Maxwell's *Treatise on Electricity and Magnetism*:
The Central Argument

Maxwell's *Treatise on Electricity and Magnetism*:
The Central Argument

Selections with Commentary
by Howard J. Fisher

Green Lion Press

Santa Fe, New Mexico

Manufactured in the United States of America

Published by Green Lion Press, Santa Fe, New Mexico, USA
www.greenlion.com

Green Lion Press books are printed on fine quality acid-free paper of high opacity. Both softbound and clothbound editions feature bindings sewn in signatures. Sewn signature binding allows our books to open securely and lie flat. Pages do not loosen or fall out and bindings do not split under heavy use by students and researchers. Clothbound editions meet the guidelines for permanence and durability of the Committee on Production Guidelines for Book Longevity of the Council on Library Resources. The paper used in all Green Lion Press books meets the minimum requirements of American National Standard for Information Sciences—Permanence of Paper for Printed Library Materials, ANSI Z39.48-1984.

Set in 11-point ITC New Baskerville.
This book is printed on acid-free paper.
Printed and bound by Sheridan Books, Inc., Chelsea, Michigan.

Cover photograph by Estan Cabigas.

Cataloguing-in-Publication data:
Fisher, Howard J.
Maxwell's Treatise on Electricity and Magnetism: The Central Argument /
by Howard J. Fisher

Includes abridged text of James Clerk Maxwell's *Treatise on Electricity and Magnetism*, index, introductions, biographical sketches, and notes.

ISBN 978-1-888009-45-3 (softcover binding)

1. Maxwell, Treatise on Electricity and Magnetism. 2. History of Science.
3. Physics. 4. Electricity.

I. James Clerk Maxwell (1831–1879). II. Fisher, Howard J. (1942–). III. Title

Library of Congress Control Number: 2014917773

In his intellectual faculties there was also a balance of powers which are often opposed. His imagination was in the highest sense concrete, grasping the actual reality, and not only the relations of things. No one was ever more impatient of mere abstractions. Yet few have had so firm a hold upon ideas. Once more, while he was continually striving to reduce to greater definiteness men's conceptions of leading physical laws, he seemed habitually to live in a sort of mystical communion with the infinite.

Lewis Campbell
Campbell and Garnett, *The Life of James Clerk Maxwell*, 1884

Contents

CONTENTS

CONTENTS

The Green Lion's Preface

Maxwell's fate has been somewhat like Kepler's, in that posterity has boiled down his life's work into just four equations—magnificent equations, to be sure, but only a bit of the husk of what Maxwell offers. Far from proposing a mere extension of mathematical physics, Maxwell hoped that the *Treatise* would illustrate and exemplify a new and more revealing way of looking at the physical universe. In this regard, Maxwell still has much to teach us.

First, the *Treatise* (as Thomas Simpson has eloquently claimed(*)) is far more than an exposition of electromagnetic theory. Maxwell endeavors to show how the treatment of electricity and magnetism points the way to a new paradigm for mathematical physics as a whole. As an alternative to the rigid "Newtonian" style, he proposes the much more flexible and analogically rich Lagrangian mathematics, which operates upon "generalized quantities" rather than predetermined physical entities such as masses and forces. This feature of Maxwell's work remains insufficiently appreciated, and the present edition highlights Maxwell's efforts in the *Treatise* to commend this approach.

On a higher level, the *Treatise* provides a humane alternative to excessively formalistic views of mathematical physics and of science in general. Maxwell promotes scientific theory as driven by *thought* rather than by mere symbolic manipulation. Like his predecessor and source of inspiration, Michael Faraday, Maxwell's primary vehicle of inquiry and interpretation is English prose, even when presenting highly technical ideas. The present edition, accordingly, attends to Maxwell's vocabulary, prose style, and imagery as essential resources for defining and expressing the central concepts of electricity and magnetism.

In the ostensibly quixotic enterprise of offering Maxwell's text for study by ordinary mortals, we are not oblivious to the considerable difficulties and challenges that such a reader must confront. There are three chief sources of these difficulties, and the present edition provides ample assistance in overcoming them.

1. Maxwell's *Treatise* combines two disparate and even conflicting functions. On the one hand, Maxwell aimed to set the science of electricity and magnetism on a firm experimental as well as theoretical basis—a task calling for concise, logical organization of essential

* See *Figures of Thought* (Green Lion Press, 2005) and *Maxwell's Mathematical Rhetoric* (Green Lion Press, 2010).

material. On the other hand, he had to produce a textbook for the Cavendish Laboratory program and mathematical tripos examinations at Cambridge University. This charge required him to treat a hefty repertory of then-timely topics and practical techniques; however, the range and variety of these subjects tend to obscure the unity of Maxwell's argument. The selections in this edition follow the *Treatise*'s central line of argument as indicated by Maxwell himself in his prefatory remarks, bypassing the distractions.

2. Maxwell employs a mathematical formalism that may be unfamiliar to modern readers. The *Treatise* is famous (or infamous) for its use of quaternions, but even when Maxwell does not utilize quaternion forms, he follows notational conventions that have by now been largely abandoned. This edition, while retaining all of Maxwell's mathematical forms, provides access to their meaning by offering modern equivalents in a running commentary.

3. The sheer quantity of mathematics in the *Treatise* can be daunting, especially since Maxwell's mathematical demonstrations often leave intermediate steps unexpressed. Though the mathematics in the *Treatise* is indeed copious, it is for the most part not really hard: Maxwell seldom demands much beyond elementary calculus from his readers. This edition strives to clarify his analyses and to supply missing steps where necessary.

Anyone who has read Fisher's guide to Faraday* knows how helpful his comments are: one reviewer describes them as "readable, informative, illuminating," and concludes, "to experience Faraday with Fisher is a delight." Fisher brings the same spirit of friendly guidance to his treatment of Maxwell, and the result lends us a sure hand in rising to the Olympian heights of Maxwell's culminating work.

<div align="right">

Dana Densmore
William H. Donahue
for Green Lion Press

</div>

* *Faraday's Experimental Researches in Electricity: Guide to a First Reading* (Green Lion Press, 2001).

Acknowledgments

My sincere gratitude goes to several advance readers of these selections who kindly offered their comments, suggestions, and corrections. Thomas K. Simpson and Chester Burke read portions of the original manuscript. Ryan Twomey faithfully worked through every page of a preliminary version; his keen observations and thoughtful suggestions have helped make this a much better book than it would otherwise have been, and have also provided occasions for a delightful exchange of views on many related topics.

I have been indebted over many years to Thomas K. Simpson's writings and conversations. It was he who helped me understand that my initial "physicist's reading" of the *Treatise* overlooked Maxwell's broad and humane vision for science and human knowing. I have tried to clear the way for Maxwell's larger voice in this book.

I also wish to thank members of the summer 2008 Faculty Study Group on Maxwell, which was held at St. John's College in Santa Fe: Grant Franks (leader), Lauren Brubaker, Michael Bybee, John Cornell, Jacques Duvoisin, Frank Hunt, Fritz Monsma, Peter Pesic, Arcelia Rodriguez, Cary Stickney, Edward Walpin, and Ken Wolfe. Special thanks are due to Grant Franks for his help in unpacking Maxwell's proof of Stokes' Theorem in Article 24.

Certain photographic illustrations included in my commentary were available through the generosity of several individuals. Thomas B. Greenslade provided the photograph of his Bohnenberger electroscope (Art. 33), J.J. O'Connor the photograph of Maxwell's "Diabolo" (Art. 15). Jim Williamson furnished information and assistance regarding the splendid Clarendon Bell (Art. 242) displayed at the Clarendon Laboratory, Oxford.

My ongoing thanks and appreciation go to Dana Densmore and William Donahue, founders and keepers of the Green Lion Press, for their commitment to, encouragement in, and support of this effort throughout its various stages of preparation.

It is a special pleasure to offer thanks to my son Jason Fisher for information about solar radiation which he supplied in connection with Maxwell's chapter on the Electromagnetic Theory of Light; and to all three of my sons—Erik, Jason, and Michael—for their continuing interest in this and others of their father's projects over many years.

HJF

Note to Readers

Maxwell's *Treatise* saw three editions. The first, issued in 1873, was the only edition published during Maxwell's lifetime. Prior to his death in 1879, Maxwell extensively rewrote the first nine chapters of the *Treatise* for a projected second edition, which appeared in 1881 under the editorship of W. D. Niven. A third edition, published in 1892, was edited by J. J. Thomson. The present text holds for the most part to the third edition, but I have made occasional reference to earlier editions where the differences are of interest.

Maxwell numbered successive Articles of the *Treatise*, but (unlike Faraday) not individual paragraphs. To facilitate commentary in the present volume, I have extended Maxwell's numbering scheme; thus the paragraphs of Article 27, for example, bear the numbers 27, 27.1, 27.2, and so on. I have employed a conspicuous separator line on each page to distinguish Maxwell's text from my commentary. With the exception of the paragraph numbers just mentioned, material above the line faithfully reproduces at least one of the three editions of the *Treatise*; material below the line represents my attempt to elucidate or interpret.

It will perhaps be useful to outline the mathematical policy I have followed in preparing my comments. In Maxwell's Preface to the First Edition he draws a sharp contrast between the supposedly unmathematical Faraday and "the professed mathematicians." The latter, Maxwell rather archly implies, are in effect stymied by ordinary narrative prose; they are incapable of taking ideas seriously unless translated into symbolic form. Such translation is what Maxwell has undertaken to do; for, he sees, if the professors could only be helped to take in Faraday's thinking they would see that Faraday too is a mathematician.

Such sentiments make it clear that, for Maxwell, the heart of mathematical reasoning is something apart from formality and symbolism. In that spirit my comments will generally attempt to explicate Maxwell's mathematical reasoning without maintaining a strict formalism; I will, for example, often take the liberty of treating infinitesimals as magnitudes. In other places I will undertake to provide algebraic equivalents of some of Maxwell's prose—not because I think the equations represent

superior reasoning, but in order to articulate Maxwell's argument more compactly; for in some respects, I am afraid, we modern readers are as prose-challenged as were Maxwell's "professed" mathematicians!

Prospective readers of the *Treatise* have sometimes been deterred by its reputation for employing quaternion mathematics. In truth, Maxwell makes scant use of quaternions in the *Treatise*; and what he does include is readily expressed in the terms of today's vector calculus. I have provided some notes on quaternion algebra on pages 49–51 below; and throughout the *Treatise*, whenever Maxwell writes a quaternion expression I offer a vector equivalent in the commentary.

In one other matter of notation I have taken a free hand. Maxwell has two different forms to indicate the square of a derivative, in some places writing $\overline{\dfrac{dV}{dx}}\Big|^2$, in others $\left(\dfrac{dV}{dx}\right)^2$. The expressions are equivalent, but I have everywhere adopted the parenthesis form, as being more familiar to modern readers.

Maxwell frequently telescopes his mathematical proofs, which is perhaps to be expected in a textbook for students who will have to survive the grueling Tripos examinations at Cambridge. In my comments I have endeavored to expand the arguments, even to the point of dotting every i and crossing every t. I have chosen to do this for the benefit of readers who have some elementary calculus but who may lack the experience to recognize common transformations at first glance. With few exceptions, Maxwell's mathematical deductions are not really difficult,* but they can be complex and tiring. I have supplied the intermediate steps so that readers can see the demonstrations unfold if they so choose. To experienced readers who need no such intervention I ask forbearance and invite them to pass over whatever is not helpful.

To less mathematically savvy readers I offer the same advice, while requesting a similar measure of tolerance: If a comment goes into more detail than you have either the desire or the stomach for—or if it is for any other reason just not useful—by all means pass it by, especially on a first reading.

It will be clear from these remarks that I have not chosen to target my commentary to a particular audience. I have, in fact, tried not to make assumptions about prospective readers. I suspect that all serious students have, at one time or another, benefited from books that some

* Heinrich Hertz, a careful if somewhat narrow reader of Maxwell, wrote: "I cannot agree ... that this difficulty [in Maxwell's theory] is of a mathematical nature." See his *Electric Waves*, tr. D. E. Jones (1893); Dover reprint 1962, pp. 21–22.

well-meaning guide has declared either "too hard," or, on the contrary, "too elementary" for them. It is my conviction that *everyone* with an inter-est in doing so can study and enjoy a substantial portion of Maxwell's great book. I hope my efforts will have contributed to that end.

&

In adopting the subtitle "The Central Argument" for this book I of course mean to signal an aim similar to that of Dana Densmore in her guide to Newton's *Principia*.* As Densmore there explains, Newton him-self specified what he regarded as the central thread of that great work, which establishes both the concept and the actuality of universal gravitation. Newton's "argument" is a connected series of propositions culminating in a reasoned conclusion.

But towards what conclusion does Maxwell work in the *Treatise*? Is it the electromagnetic theory of light in Part IV, Chapter XX? Is it the ensemble of Maxwell's equations found in Part IV, Chapter IX? Or is it, as Thomas Simpson has implied, the vision of a new science of *connected systems* in Part IV, Chapter VI?† Can we even assume that a *central argu-ment* necessarily implies a *chief conclusion*?

Unlike Newton, Maxwell is by no means explicit in charting a central course through the *Treatise*. But he does reveal what he regards as an "elementary" pathway. In his Preface to the First Edition, Maxwell notes:

The description of the phenomena, and the elementary parts of the theory of each subject, will be found in the earlier chapters of each of the four Parts into which this treatise is divided. The stu-dent will find in these chapters enough to give him an elementary acquaintance with the whole science.

I have chosen the present selections in general accordance with Maxwell's criterion, adding to his scheme only two late chapters from Part IV, whose topics—the ratio of electromagnetic and electrostatic units (Chapter XIX) and the electromagnetic theory of light (Chapter XX)—have become even more fundamental for readers in our day than for Maxwell's readers during his lifetime.

The *Treatise* is not a single-minded work, however; and neither is this pathway through it. Oliver Heaviside regretted that Maxwell tried to accommodate diverse viewpoints in the *Treatise*, keeping his own ideas

*Densmore, Dana. *Newton's Principia, The Central Argument*, 3rd ed. Santa Fe, New Mexico: Green Lion Press (2003).

†Simpson, Thomas K. *Figures of Thought, a Literary Appreciation of Maxwell's Treatise on Electricity and Magnetism*. Santa Fe, New Mexico: Green Lion Press (2005); see especially Part Two, Chapter 5.

"too much in the background." He was confident that, had Maxwell lived, he would "in some future edition, have brought his views prominently forth *ab initio*, and developed the whole treatise on their basis exclusively."* I think Maxwell's intentions were very different, as we may surmise both from the work he actually produced, and from his commendation of Faraday for revealing *all* the stages of his thinking, "his unsuccessful as well as his successful experiments, and his crude ideas as well as his developed ones."† Maxwell, I suggest, is offering his readers not "Maxwell's ideas" but an *experience*—the experience of a complete act of thought.

Plato defined the paradigmatic act of human thought in his allegory of the Cave,‡ with that moment when the mind is wrenched away from shadows towards the light—first the light of fire, ultimately the light of the Sun itself. The "complete act" that is Maxwell's *Treatise* is similarly defined—even in the reduced form of the central argument—by a deep transformation. The transformation is marked by Maxwell's abrupt turn from the phenomena, together with the tussle of competing hypotheses and symbolic languages that cluster about them, to a vision of the most universal form of human understanding;§ and it eventuates, like Plato's, in the discovery of Light. ‖

*Heaviside, Oliver. *Electrical Papers*, vol. 2 (1892), p. 167. See also Bruce J. Hunt's characterization of the *Treatise* in *The Maxwellians*, Cornell University Press (1991), p. 13, where Heaviside's speculation about a future edition is cited.

†See below, Part IV, Chapter III, Art. 528. ‡ *Republic*, Book VII

§Part IV, Chapter V ‖Part IV, Chapter XX

A TREATISE ON
ELECTRICITY AND MAGNETISM

BY

JAMES CLERK MAXWELL

PREFACE TO THE FIRST EDITION

T HE fact that certain bodies, after being rubbed, appear to attract other bodies, was known to the ancients. In modern times, a great variety of other phenomena have been observed, and have been found to be related to these phenomena of attraction. They have been classed under the name of electric phenomena, amber, *ἤλεκτρον*, having been the substance in which they were first described.

0.1 Other bodies, particularly the loadstone, and pieces of iron and steel which have been subjected to certain processes, have also been long known to exhibit phenomena of action at a distance. These phenomena, with others related to them, were found to differ from the electric phenomena, and have been classed under the name of Magnetic phenomena, the loadstone, *μάγνης*, being found in the Thessalian Magnesia.

0.2 These two classes of phenomena have since been found to be related to each other, and the relations between the various phenomena of both classes, so far as they are known, constitute the science of Electromagnetism.

0.1 *Magnesia*: In modern Greece, one of the four prefectures of Thessaly.

0.3 In the following Treatise I propose to describe the most important of these phenomena, to shew how they may be subjected to measurement, and to trace the mathematical connexions of the quantities measured. Having thus obtained the data for a mathematical theory of electromagnetism, and having shewn how this theory may be applied to the calculation of phenomena, I shall endeavour to place in as clear a light as I can the relations between the mathematical form of this theory and that of the fundamental science of Dynamics, in order that we may be in some degree prepared to determine the kind of dynamical phenomena among which we are to look for illustrations or explanations of the electromagnetic phenomena.

0.4 In describing the phenomena, I shall select those which most clearly illustrate the fundamental ideas of the theory, omitting others, or reserving them till the reader is more advanced.

0.5 The most important aspect of any phenomenon from a mathematical point of view is that of a measurable quantity. I shall therefore consider electrical phenomena chiefly with a view to their measurement, describing the methods of measurement, and defining the standards on which they depend.

0.6 In the application of mathematics to the calculation of electrical quantities, I shall endeavour in the first place to deduce the most general conclusions from the data at our disposal, and in the next place to apply the results to the simplest cases that can be chosen. I shall avoid, as much as I can, those questions which, though they have elicited the skill of mathematicians, have not enlarged our knowledge of science.

0.7 The internal relations of the different branches of the science which we have to study are more numerous and complex than those of any other science hitherto developed. Its external relations, on the one hand to dynamics, and on the other to heat, light, chemical action, and the constitution of bodies, seem to indicate the special importance of electrical science as an aid to the interpretation of nature.

0.3 *phenomena ... measurement ... mathematical connexions ... theory*: These four stages of investigation provide a concise outline of Maxwell's approach in the *Treatise*.

I shall endeavour to place in as clear a light as I can the relations between the mathematical form of this theory and that of the fundamental science of Dynamics: Thus Electricity and Magnetism, as a special science, is not the exclusive focus of this work; as we shall see, the *Treatise* aims far higher than the forging of a single science! Moreover Dynamics itself, usually defined as the science of moving bodies subject to forces, will in Maxwell's conception become far more universal in scope, as Maxwell will reveal in Part IV, Chapter V.

0.8 It appears to me, therefore, that the study of electromagnetism in all its extent has now become of the first importance as a means of promoting the progress of science.

0.9 The mathematical laws of the different classes of phenomena have been to a great extent satisfactorily made out.

0.10 The connexions between the different classes of phenomena have also been investigated, and the probability of the rigorous exactness of the experimental laws have been greatly strengthened by a more extended knowledge of their relations to each other.

0.11 Finally, some progress has been made in the reduction of electromagnetism to a dynamical science, by shewing that no electro-magnetic phenomenon is contradictory to the supposition that it depends on purely dynamical action.

0.12 What has been hitherto done, however, has by no means exhausted the field of electrical research. It has rather opened up that field, by pointing out subjects of enquiry, and furnishing us with means of investigation.

0.13 It is hardly necessary to enlarge upon the beneficial results of magnetic research on navigation, and the importance of a knowledge of the true direction of the compass, and of the effect of the iron in a ship. But the labours of those who have endeavoured to render navigation more secure by means of magnetic observations have at the same time greatly advanced the progress of pure science.

0.14 Gauss, as a member of the German Magnetic Union, brought his powerful intellect to bear on the theory of magnetism, and on the methods of observing it, and he not only added greatly to our knowledge of the theory of attractions, but reconstructed the whole of magnetic science as regards the instruments used, the methods of observation, and the calculation of the results, so that his memoirs on Terrestrial Magnetism may be taken as models of physical research by all those who are engaged in the measurement of any of the forces in nature.

0.15 The important applications of electromagnetism to telegraphy have also reacted on pure science by giving a commercial value to accurate electrical measurements, and by affording to electricians the use of apparatus on a scale which greatly transcends that of any ordinary labo-ratory. The consequences of this demand for electrical knowledge, and of these experimental opportunities for acquiring it, have been already very great, both in stimulating the energies of advanced electricians, and in diffusing among practical men a degree of accurate knowledge which is likely to conduce to the general scientific progress of the whole engineering profession.

0.16 There are several treatises in which electrical and magnetic phenomena are described in a popular way. These, however, are not what is wanted by those who have been brought face to face with quantities to be measured, and whose minds do not rest satisfied with lecture-room experiments.

0.17 There is also a considerable mass of mathematical memoirs which are of great importance in electrical science, but they lie concealed in the bulky Transactions of learned societies; they do not form a connected system; they are of very unequal merit, and they are for the most part beyond the comprehension of any but professed mathematicians.

0.18 I have therefore thought that a treatise would be useful which should have for its principal object to take up the whole subject in a methodical manner, and which should also indicate how each part of the subject is brought within the reach of methods of verification by actual measurement.

0.19 The general complexion of the treatise differs considerably from that of several excellent electrical works, published, most of them, in Germany, and it may appear that scant justice is done to the speculations of several eminent electricians and mathematicians. One reason of this is that before I began the study of electricity I resolved to read no mathematics on the subject till I had first read through Faraday's *Experimental Researches in Electricity*. I was aware that there was supposed to be a difference between Faraday's way of conceiving phenomena and that of the mathematicians, so that neither he nor they were satisfied with each other's language. I had also the conviction that this discrepancy did not arise from either party being wrong. I was first convinced of this by Sir William Thomson,* to whose advice and assistance, as well as to his published papers, I owe most of what I have learned on the subject.

0.20 As I proceeded with the study of Faraday, I perceived that his method of conceiving the phenomena was also a mathematical one,

* I take this opportunity of acknowledging my obligations to Sir W. Thomson and to Professor Tait for many valuable suggestions made during the printing of this work.

0.20 *[Faraday's] method of conceiving the phenomena was also a mathematical one, though not exhibited in the convention form of mathematical symbols*: With this observation Maxwell distinguishes the idea of mathematics from that of its symbolic garb. When in the next few paragraphs Maxwell singles out Faraday's lines of force for special appreciation, he seems to acknowledge them as real mathematical objects whose essential intelligibility can be can be expressed not only in Faraday's natural, untechnical language, but also (though perhaps not quite so well) in the formalisms of the "professed" mathematicians.

though not exhibited in the conventional form of mathematical symbols. I also found that these methods were capable of being expressed in the ordinary mathematical forms, and thus compared with those of the professed mathematicians.

0.21 For instance, Faraday, in his mind's eye, saw lines of force traversing all space where the mathematicians saw centres of force attracting at a distance: Faraday saw a medium where they saw nothing but distance: Faraday sought the seat of phenomena in real actions going on in the medium, they were satisfied that they had found it in a power of action at a distance impressed on the electric fluids.

0.22 When I had translated what I considered to be Faraday's ideas into a mathematical form, I found that in general the results of the two methods coincided, so that the same phenomena were accounted for, and the same laws of action deduced by both methods, but that Faraday's methods resembled those in which we begin with the whole and arrive at the parts by analysis, while the ordinary mathematical methods were founded on the principle of beginning with the parts and building up the whole by synthesis.

0.23 I also found that several of the most fertile methods of research discovered by the mathematicians could be expressed much better in terms of ideas derived from Faraday than in their original form.

0.24 The whole theory, for instance, of the potential, considered as a quantity which satisfies a certain partial differential equation, belongs essentially to the method which I have called that of Faraday. According to the other method, the potential, if it is to be considered at all, must be regarded as the result of a summation of the electrified particles

0.24 *"The whole theory ... of the potential ...":* Maxwell's discussion in this paragraph is highly anticipatory; he will discuss the potential in Arts. 14–16, below, and will there offer a specific example of the method he calls Faradayan—in which the potential is viewed as a quantity that solves a differential equation. "The other method," to which Maxwell opposes it, derives from the inverse-square law for electrostatic force, F.

$$F = \frac{q_i q}{r^2},$$

applied to each of a multitude of charged particles q_i and integrated to obtain the potential P_i, due to each particle, as

$$P_i = \int_{\infty}^b \frac{F}{q} \cdot dr = q_i \int_{\infty}^b r^{-2} \, dr = -\frac{q_i}{r}.$$

The contrast is essentially between a differential equation, which (as we will see) applies to *all space treated as a continuum,* and a summation of *individual centers of action.*

divided each by its distance from a given point. Hence many of the mathematical discoveries of Laplace, Poisson, Green and Gauss find their proper place in this treatise, and their appropriate expressions in terms of conceptions mainly derived from Faraday.

0.25 Great progress has been made in electrical science, chiefly in Germany, by cultivators of the theory of action at a distance. The valuable electrical measurements of W. Weber are interpreted by him according to this theory, and the electromagnetic speculation which was originated by Gauss, and carried on by Weber, Riemann, J. and C. Neumann, Lorenz, &c., is founded on the theory of action at a distance, but depending either directly on the relative velocity of the particles, or on the gradual propagation of something, whether potential or force, from the one particle to the other. The great success which these eminent men have attained in the application of mathematics to electrical phenomena, gives, as is natural, additional weight to their theoretical speculation, so that those who, as students of electricity, turn to them as the greatest authorities in mathematical electricity, would probably imbibe, along with their mathematical methods, their physical hypotheses.

0.26 These physical hypotheses, however, are entirely alien from the way of looking at things which I adopt, and one object which I have in view is that some of those who wish to study electricity may, by reading this treatise, come to see that there is another way of treating the subject, which is no less fitted to explain the phenomena, and which, though in some parts it may appear less definite, corresponds, as I think, more faithfully with our actual knowledge, both in what it affirms and in what it leaves undecided.

0.27 In a philosophical point of view, moreover, it is exceedingly important that two methods should be compared, both of which have succeeded in explaining the principal electromagnetic phenomena, and both of which have attempted to explain the propagation of light as an electromagnetic phenomenon and have actually calculated its velocity, while at the same time the fundamental conceptions of what actually takes place, as well as most of the secondary conceptions of the quantities concerned, are radically different.

0.25 *would probably imbibe ... their physical hypotheses*: Note Maxwell's implicit warning lest the power of the professed mathematicians' methods induce us to adopt hypotheses such as physical electric fluids or magnetic poles.

0.28 *an advocate rather than ... a judge*: Maxwell will play the role of *advocate* partly by showing that Faraday's thinking can compete *mathematically* with the accepted action-at-a-distance approach. But, as he noted in the previous paragraph, *both*

0.28 I have therefore taken the part of an advocate rather than that of a judge, and have rather exemplified one method than attempted to give an impartial description of both. I have no doubt that the method which I have called the German one will also find its supporters, and will be expounded with a skill worthy of its ingenuity.

0.29 I have not attempted an exhaustive account of electrical phenomena, experiments, and apparatus. The student who desires to read all that is known on these subjects will find great assistance from the *Traité d'Electricité* of Professor A. de la Rive, and from several German treatises, such as Wiedemann's *Galvanismus*, Riess' *Reibungselektricität*, Beer's *Einleitung in die Elektricität*, &c.

0.30 I have confined myself almost entirely to the mathematical treatment of the subject, but I would recommend the student, after he has learned, experimentally if possible, what are the phenomena to be observed, to read carefully Faraday's *Experimental Researches in Electricity*. He will find there a strictly contemporary historical account of some of the greatest electrical discoveries and investigations, carried on in an order and succession which could hardly have been improved if the results had been known from the first, and expressed in the language of a man who devoted much of his attention to the methods of accurately describing scientific operations and their results.*

0.31 It is of great advantage to the student of any subject to read the original memoirs on that subject, for science is always most completely assimilated when it is in the nascent state, and in the case of Faraday's *Researches* this is comparatively easy, as they are published in a separate form, and may be read consecutively. If by anything I have here written I may assist any student in understanding Faraday's modes of thought and expression, I shall regard it as the accomplishment of one of my

methods have "succeeded in explaining the principal electromagnetic phenomena." Therefore *judgment* between them will not be a simple matter of agreement or disagreement with experiment but must involve larger philosophical questions. Another aspect of Maxwell's advocacy, then, will be to legitimize philosophical interpretation as integral to scientific activity. Maxwell will speak to this concern again in Chapter I; see paragraphs 36.4 and 37.5.

0.31 *science is always most completely assimilated when it is in the nascent state:* When we are learning a subject, it is advantageous to read writers who are themselves trying to think through the most basic questions in their field. Maxwell greatly admires Faraday's example in this; we should not let Maxwell's sophisticated mathematics blind us to the fundamentally *elementary* character of his aims.

principal aims—to communicate to others the same delight which I have found myself in reading Faraday's *Researches*.

0.32 The description of the phenomena, and the elementary parts of the theory of each subject, will be found in the earlier chapters of each of the four Parts into which this treatise is divided. The student will find in these chapters enough to give him an elementary acquaintance with the whole science.

0.33 The remaining chapters of each Part are occupied with the higher parts of the theory, the processes of numerical calculation, and the instruments and methods of experimental research.

0.34 The relations between electromagnetic phenomena and those of radiation, the theory of molecular electric currents, and the results of speculation on the nature of action at a distance, are treated of in the last four chapters of the second volume.

JAMES CLERK MAXWELL

Feb. 1, 1873.

[Note: In the following chapters, material appearing within curly brackets { } was added by W. D. Niven and J. J. Thomson in editions of the *Treatise* published after Maxwell's death. Text within square brackets has been supplied by the present editor.]

Editor's Introduction to
Preliminary

"Preliminary" material is generally preparatory in nature; it makes ready for the main work or topic but is not itself part of that main work. Accordingly, Maxwell's *Preliminary* chapter opens with matters which, though fundamental to any physical science, are by no means specific to the science of electricity and magnetism. The *nature of measurement* is the first of these, including the distinction between *fundamental* and *derived units*. Next follows a discussion of *vector* and *scalar quantities* and, within vectors, the distinction—still novel in Maxwell's day—between *intensities and fluxes*.

In this chapter too Maxwell introduces certain mathematical methods he will call upon frequently; these include the *line-integral* and the *surface-integral*. In connection with these he will define the *resolved part* of a vector quantity and the concept of the *potential*. These and similar matters occupy the first 23 Articles of this chapter. They should by no means be neglected, as in several cases they have been honed to facilitate Maxwell's distinctive approach to this new science—an approach shaped largely (as Maxwell wrote in his Preface) by his profound admiration for Faraday's style of thought. But while Maxwell's survey is indeed *preliminary* in character, it is not *introductory*; Maxwell assumes a degree of familiarity with several mathematical topics that not all readers possess. Some readers, then, may choose to proceed immediately to Maxwell's Chapters I and II, referring back to this Preliminary section only in connection with Chapter II, which provides a fuller context for several of Maxwell's "preliminary" concepts.

The 24th article presents a special challenge. Here Maxwell undertakes to develop what is now known as Stokes' Theorem. His argument appears at first to be quite manageable—until the reader discovers that Maxwell has merely synopsized the most complicated of its steps! I have supplied the missing maneuvers, but they are certain to prove unattractively dense to some readers. Nevertheless, the subject of Stokes' Theorem is essential to Maxwell's reasoning and cannot be neglected. For that reason I have taken the liberty of treating Article 24 as though it were a separate chapter. That has allowed me to provide an alternative, simpler derivation of Stokes' Theorem, which some readers may prefer.

I have, similarly, treated Articles 25 and 26 as a chapter in their own right. This is where Maxwell introduces *quaternion calculus*, a forerunner of our present system of vector analysis. Though quaternions had been originated (by William Rowan Hamilton) some thirty years before, they were viewed with widespread suspicion and incomprehension even as Maxwell composed the *Treatise*. Maxwell had his own ideas concerning the resistance that quaternions encountered. In November 1870 he wrote:

> The limited use which has up to the present time been made of Quaternions must be attributed partly to the repugnance of most mature minds to new methods involving the expenditure of thought...*

Trusting that no such repugnance is to be found among readers of this book, I shall offer a brief guide to quaternion calculus in the editor's introduction to Articles 25 and 26 and in subsequent comments. In addition, wherever Maxwell writes equations in quaternion notation, I will provide the equivalent vector calculus expressions.

*"On the Application of Quaternions to Electromagnetism" (manuscript), in P M. Harman, ed. *The Scientific Letters and Papers of James Clerk Maxwell*, vol. II, pp. 570–576.

ELECTRICITY AND MAGNETISM.

❧

PRELIMINARY

ON THE MEASUREMENT OF QUANTITIES.

1.] EVERY expression of a Quantity consists of two factors or compo-
nents. One of these is the name of a certain known quantity of the same
kind as the quantity to be expressed, which is taken as a standard of
reference. The other component is the number of times the standard is
to be taken in order to make up the required quantity. The standard
quantity is technically called the Unit, and the number is called the
Numerical Value of the quantity.

1.1 There must be as many different units as there are different kinds
of quantities to be measured, but in all dynamical sciences it is possible
to define these units in terms of the three fundamental units of Length,
Time, and Mass. Thus the units of area and of volume are defined respec-
tively as the square and the cube whose sides are the unit of length.

1.2 Sometimes, however, we find several units of the same kind
founded on independent considerations. Thus the gallon, or the volume
of ten pounds of water, is used as a unit of capacity as well as the cubic
foot. The gallon may be a convenient measure in some cases, but it is not
a systematic one, since its numerical relation to the cubic foot is not a
round integral number.

2.] In framing a mathematical system we suppose the fundamental
units of length, time, and mass to be given, and deduce all the derivative
units from these by the simplest attainable definitions.

1. *The standard quantity is … called the Unit, and the number is called the Numerical Value:*
If a length, for example, is expressed as "14 feet," the foot is the unit and 14 the
numerical value. Similarly for an area of "23 acres," or a volume of "19 gallons,"
Maxwell will discuss this form of expression more fully in paragraph 3.4.

1.2 *its relation to the cubic foot is not a round integral number:* Perhaps this is not the
essential reason why the gallon is not a "systematic" unit. In the preceding para-
graph Maxwell argued that the unit of volume ought to be defined as the cube
whose side is the unit of length. But the gallon is defined independently of the
cubic foot. It is their logical independence, rather than their nonintegral rela-
tion, that prevents the gallon and the cubic foot from belonging to the same
system.

2.1 The formulae at which we arrive must be such that a person of any nation, by substituting for the different symbols the numerical values of the quantities as measured by his own national units, would arrive at a true result.

2.2 Hence, in all scientific studies it is of the greatest importance to employ units belonging to a properly defined system, and to know the relations of these units to the fundamental units, so that we may be able at once to transform our results from one system to another.

2.3 This is most conveniently done by ascertaining the *dimensions* of every unit in terms of the three fundamental units. When a given unit varies as the *n*th power of one of these units, it is said to be of *n dimensions* as regards that unit.

2.4 For instance, the scientific unit of volume is always the cube whose side is the unit of length. If the unit of length varies, the unit of volume will vary as its third power, and the unit of volume is said to be of three dimensions with respect to the unit of length.

2.5 A knowledge of the dimensions of units furnishes a test which ought to be applied to the equations resulting from any lengthened investigation. The dimensions of every term of such an equation, with respect to each of the three fundamental units, must be the same. If not, the equation is absurd, and contains some error, as its interpretation would be different according to the arbitrary system of units which we adopt.*

The Three Fundamental Units.

3.] (1) *Length.* The standard of length for scientific purposes in this country is one foot, which is the third part of the standard yard preserved in the Exchequer Chambers.

3.1 In France, and other countries which have adopted the metric system, it is the metre. The metre is theoretically the ten millionth part of the length of a meridian of the earth measured from the pole to the equator; but practically it is the length of a standard preserved in Paris, which was constructed by Borda to correspond, when at the temperature of melting ice, with the value of the preceding length as measured by Delambre. The metre has not been altered to correspond with new and more accurate measurements of the earth, but the arc of the meridian is estimated in terms of the original metre.

3.2 In astronomy the mean distance of the earth from the sun is sometimes taken as a unit of length.

*The theory of dimensions was first stated by Fourier, *Théorie de Chaleur*, §160.

3.3 In the present state of science the most universal standard of length which we could assume would be the wave length in vacuum of a particular kind of light, emitted by some widely diffused substance such as sodium, which has well-defined lines in its spectrum. Such a standard would be independent of any changes in the dimensions of the earth, and should be adopted by those who expect their writings to be more permanent than that body.

3.4 In treating of the dimensions of units we shall call the unit of length $[L]$. If l is the numerical value of a length, it is understood to be expressed in terms of the concrete unit $[L]$, so that the actual length would be fully expressed by $l [L]$.

4.] (2) *Time.* The standard unit of time in all civilized countries is deduced from the time of rotation of the earth about its axis. The sidereal day, or the true period of rotation of the earth, can be ascertained with great exactness by the ordinary observations of astronomers; and the mean solar day can be deduced from this by our knowledge of the length of the year.

4.1 The unit of time adopted in all physical researches is one second of mean solar time.

4.2 In astronomy a year is sometimes used as a unit of time. A more universal unit of time might be found by taking the periodic time of vibration of the particular kind of light whose wave length is the unit of length.

4.3 We shall call the concrete unit of time $[T]$, and the numerical measure of time t.

5.] (3) *Mass.* The standard unit of mass is in this country the avoirdupois pound preserved in the Exchequer Chambers. The grain, which is often used as a unit, is defined to be the 7000th part of this pound.

3.3 *those who expect their writings to be more permanent than* [*the earth*]: Does such an expectation spring from noble universalism or rank egotism? Let us take a moment to appreciate the ironic edge that occasionally sharpens Maxwell's humor. That edge does not often appear, but it is quite delicious when it does.

3.4 *so that the actual length would be fully expressed by* $l [L]$: More formally stated,

$$1 \times \text{actual length} = l [L].$$

Taking these products as means and extremes, we may form the proportion

$$\text{actual length} : [L] :: l : 1.$$

Thus the numerical value l is a number whose ratio to 1 is the same as that of the actual length to the unit length. To measure, then, does not mean to disclose an attribute of the measured object—"being eight feet long" is not an attribute—but a *relation.* We measure an object when we find a number ratio that is equal to the ratio between the object and the unit.

5.1 In the metrical system it is the gramme, which is theoretically the mass of a cubic centimetre of distilled water at standard temperature and pressure, but practically it is the thousandth part of the standard kilogramme preserved in Paris.

5.2 The accuracy with which the masses of bodies can he compared by weighing is far greater than that hitherto attained in the measurement of lengths, so that all masses ought, if possible, to be compared directly with the standard, and not deduced from experiments on water.

5.3 In descriptive astronomy the mass of the sun or that of the earth is sometimes taken as a unit, but in the dynamical theory of astronomy the unit of mass is deduced from the units of time and length, combined with the fact of universal gravitation. The astronomical unit of mass is that mass which attracts another body placed at the unit of distance so as to produce in that body the unit of acceleration.

5.4 In framing a universal system of units we may either deduce the unit of mass in this way from those of length and time already defined, and this we can do to a rough approximation in the present state of science; or, if we expect soon to be able to determine the mass of a single molecule of a standard substance, we may wait for this determination before fixing a universal standard of mass.

5.5 We shall denote the concrete unit of mass by the symbol $[M]$ in treating of the dimensions of other units. The unit of mass will be taken as one of the three fundamental units. When, as in the French system, a particular substance, water, is taken as a standard of density, then the unit of mass is no longer independent, but varies as the unit of volume, or as $[L^3]$.

5.6 If, as in the astronomical system, the unit of mass is defined with respect to its attractive power, the dimensions of $[M]$ are $[L^3T^{-2}]$.

5.7 For the acceleration due to the attraction of a mass m at a distance r is by the Newtonian Law m/r^2. Suppose this attraction to act for a very small time t on a body originally at rest, and to cause it to describe a space s, then by the formula of Galileo,

$$s = \frac{1}{2}ft^2 = \frac{1}{2}\frac{m}{r^2}t^2;$$

whence

$$m = 2\frac{r^2s}{t^2}.$$

Since r and s are both lengths, and t is a time, this equation cannot be true unless the dimensions of m are $[L^3T^{-2}]$. The same can be shewn from any astronomical equation in which the mass of a body appears in some but not in all of the terms.

Derived Units.

6.] The unit of **Velocity** is that velocity in which unit of length is described in unit of time. Its dimensions are $[LT^{-1}]$.

6.1 If we adopt the units of length and time derived from the vibrations of light, then the unit of velocity is the velocity of light.

6.2 The unit of **Acceleration** is that acceleration in which the velocity increases by unity in unit of time. Its dimensions are $[LT^{-2}]$.

6.3 The unit of **Density** is the density of a substance which contains unit of mass in unit of volume. Its dimensions are $[ML^{-3}]$.

6.4 The unit of **Momentum** is the momentum of unit of mass moving with unit of velocity. Its dimensions are $[MLT^{-1}]$.

6.5 The unit of **Force** is the force which produces unit of momentum in unit of time. Its dimensions are $[MLT^{-2}]$.

6.6 This is the absolute unit of force, and this definition of it is implied in every equation in Dynamics. Nevertheless, in many text books in which these equations are given, a different unit of force is adopted, namely, the weight of the national unit of mass; and then, in order to satisfy the equations, the national unit of mass is itself abandoned, and an artificial unit is adopted as the dynamical unit, equal to the national unit divided by the numerical value of the intensity of gravity at the place. In this way both the unit of force and the unit of mass are made to depend on the value of the intensity of gravity, which varies from place to place, so that statements involving these quantities are not complete without a knowledge of the intensity of gravity in the places where these statements were found to be true.

6.7 The abolition, for all scientific purposes, of this method of measuring forces is mainly due to the introduction by Gauss of a general system of making observations of magnetic force in countries in which the intensity of gravity is different. All such forces are now measured according to the strictly dynamical method deduced from our definitions, and the numerical results are the same in whatever country the experiments are made.

6.8 The unit of **Work** is the work done by the unit of force acting through the unit of length measured in its own direction. Its dimensions are $[ML^2T^{-2}]$.

6.9 The **Energy** of a system, being its capacity of performing work, is measured by the work which the system is capable of performing by the expenditure of its whole energy.

6.10 The definitions of other quantities, and of the units to which they are referred, will be given when we require them.

6.11 In transforming the values of physical quantities determined in terms of one unit, so as to express them in terms of any other unit of the same kind, we have only to remember that every expression for the quantity consists of two factors, the unit and the numerical part which expresses how often the unit is to be taken. Hence the numerical part of the expression varies inversely as the magnitude of the unit, that is, inversely as the various powers of the fundamental units which are indicated by the dimensions of the derived unit.

On Physical Continuity and Discontinuity.

7.] A quantity is said to vary continuously if, when it passes from one value to another, it assumes all the intermediate values.

7.1 We may obtain the conception of continuity from a consideration of the continuous existence of a particle of matter in time and space. Such a particle cannot pass from one position to another without describing a continuous line in space, and the coordinates of its position must be continuous functions of the time.

7.2 In the so-called 'equation of continuity,' as given in treatises on Hydrodynamics, the fact expressed is that matter cannot appear in or disappear from an element of volume without passing in or out through the sides of that element.

7.3 A quantity is said to be a continuous function of its variables if, when the variables alter continuously, the quantity itself alters continuously.

* * *

On the Relation of Physical Quantities to Directions in Space.

10.] In distinguishing the kinds of physical quantities, it is of great importance to know how they are related to the directions of those coordinate axes which we usually employ in defining the positions of things. The introduction of coordinate axes into geometry by Des Cartes was one of the greatest steps in mathematical progress, for it reduced the methods of geometry to calculations performed on numerical quantities. The position of a point is made to depend on the lengths of three lines which are always drawn in determinate directions, and the line joining two points is in like manner considered as the resultant of three lines.

10. *the resultant of three lines*: In Cartesian geometry a line, as OP, may be projected perpendicularly upon the coordinate axes Ox, Oy, and Oz to determine lines OX, OY, and OZ respectively. These projections are said to be the *components* of OP, while OP is said to be their *resultant*.

10.1 But for many purposes of physical reasoning, as distinguished from calculation, it is desirable to avoid explicitly introducing the Cartesian coordinates, and to fix the mind at once on a point of space instead of its three coordinates, and on the magnitude and direction of a force instead of its three components. This mode of contemplating geometrical and physical quantities is more primitive and more natural than the other, although the ideas connected with it did not receive their full development till Hamilton made the next great step in dealing with space, by the invention of his Calculus of Quaternions.*

10.2 As the methods of Des Cartes are still the most familiar to students of science, and as they are really the most useful for purposes of calculation, we shall express all our results in the Cartesian form. I am convinced, however, that the introduction of the ideas, as distinguished from the operations and methods of Quaternions, will be of great use to us in the study of all parts of our subject, and especially in electrodynamics, where we have to deal with a number of physical quantities, the relations of which to each other can be expressed far more simply by a few expressions of Hamilton's, than by the ordinary equations.

11.] One of the most important features of Hamilton's method is the division of quantities into Scalars and Vectors.

11.1 A Scalar quantity is capable of being completely defined by a single numerical specification. Its numerical value does not in any way depend on the directions we assume for the coordinate axes.

11.2 A Vector, or Directed quantity, requires for its definition three numerical specifications, and these may most simply be understood as having reference to the directions of the coordinate axes.

*(For an elementary account of Quaternions, the reader may be referred to Kelland and Tait's 'Introduction to Quaternions,' Tait's 'Elementary Treatise on Quaternions,' and Hamilton's 'Elements of Quaternions.')

10.1 *magnitude and direction of a force instead of its three components:* Refer to the diagram accompanying the comment on paragraph 10. For purposes of calcula-tion it is often convenient the think of line OP solely in terms of its components OX, OY, OZ. But in physical reasoning, Maxwell holds, it may be "more primi-tive and more natural"—and hence more desirable—to think of OP as a directed magnitude, without regard to its components.

10.2 *the ideas, as distinguished from the operations and methods of Quaternions:* We will make ample use of the operations of quaternions in our study; but since many of the *ideas* Maxwell regards as being "of great use to us"—are today embodied in the system of vector analysis, I will regularly provide translations of Maxwell's theorems into vector algebra form.

11.3 Scalar quantities do not involve direction. The volume of a geometrical figure, the mass and the energy of a material body, the hydrostatical pressure at a point in a fluid, and the potential at a point in space, are examples of scalar quantities.

11.4 A vector quantity has direction as well as magnitude, and is such that a reversal of its direction reverses its sign. The displacement of a point, represented by a straight line drawn from its original to its final position, may be taken as the typical vector quantity, from which indeed the name of Vector is derived.

11.5 The velocity of a body, its momentum, the force acting on it, an electric current, the magnetization of a particle of iron, are instances of vector quantities.

11.6 There are physical quantities of another kind which are related to directions in space, but which are not vectors. Stresses and strains in solid bodies are examples of these, and so are some of the properties of bodies considered in the theory of elasticity and in the theory of double refraction. Quantities of this class require for their definition nine numerical specifications. They are expressed in the language of quaternions by linear and vector functions of a vector.

11.7 The addition of one vector quantity to another of the same kind is performed according to the rule given in Statics for the composition of forces. In fact, the proof which Poisson gives of the 'parallelogram of forces' is applicable to the composition of any quantities such that turning them end for end is equivalent to a reversal of their sign.

11.8 When we wish to denote a vector quantity by a single symbol, and to call attention to the fact that it is a vector, so that we must consider its direction as well as its magnitude, we shall denote it by a German capital letter, as 𝔄, 𝔅, &c. ...

11.9 In the calculus of quaternions, the position of a point in space is defined by the vector drawn from a fixed point, called the origin, to that point. If we have to consider any physical quantity whose value depends on the position of the point, that quantity is treated as a function of the

11.4 *from which indeed the name of Vector is derived*: *Vector*, from Latin *vectus* = carrier, past participle of *vehere* = to convey.

11.7 *the rule given in Statics for the composition of forces*: See Newton, *Principia*, Corollaries 1 and 2 to the Laws of Motion.

11.8 *call attention to the fact that it is a vector*: Maxwell uses the distinctive vector notation only selectively and sometimes even employs unrelated symbols to represent a vector and its magnitude. Modern typographic practice, in contrast, demands a consistent formality. Today a common style uses boldface **A**, **B**, etc. for vectors and the corresponding unbolded letters *A*, *B* for their magnitudes.

vector drawn from the origin. The function may be itself either scalar or vector. The density of a body, its temperature, its hydrostatical pressure, the potential at a point, are examples of scalar functions. The resultant force at a point, the velocity of a fluid at a point, the velocity of rotation of an element of the fluid, and the couple producing rotation, are examples of vector functions.

12.] Physical vector quantities may be divided into two classes, in one of which the quantity is defined with reference to a line, while in the other the quantity is defined with reference to an area.

12.1 For instance, the resultant of an attractive force in any direction may be measured by finding the work which it would do on a body if the body were moved a short distance in that direction and dividing it by that short distance. Here the attractive force is defined with reference to a line.

12.2 On the other hand, the flux of heat in any direction at any point of a solid body may be defined as the quantity of heat which crosses a small area drawn perpendicular to that direction divided by that area and by the time. Here the flux is defined with reference to an area.

12.3 There are certain cases in which a quantity may be measured with reference to a line as well as with reference to an area.

12.4 Thus, in treating of the displacements of elastic solids, we may direct our attention either to the original and the actual positions of a particle, in which case the displacement of the particle is measured by the line drawn from the first position to the second, or we may consider a small area fixed in space, and determine what quantity of the solid passes across that area during the displacement.

12.1 the attractive force is defined with reference to a line: Since the product of force and distance is *work*, a force may be defined as the quotient *work ÷ distance*. Such a force will have been "defined with reference to a line" and will have the form which Maxwell will shortly call an *intensity*.

12.2 the flux is defined with reference to an area: The quantity—not just of heat but of anything that flows—which crosses a specified area perpendicularly is a quantity "defined with reference to an area" and has the form of a *flux*.

12.4 displacements of elastic solids: In any instance of flow or deformation, we can conceive the "displacement" of the material either (i) in terms of the changes of position which its particles individually undergo, or (ii) as the quantity of material that crosses a given boundary surface. Displacement in sense (i) is its kinematic use: the displacement of a particle is its change of position in a specified direction. An example of sense (ii) is the nautical use of the term: a ship of 100 tons displacement is one whose immersed portion occupies a space that would otherwise have been filled by 100 tons of water. If we imagine that space as being bounded by a surface, then 100 tons of water would have flowed through that surface when the ship was lowered into the water.

12.5 In the same way the velocity of a fluid may be investigated either with respect to the actual velocity of the individual particles, or with respect to the quantity of the fluid which flows through any fixed area.

12.6. But in these cases we require to know separately the density of the body as well as the displacement or velocity, in order to apply the first method, and whenever we attempt to form a molecular theory we have to use the second method.

12.7 In the case of the flow of electricity we do not know anything of its density or its velocity in the conductor, we only know the value of what, on the fluid theory, would correspond to the product of the density and the velocity. Hence in all such cases we must apply the more general method of measurement of the flux across an area.

12.5 *the velocity of a fluid*: Like displacement, velocity too has a double sense, referring on the one hand to the (average) velocity of individual particles, and on the other hand to the *rate in time* at which a quantity of fluid flows across a specified surface.

12.6 *we require to know ... the density ... as well as the displacement or velocity, in order to apply the first method*: Consider the simple case of uniform flow in a tube of cross-section area A. Define a volume Al as in the sketch and suppose that it contains n particles of fluid, each having mass m. If all the particles together travel distance l we shall say that all the fluid that previously occupied volume Al has crossed the bounding surface; in that sense, n particles have been displaced. And since each particle has mass m, the *total mass displacement* will be nm. Moreover, if this happens in time t, the *mass flow rate* will be nm/t.

Now we cannot realistically count the number of particles n, but if we know the density ρ of the fluid we can express $\rho = nm/Al$, or $nm = \rho Al$. Then we can write
$$\text{total mass displacement} = nm = \rho Al$$
and
$$\text{mass flow rate} = \rho Al/t = \rho Av, \qquad (i)$$
where v, that is l/t, is the velocity of a particle. Thus to compute these quantities we would need to know density ρ and particle displacement l, on the one hand, or density ρ and particle velocity v, on the other; the "first method" treats l and v as *known* quantities, and mass displacement and flow rate as *calculated* quantities.

whenever we attempt to form a molecular theory we have to use the second method: Since molecules are too small to observe individually, we cannot measure their velocities; thus the "second method" treats *mass displacement and flow rate* as known, and *molecular velocities* as calculated from eq. (i) above; even from that equation, we can at best calculate an *average* molecular velocity v; individual velocities must remain unknown.

12.7 *the product of the density and the velocity*: In our treatment of fluid flow we expressed the mass flow rate (mass per unit time) as ρAv. In a comparable treatment of electric current the same expression would represent charge per unit time. Maxwell is pointing out that our sole measure of electric current is in terms of charge per unit time, and therefore we do not have measures of ρ and

12.8 In electrical science, electromotive and magnetic intensity belong to the first class, being defined with reference to lines. When we wish to indicate this fact, we may refer to them as Intensities.

12.9 On the other hand, electric and magnetic induction, and electric currents, belong to the second class, being defined with reference to areas. When we wish to indicate this fact, we shall refer to them as Fluxes.

12.10 Each of these intensities may be considered as producing, or tending to produce, its corresponding flux. Thus, electromotive intensity produces electric currents in conductors, and tends to produce them in dielectrics. It produces electric induction in dielectrics, and probably in conductors also. In the same sense, magnetic intensity produces magnetic induction.

13.] In some cases the flux is simply proportional to the intensity and in the same direction, but in other cases we can only affirm that the direction and magnitude of the flux are functions of the direction and magnitude of the intensity.

* * *

14.] There are two mathematical operations of frequent occurrence which are appropriate to these two classes of vectors, or directed quantities.

14.1 In the case of intensity, we have to take the integral along a line of the product of an element of the line, and the resolved part of the intensity along that element. The result of this operation is called the

v independently but only of the whole product $\rho A v$. Why then does Maxwell name only density and velocity, neglecting to mention area A? Perhaps he is here tacitly assuming unit area, as he did in effect in paragraph 12.2.

12.10 Each of these intensities [produces, or tends to produce] its corresponding flux: This remark is anticipatory. Maxwell will discuss the relation of electromotive intensity to electric displacement, a flux, in Arts. 60–62 below. He discusses magnetic intensity and magnetic induction in Art. 395.

14.1 an element of the line: An element of a line is an infinitesimal part of the line. In the drawing, ds is an element of curved line AB. Its direction is necessarily the direction of the tangent PT.

resolved part of the intensity: The "resolved part" of a directed quantity is its component parallel to a specified direction, when its other component is perpendicular to that direction. In the sketch, vector R, representing an intensity, has been resolved into components that are respectively parallel to and perpendicular to the tangent PT. The component parallel to the tangent is the resolved part of R along element ds. If the angle between R and ds is ε, the drawing shows that the magnitude of the resolved part of R is $R \cos \varepsilon$.

Line-integral of the intensity. It represents the work done on a body carried along the line. In certain cases in which the line-integral does not depend on the form of the line, but only on the positions of its extremities, the line-integral is called the Potential.

14.2 In the case of fluxes, we have to take the integral, over a surface, of the flux through every element of the surface. The result of this operation is called the Surface-integral of the flux. It represents the quantity which passes through the surface.

14.3 There are certain surfaces across which there is no flux. If two of these surfaces intersect, their line of intersection is a line of flux. In those cases in which the flux is in the same direction as the force, lines of this kind are often called Lines of Force. It would be more correct, however, to speak of them in electrostatics and magnetics as Lines of Induction, and in electrokinematics as Lines of Flow.

15.] There is another distinction between different kinds of directed quantities, which, though very important from a physical point of view, is not so necessary to be observed for the sake of the mathematical methods. This is the distinction between longitudinal and rotational properties.

14.1, continued. *the Line-integral of the intensity*: Maxwell will offer a detailed discussion of the line-integral $L = \int R \cos \varepsilon \, ds$ in Art. 16.

it represents the work done: Suppose R represents *electric intensity*. Maxwell will explain in Art. 68 that electric intensity measures the force exerted on unit electrical charge. Then L is the integral of force × distance, and it represents the work done on a body (having unit electric charge) carried along that distance.

in certain cases ... the line-integral is called the Potential: Maxwell will explain these cases and discuss the Potential in Art. 16.

14.3 *surfaces across which there is no flux*: Maxwell considers flux to be "across" a surface insofar as it *penetrates the surface perpendicularly*. This is in contrast to the familiar phrase "across the water," which generally suggests a direction parallel to the water surface.

15. *different kinds of directed quantities ... longitudinal and rotational*: How can a *directed* quantity be rotational? For what is the "direction" of a rotational quantity—is not every point on a rotating body constantly *changing* its direction? Generally, however, any rotation can be assigned a direction parallel to the direction of its axis. Here is a photograph of a spinning toy, the "diabolo," actually owned by the young Maxwell. In the right-handed system of coordinates (called 'the system of the vine' in Art. 23 below), if the diabolo spins in the direction of the curved arrow, its rotational velocity can be represented by a vector **R** in the direction shown.

Photo courtesy John O'Connor

15.1 The direction and magnitude of a quantity may depend upon some action or effect which takes place entirely along a certain line, or it may depend upon the nature of rotation about that line as an axis. The laws of combination of directed quantities are the same whether they are longitudinal or rotational, so that there is no difference in the mathematical treatment of the two classes, but there may be physical circumstances which indicate to which class we must refer a particular phenomenon. Thus, electrolysis consists of the transfer of certain substances along a line in one direction, and of certain other substances in the opposite direction, which is evidently a longitudinal phenomenon, and there is no evidence of any rotational effect about the direction of the force. Hence we infer that the electric current which causes or accompanies electrolysis is a longitudinal, and not a rotational phenomenon.

15.2 On the other hand, the north and south poles of a magnet do not differ as oxygen and hydrogen do, which appear at opposite places during electrolysis, so that we have no evidence that magnetism is a longitudinal phenomenon, while the effect of magnetism in rotating the plane of polarization of plane polarized light distinctly shews that magnetism is a rotational phenomenon.

On Line-integrals.

16.] The operation of integration of the resolved part of a vector quantity along a line is important in physical science generally, and should be clearly understood.

16.1 Let x, y, z be the coordinates of a point P on a line whose length, measured from a certain point A, is s. These coordinates will be functions of a single variable s.

16.2 Let R be the numerical value of the vector quantity at P, and let the tangent to the curve at P make with the direction of R the angle ε, then R cos ε is the resolved part of R along the line, and the integral

$$L = \int_s^0 R \cos \varepsilon \, ds$$

is called the line-integral of R along the line s. We may write this expression

16.1-2 Maxwell is describing the situation depicted in the sketch. Point A is fixed while point P is free to move along the curved line shown. Distance s is measured from A to P along this line, so that the integral L is being taken along the same path from A to P.

$$L = \int_0^s \left(X\frac{dx}{ds} + Y\frac{dy}{ds} + Z\frac{dz}{ds} \right) ds,$$

where X, Y, Z are the components of R parallel to x, y, z respectively.

16.3 This quantity is, in general, different for different lines drawn between A and P. When, however, within a certain region, the quantity

$$Xdx + Ydy + Zdz = -D\Psi,$$

that is, when it is an exact differential within that region, the value of L becomes

$$L = \Psi_A - \Psi_P,$$

and is the same for any two forms of the path between A and P, provided the one form can be changed into the other by continuous motion without passing out of this region.

16.2 *X, Y, Z are the components of R parallel to x, y, z respectively*: Similarly, let dx, dy, dz be the x-, y-, and z-components of the element ds. We can interpret the form of the integral most readily in the special case for which R represents a force, as Maxwell tacitly assumed in paragraph 14.1. For if R were wholly in the x-direction, then R would be equal to X. And it could do no work along distances dy and dz, to which it would be perpendicular; the integrand would then be simply Xdx. Similarly if R were wholly in the y- or z-direction, the integrand would be Ydy or Zdz. But since R will in general have all three components, it will do work in all three directions and the integrand will be the sum $(Xdx + Ydy + Zdz)$. Then if we may both multiply and divide this quantity by the infinitesimal element ds, the integrand will become

$$\left(X\frac{dx}{ds} + Y\frac{dy}{ds} + Z\frac{dz}{ds} \right) ds \,.$$

16.3 *when it is an exact differential*: The value obtained for integral L over some path 1 will generally be different from the value obtained over another path 2. But it can happen that the integral is independent of path, so that L will have the same value for all paths between the same two endpoints. In that case,

$$L = \int_A^P \left(X\frac{dx}{ds} + Y\frac{dy}{ds} + Z\frac{dz}{ds} \right) ds, \text{ that is, } \int_A^P (Xdx + Ydy + Zdz),$$

will be a function only of points A and P. If we call that function Ψ, then we may consider $(Xdx + Ydy + Zdz)$ as the differential of Ψ, which differential Maxwell writes $-D\Psi$ (modern notation would be $-d\Psi$). Thus for Maxwell,

$$L = \int_A^P (Xdx + Ydy + Zdz) = \int_A^P (-D\Psi).$$

(The negative sign is conventional, and remains standard today.) But what is "exact" about $D\Psi$? Certainly we could write the line integral as $L = \int_A^P (-D\Psi)$

On Potentials.

16.4 The quantity Ψ is a scalar function of the position of the point, and is therefore independent of the directions of reference. It is called the Potential Function, and the vector quantity whose components are X, Y, Z is said to have a potential Ψ, if

$$X = -\frac{d\Psi}{dx}, \quad Y = -\frac{d\Psi}{dy}, \quad Z = -\frac{d\Psi}{dz}.$$

whether it is the same for all paths or not. But if it does depend on the path, the statement would not be *complete*, since there is some other, unexpressed, factor or relation which causes the path to make a difference. If nevertheless we insisted on writing $(Xdx + Ydy + Zdz) = -D\Psi$ we would have to admit that $D\Psi$ was an *incomplete* or, more commonly, an *inexact* differential. Only when the line integral is path-independent does $D\Psi$ deserve to be called an *exact* differential; Maxwell reserves the notation $D\Psi$ for the exact case.

The integral L is evaluated like any other and yields

$$L = \int_A^P (-D\Psi) = (-\Psi)\big|_P - (-\Psi)\big|_A = \Psi_A - \Psi_P,$$

as Maxwell states. Note that this is a line-integral whose value depends only on the endpoints—the criterion Maxwell named in paragraph 14.1 that identifies a line-integral as a *potential*. Note too that the integral from A to P must be the negative of the integral from P to A—and thus the integral over the round-trip path APA will be zero. Maxwell will make use of this fact in Art. 45 below.

16.4 *the vector quantity whose components are X, Y, Z is said to have a potential if …*: We saw in the previous paragraph that the vector will have potential Ψ if its components satisfy the relation $(Xdx + Ydy + Zdz) = -D\Psi$. But this relation can be restated by reversing the reasoning of paragraph 16.2. For if we attend only to the work done in the x-direction, Xdx will equal whatever portion of $-D\Psi$ is due exclusively to action along dx. Calling that portion $-d\Psi$, we shall have

$$Xdx = -d\Psi, \quad \text{from which} \quad X = -\frac{d\Psi}{dx}. \quad \text{Similarly,} \quad Y = -\frac{d\Psi}{dy} \quad \text{and} \quad Z = -\frac{d\Psi}{dz},$$

with $-d\Psi$ in each case representing the portion of $-D\Psi$ that is due exclusively to action in the corresponding direction. Thus the three expressions for X, Y, and Z are together equivalent to the relation $(Xdx + Ydy + Zdz) = -D\Psi$ and like it they imply the existence of potential Ψ.

The three expressions resemble derivatives; but they are only *partial* derivatives because, as was mentioned, their numerators represent in each case only a portion of the whole $D\Psi$. Maxwell has no special notation for partial derivatives; readers are expected to remember when a derivative is partial and when it is not. Today we distinguish a partial derivative by using the notation $X = -\frac{\partial \Psi}{\partial y}$. Otherwise stated, the components of the intensity are the (nega-tive) *partial derivatives of the potential*, respectively. As Maxwell will put it in Art. 395, "the force … is deduced from the potential."

16.5 When a potential function exists, surfaces for which the potential is constant are called Equipotential surfaces. The direction of R at any point of such a surface coincides with the normal to the surface, and if n be a normal at the point P, then

$$R = -\frac{d\Psi}{dn}.$$

16.6 The method of considering the components of a vector as the first derivatives of a certain function of the coordinates with respect to these coordinates was invented by Laplace* in his treatment of the theory of attractions. The name of Potential was first given to this function by Green,† who made it the basis of his treatment of electricity. Green's essay was neglected by mathematicians till 1846, and before that time most of its important theorems had been rediscovered by Gauss, Chasles, Sturm, and Thomson.‡

16.7 In the theory of gravitation the potential is taken with the opposite sign to that which is here used, and the resultant force in any

* *Méc. Céleste*, liv. iii.

† Essay on the Application of Mathematical Analysis to the Theories of Electricity and Magnetism, Nottingham, 1828. Reprinted in *Crelle's Journal*, and in Mr. Ferrers' edition of Green's Works.

‡ Thomson and Tait, *Natural Philosophy*, § 483.

16.5 *The direction of R at any point of such a surface coincides with the normal to the surface*: If R were not normal to the equipotential surface, it would have a component parallel to the surface. Then the line-integral of that component along the surface would not be zero, and the surface would not be equipotential—which is contrary to the supposition.

if n be a normal at the point P, then $R = -\dfrac{d\Psi}{dn}$: Just as the x-component of R is the derivative of Ψ with respect to x, and the y-component of R is the derivative of Ψ with respect to y, etc., so R itself will be the derivative of Ψ with respect to R's own direction—which is the direction of the normal, n.

16.7 *In the theory of gravitation the potential is taken with the opposite sign*: As Maxwell will note later in this paragraph, he has so defined potential that the force derived from it has the direction in which the potential decreases; that is the meaning of the negative sign in the equations $X = -d\Psi/dx$, etc., in paragraph 16.4 above. When the force is attractive, then, the potential decreases in the direction of decreasing distance from the center of attraction; thus $d\Psi/dx$ will be positive and X will be negative. Gravitational forces are attractive and therefore ought to be negative according to this rule. But the gravitational force on a body is its *weight*—which Maxwell's contemporaries were reluctant to call negative! In the special case of gravitation, therefore, they reversed the sign of the potential; thus gravitational potential was taken as increasing in the downward direction (Thomson and Tait, *Treatise on Natural Philosophy* §483). Today we

direction is then measured by the rate of *increase* of the potential function in that direction. In electrical and magnetic investigations the potential is defined so that the resultant force in any direction is measured by the *decrease* of the potential in that direction. This method of using the expression makes it correspond in sign with potential energy, which always decreases when the bodies are moved in the direction of the forces acting on them.

17.] The geometrical nature of the relation between the potential and the vector thus derived from it receives great light from Hamilton's discovery of the form of the operator by which the vector is derived from the potential.

17.1 The resolved part of the vector in any direction is, as we have seen, the first derivative of the potential with respect to a coordinate drawn in that direction, the sign being reversed.

17.2 Now if i, j, k are three unit vectors at right angles to each other, and if X, Y, Z are the components of the vector F resolved parallel to these vectors, then

$$\mathfrak{F} = iX + jY + kZ; \tag{1}$$

and by what we have said above, if Ψ is the potential,

$$\mathfrak{F} = -\left(i\,\frac{d\Psi}{dx} + j\,\frac{d\Psi}{dy} + k\,\frac{d\Psi}{dz} \right). \tag{2}$$

17.3 If we now write ∇ for the operator

$$i\,\frac{d}{dx} + j\,\frac{d}{dy} + k\,\frac{d}{dz}, \tag{3}$$

[then] $$F = -\nabla\Psi. \tag{4}$$

are less squeamish about negative quantities and generally treat gravitation no differently from other centripetal forces.

17.2 *Unit vectors*: Vectors i, j, and k have unit length and are directed in the x-, y-, and z-directions, respectively. Note that the vector whose magnitude Maxwell has up to now called F is here expressed as \mathfrak{F} when he writes it explicitly as a vector—an example of the practice pointed out in the comment on paragraph 11.8.

17.3 *If we now write* ∇: The mathematical meaning of ∇ as set forth in eq. (3) was originated by Hamilton; but the symbol, unnamed, was introduced by Tait. Maxwell adopted for it the name *nabla*, at first playfully and later in earnest, on the basis of a suggestion that the inverted delta shape resembled that of the harplike instrument called נבל (nebhel) in the Hebrew Bible. You can see that equation (2) may be obtained by applying the operator −∇, term by term, to the potential Ψ. That is what equation (4) expresses.

In the present discussion, nabla has been applied to the scalar quantity Ψ. Maxwell will treat the effects of nabla on a vector quantity in Art. 25.

17.4 The symbol of operation ∇ may be interpreted as directing us to measure, in each of three rectangular directions, the rate of increase of Ψ, and then, considering the quantities thus found as vectors, to compound them into one. This is what we are directed to do by the expression (3). But we may also consider it as directing us first to find out in what direction Ψ increases fastest, and then to lay off in that direction a vector representing this rate of increase.

17.5 M. Lamé, in his *Traité des Fonctions Inverses*, uses the term Differential Parameter to express the magnitude of this greatest rate of increase, but neither the term itself, nor the mode in which Lamé uses it, indicates that the quantity referred to has direction as well as magnitude. On those rare occasions in which I shall have to refer to this relation as a purely geometrical one, I shall call the vector \mathfrak{F} the space-variation of the scalar function Ψ, using the phrase to indicate the direction, as well as the magnitude, of the most rapid decrease of Ψ.

* * *

On Surface-Integrals.

21.] Let dS be the element of a surface, and ε the angle which a normal to the surface drawn towards the positive side of the surface makes with the direction of the vector quantity R, then

$$\int R \cos \varepsilon \, dS$$

is called the surface-integral of R over the surface S.*

21.1 THEOREM III. *The surface-integral of the flux inwards through a elosed surface may be expressed as the volume-integral of its convergence taken within the surface.* (See Art. 25.)

*{In the following investigations the positive direction of the normal is out-wards from the surface.}

17.5 *the space-variation of ... Ψ*: This will be Maxwell's term for $\nabla\Psi$. Modern writers refer to $\nabla\Psi$ as the *gradient* of Ψ, abbreviated grad Ψ.

21.1 *The surface-integral of the flux inwards*: The expression $\iint R\cos\varepsilon \, dS$ in the preceding paragraph is the surface-integral of a *flux*, as discussed in paragraph 14.2. The theorem Maxwell is about to prove is known today as Gauss's Theorem, except that Maxwell takes account of flux *inwards* through a closed surface, while modern practice considers flux in the outward direction. Note that if R is directed inwards, cos ε will be negative and therefore the integral $\iint R\cos\varepsilon \, dS$ will likewise be negative.

the volume-integral of its convergence: "Convergence" is in effect defined by the present theorem, being that quantity whose volume-integral is equal to the surface-integral of flux inwards through a closed surface. In paragraph 21.6 below,

21.2 Let X, Y, Z be the components of R, and let l, m, n be the direc-
tion-cosines of the normal to S measured outwards. Then the
surface-integral of R over S is

$$\iint R\cos\varepsilon\,dS = \iint Xl\,dS + \iint Ym\,dS + \iint Zn\,dS,\qquad(1)$$

the values of X, Y, Z being those at a point in the surface, and the inte-
grations being extended over the whole surface.

that quantity turns out to be $\left(\dfrac{dX}{dx} + \dfrac{dY}{dy} + \dfrac{dZ}{dz}\right)$, a scalar; it will be negative since,
as we saw, the surface-integral of flux inwards is negative. Maxwell's term *conver-
gence* is meant to express what is happening to the flux within the enclosed
volume: if flux continues to pour into a finite volume it must converge there.

On the other hand, when $\left(\dfrac{dX}{dx} + \dfrac{dY}{dy} + \dfrac{dZ}{dz}\right)$ is positive flux passes *out* of the
volume and must diverge; accordingly, modern terminology names this same
mathematical expression the *divergence* of the flux.

21.2 let l, m, n be the direction-cosines: In the diagram,
n is normal to the surface and makes angles λ, μ, ν
with the x-, y-, and z-axes, respectively. Then let
$l = \cos\lambda$, $m = \cos\mu$, and $n = \cos\nu$. These quantities
are called "direction-cosines" because together they
express the direction of **n** with respect to the
coordinate axes.

It is convenient to consider at this point a
little lemma that will be useful later. Consider a
rectangle of area A inclined at an angle ε to the
xy-plane. Then its normal, **n**, will also make
angle ε with the z-axis, since the z-axis is itself
normal to the xy-plane. It is evident that the
projection of area A onto the xy-plane will be
A cos ε.

But since ε is the angle between **n** and the z-axis, cos ε is one of the *direction-
cosines* of the normal to the surface, namely n.
Thus the projection of an area A onto the xy-plane is An. Similarly its pro-
jection onto the yz-plane is Al, and its projection onto the xz-plane is Am.

Equation (1): Note that by substituting for l, m, and n the equation can be re-
written $\iint R\cos\varepsilon\,dS = \iint X\cos\lambda\,dS + \iint Y\cos\mu\,dS + \int Z\cos\nu\,dS$. Since we
have let λ, μ, and ν be the angles made by the normal with the x-, y-, and z-axes
respectively, they also express the angles made by the normal with X, Y, and Z,
the respective components of R. Then just as $R\cos\varepsilon$ expresses the component
of R that is normal to dS, so also $X\cos\lambda$ expresses the normal component of X,
and similarly for the remaining righthand terms. Thus every term of the equa-
tion is a surface-integral; and the equation simply states that the surface-integral
of R is the sum of the surface-integrals of its three components.

21.3 If the surface is a closed one, then, when y and z are given, the coordinate x must have an even number of values, since a line parallel to x must enter and leave the enclosed space an equal number of times provided it meets the surface at all.

21.4 At each entrance

$$l \, dS = -dy \, dz,$$

and at each exit

$$l \, dS = dy \, dz.$$

21.5 Let a point travelling from $x = -\infty$ to $x = +\infty$ first enter the space when $x = x_1$, then leave it when $x = x_2$, and so on; and let the values of X at these points be X_1, X_2, &c., then

$$\iint Xl \, dS = -\iint \left[\left(X_1 - X_2 \right) + \left(X_3 - X_4 \right) + \text{&c.} + \left(X_{2n-1} - X_{2n} \right) \right] dy \, dz. \quad (2)$$

21.6 If X is a quantity which is continuous, and has no infinite values between x_1 and x_2, then

$$X_2 - X_1 = \int_{x_1}^{x_2} \frac{dX}{dx} \, dx; \quad (3)$$

where the integration is extended from the first to the second intersection, that is, along the first segment of x which is within the closed

21.3 *when y and z are given, the coordinate x must have an even number of values*: First let the z-coordinate be given; this is equivalent to taking a section through the solid, parallel to the xy-plane. But the intersection of a plane and a closed solid surface is a closed curve, as shown in the drawing. If then the y-coordinate in the plane is also given, the remaining x-coordinate must have an even number of values, because a straight line that enters the curve any number of times must exit from it the same number of times.

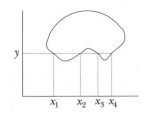

21.4 *At each entrance...*: Since λ is the angle between X and the normal \mathbf{n} to the surface, $l \, dS$ or $dS \cos \lambda$ is that aspect of area dS that presents itself perpendicularly to X and is therefore the area $-dy \, dz$. The sign is negative because Maxwell is taking the normal to be exiting the closed surface, while X is entering the surface. Thus angle λ is obtuse, and its cosine will be negative. But the sign will be positive wherever X exits the surface, for then angle λ will be acute and its cosine positive.

21.5 *Equation* (2): Maxwell is adding up the products $X \, dy \, dz$ at every entrance and exit of X, taking into account that X will in general have different values X_1, X_2, etc., at different locations. Notice that he uses the double-integral sign \iint to acknowledge that the integration is carried out over a surface and therefore over two variables, in this case y and z.

surface. Taking into account all the segments which lie within the closed surface, we find

$$\int X\, l\, dS = \iiint \frac{dX}{dx} dx\,dy\,dz, \tag{4}$$

the double integration being confined to the closed surface, but the triple integration being extended to the whole enclosed space. Hence, if X, Y, Z are continuous and finite within a closed surface S, the total surface-integral of R over that surface will be

$$\iint R \cos \varepsilon\, dS = \iiint \left(\frac{dX}{dx} + \frac{dY}{dy} + \frac{dZ}{dz} \right) dx\,dy\,dz, \tag{5}$$

the triple integration being extended over the whole space within S. ...

21.7 If at every point where X, Y, Z are continuous

$$\frac{dX}{dx} + \frac{dY}{dy} + \frac{dZ}{dz} = 0, \tag{6}$$

then the surface-integral over every closed surface is zero, and the distribution of the vector quantity is said to be Solenoidal.

21.6 *Equation (3)*: The difference in X between the first entrance and first exit can be expressed as the integral $X_2 - X_1 = \int_{x_1}^{x_2} dX$ since X is continuous and finite. To obtain Maxwell's form, multiply and divide the integrand by dx.

Equation (4): When each difference of terms in eq. (2) is transformed on the model of eq. (3), their sum retains the form $\int dX \frac{dx}{dx}$. Substitute this for the quantity in square brackets in eq. (2) to obtain Maxwell's form. Notice that since a single integral has been inserted into a double integral we now have a triple integral; it is an integral over the enclosed volume and thus over the three variables x, y, z.

Equation (5): Repeat the previous argument for Y and Z to obtain $\iint Y m\, dS = \iiint \frac{dY}{dy} dx\,dy\,dz$ and $\iint Z n\, dS = \iiint \frac{dZ}{dz} dx\,dy\,dz$. Substitute these along with eq. (4) into eq. (1) to obtain eq. (5), which completes the proof of Theorem III (21.1). We will pass over Maxwell's consideration of the case where X, Y, and Z are not continuous.

21.7 *the surface-integral over every closed surface is zero*: The left-hand side of eq. (9) measures the convergence (if negative) or divergence (if positive) of flux within the enclosed volume. If it is equal to zero, then as much flux exits the surface as enters it; and the overall flux through the enclosing surface must be zero.

Solenoidal: From Greek *σωλήν*, tube or pipe. In the following paragraphs, Maxwell will show that if eq. (9) applies throughout a region, the flux in that region can be understood as forming a pattern of adjacent and parallel tubes. Hence "the distribution of the vector quantity is said to be Solenoidal," that is, tubelike.

21.8 We shall refer to equation (9) as the General solenoidal condition...

22.] Let us now consider the case in which at every point within the surface S the equation

$$\frac{dX}{dx} + \frac{dY}{dy} + \frac{dZ}{dz} = 0, \tag{11}$$

is satisfied. We have as a consequence of this the surface-integral over the closed surface equal to zero.

22.1 Now let the closed surface S consist of three parts S_1, S_0, and S_2. Let S_1 be a surface of any form bounded by a closed line L_1. Let S_0 be formed by drawing lines from every point of L_1 always coinciding with the direction of R. If l, m, n are the direction-cosines of the normal at any point of the surface S_0, we have

$$R\cos\varepsilon = Xl + Ym + Zn = 0. \tag{12}$$

Hence this part of the surface contributes nothing towards the value of the surface-integral.

22.2 Let S_2 be another surface of any form bounded by the closed curve L_2 in which it meets the surface S_0.

22.3 Let Q_1, Q_0, Q_2 be the surface-integrals of the surfaces S_1, S_0, S_2, and let Q be the surface-integral of the closed surface S. Then

22. *at every point within the surface S*: That is, at every point in the *volume* enclosed by surface S.

22.1 *let the closed surface S consist of three parts S_1, S_0, and S_2* : In this and the next three paragraphs, Maxwell discusses a surface like the one shown in the diagram. The lines leading from S_1 to S_2 coincide everywhere with the direction of R.

Equation (12): Maxwell is using the form of eq. (1) in paragraph 21.2 to express the integrands that pertain to surface S_0. But it is clear in any case that the surface-integral of R over S_0 is zero, because surface S_0 everywhere coincides with the direction of R, so that R nowhere crosses the surface. Thus "this part of the surface [that is, S_0,] contributes nothing towards the value of the surface-integral" over the whole surface S.

22.3 *equal and opposite*: By thus restating eq. (15), Maxwell is emphasizing a more visual understanding: All the flux entering S_1 must exit S_2 since no flux crosses the tubular walls, and there is no convergence of flux within the enclosed volume. The quantities are *opposite* because entrances and exits have opposite directions with respect to the normals to S_1 and S_2; these are always taken to point outward.

surface ... for which R is always tangential: Surface S_0 is such a surface because at any point on it, a line having the direction of R will be tangent to the surface.

$$Q_1 + Q_0 + Q_2 = 0; \qquad (13)$$

and we know that

$$Q_0 = 0; \qquad (14)$$

therefore

$$Q_2 = -Q_1; \qquad (15)$$

or, in other words, the surface-integral over the surface S_2 is equal and opposite to that over S_1, whatever be the form and position of S_2, provided that the intermediate surface S_0 is one for which R is always tangential.

22.4 If we suppose L_1 a closed curve of small area, S_0 will be a tubular surface having the property that the surface-integral over every complete section of the tube is the same.

22.5 Since the whole space can be divided into tubes of this kind provided

$$\frac{dX}{dx} + \frac{dY}{dy} + \frac{dZ}{dz} = 0, \qquad (16)$$

a distribution of a vector quantity consistent with this equation is called a Solenoidal Distribution.

On Tubes and Lines of Flow.

22.6 If the space is so divided into tubes that the surface-integral for every tube is unity, the tubes are called Unit tubes, and the surface-integral over any finite surface S bounded by a closed curve L is equal to the *number* of such tubes which pass through it in the positive

22.4 *the surface-integral over every complete section of the tube is the same*: Refer to the diagram in the comment on paragraph 22.1. A "complete" section is one which, like S_2 in that drawing, intersects all the lines that cut L_1. If any additional sections are made, the argument of paragraph 22.3 will apply to them, showing that their surface-integrals are equal to one another and to the surface-integral over S_1.

22.5 *the whole space can be divided into tubes of this kind* ...: Tubes "of this kind" are tubes for which the surface-integral over every complete section is the same. That equality was established by the argument of paragraph 22.3, which in turn required the convergence to be zero—the condition expressed by eq. (16)—in order to affirm that the integral over the whole surface S is zero.

22.6 *the surface-integral over any finite surface S ... is equal to the number of such tubes which pass through it*: Maxwell might almost be speaking to Faraday here. Faraday had regarded his Moving Wire as able, in effect, to "count" magnetic lines of force—at the same time knowing full well that lines of force can no more be counted than can the points on a surface. By replacing Faraday's lines with *unit tubes*, Maxwell has defined an entity that occupies finite area and thus really is countable in the way Faraday had intended.

direction, or, what is the same thing, the number which pass through the closed curve L.

22.7 Hence the surface-integral of S depends only on the form of its boundary L, and not on the form of the surface within its boundary.

<p style="text-align:center">* * *</p>

On Right-handed and Left-handed Relations in Space.

23.] In this treatise the motions of translation along any axis and of rotation about that axis will be assumed to be of the same sign when their directions correspond to those of the translation and rotation of an ordinary or right-handed screw.*

> *The combined action of the muscles of the arm when we turn the upper side of the right hand outwards, and at the same time thrust the hand forwards, will impress the right-handed screw motion on the memory more firmly than any verbal definition. A common corkscrew may be used as a material symbol of the same relation.
>
> Professor W. H. Miller has suggested to me that as the tendrils of the vine are right-handed screws and those of the hop left-handed, the two systems of relations in space might be called those of the vine and the hop respectively.
>
> The system of the vine, which we adopt, is that of Linnæus, and of screw-makers in all civilized countries except Japan. De Candolle was the first who called the hop-tendril right-handed, and in this he is followed by Listing, and by most writers on the circular polarization of light. Screws like the hop-tendril are made for the couplings of railway-carriages, and for the fittings of wheels on the left side of ordinary carriages, but they are always called left-handed screws by those who use them.

22.7, continued. *the number which pass through the closed curve L*: That is, the number that are *encircled* by closed curve L, since any tubes cut by a surface will necessarily be encircled by that surface's boundary, as indicated in the sketch.

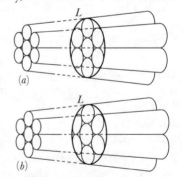

Notice that unit tubes need not have uniform cross-section; it is the *surface-integral*, not the *surface area*, that has unit value throughout.

22.7 *the surface-integral of S depends only on the form of its boundary L, and not on the form of the surface*: The sketch shows two cases of a surface intersecting a number of unit tubes. Clearly, the same number of tubes will be intersected whether the surface is plane in shape (case a) or convex (case b), so long as it is bounded by the same curve L.

Maxwell could have drawn this inference directly from paragraph 22.4, since equality of the surface-integrals for all complete sections depends in no way on the surfaces' shapes. Instead, he has waited until the highly visual notion of *unit tubes* was defined. He will state a formal relation between the surface-integral and the surface boundary as Theorem IV in Art. 24.

23.1 For instance, if the actual rotation of the earth from west to east is taken positive, the direction of the earth's axis from south to north will be taken positive, and if a man walks forward in the positive direction, the positive rotation is in the order, head, right-hand, feet, left-hand.

23.2 If we place ourselves on the positive side of a surface, the positive direction along its bounding curve will be opposite to the motion of the hands of a watch with its face towards us.

23.3 This is the right-handed system which is adopted in Thomson and Tait's *Natural Philosophy*, and in Tait's *Quaternions*. The opposite, or left-handed system, is adopted in Hamilton's *Quaternions*... The operation of passing from the one system to the other is called by Listing, Perversion.

23.4 The reflexion of an object in a mirror is a perverted image of the object.

23.5 When we use the Cartesian axes of x, y, z, we shall draw them so that the ordinary conventions about the cyclic order of the symbols lead to a right-handed system of directions in space. Thus, if x is drawn east-ward and y northward, z must be drawn upward.

* * *

23.2 *If we place ourselves on the positive side of a surface*: First imagine lying face-up upon the surface; we shall be oriented in the positive direction, that is, away from the surface. Then the direction of positive rotation will be the same as the example offered in the previous paragraph: head, right hand, feet, left hand; it will be clockwise. But if instead we lie face-down, that is, oriented in the negative direction, the same positive rotation will appear to us in the reverse order, that is, counter-clockwise. Maxwell might have been more explicit about what he means by "placing ourselves" on the positive side, but it is clear that in the present example he intends the second case, facing the surface.

23.5 *if x is drawn eastward and y northward, z must be drawn upward*: If the book is placed on a horizontal reading surface, "upward" means vertically out of the page. The sketch shows another perspective on the mutual relation of the axes.

Editor's Introduction to
Article 24 (Stokes' Theorem)

In Art. 24 of "Preliminaries" Maxwell will prove that

A line-integral taken round a closed curve may be expressed in terms of a surface-integral taken over a surface bounded by the curve.

Today this theorem is known as Stokes' Theorem (24., *comment*); in the *Treatise* Maxwell simply calls it Theorem IV. To demonstrate it, Maxwell derives the following equation, appearing there as equation (11):

$$\int \left(X \frac{dx}{ds} + Y \frac{dy}{ds} + Z \frac{dz}{ds} \right) ds = \iint (l\xi + m\eta + n\zeta)\, dS.$$

The left-hand member of the equation expresses a line-integral, the right-hand member a surface-integral.

Maxwell's proof, while quite beautiful, is lengthy and dense. Here, for readers who may prefer it, is a simpler alternative; similar proofs can be found in many modern textbooks. Maxwell's own treatment begins on page 41 below.

1. The form of the line integral

Maxwell introduced the line-integral of vector R in Art. 16 and noted (16.2, *comment*) that if the x-, y-, and z-components of R are X, Y, and Z, the line-integral takes the form

$$L = \int \left(X \frac{dx}{ds} + Y \frac{dy}{ds} + Z \frac{dz}{ds} \right) ds. \qquad (i)$$

The present Theorem applies when the path of integration is a *closed loop*. Modern practice employs the special "loop integral" symbol \oint for this case, but Maxwell uses no distinctive notation. Differentiate both sides of the above equation to obtain the *element* of L,

$$dL = \left(X \frac{dx}{ds} + Y \frac{dy}{ds} + Z \frac{dz}{ds} \right) ds = X\,dx + Y\,dy + Z\,dz. \qquad (ii)$$

2. Infinitesimal loop in coordinate plane

Consider an infinitesimal rectangular loop in the yz-plane, with sides dy and dz. The values of Y and Z (components of R) are indicated at relevant locations in the neighborhood of the loop. Now apply eq. (ii) to each leg of the loop in turn; we obtain

$$dL = Y_1\,dy + (Z_1 + dZ)\,dz - (Y_1 + dY)\,dy - Z_1\,dz.$$

(Note that each leg dy or dz is taken as positive in the *counter-clockwise* direction about the loop; see Art. 23.). Collecting like terms, we have

$$dL = dZ\,dz - dY\,dy. \qquad\qquad (iii)$$

Next, divide and multiply the right-hand side of eq. (iii) by $dy\,dz$, the area of the loop, to obtain

$$dL = \left(\frac{dZ}{dy} - \frac{dY}{dz}\right)dy\,dz. \qquad\qquad (iv)$$

A similar treatment of loops in the xz- and xy-planes yields

$$dL = \left(\frac{dX}{dz} - \frac{dZ}{dx}\right)dx\,dz \quad \text{and} \quad dL = \left(\frac{dY}{dx} - \frac{dX}{dy}\right)dx\,dy.$$

3. Infinitesimal loop in any direction

Consider a loop dS having any orientation whatsoever, and suppose that the differential loops $dy\,dz$, $dx\,dz$, and $dx\,dy$ which we have been discussing are the projections of dS onto the yz-, xz-, and xy-planes, respectively; the sketch depicts one such example. And let l, m, n be the direction-cosines (21.2) of the normal to dS. Then by the lemma offered in (21.2, *comment*), the respective projections will be $l\,dS$, $m\,dS$, and $n\,dS$.

Thus we shall have, in the yz-plane,

$$dy\,dz = l\,dS.$$

When this is substituted into eq. (iv) it yields the x-component of dL for the loop dS:

$$dL_x = \left(\frac{dZ}{dy} - \frac{dY}{dz}\right)l\,dS,$$

and, similarly, the y- and z-components are

$$dL_y = \left(\frac{dX}{dZ} - \frac{dZ}{dX}\right) m\, dS \quad \text{and} \quad dL_z = \left(\frac{dY}{dX} - \frac{dX}{dY}\right) n\, dS.$$

Adding these components, we have for the loop dS

$$dL = \left(\frac{dZ}{dY} - \frac{dY}{dZ}\right) l\, dS + \left(\frac{dX}{dZ} - \frac{dZ}{dX}\right) m\, dS + \left(\frac{dX}{dY} - \frac{dY}{dX}\right) n\, dS.$$

In (24.2) Maxwell will define the quantities

$$\xi = \frac{dZ}{dY} - \frac{dY}{dZ}, \quad \eta = \frac{dX}{dZ} - \frac{dZ}{dX}, \quad \zeta = \frac{dY}{dX} - \frac{dX}{dY}.$$

Using Maxwell's own symbols, then, we may write

$$dL = \left(l\xi + m\eta + n\zeta\right) dS. \qquad (v)$$

But the right-hand side of this equation is an element of *area*. We have thus shown that a differential element of a *line-integral* (the left-hand side) can be expressed as a differential element of a *surface-integral* (the right-hand side). This represents the infinitesimal case, as it were, of Stokes' Theorem. In the next section we shall extend the Theorem to finite lines and areas.

4. Extension to a finite loop

The following argument was originated by Ampère. Although Maxwell does not use it explicitly in his own proof of Stokes' Theorem, he will put it to another use in Art. 483.

Consider two adjacent loops, and for some vector R, form the line-integral about each of them, keeping consistently to a single direction; in the sketch, for example, that direction is counter-clockwise.

Leg *be* is common to both loops. Clearly the magnitude and orientation of R (or magnitudes and orientations, if it is not uniform) along *be* are the same, no matter to which loop we assign it. But it is evident that if the direction *be* is positive for the left-hand loop, that same direction is negative for the adjacent loop. When the line-integrals are added, therefore, the two contributions arising from leg *be* cancel each other; and the resulting line-integral is equivalent to the line-integral around a single bounding loop *acdf*. And it will be the sum of two terms, each evaluated for its corresponding loop, and each having the form

$(l\xi + m\eta + n\zeta) dS.$

Since this is the element of a surface-integral, the integral of such terms will constitute a surface-integral.

Then to evaluate the line-integral about any finite loop *abcd*, let it be subdivided into indefinitely many elements; and since eq. (*v*) applies to each element, integrate eq. (*v*) over the entire surface enclosed by the loop to obtain

$$L = \iint (l\xi + m\eta + n\zeta) dS.$$

At last, substitute this expression for L into eq. (*i*). We have

$$\int \left(X\frac{dx}{ds} + Y\frac{dy}{ds} + Z\frac{dz}{ds} \right) ds = \iint (l\xi + m\eta + n\zeta) dS,$$

which is the relationship that was to be proved. This equation is the equivalent of Maxwell's eq. (11) in Article 24, which follows.

23.9 We now proceed to prove a theorem which is useful as establishing a connection between the surface-integral taken over a finite surface and a line-integral taken round its boundary.

24.] THEOREM IV. *A line-integral taken round a closed curve may be expressed in terms of a surface-integral taken over a surface bounded by the curve.*

24.1 Let X, Y, Z be the components of a vector quantity \mathfrak{A} whose line-integral is to be taken round a closed curve s.

24.2 Let S be any continuous finite surface bounded entirely by the closed curve s, and let ξ, η, ζ be the components of another vector quantity \mathfrak{B}, related to X, Y, Z by the equations

$$\xi = \frac{dZ}{dy} - \frac{dY}{dz}, \quad \eta = \frac{dX}{dz} - \frac{dZ}{dx}, \quad \zeta = \frac{dY}{dx} - \frac{dX}{dy}. \quad (1)$$

Then the surface-integral of \mathfrak{B} taken over the surface S is equal to the line-integral of \mathfrak{A} taken round the curve s. It is manifest that ξ, η, ζ satisfy of themselves the solenoidal condition,

$$\frac{d\xi}{dx} + \frac{d\eta}{dy} + \frac{d\zeta}{dz} = 0.$$

23.9 *We now proceed to prove a theorem:* As previously explained, I am treating Art. 24 as a separate chapter. The chapter heading that appears at the top of this page did not, therefore, attach to it in the original.

24. *THEOREM IV:* George Stokes set the proof of this theorem as Question 8 of the Smith's Prize Examination in 1854—the year in which Maxwell, then an undergraduate at Trinity College, became First Smith's Prizeman jointly with E. J. Routh. Although William Thomson actually discovered the relation and communicated it to Stokes in 1850 (Darrigol, p.146n), it continues to be known as Stokes' Theorem.

24.2 *It is manifest that ξ, η, ζ satisfy of themselves the solenoidal condition:* With ξ, η, ζ defined according to eqs. (1), differentiate:

$$\frac{d\xi}{dx} = \frac{d}{dx}\left(\frac{dZ}{dy}\right) - \frac{d}{dx}\left(\frac{dY}{dz}\right) = \frac{d^2 Z}{dx\,dy} - \frac{d^2 Y}{dz\,dx},$$

$$\frac{d\eta}{dy} = \frac{d}{dy}\left(\frac{dX}{dz}\right) - \frac{d}{dy}\left(\frac{dZ}{dx}\right) = \frac{d^2 X}{dy\,dz} - \frac{d^2 Z}{dx\,dy},$$

$$\frac{d\zeta}{dz} = \frac{d}{dz}\left(\frac{dY}{dx}\right) - \frac{d}{dz}\left(\frac{dX}{dy}\right) = \frac{d^2 Y}{dz\,dx} - \frac{d^2 X}{dy\,dz}.$$

The terms cancel pairwise and thus sum to zero, thereby satisfying the solenoidal condition.

24.3 Let l, m, n be the direction-cosines of the normal to an element of the surface dS, reckoned in the positive direction. Then the value of the surface-integral of \mathfrak{B} may be written

$$\iint (l\xi + m\eta + n\zeta)\,dS. \tag{2}$$

24.4 In order to form a definite idea of the meaning of the element dS, we shall suppose that the values of the coordinates x, y, z for every point of the surface are given as functions of two independent variables α and β. If β is constant and α varies, the point (x, y, z) will describe a curve on the surface, and if a series of values is given to β, a series of such curves will be traced, all lying on the surface S. In the same way, by giving a series of constant values to α, a second series of curves may be traced, cutting the first series, and dividing the whole surface into elementary portions, any one of which may be taken as the element dS.

24.5 The projection of this element on the plane of yz is, by the ordinary formula,

$$l\,dS = \left(\frac{dy}{d\alpha}\frac{dz}{d\beta} - \frac{dy}{d\beta}\frac{dz}{d\alpha} \right) d\beta\,d\alpha. \tag{3}$$

The expressions for $m\,dS$ and $n\,dS$ are obtained from this by substituting x, y, z in cyclic order.

24.3 *direction-cosines ... the positive direction*: The direction-cosines of a vector were defined in paragraph 21.2. The positive direction is *outward* from a closed surface; compare the footnote to Art. 21.

24.4 *a series of such curves will be traced, all lying on the surface S*: The drawing shows a network of curves obtained by choosing a succession of values of α, letting β vary for each α, followed by a succession of values of β, letting α vary for each β. These lines divide the surface into areas like the ones shown; if the successive values of α and β differ by the infinitesimal amounts $d\alpha$ and $d\beta$ respectively, any one of those areas may be taken as an element, dS, of the surface.

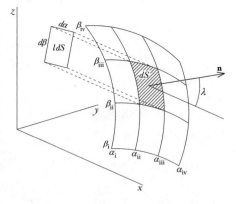

24.5 *the ordinary formula*: Maxwell means the formula for the area of a parallelogram, namely, the product of two adjacent sides together with the sine of the angle between them. It contributes to the derivation of eq. (3) in this way:

Since l is the direction-cosine of the normal to dS with respect to the x-axis, product $l\,dS$ represents the projection of surface-element dS onto the yz-plane. As the diagrams on this and the facing page show, this projection will be an infinitesimal parallelogram with sides $d\alpha$ and $d\beta$, making angles a and b respectively

24.6 The surface-integral which we have to find is

$$\iint \left(l\xi + m\eta + n\zeta \right) dS;$$ (4)

or, substituting the values of ξ, η, ζ in terms of X, Y, Z,

$$\iint \left(m\frac{dX}{dz} - n\frac{dX}{dy} + n\frac{dY}{dx} - l\frac{dY}{dz} + l\frac{dZ}{dy} - m\frac{dZ}{dx} \right) dS.$$ (5)

The part of this which depends on X may be written

$$\iint \left[\frac{dX}{dz}\left(\frac{dz}{d\alpha}\frac{dx}{d\beta} - \frac{dz}{d\beta}\frac{dx}{d\alpha} \right) - \frac{dX}{dy}\left(\frac{dx}{d\alpha}\frac{dy}{d\beta} - \frac{dx}{d\beta}\frac{dy}{d\alpha} \right) \right] d\beta\, d\alpha;$$ (6)

with the z-axis. The angle between the sides is thus $a-b$ and the area of the parallelogram is therefore

$$l\, dS = d\alpha\, d\beta \sin(a-b)$$
$$= d\alpha\, d\beta (\sin a \cos b - \sin b \cos a).$$

But it is also evident from the diagram that

$$\sin a = \frac{\frac{\partial y}{\partial \alpha} d\alpha}{d\alpha} = \frac{\partial y}{\partial \alpha};$$

and similarly

$$\cos b = \partial z/\partial\beta, \quad \sin b = \partial y/\partial\beta, \quad \cos a = \partial z/\partial\alpha.$$

Substituting, we have

$$l\, dS = \left(\frac{\partial y}{\partial \alpha}\frac{\partial z}{\partial \beta} - \frac{\partial y}{\partial \beta}\frac{\partial z}{\partial \alpha} \right) d\alpha\, d\beta,$$

which is Maxwell's eq. (3). Note that the derivatives are *partial* derivatives; otherwise the terms in the parentheses would appear to cancel one another. As mentioned earlier, Maxwell has no special notation for partial derivatives.

The expressions for mdS and ndS are obtained from this...: We obtain

$$m\, dS = \left(\frac{\partial z}{\partial \alpha}\frac{\partial x}{\partial \beta} - \frac{\partial z}{\partial \beta}\frac{\partial x}{\partial \alpha} \right) d\alpha\, d\beta \quad \text{and} \quad n\, dS = \left(\frac{\partial x}{\partial \alpha}\frac{\partial y}{\partial \beta} - \frac{\partial x}{\partial \beta}\frac{\partial y}{\partial \alpha} \right) d\alpha\, d\beta,$$

which are the projections of dS onto the zx- and xy-planes, respectively.

24.6 *The surface-integral which we have to find*: Eq. (4) repeats eq. (2), which was identified in paragraph 24.3 as the integral of \mathfrak{B} over the surface-element dS.

the values of ξ, η, ζ: These were defined in eq. (1), paragraph 24.2.

The part of this which depends on X: In eq. (5), ignore terms containing Y and Z, then substitute $m\, dS$ and $n\, dS$ just obtained to yield eq. (6). Maxwell is about to transform the surface-integral of \mathfrak{B} into a line-integral of \mathfrak{A}, whose components are X, Y, and Z. By considering these components separately, he will be able to use the treatment of X as a pattern to be applied to Y and to Z—just as, in the previous step, he used the argument for projecting dS onto the yz-plane as a pattern for projecting it also onto the zx- and xy-planes.

adding and subtracting $\dfrac{dX}{dx}\dfrac{dx}{d\alpha}\dfrac{dx}{d\beta}$, this becomes

$$= \iint \left\{ \frac{dx}{d\beta}\left(\frac{dX}{dx}\frac{dx}{d\alpha} + \frac{dX}{dy}\frac{dy}{d\alpha} + \frac{dX}{dz}\frac{dz}{d\alpha} \right)\right.$$
$$\left. - \frac{dx}{d\alpha}\left(\frac{dX}{dx}\frac{dx}{d\beta} + \frac{dX}{dy}\frac{dy}{d\beta} + \frac{dX}{dz}\frac{dz}{d\beta} \right) \right\} d\beta\, d\alpha; \qquad (7)$$

$$= \iint \left(\frac{dX}{d\alpha}\frac{dx}{d\beta} - \frac{dX}{d\beta}\frac{dx}{d\alpha} \right) d\beta\, d\alpha. \qquad (8)$$

24.7 Let us now suppose that the curves for which β is constant form a series of closed curves surrounding a point on the surface for which α has its minimum value, α_0, and let the last curve of the series, for which $\alpha = \alpha_1$, coincide with the closed curve s.

24.8 Let us also suppose that the curves for which β is constant form a series of lines drawn from the point at which $\alpha = \alpha_0$ to the closed curve s, the first, β_0, and the last, β_1, being identical.

24.6, continued. *adding and subtacting… this becomes*: Expand the terms in the integrand of eq. (6) to obtain

$$\frac{dX}{dz}\frac{dz}{d\alpha}\frac{dx}{d\beta} - \frac{dX}{dz}\frac{dz}{d\beta}\frac{dx}{d\alpha} - \frac{dX}{dy}\frac{dx}{d\alpha}\frac{dy}{d\beta} + \frac{dX}{dy}\frac{dx}{d\beta}\frac{dy}{d\alpha}.$$

Then when the term $\dfrac{dX}{dx}\dfrac{dx}{d\alpha}\dfrac{dx}{d\beta}$ is added and subtracted we shall have

$$\frac{dX}{dz}\frac{dz}{d\alpha}\frac{dx}{d\beta} - \frac{dX}{dz}\frac{dz}{d\beta}\frac{dx}{d\alpha} - \frac{dX}{dy}\frac{dx}{d\alpha}\frac{dy}{d\beta} + \frac{dX}{dy}\frac{dx}{d\beta}\frac{dy}{d\alpha} + \frac{dX}{dx}\frac{dx}{d\alpha}\frac{dx}{d\beta} - \frac{dX}{dx}\frac{dx}{d\alpha}\frac{dx}{d\beta}.$$

Factor out $dx/d\beta$ and $dx/d\alpha$ to obtain the integrand of eq. (7).

24.7 *the curves for which α is constant form a series of closed curves*: The curves having constant α were introduced in paragraph 24.4, but now Maxwell specifies that they are to be *closed* curves, like earth's lines of latitude. Since he has all along assumed that the whole surface is bounded by the closed curve s, he is able to make s the last curve in the series.

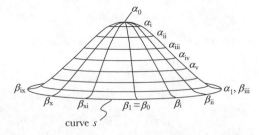

24.8 *the curves for which β is constant form a series of lines … the first, β_0, and the last, β_1, being identical*: The requirement that $\beta_0 = \beta_1$ makes these curves comparable to earth's lines of longitude, with point α_0 corresponding to the pole.

24.9 Integrating (8) by parts, the first term with respect to α and the second with respect to β, the double integrals destroy each other and the expression becomes

$$\int_{\beta_0}^{\beta_1}\left(X\frac{dx}{d\beta}\right)d\beta - \int_{\beta_0}^{\beta_1}\left(X\frac{dx}{d\beta}\right)d\beta - \int_{\alpha_0}^{\alpha_1}\left(X\frac{dx}{d\alpha}\right)d\alpha - \int_{\alpha_0}^{\alpha_1}\left(X\frac{dx}{d\alpha}\right)d\alpha\Big|_{\beta=\beta_0}. \qquad (9)$$

24.9 *Integrating (8) by parts*: To a modern reader, "integration by parts" is the technique of replacing an integral of form $\int u\,dv$ with the equivalent expression $uv - \int v\,du$. But Maxwell is using the expression "by parts" in a more colloquial sense, as though to say "step by step." Here are the steps.

In eq. (8), integrate the terms in parentheses separately, as

$$\iint\left(\frac{dX}{d\alpha}\frac{dx}{d\beta}\right)d\beta\,d\alpha - \iint\left(\frac{dX}{d\beta}\frac{dx}{d\alpha}\right)d\beta\,d\alpha.$$

the first term with respect to α and the second with respect to β: Form the inner integrals, assigning $d\alpha$ to the first and $d\beta$ to the second, as

$$\int\left(\int\frac{dX}{d\alpha}\frac{dx}{d\beta}d\alpha\right)d\beta - \int\left(\int\frac{dX}{d\beta}\frac{dx}{d\alpha}d\beta\right)d\alpha.$$

Then since $d\alpha$ and $d\beta$ are infinitesimally small, $dx/d\alpha$ and $dx/d\beta$ may be considered constant and may be removed from their respective integrands. Then since the limits of integration are from α_0 to α_1 and from β_0 to β_1 we have on the one hand

$$\int_{\alpha_0}^{\alpha_1}\frac{dX}{d\alpha}\frac{dx}{d\beta}d\alpha = \frac{dx}{d\beta}\int_{\alpha_0}^{\alpha_1}\frac{dX}{d\alpha}d\alpha = \frac{dx}{d\beta}\int_{\alpha_0}^{\alpha_1}dX = \frac{dx}{d\beta}X\Big|_{\alpha_0}^{\alpha_1},$$

and on the other hand

$$\int_{\beta_0}^{\beta_1}\frac{dX}{d\beta}\frac{dx}{d\alpha}d\beta = \frac{dx}{d\alpha}\int_{\beta_0}^{\beta_1}\frac{dX}{d\beta}d\beta = \frac{dx}{d\alpha}\int_{\beta_0}^{\beta_1}dX = \frac{dx}{d\alpha}X\Big|_{\beta_0}^{\beta_1}.$$

Now we may take the outer integrals, integrating the first of the above expressions with respect to β and the second with respect to α to obtain

$$\int_{\beta_0}^{\beta_1}\left(X\frac{dx}{d\beta}\right)d\beta - \int_{\beta_0}^{\beta_1}\left(X\frac{dx}{d\beta}\right)d\beta - \int_{\alpha_0}^{\alpha_1}\left(X\frac{dx}{d\alpha}\right)d\alpha + \int_{\alpha_0}^{\alpha_1}\left(X\frac{dx}{d\alpha}\right)d\alpha\Big|_{\beta=\beta_0},$$

which is Maxwell's eq. (9).

the double integrals destroy one another: Maxwell means this remark to apply to the previous step, which yielded his eq. (9). But his choice of expression is rather odd, since that procedure did not involve any sort of destruction or mutual cancellation. Perhaps he means only that when the double integrals are treated as concatenations of single integrals they cease to *be* double integrals.

24.10 Since the point (α, β_1) is identical with the point (α, β_0), the third and fourth terms destroy each other; and since there is but one value of x at the point where $\alpha = \alpha_0$, the second term is zero, and the expression is reduced to the first term.

24.11 Since the curve $\alpha = \alpha_1$ is identical with the closed curve s, we may write the expression in the form

$$\int X \frac{dX}{ds} ds, \tag{10}$$

where the integration is to be performed round the curve s. We may treat in the same way the parts of the surface-integral which depend upon Y and Z, so that we get finally,

24.10 *the third and fourth terms destroy each other*: Examine eq. (9),

$$\int_{\beta_0}^{\beta_1}\left(X\frac{dx}{d\beta}\right)_{\alpha=\alpha_1} d\beta - \int_{\beta_0}^{\beta_1}\left(X\frac{dx}{d\beta}\right)_{\alpha=\alpha_0} d\beta - \int_{\alpha_0}^{\alpha_1}\left(X\frac{dx}{d\alpha}\right)_{\beta=\beta_1} d\alpha + \int_{\alpha_0}^{\alpha_1}\left(X\frac{dx}{d\alpha}\right)_{\beta=\beta_0} d\alpha .$$

Since $\beta_0 = \beta_1$, the third and fourth terms are identical but carry opposite signs; they therefore cancel one another.

the second term is zero: All the values of α other than α_0 determine curves that extend over a range of points and therefore over varying coordinate x. But α_0 corresponds to a single point and therefore to a fixed value of x. Then since the second term is to be evaluated at α_0, dx will be 0 in that term and the integral will likewise be zero, "and the expression is reduced to the first term." We can now appreciate Maxwell's foresight in paragraphs 24.7 and 24.8; his artful choice of coordinates there has caused three of four integrals to disappear!

Recall that this whole line of reasoning, beginning with eq. (6) in paragraph 24.6, has concerned the contribution of component X to the surface-integral of \mathfrak{B}. Thus that contribution amounts to the single term

$$\int_{\beta_0}^{\beta_1}\left(X\frac{dx}{d\beta}\right)_{\alpha=\alpha_1} d\beta,$$

and it has the form of a line integral—the form Maxwell is seeking.

24.11 *the curve $\alpha = \alpha_1$ is identical with the closed curve s*: Maxwell specified this in paragraph 24.7 above. Then $d\beta$ will be identical to ds in the above integral, which may therefore be rewritten as

$$\int\left(X\frac{dx}{ds}\right) ds.$$

Note that this integral is taken over a *closed* path. Modern notation employs the special symbol \oint to indicate this case, but Maxwell uses no distinctive integral sign.

We may treat in the same way the parts of the surface-integral which depend upon Y and Z: Just as Maxwell has obtained the above expression for the contribution of X to the surface-integral of \mathfrak{B}, we can obtain similar expressions containing $Y dy/ds$ and $Z dz/ds$ for the contributions of Y and Z.

$$\iint (l\xi + m\eta + n\zeta)\,dS = \int \left(X\frac{dx}{ds} + Y\frac{dy}{ds} + Z\frac{dz}{ds} \right) ds; \qquad (11)$$

where the first integral is extended over the surface S, and the second round the bounding curve s.

Equation (11): The left-hand member is the surface-integral of \mathfrak{B}, the components of which are ξ, η, ζ, while the right-hand member is the line-integral of \mathfrak{A}, whose components are X, Y, Z—as was required to prove.

Editor's Introduction to
Articles 25 and 26

A brief note on quaternions

In these Articles Maxwell explores some fundamentals of the formalism of quaternions. But to Maxwell the quaternion approach represented far more than a formalism; he found it to be, rather, "a method for thinking [that] calls upon us at every step to form a mental image of the geometrical features represented by the symbols...."*

The following remarks offer a rough and ready guide comparing quaternion operations to those of our modern vector algebra, and I hope readers will find them helpful. But the history of quaternions, and indeed of vectorial systems generally, is a fascinating topic in itself; and readers who wish to delve into it will not find a better or more engaging treatment than Michael J. Crowe's splendid history of that subject.†

A quaternion consists of two parts, a scalar and a vector. A convenient, though historically anachronistic, way to represent a quaternion Q is \tilde{Q} (q, \mathbf{q}), where the italicized letter signifies the scalar part, the bolded letter the vector part. Quaternion addition and subtraction are straightforward; one simply adds or subtracts the vector and scalar parts separately. Multiplication is more complicated; the product of quaternions A and B can be expressed, however anachronistically, this way:

$$AB = (a, \mathbf{a})(b, \mathbf{b}) = (ab - \mathbf{a} \cdot \mathbf{b}, \mathbf{a} \times \mathbf{b} + a\mathbf{b} + b\mathbf{a}),$$

where $\mathbf{a} \cdot \mathbf{b}$ and $\mathbf{a} \times \mathbf{b}$ will be interpreted momentarily. From a quaternion point of view, vectors are simply quaternions that lack their scalar parts, so the product of two vectors will be

$$\mathbf{ab} = (0, \mathbf{a})(0, \mathbf{b}) = (0 - \mathbf{a} \cdot \mathbf{b}, \mathbf{a} \times \mathbf{b} + 0 + 0)$$
$$= (-\mathbf{a} \cdot \mathbf{b}, \mathbf{a} \times \mathbf{b}).$$

It is evident that $-\mathbf{a} \cdot \mathbf{b}$ occupies the scalar position, and $\mathbf{a} \times \mathbf{b}$ the vector position, in the quaternion product notation. Maxwell will express

* Maxwell, James Clerk, "Quaternions," in *Nature*, 9 (1873).
† Michael J. Crowe, *A History of Vector Analysis: The Evolution of the Idea of a Vectorial System*, University of Notre Dame Press (1967), reprinted Dover Publications (1994).

this fact in paragraph 25.3 by saying that the product of two vectors has "two parts, one scalar and the other vector." In modern vector algebra, **a•b** (without the negative sign) has been independently defined as the *scalar* (or *dot*) product of two vectors; it is a scalar quantity equal to the product of the magnitudes a and b times the cosine of the angle between vectors **a** and **b**. On the other hand, **a**×**b** is called the *vector* (or *cross*) product of two vectors; it is a vector quantity having magnitude equal to the product of a and b times the sine of the angle between vectors **a** and **b**, and direction perpendicular to both **a** and **b**.

Dot product and cross product

In (16.2) Maxwell calculated the line-integral of intensity R over path AB by integrating $R\cos\varepsilon\,ds$ over that path. Notice that the integrand is an example of the dot product **R•ds**, where **R** and **ds** are the vectors whose magnitudes are R and ds, respectively.

In (596.1) Maxwell will consider a parallelogram with sides AB and AM. If the corner angle is ε, the parallelogram's area will be $AB\,AM\sin\varepsilon$. Notice that this is the magnitude of the cross product $AB\times AM$, while the direction of the cross product is perpendicular to the plane of the parallelogram, and so expresses the parallelogram's orientation.

Multiplication of i, j, k

As was described in paragraph 17.2, **i**, **j**, and **k** are unit vectors in the x-, y-, and z-directions, respectively. Applying the definitions of dot and cross products, we may calculate, first, the products of two identical unit vectors—for example

$$\mathbf{i}\boldsymbol{\cdot}\mathbf{i} = 1\cos 0° \text{ or } 1$$

and
$$\mathbf{i}\times\mathbf{i} = 1\sin 0° \text{ or } 0.$$

The same results will obtain for the dot or cross product of *any* unit vector with itself.

Similarly, we take the products of two nonidentical unit vectors, such as **i** and **j**. But what is the angle between **i** and **j**—is it 90° or is it 270°? For the dot product it doesn't matter, since in either case the cosine of the angle will be zero. Thus

$$\mathbf{i}\boldsymbol{\cdot}\mathbf{j} = 1\cos 90° = 1\cos 270° = 0.$$

But we cannot ignore the question in the case of the cross product, since $\sin 90° = 1$, while $\sin 270° = -1$. If we adopt the convention that the angle between vectors taken in cyclical order (**i**→**j, j**→**k, k**→**i**) shall be

$90°$, then that between vectors taken in reverse order will be $270°$. Then $\mathbf{i} \times \mathbf{j}$ will have magnitude 1 in the \mathbf{k}-direction; in other words,

$$\mathbf{i} \times \mathbf{j} = \mathbf{k}.$$

On the other hand, $\mathbf{j} \times \mathbf{i}$ will have magnitude -1 in the \mathbf{k}-direction, that is,

$$\mathbf{j} \times \mathbf{i} = -\mathbf{k}.$$

Evidently multiplication of nonparallel vectors is not commutative! The remaining combinations are easily inferred, but I will summarize them for convenience:

$$\begin{aligned} \mathbf{i} \times \mathbf{j} &= \mathbf{k} & \mathbf{j} \times \mathbf{i} &= -\mathbf{k}, \\ \mathbf{j} \times \mathbf{k} &= \mathbf{i} & \mathbf{k} \times \mathbf{j} &= -\mathbf{i}, \\ \mathbf{k} \times \mathbf{i} &= \mathbf{j} & \mathbf{i} \times \mathbf{k} &= -\mathbf{j}. \end{aligned}$$

Multiplication of vectors in component form

The x-, y-, and z-components (also called *rectangular* components) of a vector were defined in (10.) and its comment; and any vector may be expressed as the sum of its components. Consider, then, vectors \mathbf{A} and \mathbf{B}, expressed in component form as

$$\mathbf{A} = a\mathbf{i} + b\mathbf{j} + c\mathbf{k}$$

and

$$\mathbf{B} = d\mathbf{i} + e\mathbf{j} + f\mathbf{k}.$$

We form their products by taking the products of successive pairs of terms; thus

$$\begin{aligned} \mathbf{A} \boldsymbol{\cdot} \mathbf{B} &= (a\mathbf{i} + b\mathbf{j} + c\mathbf{k}) \boldsymbol{\cdot} (d\mathbf{i} + e\mathbf{j} + f\mathbf{k}) \\ &= ad\mathbf{i} \boldsymbol{\cdot} \mathbf{i} + ae\mathbf{i} \boldsymbol{\cdot} \mathbf{j} + af\mathbf{i} \boldsymbol{\cdot} \mathbf{k}, + \text{etc.} \end{aligned}$$

But we saw that the dot product of a unit vector with itself is just 1, while the dot products of unlike unit vectors disappear. Thus we have

$$\mathbf{A} \boldsymbol{\cdot} \mathbf{B} = ad + be + cf$$

or, the dot product of two vectors is the sum of the products of their x-, y-, and z-components, respectively.

Similarly for the cross product,

$$\begin{aligned} \mathbf{A} \times \mathbf{B} &= (a\mathbf{i} + b\mathbf{j} + c\mathbf{k}) \times (d\mathbf{i} + e\mathbf{j} + f\mathbf{k}) \\ &= ad\mathbf{i} \times \mathbf{i} + ae\mathbf{i} \times \mathbf{j} + af\mathbf{i} \times \mathbf{k}, + \text{etc.} \end{aligned}$$

But the cross products of unit vectors with themselves vanish, while the cross products of unlike unit vectors depend on their order of multiplication. We shall have

$$\mathbf{A} \times \mathbf{B} = ae\mathbf{k} - af\mathbf{j} + bf\mathbf{i} - bd\mathbf{k} + cd\mathbf{j} - ce\mathbf{i}$$

or, collecting terms,

$$\mathbf{A} \times \mathbf{B} = (bf - ce)\mathbf{i} + (cd - af)\mathbf{j} + (ae - bd)\mathbf{k}.$$

We will find these modern vector forms arising again and again throughout the *Treatise*, even when Maxwell is employing quaternion notation. I will regularly call attention to them when they do.

The "resolved part" of a vector

As noted previously in (14.1, *comment*), The "resolved part" of a directed quantity is its component parallel to a specified direction, when its other component is perpendicular to that direction. Thus in the sketch, the resolved part of vector **A** in the direction of vector **B** is $A\cos\varepsilon$, where A is the magnitude of vector **A**.

But the dot product $\mathbf{A}\cdot\mathbf{B} = AB\cos\varepsilon$, where A and B are the magnitudes of vectors **A** and **B**, respectively (page 50, above). Thus we can also express the resolved part of vector **A** in the direction of vector **B** as:

$$\frac{\mathbf{A}\cdot\mathbf{B}}{B}.$$

And if, as before, the rectangular components of **A** and **B** are a, b, c and d, e, f, respectively, an equivalent expression for the resolved part of **A** in the direction of **B** is:

$$\frac{ad+be+cf}{B},$$

where B is the magnitude of vector **B**.

[PRELIMINARY, ARTS. 25 and 26]

On the effect of the operator ∇ on a vector function.

25.] We have seen that the operation denoted by ∇ is that by which a vector quantity is deduced from its potential. The same operation, however, when applied to a vector function, produces results which enter into the two theorems we have just proved (III and IV). The extension of this operator to vector displacements, and most of its further development, are due to Professor Tait.

25.1 Let σ be a vector function of ρ, the vector of a variable point. Let us suppose, as usual, that

$$\rho = \mathbf{i}x + \mathbf{j}y + \mathbf{k}z, \quad \text{and} \quad \sigma = \mathbf{i}X + \mathbf{j}Y + \mathbf{k}Z;$$

where X, Y, Z are the components of σ in the directions of the axes.

25.2 We have to perform on σ the operation

$$\nabla = \mathbf{i}\frac{d}{dx} + \mathbf{j}\frac{d}{dy} + \mathbf{k}\frac{d}{dz}.$$

25.3 Performing this operation, and remembering the rules for the multiplication of $\mathbf{i}, \mathbf{j}, \mathbf{k}$, we find that $\nabla\sigma$ consists of two parts, one scalar and the other vector.

25. *the operation denoted by ∇ is that by which a vector quantity is deduced from its potential*: Maxwell is not defining ∇ ("nabla") with this statement; rather he is reminding us that we have already seen one example of its operation (17.3).

25.3 *the rules for the multiplication of* $\mathbf{i}, \mathbf{j}, \mathbf{k}$: that is, the rules for their *quaternion* multiplication. We can formulate them in terms of the modern vector products outlined in the editor's introduction—first, the quaternion product of identical unit vectors, for example

$$\mathbf{ii} = (0, \mathbf{i})(0, \mathbf{i}) = (-\mathbf{i} \cdot \mathbf{i}, \mathbf{i} \times \mathbf{i}).$$

The scalar part of this quaternion product will be -1. The vector part will vanish. Thus $\mathbf{ii} = -1$, a pure scalar; and it is clear that the same results will obtain for the quaternion product of *any* unit vector with itself.

Similarly, we find the product of two nonidentical unit vectors, such as

$$\mathbf{ij} = (0, \mathbf{i})(0, \mathbf{j}) = (-\mathbf{i} \cdot \mathbf{j}, \mathbf{i} \times \mathbf{j}).$$

The scalar part will be 0. The vector part will equal \mathbf{k} in this example, but it would be $-\mathbf{k}$ if the unit vectors were not being multiplied in cyclical order. Thus, $\mathbf{ij} = \mathbf{k}$, but $\mathbf{ji} = -\mathbf{k}$; the remaining cases are easily inferred.

we find that $\nabla\sigma$ consists of two parts: Since both ∇ and σ are vectors, their product will exhibit the scalar and vector parts we have been discussing.

25.4 The scalar part is

$$S\nabla\sigma = -\left(\frac{dX}{dx} + \frac{dY}{dy} + \frac{dZ}{dz}\right), \text{ see Theorem III,}$$

and the vector part is

$$V\nabla\sigma = \mathbf{i}\left(\frac{dZ}{dy} - \frac{dY}{dz}\right) + \mathbf{j}\left(\frac{dX}{dz} - \frac{dZ}{dx}\right) + \mathbf{k}\left(\frac{dY}{dx} - \frac{dX}{dy}\right).$$

25.4 *The scalar part is...*: $S\nabla\sigma$ is quaternion notation for the scalar part of the product $\nabla\sigma$. From the discussion in the editor's introduction to this Article, we know that $S\nabla\sigma = -\nabla\bullet\sigma$; let us proceed therefore to evaluate this expression.

In paragraph 25.1, Maxwell defined ∇ and σ in terms of their components. We therefore write

$$S\nabla\sigma = -\nabla\bullet\sigma = -\left(\mathbf{i}\frac{d}{dx} + \mathbf{j}\frac{d}{dy} + \mathbf{k}\frac{d}{dz}\right)\bullet(\mathbf{i}X + \mathbf{j}Y + \mathbf{k}Z).$$

Multiplying through term-by-term, products like

$$-\mathbf{i}\frac{d}{dx}\bullet\mathbf{i}X = -\mathbf{i}\bullet\mathbf{i}\frac{dX}{dx} = -\frac{dX}{dx},$$

involving identical unit vectors, will stand; while products involving unlike unit vectors, such as

$$-\mathbf{i}\frac{d}{dx}\bullet\mathbf{j}Y = -\mathbf{i}\bullet\mathbf{j}\frac{dY}{dx} = 0,$$

will vanish. Summing, we shall have Maxwell's expression.

see Theorem III: Specifically, eq. (5) in paragraph 21.6; we will recognize $S\nabla\sigma$ as the volume integrand in that theorem.

and the vector part is...: $V\nabla\sigma$ denotes the vector part of the product $\nabla\sigma$, so that

$$V\nabla\sigma = \nabla\times\sigma = \left(\mathbf{i}\frac{d}{dx} + \mathbf{j}\frac{d}{dy} + \mathbf{k}\frac{d}{dz}\right)\bullet(\mathbf{i}X + \mathbf{j}Y + \mathbf{k}Z).$$

This time the products involving identical unit vectors will disappear, and when we evaluate the six remaining products we shall have

$$\mathbf{i}\frac{d}{dx}\times\mathbf{j}Y + \mathbf{i}\frac{d}{dx}\times\mathbf{k}Z + \mathbf{j}\frac{d}{dy}\times\mathbf{i}X + \mathbf{j}\frac{d}{dy}\times\mathbf{k}Z + \mathbf{k}\frac{d}{dz}\times\mathbf{i}X + \mathbf{k}\frac{d}{dz}\times\mathbf{j}Y =$$

$$(\mathbf{i}\times\mathbf{j})\frac{dY}{dx} + (\mathbf{i}\times\mathbf{k})\frac{dZ}{dx} + (\mathbf{j}\times\mathbf{i})\frac{dX}{dy} + (\mathbf{j}\times\mathbf{k})\frac{dZ}{dy} + (\mathbf{k}\times\mathbf{i})\frac{dX}{dz} + (\mathbf{k}\times\mathbf{j})\frac{dY}{dz}.$$

Apply the rules for multiplying $\mathbf{i}, \mathbf{j}, \mathbf{k}$—and remember to observe the order of multiplication—to obtain Maxwell's expression.

Note: Maxwell occasionally writes $S.\nabla\sigma$ and $V.\nabla\sigma$ rather than $S\nabla\sigma$ and $V\nabla\sigma$. In such instances the period is merely a separator added for clarity; it does not alter the meanings of the expressions.

25.5 If the relation between X, Y, Z and ξ, η, ζ is that given by equation (1) of the last theorem, we may write

$$V \nabla \sigma = \mathbf{i}\xi + \mathbf{j}\eta + \mathbf{k}\zeta. \text{ See Theorem IV.}$$

25.6 It appears therefore that the functions of X, Y, Z which occur in the two theorems are both obtained by the operation ∇ on the vector whose components are X, Y, Z. The theorems themselves may be written

$$\iiint S \nabla \sigma \, d\varsigma = \iint S . \sigma U \nu \, ds, \quad \text{(III)}$$

$$\text{and} \quad \int S \sigma \, d\rho = -\iint S . \nabla \sigma U \nu \, ds; \quad \text{(IV)}$$

where $d\varsigma$ is an element of a volume, ds of a surface, $d\rho$ of a curve, and $U\nu$ a unit-vector in the direction of the normal.

25.5 *If the relation between X, Y, Z and ξ, η, ζ is that given by equation* (1) *of the last theorem, we may write*...: Eq. (1) in paragraph 24.2 reads:

$$\xi = \frac{dZ}{dy} - \frac{dY}{dz}, \quad \eta = \frac{dX}{dz} - \frac{dZ}{dx}, \quad \zeta = \frac{dY}{dx} - \frac{dX}{dy}.$$

Substitute ξ, η, ζ into Maxwell's expression for $V\nabla\sigma$ in the previous paragraph to obtain the compact statement given here. Note the similarity between this expression for $V\nabla\sigma$ and the surface integrand of Theorem IV (paragraph 24.11).

25.6 *The theorems themselves may be written*...: Alternatively, the theorems expressed in vector calculus notation will be

$$\iiint (-\nabla \cdot \sigma) d\varsigma = \iint (-\sigma \cdot \mathbf{n}) ds \quad \text{(III)}$$

$$\int (-\sigma \cdot d\rho) = -\iint (-\sigma \cdot \mathbf{n}) ds \quad \text{(IV)}$$

where \mathbf{n} is the unit normal vector to surface-element ds. Although several of the negative signs are algebraically spurious here, I have let them stand as explicit reminders that the quaternion scalar part of the product of two vectors is the *negative* of the modern scalar or dot product. Note also that Maxwell has, for no obvious reason, abruptly changed his notation for a surface-element from dS to ds, that for a line-element from ds to $d\rho$, and has introduced $d\varsigma$ to denote the volume-element previous represented as $dx\,dy\,dz$. Such changes in notation happen pretty often in the *Treatise*. I would suggest that this is Maxwell's deliberate practice, arising out of his desire to avert merely formal thinking. He wishes to reason, like Faraday, about *physical quantities*, not their mathematical signs; therefore a certain flexibility of notation may help us break the routine analytical habits that tend to develop when we rely upon a fixed glossary of symbols. Recall his remark that Faraday's method of conceiving the phenomena "was also a mathematical one, though not exhibited in the conventional form of mathematical symbols" (page 4).

25.7 To understand the meaning of these functions of a vector, let us suppose that σ_0 is the value of σ at a point P, and let us examine the value of $\sigma - \sigma_0$ in the neighbourhood of P.

25.8 If we draw a closed surface round P, then, if the surface-integral of σ over this surface is directed inwards, $S\nabla\sigma$ will be positive, and the vector $\sigma - \sigma_0$ near the point P will be on the whole directed towards P, as in the figure (1).

25.9 I propose therefore to call the scalar part of $\nabla\sigma$ the *convergence* of σ at the point P.

Fig. 1.

25.8 *If the surface-integral of σ over this surface is directed inwards...*: In order for the surface *integral* to be inwards, σ must be directed inwards (not necessarily perpendicularly) through the surface—and not necessarily everywhere throughout the surface; we need only suppose that, when integrated, the inward direction of σ predominates.

$S\nabla\sigma$ *will be positive*: In the restated Theorem III, let σ be directed inwards. But normal **n** is outwards, so $(\sigma \cdot \mathbf{n})$ will be negative. Therefore $-(\sigma \cdot \mathbf{n})$ will be positive, so that both the right-hand and left-hand terms are positive. And therefore $-(\nabla \cdot \sigma)$ on the left-hand side, which is Maxwell's $S\nabla\sigma$, is positive.

the vector $\sigma - \sigma_0$ near the point P will be ... directed towards P: Notice that Maxwell's Figure 1 plots the *difference* between σ and σ_0. As an example of how that pattern might come about, suppose that panel (a) in the sketch below maps values of vector σ at various points in the neighborhood of P, as well as the value σ_0 which σ has at point P itself. In panel (b), vector σ_0 has been subtracted from each of the other vectors by placing its tail at the tail of each σ. The dotted arrows drawn from the head of σ_0 to the head of each σ correspond to Maxwell's radial arrows.

Maxwell's Figure 1 shows the centripetal arrows as being all equal, but equality is not a condition of their being centripetal. You can probably see how any number of modifications in the pattern of σ shown in panel (b) would result in dotted lines that are centripetal but unequal.

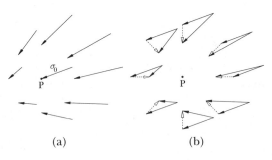

(a) (b)

25.9 *I propose ... to call the scalar part of $\nabla\sigma$ the convergence of σ at the point P*: Convergence thus represents the extent to which not σ itself, but its relative values compared to the value at point P, are centripetally directed.

As we have seen, $S\nabla\sigma$, the scalar part of $\nabla\sigma$, corresponds to $-\nabla \cdot \sigma$ in the language of vector algebra. Inasmuch, then, as $-\nabla \cdot \sigma$ was Maxwell's *convergence*, the operation $\nabla \cdot \sigma$ itself (without negative sign) came to be titled *divergence* by vector theorists. William Kingdon Clifford introduced the term in 1878.

25.10 To interpret the vector part of $\nabla\sigma$, let the direction of the vector whose components are ξ, η, ζ be upwards from the paper and at right angles to it, and let us examine the vector $\sigma-\sigma_0$ near the point P. It will appear as in the figure (2), this vector being arranged on the whole tangentially in the direction opposite to the hands of a watch.

Fig. 2.

25.11 I propose (with great diffidence) to call the vector part of $\nabla\sigma$ the *rotation* of σ at the point P.

25.10 *the vector whose components are ξ, η, ζ*: This is the vector \mathfrak{B} that Maxwell defined in paragraph 24.2. In 25.5 he drew the consequence $V.\nabla\sigma = \mathbf{i}\xi + \mathbf{j}\eta + \mathbf{k}\zeta$; thus Maxwell is here supposing that $V.\nabla\sigma$, the vector part of $\nabla\sigma$, is directed normally to the page.

this vector being arranged ... tangentially: As we have seen, $V.\nabla\sigma$ is equivalent to $\nabla\times\sigma$ and will be perpendicular to both multiplicands. Thus both ∇ and σ lie in the plane of the page. Diagrams similar to those of panels (a) and (b) (on the opposite page) can then be constructed to derive the pattern of Figure 2, which Maxwell is about to call a pattern of "rotation." Here too, equality of the tangential arrows is not a condition of their being tangential. A striking illustration of this is shown in panels (c) and (d), where the vectors σ have the same direction and differ only in magnitude. That is enough to produce counterclockwise "rotation" about point P, since the dotted arrows on opposite sides of P are opposite in direction.

(c) (d)

25.11 *with great diffidence*: Why this reserve concerning the name "rotation"? Previously, in his 1871 paper On the Mathematical Classification of Physical Quantities, Maxwell had proposed "curl" or "version" as names for the vector part of $\nabla\sigma$. Interestingly, he there uses the same phrase, "with great diffidence," in proposing "slope" to name the operation of ∇ on a *scalar* quantity—the operation he calls "space-variation" in the present *Treatise* (17.3–17.5 above). Even earlier, in an 1870 letter to Tait, Maxwell had asked for advice concerning the "rough hewn" names "twist, version, twirl, curl" (Harman, vol. ii, p. 568). Maxwell used "curl" in the first edition of the *Treatise*, but changed it to "rotation" in the second. Today, the term "curl" has become standard.

Maxwell's close attention to naming stands in marked contrast to his sometimes indifferent treatment of mathematical symbology (25.6, *comment*). In general, Maxwell regards the naming of quantities as a *fundamental* act, one that carries a degree of commitment with respect to the physicality of the quantity. That is especially true in the present case of "rotation," since Maxwell knows that $\mathfrak{A} = V.\nabla\mathfrak{B}$ will in Part IV of this work prove to represent the *real physical momentum* of electric current—the quantity that Faraday struggled with and searched for all his life under the appellation "electro-tonic state."

25.12 In Fig. 3 we have an illustration of rotation combined with convergence.

25.13 Let us now consider the meaning of the equation

$$V \nabla \sigma = 0.$$

This implies that $\nabla \sigma$ is a scalar, or that the vector σ is the space-variation of some scalar function Ψ.

Fig. 3.

26.] One of the most remarkable properties of the operator ∇ is that when repeated it becomes

$$\nabla^2 = -\left(\frac{d^2}{dx^2} + \frac{d^2}{dy^2} + \frac{d^2}{dz^2} \right),$$

25.12 *rotation combined with convergence*: Thus the method of subtracting σ_0, the value of σ at a given point, from the values of σ at neighboring points, constitutes a clear analytic test of the character of σ. If the resulting vectors are centripetal, σ is wholly convergent; if tangential, it is wholly rotational; if oblique, σ exhibits both characters in combination.

25.13 *This implies that $\nabla \sigma$ is a scalar...*: If the vector part of $\nabla \sigma$ is zero, $\nabla \sigma$ must consist wholly of a scalar part.

... or that the vector σ is the space-variation of some scalar function Ψ: That is, it implies that $\sigma = \nabla \Psi$, for Maxwell named $\nabla \Psi$ the "space-variation" of Ψ in paragraph 17.5 above. Now by definition, $\nabla \Psi = \mathbf{i} \dfrac{d\Psi}{dx} + \mathbf{j} \dfrac{d\Psi}{dy} + \mathbf{k} \dfrac{d\Psi}{dz}$ (remember that the derivatives are partials). And the vector part of this quantity, expressed as its vector algebra equivalent, will be

$$\nabla \times \nabla \Psi = \left(\mathbf{i} \frac{d}{dx} + \mathbf{j} \frac{d}{dy} + \mathbf{k} \frac{d}{dz} \right) \times \left(\mathbf{i} \frac{d\Psi}{dx} + \mathbf{j} \frac{d\Psi}{dy} + \mathbf{k} \frac{d\Psi}{dz} \right).$$

Multiply this through term by term, remembering the "rules for the multiplication of $\mathbf{i}, \mathbf{j}, \mathbf{k}$" (25.3, *comment*) to obtain:

$$\left(\mathbf{i} \frac{d}{dx} \times \mathbf{j} \frac{d\Psi}{dy} \right) + \left(\mathbf{i} \frac{d}{dx} \times \mathbf{k} \frac{d\Psi}{dz} \right) + \left(\mathbf{j} \frac{d}{dy} \times \mathbf{i} \frac{d\Psi}{dx} \right) + \left(\mathbf{j} \frac{d}{dy} \times \mathbf{k} \frac{d\Psi}{dz} \right)$$

$$+ \left(\mathbf{k} \frac{d}{dz} \times \mathbf{i} \frac{d\Psi}{dx} \right) + \left(\mathbf{k} \frac{d}{dz} \times \mathbf{j} \frac{d\Psi}{dy} \right) =$$

$$\mathbf{k} \frac{d^2\Psi}{dx\,dy} - \mathbf{j} \frac{d^2\Psi}{dx\,dz} - \mathbf{k} \frac{d^2\Psi}{dy\,dx} + \mathbf{i} \frac{d^2\Psi}{dy\,dz} + \mathbf{j} \frac{d^2\Psi}{dz\,dx} - \mathbf{i} \frac{d^2\Psi}{dz\,dy}.$$

The terms cancel pairwise and thus sum to zero. It is evident that *only* if σ has this form, the form $\nabla \Psi$, will $\nabla \times \sigma$ reduce in general to zero.

26. *when [∇ is] repeated it becomes*: By "repeated" Maxwell means multiplied *in the quaternion sense*. Recall from the formalism discussed on page 49 that

$$\nabla \nabla = (-\nabla \bullet \nabla, \nabla \times \nabla).$$

But the cross-product of a vector with itself will be zero, and therefore $\nabla \nabla$ (or

an operator occurring in all parts of Physics, which we may refer to as Laplace's Operator.

26.1 This operator is itself essentially scalar. When it acts on a scalar function the result is scalar, when it acts on a vector function the result is a vector.

26.2 If, with any point P as centre, we draw a small sphere whose radius is r, then if q_0 is the value of q at the centre, and \bar{q} the mean value of q for all points within the sphere,

$$q_0 - \bar{q} = \tfrac{1}{10} r^2 \nabla^2 q;$$

so that the value at the centre exceeds or falls short of the mean value according as $\nabla^2 q$ is positive or negative.

26.3 I propose therefore to call $\nabla^2 q$ the *concentration* of q at the point P, because it indicates the excess of the value of q at that point over its mean value in the neighbourhood of the point.

∇^2, as Maxwell represents it) is scalar and equal to

$$-\left(\mathbf{i}\frac{d}{dx} + \mathbf{j}\frac{d}{dy} + \mathbf{k}\frac{d}{dz} \right) \bullet \left(\mathbf{i}\frac{d}{dx} + \mathbf{j}\frac{d}{dy} + \mathbf{k}\frac{d}{dz} \right).$$

Multiply through term by term; products like $-\mathbf{i}\cdot\mathbf{i}$ become -1, products like $\mathbf{i}\cdot\mathbf{j}$ or $\mathbf{i}\cdot\mathbf{k}$ vanish, and thus we obtain Maxwell's expression. (Note: Modern practice defines ∇^2 simply as $\nabla\cdot\nabla$ and therefore omits Maxwell's negative sign.)

26.2 $q_0 - \bar{q} = \tfrac{1}{10} r^2 \nabla^2 q$: Maxwell cites this result as though it were evident, but the derivation actually takes some doing. He was considerably more forthcoming in an earlier (1871) paper, "On the Mathematical Classification of Physical Quantities" (*Proceedings of the London Mathematical Society*, Vol. III, No. 34), writing:

> For if Q be a quantity, either scalar or vector, which is a function of the position of a point; and if we find the integral of Q taken throughout the volume of a sphere whose radius is r; then, if we divide this by the volume of the sphere, we shall obtain \bar{Q}, the mean value of Q within the sphere. If Q_0 is the value of Q at the centre of the sphere, then, when r is small,
>
> $$Q_0 - \bar{Q} = \frac{r^2}{10} \nabla^2 Q.$$
>
> or the value of Q at the centre of the sphere exceeds the mean value of Q within the sphere by a quantity depending on the radius, and on $\nabla^2 Q$. Since, therefore, $\nabla^2 Q$ indicates the excess of the value of Q at the centre above its mean value in the sphere, I shall call it the concentration of Q.

26.3 *I propose ... to call $\nabla^2 q$ the concentration of q*: Recall that Maxwell identified ∇^2 as Laplace's Operator in paragraph 26. Although his proposed term "concentration" would appear to express a useful interpretation of that quantity, it has not been adopted; today $\nabla^2 q$ is merely called the Laplacian of q. Maxwell will use the term *concentration* again in paragraph 77.4 below.

26.4 If q is a scalar function, the method of finding its mean value is well known. If it is a vector function, we must find its mean value by the rules for integrating vector functions. The result of course is a vector.

Editor's Introduction to
Chapter I

Part I of the *Treatise* concerns electrostatics, the study of electricity at rest. As we shall see, Maxwell will found his electrical theory on a careful definition of electricity as a mathematical quantity. But a mathematically defined quantity, he warns us, is not necessarily a *substance*; and Maxwell will favor an agnostic position regarding one venerable element of traditional thinking—the conception of electricity as a fluid (or two fluids) which act upon other, distant, quantities of electric fluid.

Maxwell will more energetically revise another old chestnut in the tradition. Classical action-at-a-distance theory held that portions of electricity acted upon one another directly, with no intervening medium, regardless of how distant from one another they might be—in that way resembling the action of gravitating bodies in Newton's gravitational theory (despite Newton's own rejection of those views!*). It was Faraday who showed conclusively that the medium which fills the space between electrified bodies plays a definite role. His measurements of the specific inductive capacity of various materials† revealed that when different materials intervene between the same electrified bodies, they sustain characteristically different intensities of what Faraday regarded as *electric tension*. In Art. 59 of the present chapter Maxwell will develop Faraday's conception of a medium under stress; in Art. 60 he will associate that stress with electrical *displacement*, by analogy with the mechanical displacement of an elastic material when it is subjected to stress.

*See, for example, Newton's oft-quoted letter to Bentley of 1693: "[T]hat one body may act upon another at a distance through a vacuum without the mediation of any thing else by and through which their action or force may be conveyed from one to another is to me so great an absurdity that I believe no man who has in philosophical matters any competent faculty of thinking can ever fall into it."

† *Experimental Researches in Electricity*, Eleventh Series.

But Maxwell's elegant treatment of the medium introduces a potentially troubling ambivalence: *Which* element, the electricity or the medium, is fundamental? Is displacement in the medium merely a result of the attractive and repulsive forces exerted on it by electricity? Or might what we call electricity be simply the superficial manifestation of a more fundamental displacement in the medium?

Conventional electrical theory was in no doubt: *electricity*, conceived as endowed with attractive and repulsive force, was primary; the medium's response was secondary and passive. Faraday argued just the opposite, that electrical "charge" is secondary and arises from the more fundamental state of polarization in the medium.

We shall find that Maxwell, on the whole, regards the two views as correlatives. In *every* electrical configuration there is one region in which polarization has to be taken as fundamental, and another in which it must be regarded as derivative. But this understanding will take some time to work out. We shall see him lay the groundwork for it in the present chapter; but the full extent of its power will probably not be appreciated until we have taken up the topic of *magnetism*, in Part III.

PART I.

ELECTROSTATICS.

CHAPTER I

DESCRIPTION OF PHENOMENA.

Electrification by Friction.

27.] EXPERIMENT I. Let a piece of glass and a piece of resin, neither of which exhibits any electrical properties, be rubbed together and left with the rubbed surfaces in contact. They will still exhibit no electrical properties. Let them be separated. They will now attract each other.

27.1 If a second piece of glass be rubbed with a second piece of resin, and if the pieces be then separated and suspended in the neighbourhood of the former pieces of glass and resin, it may be observed—

(1) That the two pieces of glass repel each other.

(2) That each piece of glass attracts each piece of resin.

(3) That the two pieces of resin repel each other.

27.2 These phenomena of attraction and repulsion are called Electrical phenomena, and the bodies which exhibit them are said to be electrified, or to be charged with electricity.

27.3 Bodies may be electrified in many other ways, as well as by friction.

27.4 The electrical properties of the two pieces of glass are similar to each other but opposite to those of the two pieces of resin: the glass attracts what the resin repels and repels what the resin attracts.

27. *resin*: a viscous secretion of some plants, especially trees, that hardens upon drying. Amber (Arabic *anbar*), whose Greek name, *elektron*, has given "electricity" its name, is a fossilized resin. Amber is used in jewelry, while other resins are used in the manufacture of varnishes, lacquers, and perfumes.

27.5 If a body electrified in any manner whatever behaves as the glass does, that is, if it repels the glass and attracts the resin, the body is said to be vitreously electrified, and if it attracts the glass and repels the resin it is said to be resinously electrified. All electrified bodies are found to be either vitreously or resinously electrified.

27.6 It is the established practice of men of science to call the vitreous electrification positive, and the resinous electrification negative. The exactly opposite properties of the two kinds of electrification justify us in indicating them by opposite signs, but the application of the positive sign to one rather than to the other kind must be considered as a matter of arbitrary convention, just as it is a matter of convention in mathematical diagrams to reckon positive distances towards the right hand.

27.7 No force, either of attraction or of repulsion, can be observed between an electrified body and a body not electrified. When, in any case, bodies not previously electrified are observed to be acted on by an electrified body, it is because they have become electrified by induction.

Electrification by Induction.

28.] EXPERIMENT II.* Let a hollow vessel of metal be hung up by white silk threads, and let a similar thread be attached to the lid of the vessel so that the vessel may be opened or closed without touching it.

28.1 l.et the pieces of glass and resin be similarly suspended and electrified as before.

28.2 Let the vessel be originally unelectrified, then if an electrified piece of glass is hung up within it by its thread without touching the vessel, and the lid closed, the outside of the vessel will be found to be vitreously electrified, and it may be shewn that the electrification outside of the vessel is exactly the same in whatever part of the interior space the glass is suspended.

Fig. 4.

*This, and several experiments which follow, are due to Faraday, 'On Static Electrical Inductive Action,' ... *Exp. Res.*, vol. ii. p. 279.

27.5 *vitreous*: of, or of the nature of, glass (Latin *vitrum*).

27.6 *exactly opposite properties of the two kinds of electrification*: The two kinds are "opposite" inasmuch as they are capable of nullifying one another. Maxwell will give examples in paragraphs 29.6, 30., and 31.2.

28. *white silk threads*: Not dyed, as dyeing might impair the insulating ability of the silk.

28.3 If the glass is now taken out of the vessel without touching it, the electrification of the glass will be the same as before it was put in, and that of the vessel will have disappeared.

28.4 This electrification of the vessel, which depends on the glass being within it, and which vanishes when the glass is removed, is called electrification by Induction.

28.5 Similar effects would be produced if the glass were suspended near the vessel on the outside, but in that case we should find an electrification, vitreous in one part of the outside of the vessel and resinous in another. When the glass is inside the vessel the whole of the outside is vitreously and the whole of the inside resinously electrified.

Electrification by Conduction.

29.] EXPERIMENT III. Let the metal vessel be electrified by induction, as in the last experiment, let a second metallic body be suspended by white silk threads near it, and let a metal wire, similarly suspended, be brought so as to touch simultaneously the electrified vessel and the second body.

29.1 The second body will now be found to be vitreously electrified, and the vitreous electrification of the vessel will have diminished.

29.2 The electrical condition has been transferred from the vessel to the second body by means of the wire. The wire is called a *conductor* of electricity, and the second body is said to be *electrified by conduction*.

Conductors and Insulators.

29.3 EXPERIMENT IV. If a glass rod, a stick of resin, or gutta-percha, or a white silk thread, had been used instead of the metal wire, no transfer of electricity would have taken place. Hence these latter substances are called Non-conductors of electricity. Non-conductors are used in electrical experiments to support electrified bodies without carrying off their electricity. They are then called Insulators.

28.4 *Induction*: From Latin *inducere*, lead into or within; hence to bring about, give rise to. Note the important qualification, "which vanishes when the glass is removed." Electrification by induction is maintained only in the presence of the inducing body. Electrification by conduction, which will be described in Experiment III, remains even after the electrifying body has been withdrawn.

29.3 *gutta-percha*: a tough plastic substance obtained from the milky secretion of the Malaysian percha and other trees.

Non-conductors ... Insulators: Note that the first term names the *property*, the second names the *use* of the material. From Latin *insula*, island.

29.4 The metals are good conductors; air, glass, resins, gutta-percha, vulcanite, paraffin, &c. are good insulators; but, as we shall see afterwards, all substances resist the passage of electricity, and all substances allow it to pass, though in exceedingly different degrees. This subject will be considered when we come to treat of the motion of electricity. For the present we shall consider only two classes of bodies, good conductors, and good insulators.

29.5 In Experiment II an electrified body produced electrification in the metal vessel while separated from it by air, a non-conducting medium. Such a medium, considered as transmitting these electrical effects without conduction, has been called by Faraday a Dielectric medium, and the action which takes place through it is called Induction.

29.6 In Experiment III the electrified vessel produced electrification in the second metallic body through the medium of the wire. Let us suppose the wire removed, and the electrified piece of glass taken out of the vessel without touching it, and removed to a sufficient distance. The second body will still exhibit vitreous electrification, but the vessel, when the glass is removed, will have resinous electrification. If we now bring the wire into contact with both bodies, conduction will take place along the wire, and all electrification will disappear from both bodies, indicating that the electrification of the two bodies was equal and opposite.

29.4 *vulcanite, paraffin*: Vulcanite, also called ebonite or hard rubber, is made by treating natural rubber with sulfur for increased strength. Paraffin is an inflammable wax obtained by distillation from oil. Its name derives from Latin *parum*, little + *affinis*, related—expressing its low chemical activity.

29.5 *Dielectric*: From Greek *dia*, through. The reference to action *through* something reflects Faraday's conviction—one which Maxwell shares—that induction is not action at a distance, rather, the medium plays a role in sustaining the effect. Notice that when describing *conduction* (29.2), Maxwell described the electrical condition as having been transferred "by means of" the conductor, not "through" it. Whatever the nature of conduction, Maxwell is not adopting the locution of electricity moving "through" a conductor as water flows through a pipe.

29.6 *the vessel ... will have resinous electrification*: It would appear, then, that while the vessel and the second body were joined by the conductor they constituted a single body, resinously electrified at the vessel "end" and vitreously electrified at the other extremity.

the electrification of the two bodies was equal and opposite: This is the first example of the "opposite properties" which Maxwell ascribed to vitreous and resinous electricity in (27.6) above.

30.] EXPERIMENT V. In Experiment II it was shewn that if a piece of glass, electrified by rubbing it with resin, is hung up in an insulated metal vessel, the electrification observed outside does not depend on the position of the glass. If we now introduce the piece of resin with which the glass was rubbed into the same vessel, without touching it or the vessel, it will be found that there is no electrification outside the vessel. From this we conclude that the electrification of the resin is exactly equal and opposite to that of the glass. By putting in any number of bodies, electrified in any way, it may be shewn that the electrification of the outside of the vessel is that due to the algebraic sum of all the electrifications, those being reckoned negative which are resinous. We have thus a practical method of adding the electrical effects of several bodies without altering their electrification.

31.] EXPERIMENT VI. Let a second insulated metallic vessel, *B*, be provided, and let the electrified piece of glass be put into the first vessel *A*, and the electrified piece of resin into the second vessel *B*. Let the two vessels be then put in communication by the metal wire, as in Experiment III. All signs of electrification will disappear.

31.1 Next, let the wire be removed, and let the pieces of glass and of resin be taken out of the vessels wfthout touching them. It will be found that *A* is electrified resinously and *B* vitreously.

31.2 If now the glass and the vessel *A* be introduced together into a larger insulated metal vessel *C*, it will be found that there is no electrification outside *C*. This shews that the electrification of *A* is exactly equal and opposite to that of the piece of glass, and that of *B* may be shewn in the same way to be equal and opposite to that of the piece of resin.

31.3 We have thus obtained a method of charging a vessel with a quantity of electricity exactly equal and opposite to that of an electrified body without altering the electrification of the latter, and we may in this way charge any number of vessels with exactly equal quantities of electricity of either kind, which we may take for provisional units.

31.3 *equal quantities of electricity ... which we may take for provisional units*: But the primary function of a *unit* is to enable us to *count*; and Maxwell has not yet shown how these "equal quantities of electricity" can so serve. In the series of experiments to follow he will outline how these quantities may be accumulated and compared with other quantities of electricity, thus making true measurement possible. In contrast to such "provisional" units, Maxwell will define a systematic unit of electricity, based on the law of electrical force, in Art. 41 below.

32.] EXPERIMENT VII. Let the vessel B, charged with a quantity of positive electricity, which we shall call, for the present, unity, be introduced into the larger insulated vessel C without touching it. It will produce a positive electrification on the outside of C. Now let B be made to touch the inside of C. No change of the external electrification will be observed. If B is now taken out of C without touching it, and removed to a sufficient distance, it will be found that B is completely discharged, and that C has become charged with a unit of positive electricity.

32.1 We have thus a method of transferring the charge of B to C.

32.2 Let B be now recharged with a unit of electricity, introduced into C already charged, made to touch the inside of C, and removed. It will be found that B is again completely discharged, so that the charge of C is doubled.

32.3 If this process is repeated, it will be found that however highly C is previously charged, and in whatever way B is charged, when B is first entirely enclosed in C, then made to touch C, and finally removed without touching C, the charge of B is completely transferred to C, and B is entirely free from electrification.

32.4 This experiment indicates a method of charging a body with any number of units of electricity. We shall find, when we come to the mathematical theory of electricity, that the result of this experiment affords an accurate test of the truth of the theory.

32. *Experiment VII*: The three steps of this experiment are illustrated in the drawing. Step (i): Body B, charged with unit positive electricity, induces an equal positive electrification of the exterior of C; this step is identical to Experiment II (28.2). Step (ii): When body B touches the interior of C the exterior electrification of C does not change; but what is the condition of B at that moment? Step (iii): B is removed to a great distance and is found to be fully dis-

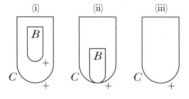

charged, while C retains unit positive electrification. It appears that B must have transferred its electricity to C by conduction, for C retains its charge even after B has been removed. And this can only have happened while B was in contact with C in step (ii). Evidently, then, B will have become discharged at the first moment of contact in step (ii); and the charge of C, which had formerly been maintained by induction so long as B and C were not in contact, now stands in its own right, just like the charge that was acquired by conduction in Experiment III (29.1–2).

32.4 *a method of charging a body with any number of units of electricity*: The whole series of experiments has been directed to this end, for we now have a practical demonstration that electricity is a countable—that is to say a measurable— quantity.

33.] Before we proceed to the investigation of the law of electrical force, let us enumerate the facts we have already established.

33.1 By placing any electrified system inside an insulated hollow conducting vessel, and examining the resultant effect on the outside of the vessel, we ascertain the character of the total electrification of the system placed inside, without any communication of electricity between the different bodies of the system.

33.2 The electrification of the outside of the vessel may be tested with great delicacy by putting it in communication with an electroscope.

33.3 We may suppose the electroscope to consist of a strip of gold leaf hanging between two bodies charged, one positively, and the other negatively. If the gold leaf becomes electrified it will incline towards the body whose electrification is opposite to its own. By increasing the electrification of the two bodies and the delicacy of the suspension, an exceedingly small electrification of the gold leaf may be detected.

33.4 When we come to describe electrometers and multipliers we shall find that there are still more delicate methods of detecting electrification and of testing the accuracy of our theories, but at present we shall suppose the testing to be made by connecting the hollow vessel with a gold leaf electroscope.

33.1 *we ascertain ... the total electrification of the system*: In the previous paragraph Maxwell invited us to "enumerate the facts we have already established." It may seem puzzling, then, that in the present paragraph he names only the one "fact" here cited; but he will continue the enumeration in the following Article.

Notice that the foregoing experiments have identified *physical operations* that are the analogs of the operations of arithmetic! Thus in (31.3) we had the unit; in (32.4) we had addition and hence multiplication; and now with the method of an "insulated hollow conducting vessel" we can manifest the overall electrification of any system and compare it to a known number of units. Electricity, therefore, is fully established as a quantity that can be represented mathematically. Maxwell will state this explicitly in (34.6) below.

33.3 *We may suppose the electroscope to consist...*: The instrument Maxwell here describes was introduced by Bohnenberger about 1814 and is of a distinctive design. The more usual electroscope consisted of a pair of gold leaves that would repel one another upon electrification, as is suggested at the bottom of the hollow vessel in Maxwell's drawing for Figure 4 (Art. 28). The charming instrument shown here, an example of Bohnenberger's design, was built by Joseph Naylor of Depauw University about 1900. The gold leaf is missing. Photograph by Thomas B. Greenslade, Jr.

33.5 This method was used by Faraday in his very admirable demonstration of the laws of electrical phenomena.*

34.] I. The total electrification of a body, or system of bodies, remains always the same, except in so far as it receives electrification from or gives electrification to other bodies.

34.1 In all electrical experiments the electrification of bodies is found to change, but it is always found that this change is due to want of perfect insulation, and that as the means of insulation are improved, the loss of electrification becomes less. We may therefore assert that the electrification of a body placed in a perfectly insulating medium would remain perfectly constant.

34.2 II. When one body electrifies another by conduction, the total electrification of the two bodies remains the same, that is, the one loses as much positive or gains as much negative electrification as the other gains of positive or loses of negative electrification.

34.3 For if the two bodies are enclosed in the hollow vessel, no change of the total electrification is observed.

34.4 III. When electrification is produced by friction, or by any other known method, equal quantities of positive and negative electrification are produced.

34.5 For the electrification of the whole system may be tested in the hollow vessel, or the process of electrification may be carried on within the vessel itself, and however intense the electrification of the parts of the system may be, the electrification of the whole, as indicated by the gold leaf electroscope, is invariably zero.

34.6 The electrification of a body is therefore a physical quantity capable of measurement, and two or more electrifications can be combined experimentally with a result of the same kind as when two quantities are added algebraically. We therefore are entitled to use language fitted to deal with electrification as a quantity as well as a quality, and to speak of any electrified body as 'charged with a certain quantity of positive or negative electricity.'

35.] While admitting electricity, as we have now done, to the rank of a physical quantity, we must not too hastily assume that it is, or is not, a substance, or that it is, or is not, a form of energy, or that it belongs to any known category of physical quantities. All that we have hitherto proved is

* Exp. Res., vol. ii. p. 279.

34.3 *For if the two bodies are enclosed in the hollow vessel*: The bodies are first to be enclosed in the vessel, and then brought into contact.

that it cannot be created or annihilated, so that if the total quantity of electricity within a closed surface is increased or diminished, the increase or diminution must have passed in or out through the closed surface.

35.1 This is true of matter, and is expressed by the equation known as the Equation of Continuity in Hydrodynamics.

35.2 It is not true of heat, for heat may be increased or diminished within a closed surface, without passing in or out through the surface, by the transformation of some other form of energy into heat, or of heat into some other form of energy.

35.3 It is not true even of energy in general if we admit the immediate action of bodies at a distance. For a body outside the closed surface may make an exchange of energy with a body within the surface. But if all apparent action at a distance is the result of the action between the parts of an intervening medium, it is conceivable that in all cases of the increase or diminution of the energy within a closed surface we may be able, when the nature of this action of the parts of the medium is clearly understood, to trace the passage of the energy in or out through that surface.

35.4 There is, however, another reason which warrants us in asserting that electricity, as a physical quantity, synonymous with the total electrification of a body, is not, like heat, a form of energy. An electrified system has a certain amount of energy, and this energy can be calculated by multiplying the quantity of electricity in each of its parts by another physical quantity, called the Potential of that part, and taking half the sum of the products. The quantities 'Electricity' and 'Potential,' when

35.] *if the total quantity ... within a closed surface is increased or diminished, the increase or diminution must have passed in or out through the closed surface*: A quantity that obeys this principle is said to be *conserved*. In succeeding paragraphs Maxwell will assert that *matter* is conserved, but *heat* is not.

35.3 *It is not true even of energy in general if we admit the immediate action of bodies at a distance*: That is, *energy* will not be conserved, if action at a distance really exists. Since the principle of conservation of energy had by this time become fully accepted among British physicists, such a statement implicitly makes a *reductio ad absurdum* argument against action at a distance.

For a body outside the closed surface may make an exchange of energy with a body within the surface: Suppose electrified bodies A and B, mutually attracted; then each will have potential energy with respect to the other. If then the electrification of B were to increase, the potential energy of A would likewise increase. But if electrical attraction is truly "at a distance," that energy increase cannot have passed through a closed surface surrounding A—contradicting the final statement in paragraph 35.

35.4 *this energy can be calculated by ... taking half the sum of the products*: Maxwell is anticipating; he will derive this calculation in Art. 84 below.

multiplied together, produce the quantity 'Energy.' It is impossible, therefore, that electricity and energy should be quantities of the same category, for electricity is only one of the factors of energy, the other factor being 'Potential.'

35.6 Energy, which is the product of these factors, may also be considered as the product of several other pairs of factors, such as

A Force	× A distance through which the force is to act.
A Mass	× Gravitation acting through a certain height.
A Mass	× Half the square of its velocity.
A Pressure	× A volume of fluid introduced into a vessel at that pressure.
A Chemical Affinity	× A chemical change, measured by the number of electro-chemical equivalents which enter into combination.

35.7 If we ever should obtain distinct mechanical ideas of the nature of electric potential, we may combine these with the idea of energy to determine the physical category in which 'Electricity' is to be placed.

36.] In most theories on the subject, Electricity is treated as a substance, but inasmuch as there are two kinds of electrification which, being combined, annul each other, and since we cannot conceive of two substances annulling each other, a distinction has been drawn between Free Electricity and Combined Electricity.

Theory of Two Fluids.

36.1 In what is called the Theory of Two Fluids, all bodies, in their unelectrified state, are supposed to be charged with equal quantities of positive and negative electricity. These quantities are supposed to be so great that no process of electrification has ever yet deprived a body of all the electricity of either kind. The process of electrification, according to

36. *In most theories on the subject, Electricity is treated as a substance*: From the standpoint of Maxwell's cautionary remarks in (35.), such treatments would appear to exhibit excessive haste. Nevertheless Maxwell will devote this Article and the next to a serious examination of the two principal fluid theories. Why do this? One reason, certainly, is that fluid thinking had already become well established in scientific discourse; no student of electricity could afford to be ignorant of its precepts. But perhaps a deeper reason is that these theories tend to generate clever contrivances that have no actual basis in the phenomena, and whose artfulness merely disguises unnecessary assumptions. One such artifice is the one Maxwell mentions at the conclusion of this paragraph: the distinction between "free" and "combined" electricity. He will call attention to its spurious character in (36.6) below.

this theory, consists in taking a certain quantity P of positive electricity from the body A and communicating it to B, or in taking a quantity N of negative electricity from B and communicating it to A, or in some combination of these processes.

36.2 The result will be that A will have $P+N$ units of negative electricity over and above its remaining positive electricity, which is supposed to be in a state of combination with an equal quantity of negative electricity. This quantity $P+N$ is called the Free electricity, the rest is called the Combined, Latent, or Fixed electricity.

36.3 In most expositions of this theory the two electricities are called 'Fluids,' because they are capable of being transferred from one body to another, and are, within conducting bodies, extremely mobile. The other properties of fluids, such as their inertia, weight, and elasticity, are not attributed to them by those who have used the theory for merely mathematical purposes; but the use of the word Fluid has been apt to mislead the vulgar, including many men of science who are not natural philosophers, and who have seized on the word Fluid as the only term in the statement of the theory which seemed intelligible to them.

36.4 We shall see that the mathematical treatment of the subject has been greatly developed by writers who express themselves in terms of the 'Two Fluids' theory. Their results, however, have been deduced entirely from data which can be proved by experiment, and which must therefore be true, whether we adopt the theory of two fluids or not. The experimental verification of the mathematical results therefore is no evidence for or against the peculiar doctrines of this theory.

36.5 The introduction of two fluids permits us to consider the negative electrification of A and the positive electrification of B as the effect of any one of three different processes which would lead to the same result. We have already supposed it produced by the transfer of P units of

36.4 *no evidence for or against ... this theory*: Experiment has amply tested the mathematical statements of fluid theorists, but not the interpretive imagery in which those writers have clothed them. We must not suppose, therefore, that successful experiments support—or, for that matter, that failed experiments refute—such imagery. Maxwell made a similar point in his Preface when he voiced his concern lest students of electricity, impressed by the achievements of mathematical physicists, should "imbibe, along with their mathematical methods, their physical hypotheses" (page 6, above).

This will be a recurring theme for Maxwell throughout the *Treatise*. Although we do not include it in our selections, he devotes the whole of Chapter II in Part IV to the presentation of Ampère's unyieldingly analytic theory—a theory Maxwell ultimately opposes!—in order to show that Ampère's treatment and Maxwell's own, very different, theory are equally capable mathematically.

positive electricity from A to B, together with the transfer of N units of negative electricity from B to A. But if $P+N$ units of positive electricity had been transferred from A to B, or if $P+N$ units of negative electricity had been transferred from B to A, the resulting 'free electricity' on A and on B would have been the same as before, but the quantity of 'combined electricity' in A would have been less in the second case and greater in the third than it was in the first.

36.6 It would appear therefore, according to this theory, that it is possible to alter not only the amount of free electricity in a body, but the amount of combined electricity. But no phenomena have ever been observed in electrified bodies which can be traced to the varying amount of their combined electricities. Hence either the combined electricities have no observable properties, or the amount of the combined electricities is incapable of variation. The first of these alternatives presents no difficulty to the mere mathematician, who attributes no properties to the fluids except those of attraction and repulsion, for he conceives the two fluids simply to annul one another, like $+e$ and $-e$, and their combination to be a true mathematical zero. But to those who cannot use the word Fluid without thinking of a substance it is difficult to conceive how the combination of the two fluids can have no properties at all, so that the addition of more or less of the combination to a body shall not in any way affect it, either by increasing its mass or its weight, or altering some of its other properties. Hence it has been supposed by some, that in every process of electrification exactly equal quantities of the two fluids are transferred in opposite directions, so that the total quantity of the two fluids in any body taken together remains always the same. By this new law they 'contrive to save appearances,' forgetting that there would have been no need of the law except to reconcile the 'Two Fluids' theory with facts, and to prevent it from predicting non-existent phenomena.

Theory of One Fluid.

37.] In the theory of One Fluid everything is the same as in the theory of Two Fluids except that, instead of supposing the two substances equal and opposite in all respects, one of them, generally the negative one, has been endowed with the properties and name of Ordinary Matter, while the other retains the name of The Electric Fluid. The particles of the fluid are supposed to repel one another according to the law

36.6 *there would have been no need of the law*: Maxwell here calls attention to the extent to which Two-Fluid partisans have allowed themselves to be carried away by their own hypotheses (36., *comment*).

of the inverse square of the distance, and to attract those of matter according to the same law. Those of matter are supposed to repel each other and attract those of electricity.

37.1 If the quantity of the electric fluid in a body is such that a particle of the electric fluid outside the body is as much repelled by the electric fluid in the body as it is attracted by the matter of the body, the body is said to be Saturated. If the quantity of fluid in the body is greater than that required for saturation, the excess is called the Redundant fluid, and the body is said to be Overcharged. If it is less, the body is said to be Undercharged, and the quantity of fluid which would be required to saturate it is sometimes called the Deficient fluid. The number of units of electricity required to saturate one gramme of ordinary matter must be very great, because a gramme of gold may be beaten out to an area of a square metre, and when in this form may have a negative charge of at least 60,000 units of electricity. In order to saturate the gold leaf when so charged, this quantity of electric fluid must be communicated to it, so that the whole quantity required to saturate it must be greater than this. The attraction between the matter and the fluid in two saturated bodies is supposed to be a very little greater than the repulsion between the two portions of matter and that between the two portions of fluid. This residual force is supposed to account for the attraction of gravitation.

37.2 This theory does not, like the Two Fluid theory, explain too much. It requires us, however, to suppose the mass of the electric fluid so small that no attainable positive or negative electrification has yet perceptibly increased or diminished either the mass or the weight of a body, and it has not yet been able to assign sufficient reasons why the

37. *the law of the inverse square of the distance*: Maxwell refers to the law which describes the attractive or repulsive force between two electrified bodies. Like Newton's gravitation force, the electric force is inversely proportional to the square of the distance between the bodies; Maxwell will describe one way to obtain that relation experimentally in Arts. 40–41.

Those [particles] of matter are supposed to repel each other: This principle might seem to contradict universal gravitation; but Maxwell will note in (37.3–37.4) that those who criticize it on this basis have overlooked some points of the theory and "ought to have reasoned better."

37.1 *a gramme of gold ... may have a negative charge of at least 60,000 units of electricity*: Maxwell calls this quantity "very great," but in comparison to what? He has not yet defined the unit of electricity, but if he means the *electrostatic unit*, to be defined in (41.1) below, 60,000 units represents about 1% of the output of Faraday's electric eel, or (very roughly) the electricity produced by one turn of his large plate electrical machine, each turn of which "will give ten or twelve sparks from the conductors, each an inch in length." See Faraday, *Experimental Researches in Electricity*, ¶¶ 290, 1770.

vitreous rather than the resinous electrification should be supposed due to an excess of electricity.

37.3 One objection has sometimes been urged against this theory by men who ought to have reasoned better. It has been said that the doctrine that the particles of matter uncombined with electricity *repel* one another, is in direct antagonism with the well-established fact that every particle of matter attracts every other particle throughout the universe. If the theory of One Fluid were true we should have the heavenly bodies repelling one another.

37.4 It is manifest however that the heavenly bodies, according to this theory, if they consisted of matter uncombined with electricity, would be in the highest state of negative electrification, and would repel each other. We have no reason to believe that they are in such a highly electrified state, or could be maintained in that state. The earth and all the bodies whose attraction has been observed are rather in an unelectrified state, that is, they contain the normal charge of electricity, and the only action between them is the residual force lately mentioned. The artificial manner, however, in which this residual force is introduced is a much more valid objection to the theory.

37.5 In the present treatise I propose, at different stages of the investigation, to test the different theories in the light of additional classes of phenomena. For my own part, I look for additional light on the nature of electricity from a study of what takes place in the space intervening between the electrified bodies. Such is the essential character of the mode of investigation pursued by Faraday in his *Experimental Researches*, and as we go on I intend to exhibit the results, as developed by Faraday,

37.3 *One objection ... urged against this theory by men who ought to have reasoned better*: Though Maxwell is a strong *critic* of the fluid theories, he is not a *foe*, ready to oppose them on any grounds. In this case he rejects a criticism he recognizes as being based on a misrepresentation. In the next paragraph he will correct the erroneous challenge—taking care, however, to replace it with "a much more valid objection to the theory."

37.5 *I look for additional light on the nature of electricity from ... what takes place in the space intervening between the electrified bodies*: From the fluid point of view, would not such hopefulness seem utterly preposterous? How could we learn anything about the electric fluid by studying the "intervening" space—precisely where the fluid *is not*? But Maxwell, with Faraday, is by no means bound by fluid theory; and he, with Faraday, understands that "electrification" may be more adequately conceived as pertaining to a body's *relation with the surrounding medium*, than to any sort of fluid, whether conceived as filling the body's bulk or painting its surface.

W. Thomson, &c., in a connected and mathematical form, so that we may perceive what phenomena are explained equally well by all the theories, and what phenomena indicate the peculiar difficulties of each theory.

Measurement of the Force between Electrified Bodies.

38.] Forces may be measured in various ways. For instance, one of the bodies may be suspended from one arm of a delicate balance, and weights suspended from the other arm, till the body, when unelectrified, is in equilibrium. The other body may then be placed at a known distance beneath the first, so that the attraction or repulsion of the bodies when electrified may increase or diminish the apparent weight of the first. The weight which must be added to or taken from the other arm, when expressed in dynamical measure, will measure the force between the bodies. This arrangement was used by Sir W. Snow Harris, and is that adopted in Sir W. Thomson's absolute electrometers. See Art. 217.

38.1 It is sometimes more convenient to use a torsion-balance, in which a horizontal arm is suspended by a fine wire or fibre, so as to be capable of vibrating about the vertical wire as an axis, and the body is attached to one end of the arm and acted on by the force in the tangential direction, so as to turn the arm round the vertical axis, and so twist the suspension wire through a certain angle. The torsional rigidity of the

what phenomena are explained equally well by all the theories: Here again is the theme last touched on in (36.4)—that since theories are generally constructed with direct reference to the phenomena, it should come as no surprise that many phenomena will be equally compatible with all of them. Even though competing theories might have their "peculiar difficulties," it is seldom to be expected that phenomena will simply *refute* a theory. The choice to affirm one theory over another will rather be based, at least in serious cases, on different criteria. As we proceed in the *Treatise*, we shall find that Maxwell's criteria emphasize *interpretive richness* more than they do simple agreement with predicted outcomes. He has perhaps already given a hint of this in (37.4), where he named the "artificiality" of some of the two-fluid concepts as an example of valid criticism. As no experiment can objectively assess the "artificiality" of concepts, perhaps no argument can be decisive. That may be one of the reasons Maxwell has decided to act as "an advocate rather than ... a judge" (Preface, page 6) in the present enterprise.

38. *dynamical measure*: Here, the expression of weights or other forces according to Newton's second law, that is, in terms of the momenta they generate. By contrast, an example of non-dynamic measure is the old English *grain*, which was originally the weight of an average barleycorn.

wire is found by observing the time of oscillation of the arm, the moment of inertia of the arm being otherwise known, and from the angle of torsion and the torsional rigidity the force of attraction or repulsion can be deduced. The torsion-balance was devised by Michell for the determination of the force of gravitation between small bodies, and was used by Cavendish for this purpose. Coulomb, working independently of these philosophers, reinvented it, thoroughly studied its action, and successfully applied it to discover the laws of electric and magnetic forces; and the torsion-balance has ever since been used in researches where small forces have to be measured. See Art. 215.

39.] Let us suppose that by either of these methods we can measure the force between two electrified bodies. We shall suppose the dimensions of the bodies small compared with the distance between them, so that the result may not be much altered by any inequality of distribution of the electrification on either body, and we shall suppose that both bodies are so suspended in air as to be at a considerable distance from other bodies on which they might induce electrification.

39.1 It is then found that if the bodies are placed at a fixed distance and charged respectively with e and e' of our provisional units of electricity, they will repel each other with a force proportional to the product of e and e'. If either e or e' is negative, that is, if one of the charges is vitreous and the other resinous, the force will be attractive, but if both e and e' are negative the force is again repulsive.

39.2 We may suppose the first body, A, charged with m units of positive and n units of negative electricity, which may be conceived separately placed within the body, as in Experiment V.

39.3 Let the second body, B, be charged with m' units of positive and n' units of negative electricity.

39.4 Then each of the m positive units in A will repel each of the m' positive units in B with a certain force, say f, making a total effect equal to $mm'f$.

39.2 *separately placed within the body, as in Experiment V*: But in Experiment V Maxwell showed that the combined effect of separate positive charges is just their algebraic sum. Why then does he consider separate charges here, instead of simply conceiving the body to have a certain overall charge, whether positive or negative? His reason will become clearer in (39.4).

39.4 *each of the m positive units ... will repel each of the m' positive units*: Experiment V did not explicitly concern attraction and repulsion, and hence we must not assume that their effects add algebraically in the present case, where attraction and repulsion are the very topics under study. In this and the following paragraphs, therefore, Maxwell takes care to treat the attractions and

39.5 Since the effect of negative electricity is exactly equal and opposite to that of positive electricity, each of the m positive units in A will attract each of the n' negative units in B with the same force f, making a total effect equal to $mn'f$.

39.6 Similarly the n negative units in A will attract the m' positive units in B with a force $nm'f$, and will repel the n' negative units in B with a force $nn'f$.

39.7 The total repulsion will therefore be $(mm'+nn')f$; and the total attraction will be $(mn'+m'n)f$.

39.8 The resultant repulsion will be

$$\left(mm'+nn'-mn'-nm'\right)f \quad \text{or} \quad (m-n)(m'-n')f.$$

39.9 Now $m-n=e$ is the algebraical value of the charge on A, and $m'-n'=e'$ is that of the charge on B, so that the resultant repulsion may be written $ee'f$, the quantities e and e' being always understood to be taken with their proper signs.

Variation of the Force with the Distance.

40.] Having established the law of force at a fixed distance, we may measure the force between bodies charged in a constant manner and placed at different distances. It is found by direct measurement that the force, whether of attraction or repulsion, varies inversely as the square of the distance, so that if f is the repulsion between two units at unit distance, the repulsion at distance r will be fr^{-2}, and the general expression for the repulsion between e units and e' units at distance r will be

$$fee'r^{-2}.$$

Definition of the Electrostatic Unit of Electricity.

41.] We have hitherto used a wholly arbitrary standard for our unit of electricity, namely, the electrification of a certain piece of glass as it happened to be electrified at the commencement of our experiments. We

repulsions of the several combinations of positive and negative charges independently. In (39.9) we will find that their additivity is derived and does not have to be assumed.

a total effect equal to $mm'f$: Since Maxwell has defined f as the force of *each* of the m units upon *each* of the m' units, the total force will be the product $mm'f$.

39.9 *the resultant repulsion may be written* $ee'f$: Having found in the preceding paragraph that the algebraic sums $m-n$ and $m'-n'$ appear in the expression for the force between two charged bodies, Maxwell is now justified in replacing each of those sums by the overall charges, e and e', of the respective bodies.

are now able to select a unit on a definite principle, and in order that this unit may belong to a general system we define it so that f may be unity, or in other words—

41.1 *The electrostatic unit of electricity is that quantity of positive electricity which, when placed at unit of distance from an equal quantity, repels it with unit of force.* *

41.2 This unit is called the Electrostatic unit to distinguish it from the Electromagnetic unit, to be afterwards defined.

41.3 We may now write the general law of electrical action in the simple form $F = ee'r^{-2}$; or,

41.4 *The repulsion between two small bodies charged respectively with e and e' units of electricity is numerically equal to the product of the charges divided by the square of the distance.*

Dimensions of the Electrostatic Unit of Quantity.

42.] If $[Q]$ is the concrete electrostatic unit of quantity itself and e, e' the numerical values of particular quantities; if $[L]$ is the unit of length, and r the numerical value of the distance; and if $[F]$ is the unit of force, and F the numerical value of the force, then the equation becomes

$$F[F] = ee'r^{-2}[Q^2][L^{-2}];$$

whence
$$[Q] = \left[LF^{\frac{1}{2}} \right]$$
$$= \left[L^{\frac{3}{2}} T^{-1} M^{\frac{1}{2}} \right]$$

42.1 This unit is called the Electrostatic Unit of electricity. Other units may be employed for practical purposes, and in other departments of electrical science, but in the equations of electrostatics quantities of electricity are understood to be estimated in electrostatic units, just as in physical astronomy we employ a unit of mass which is founded on the phenomena of gravitation, and which differs from the units of mass in common use.

Proof of the Law of Electrical Force.

43.] The experiments of Coulomb with the torsion-balance may be considered to have established the law of force with a certain approximation to accuracy. Experiments of this kind, however, are rendered difficult, and in some degree uncertain, by several disturbing causes, which must be carefully traced and corrected for.

* {In this definition and in the enunciation of the law of electrical action the medium surrounding the electrified bodies is supposed to be air. See Art. 94.}

42. [F] *is the unit of force*: with dimensions $[MLT^{-2}]$, as Maxwell noted in (6.5).

43.1 In the first place, the two electrified bodies must be of sensible dimensions relative to the distance between them, in order to be capable of carrying charges sufficient to produce measurable forces. The action of each body will then produce an effect on the distribution of electricity on the other, so that the charge cannot be considered as evenly distributed over the surface, or collected at the centre of gravity; but its effect must be calculated by an intricate investigation. This, however, has been done as as regards two spheres by Poisson in an extremely able manner, and the investigation has been greatly simplified by W. Thomson in his *Theory of Electrical Images*. See Arts. 172–175.

43.2 Another difficulty arises from the action of the electricity induced on the sides of the case containing the instrument. By making the inner surface of the instrument of metal, this effect can be rendered definite and measurable.

43.3 An independent difficulty arises from the imperfect insulation of the bodies, on account of which the charge continually decreases. Coulomb investigated the law of dissipation, and made corrections for it in his experiments.

43.4 The methods of insulating charged conductors, and of measuring electrical effects, have been greatly improved since the time of Coulomb, particularly by Sir W. Thomson; but the perfect accuracy of Coulomb's law of force is established, not by any direct experiments and measurements (which may be used as illustrations of the law), but by a mathematical consideration of the phenomenon described as Experiment VII, namely, that an electrified conductor B, if made to touch the inside of a hollow closed conductor C and then withdrawn without touching C, is perfectly discharged, in whatever manner the outside of C may be electrified. By means of delicate electroscopes it is easy to show that no electricity remains on B after the operation, and by the mathematical theory given at [Art. 74], this can only be the case if the force varies inversely as the square of the distance, for if the law were of any different form B would be electrified.

43.1 *so that the charge cannot be considered as ... collected at the centre*: Maxwell will show this by a graphical method in Art. 119.

43.4 *Coulomb's law of force is established, not by any direct experiments ... but by a mathematical consideration...*: Measurements with the torsion-balance can serve, as Maxwell says, as "illustrations" of Coulomb's law (the inverse-square law stated in Art. 40 above), but they are too error-prone to "establish" it. Instead Maxwell will describe a far more adequate experiment in Art. 74. So elegant is this method—in effect a refined version of the Experiment VII described earlier in Art. 32—that its connection with a law of force might not be evident except through such a mathematical analysis as Maxwell will there provide.

The Electric Field.

44.] The Electric Field is the portion of space in the neighbourhood of electrified bodies, considered with reference to electric phenomena. It may be occupied by air or other bodies, or it may be a so-called vacuum, from which we have withdrawn every substance which we can act upon with the means at our disposal.

44.1 If an electrified body be placed at any part of the electric field it will, in general, produce a sensible disturbance in the electrification of the other bodies.

44.2 But if the body is very small, and its charge also very small, the electrification of the other bodies will not be sensibly disturbed, and we may consider the position of the body as determined by its centre of mass. The force acting on the body will then be proportional to its charge, and will be reversed when the charge is reversed.

44.3 Let e be the charge of the body, and F the force acting on the body in a certain direction, then when e is very small F is proportional to e, or

$$F = Re$$

where R depends on the distribution of electricity on the other bodies in the field. If the charge e could be made equal to unity without disturbing the electrification of other bodies we should have $F = R$.

44.4 We shall call R the Resultant Electromotive Intensity at the given point of the field. When we wish to express the fact that this quantity is a vector we shall denote it by the German letter \mathfrak{E}.

Total Electromotive Force and Potential.

45.] If the small body carrying the small charge e be moved from one given point, A, to another B, along a given path, it will experience at each point of its course a force Re, where R varies from point to point of the course. Let the whole work done on the body by the electrical force

44. *The Electric Field is the portion of space ... considered with reference to electric phenomena*: Notice that for Maxwell (as previously for Faraday) the Field is a "portion" of space and not, as is common parlance in our time, an entity conceived as *occupying* space.

44.4 *We shall call R the Resultant Electromotive Intensity ... the German letter* \mathfrak{E}: The German letter \mathfrak{E} corresponds to the Roman E. Confusingly, Maxwell has chosen to represent the vector (\mathfrak{E}) and its scalar value (R) with different letters! Fortunately this seldom happens. Note that according to the relation $F = Re$ in the preceding paragraph, R (and therefore also \mathfrak{E}) will have the dimensions of *force per unit charge*. Maxwell will offer a more formal definition of Resultant Electromotive Intensity in Art. 68 below.

be *Ee*, then *E* is called the Total Electromotive Force along the path *AB*. If the path forms a complete circuit, and if the total electromotive force round the circuit does not vanish, the electricity cannot be in equilibrium but a current will be produced. Hence in Electrostatics the total electromotive force round any closed circuit must be zero, so that if *A* and *B* are two points on the circuit, the total electromotive force from *A* to *B* is the same along either of the two paths into which the circuit is broken, and since either of these can he altered independently of the other, the total electromotive force from *A* to *B* is the same for all paths from *A* to *B*.

45.1 If *B* is taken as a point of reference for all other points, then the total electromotive force from *A* to *B* is called the Potential of *A*. It

45. *then E is called the Total Electromotive Force*: Since the "whole work" done on the body is *Ee*, then *E*, the total electromotive force, signifies the *work done per unit charge*—that is, the "resolved part" of the *force per unit charge*, integrated over distance. But the electromotive intensity *R* is the force exerted per unit charge; hence the total electromotive force along *AB* will be the line-integral of electromotive intensity over the path *AB*. Maxwell will discuss this line-integral at length in Art. 69.

Previously, *electromotive force* was the conventional term for a hypothetical Newtonian force that was supposed to move electricity as gravitational force moves matter. But notice that by Maxwell's definition, electromotive force is *not a force* in the Newtonian sense, since it is the *product* of a force and a distance, per unit charge.

if the total electromotive force round the circuit does not vanish, the electricity cannot be in equilibrium but a current will be produced: In (16.3) Maxwell noted that if a line-integral is the integral of an exact differential, then it will depend only on the endpoints of the path of integration and can be expressed as a difference of potentials in this form:

$$\int_A^B \left(X\,dx + Y\,dy + Z\,dz \right) = \Psi_A - \Psi_B.$$

Thus the total electromotive force along a path will be equal to the potential difference between the endpoints. As Maxwell is about to explain (45.3), electricity tends to move in the direction that lessens its electrical potential energy (just as gravitating matter tends to move in the direction that reduces its gravitational potential energy). Insofar as the result of integration over a given path depends solely on the potentials at its endpoints, the line integral around a *closed* path (for which the beginning and end points are the same) must be zero. But in cases where the line-integral is *not* reducible to a difference of potentials, the integral round a closed path will not be zero—will not "vanish," as Maxwell puts it. This means that electricity could perpetually lose energy by circulating continuously about that path! Such are all non-transient *electric currents*, and their paths are suitably called *circuits*.

45.1 *the Potential of A*: Maxwell will offer a more formal definition of the potential at a point in Art. 70.

depends only on the position of A. In mathematical investigations, B is generally taken at an infinite distance from the electrified bodies.

45.2 A body charged positively tends to move from places of greater positive potential to places of smaller positive, or of negative, potential, and a body charged negatively tends to move in the opposite direction.

45.3 In a conductor the electrification is free to move relatively to the conductor. If therefore two parts of a conductor have different potentials, positive electricity will move from the part having greater potential to the part having less potential as long as that difference continues. A conductor therefore cannot be in electrical equilibrium unless every point in it has the same potential. This potential is called the Potential of the Conductor.

Equipotential Surfaces.

46.] If a surface described or supposed to be described in the electric field is such that the electric potential is the same at every point of the surface it is called an Equipotential surface.

46.1 An electrified particle constrained to rest upon such a surface will have no tendency to move from one part of the surface to another, because the potential is the same at every point. An equipotential surface is therefore a surface of equilibrium or a level surface.

46.2 The resultant force at any point of the surface is in the direction of the normal to the surface, and the magnitude of the force is such that the work done on an electrical unit in passing from the surface V to the surface V' is $V - V'$.

45.1, continued. *In mathematical investigations, B is generally taken at an infinite distance from the electrified bodies*: In practical electrical measurements *the earth*, rather than a place infinitely distant, is usually taken as the common reference point.

46.1 *An electrified particle constrained to rest upon such a surface will have no tendency to move from one part of the surface to another*: Such a particle will tend to move to *another* surface—one having lower potential—and would do so if not constrained; but it has no tendency to move *along* the surface.

a level surface: Maxwell is of course speaking metaphorically here. An electrically "level" surface need not be *flat*! It is a surface along which an electrified body will have no tendency to move; in the same way, a material body on a gravitationally level surface has no tendency to move along the surface. (Note that, strictly speaking, a gravitationally level surface is not flat either; it is everywhere equidistant from the center of the earth.

46.2 *The resultant force ... is in the direction of the normal to the surface*: If it were not, the force would have a component tangent to the surface; and an electrified particle constrained to occupy the surface could not then be in equilibrium, contrary to (46.1).

46.3 No two equipotential surfaces having different potentials can meet one another, because the same point cannot have more than one potential, but one equipotential surface may meet itself, and this takes place at all points and along all lines of equilibrium.

46.4 The surface of a conductor in electrical equilibrium is necessarily an equipotential surface. If the electrification of the conductor is positive over the whole surface, then the potential will diminish as we move away from the surface on every side, and the conductor will be surrounded by a series of surfaces of lower potential.

46.5 But if (owing to the action of external electrified bodies) some regions of the conductor are charged positively and others negatively, the complete equipotential surface will consist of the surface of the conductor itself together with a system of other surfaces, meeting the surface of the conductor in the lines which divide the positive from the negative regions.* These lines will be lines of equilibrium, and an electrified particle placed on one of these lines will experience no force in any direction.

* {See Arts. 80, 114}

46.4 *The surface of a conductor in electrical equilibrium is necessarily an equipotential surface*: since the whole conductor has a uniform potential, as Maxwell deduced in (45.3).

the potential will diminish as we move away from the surface: For the resultant electromotive force is in the direction of the normal (46.2), away from a positive surface (45.2); and from higher to lower potential (45., *comment*).

46.5 *But if … some regions of the conductor are charged positively and others negatively*: The diagram shows a conductor B, whose surface has developed positive and negative regions "owing to the action of" an external, positively-electrified body A—an example of what Maxwell called "electrification by induction" in paragraph 28.4.

… the complete equipotential system will consist of the conductor itself together with a system of other surfaces: In the diagram, body A is surrounded by the series of equipotential surfaces described in (46.4). One of these surfaces has the same potential as the conductor B and meets it; and thus the "complete equipotential system" consists of the conductor together with that surface. Although Maxwell allows for a "system" of such surfaces, there is only one in the simple case diagrammed here.

… meeting the surface of the conductor in the lines which divide the positive from the negative regions: The one such line in this diagram is shown dotted.

These lines will be lines of equilibrium: As Maxwell remarked in (46.3), "one equipotential surface may meet itself, and this takes place at all points and along all lines of equilibrium."

46.6 When the surface of a conductor is charged positively in some parts and negatively in others, there must be some other electrified body in the field besides itself. For if we allow a positively electrified particle, starting from a positively charged part of the surface, to move always in the direction of the resultant force upon it, the potential at the particle will continually diminish till the particle reaches either a negatively charged surface at a potential less than that of the first conductor, or moves off to an infinite distance. Since the potential at an infinite distance is zero, the latter case can only occur when the potential of the conductor is positive.

46.7 In the same way a negatively electrified particle, moving off from a negatively charged part of the surface, must either reach a positively charged surface, or pass off to infinity, and the latter case can only happen when the potential of the conductor is negative.

46.8 Therefore, if both positive and negative charges exist on a conductor, there must be some other body in the field whose potential has the same sign as that of the conductor but a greater numerical value, and if a conductor of any form is alone in the field the charge of every part is of the same sign as the potential of the conductor.

46.9 The interior surface of a hollow conducting vessel containing no charged bodies is entirely free from charge. For if any part of the surface

46.6 *For if we allow a positively electrified particle ... to move always in the direction of the resultant force upon it*: The moving particle is describing what Maxwell will shortly call a Line of Force (47.).

till the particle reaches ... a negatively charged surface at a potential less than that of the first conductor: Potential diminishes along a line of force (45.2); hence the surface at which the line terminates must be at a lower potential than that of the conductor from which it originated. But since that conductor is equipotential, the terminating surface must be that of some *other* electrified body. Alternatively, the line of force may extend to infinite distance—but only if the potential of the originating conductor is greater than zero, that is, positive.

In the next paragraph Maxwell will construct the analogous argument for a negatively electrified particle.

46.8 In this paragraph Maxwell combines the arguments of the two previous paragraphs to conclude that an *isolated* conductor cannot exhibit regions of unlike electrification; conversely, if a conductor does exhibit such regions, it cannot be isolated but must be under the influence of some other electrified body.

46.9 *The interior surface of a hollow conducting vessel containing no charged bodies is entirely free from charge*: This was the result of Experiment VII, described in Art. 32. In a later Article (which could not be included in the present selections) Maxwell describes a more precise experimental test of this result that was carried out by Cavendish in 1773, exactly 100 years before publication of this *Treatise*.

were charged positively, a positively electrified particle moving in the direction of the force upon it, must reach a negatively charged surface at a lower potential. But the whole interior surface has the same potential. Hence it can have no charge.

46.10 A conductor placed inside the vessel and communicating with it, may be considered as bounded by the interior surface. Hence such a conductor has no charge.

Lines of Force.

47.] The line described by a point moving always in the direction of the resultant intensity is called a Line of Force. It cuts the equipotential surfaces at right angles. The properties of lines of force will be more fully explained afterwards, because Faraday has expressed many of the laws of electrical action in terms of his conception of lines of force drawn in the electric field, and indicating both the direction and the intensity at every point.

Electric Tension.

48.] Since the surface of a conductor is an equipotential surface, the resultant intensity is normal to the surface, and it will be shewn in Art. 80 that it is proportional to the superficial density of the electrification. Hence the electricity on any small area of the surface will be acted on by a force tending from the conductor and proportional to the product of the resultant intensity and the density, that is, proportional to the square of the resultant intensity.

48.1 This force, which acts outwards as a tension on every part of the conductor, will be called electric Tension. It is measured like ordinary mechanical tension, by the force exerted on unit of area.

48.2 The word tension has been used by electricians in several vague senses, and it has been attempted to adopt it in mathematical language

47. *It cuts the equipotential surfaces at right angles*: As we saw in (46.2), the resultant electromotive force at any point on the surface is normal to the surface.

48. *superficial density*: Superficial density of electrification, also called "surface-density," is the quotient of quantity of electricity by the surface area over which the electricity is distributed; Maxwell will formally define it in Art. 64 and will state its proportionality to resultant intensity in paragraphs 80.3–80.4 below.

a force ... proportional to the product of the resultant intensity and the density: In a region characterized by electromotive intensity R, the force acting upon a quantity of electricity e will be $F = Re$ (44.3). If e resides on a "small area of the surface," it will in turn be proportional to the surface-density σ; and thus F will be proportional to $R\sigma$, the product of the resultant intensity and the density.

as a synonym for Potential; but on examining the cases in which the word has been used, I think it will be more consistent with usage and with mechanical analogy to understand by tension a pulling force of so many pounds weight per square inch exerted on the surface of a conductor or elsewhere. We shall find that the conception of Faraday, that this electric tension exists not only at the electrified surface but all along the lines of force, leads to a theory of electric action as a phenomenon of stress in a medium.

Electromotive Force.

49.] When two conductors at different potentials are connected by a thin conducting wire, the tendency of electricity to flow along the wire is measured by the difference of the potentials of the two bodies. The difference of potentials between two conductors or two points is therefore called the Electromotive force between them.

48.2 *stress in a medium*: Faraday's notion of electric action as involving stress in a medium put him at odds with action-at-a-distance theorists, who recognized no role for any medium that might subsist between the acting bodies. As Maxwell stated in his Preface, "Faraday saw a medium where they saw nothing but distance."

49. *the tendency of electricity to flow ... is measured by the difference of the potentials*: This would appear to be an implicit definition, for it is not clear what would, in general, constitute a measure of "tendency": Does a body have "greater tendency to fall" if its weight increases? Does a fluid manifest "greater tendency to flow" if its velocity increases? In contrast, a difference of potentials is clearly measurable; and by equating the two Maxwell has attempted to clarify the idea of "tendency" in the electrostatic case. Notice, however, how different this is from our more ordinary usage pertaining to heavy bodies—where to declare that "tendency" shall be measured by "difference of potential" would be to say that a body has a greater tendency to fall into a deep abyss than into a shallow ditch! Notwithstanding that such may actually characterize the perception of acrophobes, we do not commonly speak that way regarding inanimate bodies.

The difference of potentials ... is therefore called the Electromotive force: By capitalizing *Electromotive*, Maxwell identifies it as the term being defined—the name evidently deriving from "the tendency of *electricity* to *flow* [move]." But surely the term "force" is the more questionable since, as noted in the comment to (45.), this "force" is not a force in the Newtonian sense. It might be surprising at this point that Maxwell appears unconcerned with such a discrepancy in terminology, but in Part IV he will assert that the electromotive sense of force is actually a more universal one than the Newtonian mechanical sense! See Art. 501.

This identification of potential difference with "tendency to flow" and "electromotive force" is useful and valid for the electrostatic cases discussed in Part I of the Treatise. In the following paragraph, however, Maxwell will caution that it is not accurate in general.

49.1 Electromotive force cannot in all cases be expressed in the form of a difference of potentials. These cases, however, are not treated of in Electrostatics. We shall consider them when we come to heterogeneous circuits, chemical actions, motions of magnets, inequalities of temperature, &c.

Capacity of a Conductor.

50.] If one conductor is insulated while all the surrounding conductors are kept at the zero potential by being put in communication with the earth, and if the conductor, when charged with a quantity E of electricity, has a potential V, the ratio of E to V is called the Capacity of the conductor. If the conductor is completely enclosed within a conducting vessel without touching it, then the charge on the inner conductor will be equal and opposite to the charge on the inner surface of the outer conductor, and will be equal to the capacity of the inner conductor multiplied by the difference of the potentials of the two conductors.

49.1 *Electromotive force cannot in all cases be expressed in the form of a difference of potentials*: Maxwell has in mind cases like that mentioned in (45.), in which the electromotive force about a closed path does not vanish, and therefore to which consistent potentials cannot be assigned.

50. *the ratio of E to V is called the Capacity of the conductor*: Note that Maxwell is ascribing "capacity" to the *insulated conductor* specifically. The charge E is the charge of the insulated conductor, and the potential V is the potential of the insulated conductor.

If the conductor is completely enclosed within a conducting vessel...: Unlike the "surrounding conductors" mentioned in the first sentence, the "conducting vessel" here introduced is not necessarily grounded and therefore may have *any* potential V'.

the charge on the inner conductor will be equal and opposite to the charge on the inner surface of the outer conductor...: Maxwell drew this conclusion from Experiment II; see (28.5) above.

... and will be equal to the capacity of the inner conductor multiplied by the difference of the potentials of the two conductors: Why the *difference* of the potentials? By definition, the capacity C of the inner conductor is the quotient of its charge E by its potential V *when the outer conductor is at zero potential*. Let the outer conductor be raised to potential V', and suppose the inner conductor carries no charge. Then within the space enclosed by the outer conductor there can be no electromotive intensity, so the potential will be uniform throughout and equal to the potential of the outer conductor, V'. If we then add charge E to the inner conductor its potential will, by definition, rise by E/C, to become $V = V' + E/C$. Solve this equation for E to obtain $E = C(V - V')$ or $E = C\Delta V$, the relation Maxwell asserts.

Electric Accumulators.

50.1 A system consisting of two conductors whose opposed surfaces are separated from each other by a thin stratum of an insulating medium is called an electric Accumulator. The two conductors are called the Electrodes and the insulating medium is called the Dielectric. The capacity of the accumulator is directly proportional to the area of the opposed surfaces and inversely proportional to the thickness of the stratum between them. A Leyden jar is an accumulator in which glass is the insulating medium. Accumulators are sometimes called Condensers, but I prefer to restrict the term 'condenser' to an instrument which is used not to hold electricity but to increase its superficial density.

PROPERTIES OF BODIES IN RELATION TO STATICAL ELECTRICITY.

Resistance to the Passage of Electricity through a Body.

51.] When a charge of electricity is communicated to any part of a mass of metal the electricity is rapidly transferred from places of high to places of low potential till the potential of the whole mass becomes the same. In the case of pieces of metal used in ordinary experiments this process is completed in a time too short to be observed, but in the case

50.1 *directly proportional to the area of the opposed surfaces and inversely proportional to the thickness of the stratum between them*: Here Maxwell is anticipating the results of later inferences. The charge on either surface equals the surface-density (48., *comment*) multiplied by the surface area; while the potential difference equals the electric intensity multiplied by the very small distance between the plates. But as Maxwell has already asserted in (48.), and will confirm in paragraphs 80.3–80.4 below, the electric intensity adjacent to a charged surface is proportional to the surface-density of the charge. Hence the capacity—the quotient of charge by potential—will be proportional to the surface area and inversely proportional to separation distance.

Leyden jar: It consists of a glass jar coated both inside and outside with metal foil; the two opposing conducting surfaces are thus separated by a glass insulating medium. The central post extends to contact the inner coating.

Accumulators are sometimes called Condensers, but I prefer to restrict the term 'condenser'...: Notwithstanding Maxwell's expressed preference, the term 'condenser' effectively became the standard electrical term for the device Maxwell calls Accumulator; and he himself uses it in that sense as early as Art. 62 below. Later, towards the middle of the twentieth century, the term 'condenser' would itself be superseded by 'capacitor.'

51. *electricity is rapidly transferred from places of high to places of low potential till the potential of the whole mass becomes the same*: Maxwell has noted already (45.3, 36.4) that a conductor will not be in equilibrium until it attains uniform potential, but he did not there call attention to the *time* this process requires.

of very long and thin wires, such as those used in telegraphs, the potential does not become uniform till after a sensible time, on account of the resistance of the wire to the passage of electricity through it.

51.1 The resistance to the passage of electricity is exceedingly diffferent in different substances, as may be seen from the tables at Arts. 362, 364, and 367, which will be explained in treating of Electric Currents.

51.2 All the metals are good conductors, though the resistance of lead is 12 times that of copper or silver, that of iron 6 times, and that of mercury 60 times that of copper. The resistance of all metals increases as their temperature rises.

51.3 Many liquids conduct electricity by electrolysis. This mode of conduction will be considered in Part II. For the present, we may regard all liquids containing water and all damp bodies as conductors, far inferior to the metals but incapable of insulating a charge of of electricity for a sufficient time to be observed. The resistance of electrolytes diminishes as the temperature rises.

51.4 On the other hand, the gases at the atmospheric pressure, whether dry or moist, are insulators so nearly perfect when the electric tension is small that we have as yet obtained no evidence of electricity passing through them by ordinary conduction. The gradual loss of charge by electrified bodies may in every case be traced to imperfect insulation in the supports, the electricity either passing through the substance of the support or creeping over its surface. Hence, when two charged bodies are hung up near each other, they will preserve their charges longer if they are electrified in opposite ways, than if they are electrified in the same way. For though the electromotive force tending to make electricity pass through the air between them is much greater when they are oppositely electrified, no perceptible loss occurs in this way. The actual loss takes place through the supports, and the electromotive force through the supports is greatest when the bodies are electrified in the same way. The result appears anomalous only when we expect the loss to occur by the passage of electricity through the air between the bodies. The passage of electricity through gases takes place, in general, by disruptive discharge, and does not begin till the

51.3 *electrolysis*: Substances, the chemical decomposition of which is essential to their electrical conduction, are said to conduct by electrolysis (*electro* + *lysis* from Greek λύω, *loosen*) and are accordingly called electrolytes. The only known electrolytes are liquids. Water containing a small quantity of dissolved salt or acid is an electrolyte, but pure water conducts electricity so feebly it is not usually ranked as an electrolyte.

51.4 *disruptive discharge*: Examples are spark, arc, and glow. Maxwell will describe disruptive discharge more fully at (55.1) below.

electromotive intensity has reached a certain value. The value of the electromotive intensity which can exist in a dielectric without a discharge taking place is called the Electric Strength of the dielectric. The electric strength of air diminishes as the pressure is reduced from the atmospheric pressure to that of about three millimetres of mercury.* When the pressure is still further reduced, the electric strength rapidly increases; and when the exhaustion is carried to the highest degree hitherto attained, the electromotive intensity required to produce a spark of a quarter of an inch is greater than that which will give a spark of eight inches in air at the ordinary pressure.

51.5 A vacuum, that is to say, that which remains in a vessel after we have removed everything which we can remove from it, is therefore an insulator of very great electric strength.

51.6 The electric strength of hydrogen is much less than that of air at the same pressure.

51.7 Certain kinds of glass when cold are marvellously perfect insulators, and Sir W. Thomson has preserved charges of electricity for years in bulbs hermetically sealed. The same glass, however, becomes a conductor at a temperature below that of boiling water.

51.8 Gutta-percha, caoutchouc, vulcanite, paraffin, and resins are good insulators, the resistance of gutta-percha at 75° F. being about 6×10^{19} times that of copper.

51.9 Ice, crystals, and solidified electrolytes, are also insulators.

51.10 Certain liquids, such as naphtha, turpentine, and some oils, are insulators, but inferior to the best solid insulators.

DIELECTRICS.

Specific Inductive Capacity.

52.] All bodies whose insulating power is such that when they are placed between two conductors at different potentials the electromotive force acting on them does not immediately distribute their electricity so as to reduce the potential to a constant value, are called by Faraday Dielectrics.

> * {The pressure at which the electric strength is a minimum depends on the shape and size of the vessel in which the gas is contained.}

51.8 *caoutchouc*: The term refers to natural rubber; somewhat sticky and not very durable compared to modern vulcanized rubber.

52. *All bodies whose insulating power is such that ... the electromotive force acting on them does not immediately distribute their electricity so as to reduce the potential to a constant value*: In other words, all bodies that are not excellent *conductors* (45.3, 51.).

52.1 It appears from the hitherto unpublished researches of Cavendish that he had, before 1773, measured the capacity of plates of glass, resin, bees-wax, and shellac, and had determined the ratios in which their capacities exceeded that of plates of air of the same dimensions.

52.2 Faraday, to whom these researches were unknown, discovered that the capacity of an accumulator depends on the nature of the insulating medium between the two conductors, as well as on the dimensions and relative position of the conductors themselves. By substituting other insulating media for air as the dielectric of the accumulator, without altering it in any other respect, he found that when air and other gases were employed as the insulating medium the capacity of the accumulator remained sensibly the same, but that when shellac, sulphur, glass, &c. were substituted for air, the capacity was increased in a ratio which was different for each substance.

52.3 By a more delicate method of measurement Boltzmann succeeded in observing the variation of the inductive capacities of gases at different pressures.

52.4 This property of dielectrics, which Faraday called Specific Inductive Capacity, is also called the Dielectric Constant of the substance. It is defined as the ratio of the capacity of an accumulator when its dielectric is the given substance, to its capacity when the dielectric is a vacuum.

52.5 If the dielectric is not a good insulator, it is difficult to measure its inductive capacity, because the accumulator will not hold a charge for a sufficient time to allow it to be measured; but it is certain that inductive capacity is a property not confined to good insulators, and it is probable that it exists in all bodies.

52.1 *the hitherto unpublished researches of Cavendish*: Cavendish's astonishingly extensive researches remained largely unpublished when the first edition of the *Treatise* appeared in 1873, and when these words were written. Subsequently, from 1874 until 1879, Maxwell edited Cavendish's manuscripts with dedication and devotion, publishing them as *The Electrical Researches of the Honourable Henry Cavendish*. Maxwell sent the book to press in June 1879, five months before his death.

52.4 *It is defined as the ratio of the capacity of an accumulator when its dielectric is the given substance, to its capacity when the dielectric is a vacuum*: The capacity C of an accumulator utilizing any dielectric whatsoever was defined as $C = E/\Delta V$ (50., *comment*); when the dielectric is vacuum let us write $C_0 = E_0/\Delta V_0$. The definition of specific inductive capacity K can therefore be expressed $K = C/C_0 = \Delta V_0/\Delta V$ if the charge E is held constant. Finally, since the physical dimensions of the accumulator do not change, the potential difference will vary directly as the resultant intensity; thus $\Delta V_0/\Delta V = R_0/R$. Substituting, $K = R_0/R$ or $R_0 = KR$. This expression will prove useful at (62.3) below.

Absorption of Electricity.

53.] It is found that when an accumulator is formed of certain dielectrics, the following phenomena occur.

53.1 When the accumulator has been for some time electrified and is then suddenly discharged and again insulated, it becomes recharged in the same sense as at first, but to a smaller degree, so that it may be discharged again several times in succession, these discharges always diminishing. This phenomenon is called that of the Residual Discharge.

53.2 The instantaneous discharge appears always to be proportional to the difference of potentials at the instant of discharge, and the ratio of these quantities is the true capacity of the accumulator; but if the contact of the discharger is prolonged so as to include some of the residual discharge, the apparent capacity of the accumulator, calculated from such a discharge, will be too great.

53.3 The accumulator if charged and left insulated appears to lose its charge by conduction, but it is found that the proportionate rate of loss is much greater at first than it is afterwards, so that the measure of conductivity, if deduced from what takes place at first, would be too great. Thus, when the insulation of a submarine cable is tested, the insulation appears to improve as the electrification continues.

53.1 *the Residual Discharge*: Charge a Leyden jar to a high potential, perhaps using an electric machine as depicted in the drawing (*a*). (Maxwell will discuss several forms of electric machine in Arts. 207ff.) After the jar has remained charged for some time, discharge it with a spark, as in drawing (*b*). One might think that such discharge would have exhausted the jar's electrification; and indeed, if we attempt a second discharge immediately, nothing happens. But after a few minutes' wait we find that the jar *will* produce another spark, though one noticeably weaker than the first. Often several more discharges, after similar delays, can be obtained. It is as though the jar had successively regained portions of its original charge—or, alternatively, as though it had initially held back some of its charge, giving it up only later, in stages. This second interpretation is expressed in the name "residual discharge."

(a)

(b)

53.2 *the apparent capacity ... will be too great*: The initial discharge will be proportional to the initial potential difference between the inner and outer conductors; and the measured quantity of that discharge, divided by the measured potential difference, will be an accurate measure of the jar's electrical capacity (50.). But if the discharge is prolonged, so that we inadvertently measure the sum of the initial discharge and any part of the residual discharge, the measured charge will be greater, with no corresponding increase in initial potential difference; hence the calculated capacity will be too great.

53.4 Thermal phenomena of a kind at first sight analogous take place in the case of the conduction of heat when the opposite sides of a body are kept at different temperatures. In the case of heat we know that they depend on the heat taken in and given out by the body itself. Hence, in the case of the electrical phenomena, it has been supposed that electricity is absorbed and emitted by the parts of the body. We shall see, however, in Art. 329, that the phenomena can be explained without the hypothesis of absorption of electricity, by supposing the dielectric in some degree heterogeneous.

53.4 *Thermal phenomena of a kind at first sight analogous*: Suppose the two ends of a metal rod held at constant but unequal temperatures (sketch *a*); this might be done by means of large water reservoirs whose temperatures can be independently controlled. In practice we would also wrap the rod with insulation to minimize extraneous heat loss.

(*b*) After some time, place the rod between other reservoirs having *equal* and intermediate temperatures—a move that might seem to resemble the initial discharge of a Leyden jar, which brings the inner and outer conductors to equal potential.

(*c*) Remove the rod from the reservoirs; after a time the initially hot end begins to grow warm again, and the initially cold end begins again to cool; though the temperature difference never becomes as great as it had been at first.

If the rod is equipped with thermometers as shown, the story is clear: At first, heat was passing slowly but steadily from the hot end to the cold. In the second step, each end exchanged heat both with its corresponding reservoir *and its adjoining region* until the ends attained the termperature of the reservoirs—that is, equal temperature.

But when the rod was removed from all reservoirs, the initially hot end continued to receive heat from the parts adjoining it, and similarly the initially cold end continued to lose heat to its neighboring precinct; in this way a portion of the original temperature difference was regained at the expense of heat that had been *stored in the center regions* of the rod. Similarly, it was supposed, the Leyden jar might be thought to store, or *absorb*, electric fluid.

53.5 That the phenomena called Electric Absorption are not an actual absorption of electricity by the substance may be shewn by charging the substance in any manner with electricity while it is surrounded by a closed metallic insulated vessel. If, when the substance is charged and insulated, the vessel be instantaneously discharged and then left insulated, no charge is ever communicated to the vessel by the gradual dissipation of the electrification of the charged substance within it.

54.] This fact is expressed by the statement of Faraday that it is impossible to charge matter with an absolute and independent charge of one kind of electricity.*

54.1 In fact it appears from the result of every experiment which has been tried that in whatever way electrical actions may take place among a system of bodies surrounded by a metallic vessel, the charge on the outside of that vessel is not altered.

54.2 Now if any portion of electricity could be forced into a body so as to be absorbed in it, or to become latent, or in any way to exist in it, without being connected with an equal portion of the opposite electricity by lines of induction, or if, after having been absorbed, it could gradually emerge and return to its ordinary mode of action, we should find some change of electrification in the surrounding vessel.

* *Exp. Res.*, vol. i. series xi. ¶ ii. 'On the Absolute
Charge of Matter,' and § 1244.

53.5 *no charge is ever communicated to the vessel by the gradual dissipation of the electrification*: If the substance did actually hold back a portion of its previous electrification in an "absorbed" or latent condition, then as that portion gradually became manifest at the substance's surface, the outside of the surrounding vessel ought to become electrified in the manner of Experiment V (30.). Since that does not happen, Maxwell argues, there is no "actual absorption."

This argument does not seem particularly convincing. While it might be admissible on the single-fluid theory of electricity (Art. 37), adherents of the two-fluid theory (Art. 36) would reply that, as equal quantities of positive and negative fluid are "absorbed," so equal quantities are set free to develop the residual discharge; the quantities of liberated charge thus sum to zero, and no effect on the surrounding vessel would be expected. But Maxwell really has no serious interest in fluid theories anyway, as we shall see in (54.2) below.

54.2 *lines of induction*: Here Maxwell names what he regards as truly central, not only to his reasoning about the residual discharge but to the view that informs his entire Treatise: *Induction*, not electric fluid, is the essential basis of electric action. He will quote Faraday to that effect in the following paragraph, and he will develop and outline the consequences of this view in the final four Articles of the present chapter.

54.3 As this is never found to be the case, Faraday concluded that it is impossible to communicate an absolute charge to matter, and that no portion of matter can by any change of state evolve or render latent one kind of electricity or the other. He therefore regarded induction as 'the essential function both in the first development and the consequent phenomena of electricity.' His 'induction' is (1298) a polarized state of the particles of the dielectric, each particle being positive on one side and negative on the other, the positive and the negative electrification of each particle being always exactly equal.

*Disruptive Discharge.**

55.] If the electromotive intensity at any point of a dielectric is gradually increased, a limit is at length reached at which there is a sudden electrical discharge through the dielectric, generally accompanied with light and sound, and with a temporary or permanent rupture of the dielectric.

55.1 The electromotive intensity when this takes place is a measure of what we may call the electric strength of the dielectric. It depends on the nature of the dielectric, and is greater in dense air than in rare air, and greater in glass than in air, but in every case, if the electromotive force be made great enough, the dielectric gives way and its insulating power

* See Faraday, *Exp. Res.*, vol. i., series xii. and xiii.

54.3 *As this is never found to be the case, Faraday concluded that it is impossible to communicate an absolute charge to matter*: An "absolute" charge is one that has no relation to other, opposite, charge—in other words, charge that does not participate in *induction*. Maxwell's wording might suggest that Faraday inferred the impossibility of absolute charge on the basis of one phenomenon alone, as though the experiment described in (53.5) were by itself decisive. On the contrary, as I suggested in my comment, the experiment admits of alternate interpretations. Faraday actually drew his conclusion after studying a rich variety of phenomena; his discussion occupies some 82 paragraphs in the Eleventh Series (Faraday, *Experimental Researches in Electricity*, paragraphs 1169–1251).

a polarized state of the particles of the dielectric: This, in contrast to the supposed action of portions of "electric fluid" on one another.

each particle being positive on one side and negative on the other: Do not read this as suggesting that "electric fluids" accumulate on the respective sides of the particles; the idea of polarization entirely dispenses with electric fluids or substances. Maxwell will discuss polarization more fully in Arts. 59–62 and in Chapter V below; but, as he will soon instruct us, "its importance cannot be fully understood till we arrive at the study of electromagnetic phenomena" (61.2).

55.1 *electric strength of the dielectric*: More simply, *dielectric strength* (51.4, *comment*).

is destroyed, so that a current of electricity takes place through it. It is for this reason that distributions of electricity for which the electromotive intensity becomes anywhere infinite cannot exist.

The Electric Glow.

55.2 Thus, when a conductor having a sharp point is electrified, the theory, based on the hypothesis that it retains its charge, leads to the conclusion that as we approach the point the superficial density of the electricity increases without limit, so that at the point itself the surface-density, and therefore the resultant electromotive intensity, would be infinite. If the air, or other surrounding dielectric, had an invincible insulating power, this result would actually occur; but the fact is, that as soon as the resultant intensity in the neighbourhood of the point has reached a certain limit, the insulating power of the air gives way, so that the air close to the point becomes a conductor. At a certain distance from the point the resultant intensity is not sufficient to break through the insulation of the air, so that the electric current is checked, and the electricity accumulates in the air round the point.

55.3 The point is thus surrounded by particles of air charged with electricity of the same kind as its own. The effect of this charged air around the point is to relieve the air at the point itself from part of the enormous electromotive intensity which it would have experienced if the conductor alone had been electrified. In fact the surface of the electrified body is no longer pointed, because the point is enveloped by a rounded mass of charged air, the surface of which, rather than that of the solid conductor, may be regarded as the outer electrified surface.

55.4 If this portion of charged air could be kept still, the electrified body would retain its charge, if not on itself at least in its neighbourhood, but the charged particles of air being free to move under the action of electrical force, tend to move away from the electrified body because it is charged with the same kind of electricity. The charged particles of air therefore tend to move off in the direction of the lines of force and to approach those surrounding bodies which are oppositely electrified. When they are gone, other uncharged particles take their place round the point, and since these cannot shield those next the point itself from the excessive electric tension, a new discharge takes place, after which the newly charged particles move off, and so on as long as the body remains electrified.

55.5 In this way the following phenomena are produced:—At and close to the point there is a steady glow, arising from the constant discharges which are taking place between the point and the air very near it.

55.6 The charged particles of air tend to move off in the same general direction, and thus produce a current of air from the point, consisting of the charged particles, and probably of others carried along by them. By artificially aiding this current we may increase the glow, and by checking the formation of the current we may prevent the continuance of the glow.

55.7 The electric wind in the neighbourhood of the point is sometimes very rapid, but it soon loses its velocity, and the air with its charged particles is carried about with the general motions of the atmosphere, and constitutes an invisible electric cloud. When the charged particles come near to any conducting surface, such as a wall, they induce on that surface a charge opposite to their own, and are then attracted towards the wall, but since the electromotive force is small they may remain for a long time near the wall without being drawn up to the surface and discharged. They thus form an electrified atmosphere clinging to conductors, the presence of which may sometimes be detected by the electrometer. The electric forces, however, acting between large masses of charged air and other bodies are exceedingly feeble compared with the ordinary forces which produce winds, and which depend on inequalities of density due to differences of temperature, so that it is very improbable that any observable part of the motion of ordinary thunder clouds arises from electrical causes.

55.8 The passage of electricity from one place to another by the motion of charged particles is called Electrical Convection or Convective Discharge.

55.9 The electrical glow is therefore produced by the constant passage of electricity through a small portion of air in which the tension is very high, so as to charge the surrounding particles of air which are continually swept off by the electric wind, which is an essential part of the phenomenon.

55.10 The glow is more easily formed in rare air than in dense air, and more easily when the point is positive than when it is negative.

55.11 This and many other differences between positive and negative electrification must be studied by those who desire to discover something about the nature of electricity. They have not, however, been satisfactorily brought to bear upon any existing theory.

The Electric Brush.

56.] The electric brush is a phenomenon which may be produced by electrifying a blunt point or small ball so as to produce an electric field

in which the tension diminishes as the distance increases, but in a less rapid manner than when a sharp point is used. It consists of a succession of discharges, ramifying as they diverge from the ball into the air, and terminating either by charging portions of air or by reaching some other conductor. It is accompanied by a sound, the pitch of which depends on the interval between the successive discharges, and there is no current of air as in the case of the glow.

The Electric Spark.

57.] When the tension in the space between two conductors is considerable all the way between them, as in the case of two balls whose distance is not great compared with their radii, the discharge, when it occurs, usually takes the form of a spark, by which nearly the whole electrification is discharged at once.

57.1 In this case, when any part of the dielectric has given way, the parts on either side of it in the direction of the electric force are put into a state of greater tension so that they also give way, and so the discharge proceeds right through the dielectric, just as when a little rent is made in the edge of a piece of paper a tension applied to the paper in the direction of the edge causes the paper to be torn through, beginning at the rent, but diverging occasionally where there are weak places in the paper. The electric spark in the same way begins at the point where the electric tension first overcomes the insulation of the dielectric, and proceeds from that point, in an apparently irregular path, so as to take in other weak points, such as particles of dust floating in air.

56. *a succession of discharges, ramifying as they diverge from the ball into the air*: The constitution of the electric brush was disclosed by Wheatstone's high speed rotating mirror. That instrument, which functioned somewhat like a stroboscope, revealed innumerable spark-like branches like those shown here. This drawing, adapted from S. P. Thomson, illustrates brush at a positive electrode.

57.1 *proceeds ... in an apparently irregular path, so as to take in other weak points*: Such irregular branching of the spark is shown in this drawing, which, like the previous one, was made with the aid of the rotating mirror and adapted from Thomson.

57.2 All these phenomena differ considerably in different gases, and in the same gas at different densities. Some of the forms of electrical discharge through rare gases are exceedingly remarkable. In some cases there is a regular alternation of luminous and dark strata, so that if the electricity, for example, is passing along a tube containing a very small quantity of gas, a number of luminous disks will be seen arranged transversely at nearly equal intervals along the axis of the tube and separated by dark strata. If the strength of the current be increased a new disk will start into existence, and it and the old disks will arrange themselves in closer order. In a tube described by Mr. Gassiot the light of each of the disks is bluish on the negative and reddish on the positive side, and bright red in the central stratum.

57.3 These, and many other phenomena of electrical discharge, are exceedingly important, and when they are better understood they will probably throw great light on the nature of electricity as well as on the nature of gases and of the medium pervading space. At present, however, they must be considered as outside the domain of the mathematical theory of electricity.

* * *

Plan of this Treatise.

59.] In the following treatise I propose first to explain the ordinary theory of electrical action, which considers it as depending only on the electrified bodies and on their relative position, without taking account of any phenomena which may take place in the intervening media. In

57.2 *luminous disks ... arranged transversely at nearly equal intervals along the axis of the tube*: as illustrated in this etching, which too is adapted from Thomson. Notice how the discharge pattern follows the "axis" of the tube, howsoever contorted.

59. *I propose first to explain the ordinary theory of electrical action*: The "ordinary" theory of action-at-a-distance represents precisely the body of thought Maxwell aims to supplant! Nevertheless he begins there, where most readers will feel on familiar ground, and only subsequently develops the (Lagrangian) "system of equations" which will prove to be in a certain sense the heart of the book. Although Maxwell is writing (as he states in his Preface) "as an advocate rather than a judge," he is no doctrinaire; instead of positing the terms of his favored theory at the outset, he is going to let our familiar (but too narrow) mode of thinking lead to a method that is deeper and more universal. It seems almost in the same spirit that he has waited until now—the 59th article (in the original, some 62 pages)—before formulating his Plan!

this way we shall establish the law of the inverse square, the theory of the potential, and the equations of Laplace and Poisson. ...

59.1 We shall then investigate certain general theorems by which Green, Gauss, and Thomson have indicated the conditions of solution of problems in the distribution of electricity. ...

59.2 In Thomson's theorem, the total energy of the system is expressed in the form of the integral of a certain quantity extended over the whole space between the electrified bodies, and also in the form of an integral extended over the electrified surfaces only. The equality of these two expressions may be thus interpreted physically. We may conceive the physical relation between the electrified bodies, either as the result of the state of the intervening medium, or as the result of a direct action between the electrified bodies at a distance. If we adopt the latter conception, we may determine the law of the action, but we can go no further in speculating on its cause. If, on the other hand, we adopt the conception of action through a medium, we are led to enquire into the nature of that action in each part of the medium.

59.3 It appears from the theorem, that if we are to look for the seat of the electric energy in the different parts of the dielectric medium, the amount of energy in any small part must depend on the square of the resultant electromotive intensity at that place multiplied by a coefficient called the specific inductive capacity of the medium.

59.4 It is better, however, in considering the theory of dielectrics from the most general point of view, to distinguish between the electromotive

59.1 *We shall then investigate certain general theorems*: Maxwell presents these theorems in Chapter IV. I do not include them in the present selections since Maxwell will offer the same theorems—in their applied forms—in still later chapters.

59.2 *If we adopt the latter conception ... we can go no further in speculating on its cause*: The "latter conception," that of direct action-at-a-distance, closes off causal inquiry since it effectively *denies* any cause of propagation over distance. It was of this denial that Newton was complaining when he wrote, in his third letter to Bentley (1692): "[T]hat one body may act upon another at a distance through the vacuum, wthout the mediation of any thing else, by and through which their action and force may be conveyed from one to another, is to me so great an absurdity, that I believe no man who has in philosophical matters any competent faculty of thinking can ever fall into it."

59.3 *It appears from the theorem, that ... the amount of energy in any small part must depend on the square of the resultant electromotive intensity at that place multiplied by ... the specific inductive capacity of the medium*: By "depend on" Maxwell here means "be proportional to," and in the next paragraph he specifies the proportionality constant as *one half*. For a derivation see (84.12, *comment*).

intensity at any point and the electric polarization of the medium at that point, since these directed quantities, though related to to one another, are not, in some solid substances, in the same direction. The most general expression for the electric energy of the medium per unit of volume is half the product of the electromotive intensity and the electric polarization multiplied by the cosine of the angle between their directions. In all fluid dielectrics the electromotive intensity and the electric polarization are in the same direction and in a constant ratio.

59.5 If we calculate on this hypothesis the total energy residing in the medium, we shall find it equal to the energy due to the electrification of the conductors on the hypothesis of direct action at a distance. Hence the two hypotheses are mathematically equivalent.

59.6 If we now proceed to investigate the mechanical state of the medium on the hypothesis that the mechanical action observed between electrified bodies is exerted through and by means of the medium, as in the familiar instances of the action of one body on another by means of the tension of a rope or the pressure of a rod, we find that the medium must be in a state of mechanical stress.

59.7 The nature of this stress is, as Faraday pointed out,* a tension along the lines of force combined with an equal pressure in all directions at right angles to these lines. The magnitude of these stresses is proportional to the energy of the electrification per unit of volume, or, in other words, to the square of the resultant electromotive intensity multiplied by the specific inductive capacity of the medium.

59.8 This distribution of stress is the only one consistent** with the observed mechanical action on the electrified bodies, and also with the observed equilibrium of the fluid dielectric which surrounds them. I have

* *Exp. Res.*, series xi. 1297.

** {This statement requires modification: the distribution of stress referred to is only one among many such distributions which will all produce the required effect.}

59.4 *electric polarization*: the state of having developed opposite electric conditions on opposite faces (60.1). Maxwell will define its measure in (60.3).

59.5 *on this hypothesis*: The hypothesis of action through a medium.

the total energy residing in the medium [will be found] equal to the energy due to the electrification of the conductors on the hypothesis of direct action at a distance: This same equality was cited earlier at (59.2) in connection with Thomson's theorem.

the two hypotheses are mathematically equivalent: Therefore both hypotheses will be equally consistent with experiment; the choice between them will have to be made not on experimental but on philosophical grounds.

therefore thought it a warrantable step in scientific procedure to assume the actual existence of this state of stress, and to follow the assumption into its consequences. Finding the phrase *electric tension* used in several vague senses, I have attempted to confine it to what I conceive to have been in the minds of some of those who have used it, namely, the state of stress in the dielectric medium which causes motion of the electrified bodies, and leads, when continually augmented, to disruptive discharge. Electric tension, in this sense, is a tension of exactly the same kind, and measured in the same way, as the tension of a rope, and the dielectric medium, which can support a certain tension and no more, may be said to have a certain strength in exactly the same sense as the rope is said to have a certain strength. Thus, for example, Thomson has found that air at the ordinary pressure and temperature can support an electric tension of 9600 grains weight per square foot before a spark passes.

60.] From the hypothesis that electric action is not a direct action between bodies at a distance, but is exerted by means of the medium between the bodies, we have deduced that this medium must be in a state of stress. We have also ascertained the character of the stress, and compared it with the stresses which may occur in solid bodies. Along the lines of force there is tension, and perpendicular to them there is pressure, the numerical magnitude of these forces being equal, and each proportional to the square of the resultant intensity at the point. Having established these results, we are prepared to take another step, and to form an idea of the nature of the electric polarization of the dielectric medium.

59.8 *Finding the phrase electric tension used in several vague senses...*: Previous writers, including Volta, Faraday, and Tyndall, had not fully developed the distinction between the quantities Maxwell has defined as *electromotive intensity* and *difference of potential*; as a result, their usage tended to conflate aspects of both those quantities.

...I have attempted to confine it to ... the state of stress in the dielectric medium which causes motion of the electrified bodies...: The theory of action through a medium has no place for supposed forces of attraction or repulsion exerted by distant bodies. Rather, the motions of electrified bodies are the manifestations of stress patterns existing in the medium immediately surrounding them. The meaning of "electric tension" must therefore be refined to reflect this view.

...and leads, when continually augmented, to disruptive discharge: conventional theory depicted *attraction and repulsion* and *disruptive discharge* as distinct classes of phenomena. In Maxwell's view they are both manifestations of stress in the medium.

60. *Having established these results*: Remember that this is still part of Maxwell's "Plan" (59.); he has not yet actually "established" all these results, though he has indeed stated them.

60.1 An elementary portion of a body may be said to be polarized when it acquires equal and opposite properties on two opposite sides. The idea of internal polarity may be studied to the greatest advantage as exemplified in permanent magnets, and it will be explained at greater length when we come to treat of magnetism.

60.2 The electric polarization of an elementary portion of a dielectric is a forced state into which the medium is thrown by the action of electromotive force, and which disappears when that force is removed. We may conceive it to consist in what we may call an electric displacement, produced by the electromotive intensity. When the electromotive force acts on a conducting medium it produces a current through it, but if the medium is a non-conductor or dielectric, the current cannot {continue to} flow through the medium, but the electricity is displaced within the medium in the direction of the electromotive intensity, the extent of this displacement depending on the magnitude of the electromotive intensity, so that if the electromotive intensity increases or diminishes, the electric displacement increases or diminishes in the same ratio.

60.3 The amount of the displacement is measured by the quantity of electricity which crosses unit of area, while the displacement increases

60.2 *The electric polarization of an elementary portion of a dielectric is a forced state*: Maxwell has already compared electric tension in a medium to mechanical tension in a stretched rope. Here he begins to carry that comparison beyond a merely instructive metaphor and puts it to use in developing an articulate theory of the electric medium.

We may conceive it to consist in what we may call an electric displacement: Maxwell's concept of electric displacement marks his first opportunity to bring forth Faraday's thinking in explicitly mathematical form and so contrast it with the conventional (or, as he says, the "ordinary") theory. It is remarkable that throughout this Article 60 Maxwell chooses to write exclusively in prose—perhaps in homage to Faraday! Nevertheless he clearly has equations in mind; I will identify them in my comments.

the electric displacement increases or diminishes in the same ratio [*as the electromotive intensity*]: Here Maxwell is implicitly *defining* displacement as being proportional to electromotive intensity. Note that since electrical displacement is a wholly theoretical, even hypothetical, quantity, there is no experimental way to establish units for it. Maxwell can therefore posit only a proportion, not an equation. He will derive the units of displacement theoretically in (60.13).

60.3 *The amount of the displacement is measured by the quantity of electricity which crosses unit of area...*: Thus "displacement" properly falls into the category of fluxes (12.9), quantities that are distributed over a surface and which are directional. But Maxwell will shortly introduce the term "whole displacement," which refers to the *summation* or *integral of displacement* over a whole surface.

from zero to its actual amount. This, therefore, is the measure of the electric polarization.

60.4 The analogy between the action of electromotive intensity in producing electric displacement and of ordinary mechanical force in producing the displacement of an elastic body is so obvious that I have ventured to call the ratio of the electromotive intensity to the corresponding electric displacement the *coefficient of electric elasticity* of the medium. This coefficient is different in different media, and varies inversely as the specific inductive capacity of each medium.

60.5 The variations of electric displacement evidently constitute electric currents. These currents, however, can only exist during the variation of the displacement, and therefore, since the displacement cannot exceed a certain value without causing disruptive discharge, they cannot be continued indefinitely in the same direction, like the currents through conductors.

* * *

60.3, continued. *This, therefore, is the measure of the electric polarization*: That is, displacement is to be conceived as *proportional* to polarization. Maxwell has explained that polarization is a "forced" state, a state of tension in the medium. Just as the tension in an elastic band arises through, and is proportional to, mechanical displacement, electric tension in a medium arises through, and is proportional to, electric displacement. Maxwell will invoke the analogy with mechanical elasticity explicitly in the following paragraph.

60.4 *the analogy between … electromotive intensity in producing electric displacement and … mechanical force in producing the displacement of an elastic body*: A mechanical force F, acting on an elastic body, will produce a deformation or displacement x proportional to that force; the relation is expressed $F = kx$, where the ratio F/x, or k, is called the *elastic coefficient* of the body. Similarly, Maxwell has defined electric displacement D as proportional to electromotive intensity R and in (62.3, 68.6) below will write $D = (1/4\pi)KR$; by comparison, the ratio R/D, which equals $4\pi/K$, may be regarded as the coefficient of electrical elasticity.

varies inversely as the specific inductive capacity: Maxwell discussed specific inductive capacity in Art. 52 above, and he will identify it with K in (62.3) below. Thus the coefficient $4\pi/K$ "varies inversely as the specific inductive capacity of each medium."

60.5 *The variations of electric displacement evidently constitute electric currents*: This language is double-edged. Insofar as electric displacement represents a *change in location* of electricity (because it "crosses" a unit area—60.3), it constitutes a current in the conventional sense of *flow*. But a variation of displacement is also a variation of *tension* (60.3, *comment*)—and this constitutes a current in a different sense, that of a continually alternating rise and collapse of tension, as Faraday had proposed and as Maxwell will explicitly maintain in (62.7).

60.7 If a charge *e* is uniformly distributed over the surface of a sphere, the resultant intensity at any point of the medium surrounding the sphere is proportional to the charge *e* divided by the square of the distance from the centre of the sphere. This resultant intensity, according to our theory, is accompanied by a displacement of electricity in a direction outwards from the sphere.

60.8 If we now draw a concentric spherical surface of radius *r*, the whole displacement, *E*, through this surface will be proportional to the resultant intensity multiplied by the area of the spherical surface. But the resultant intensity is directly as the charge *e* and inversely as the square of the radius, while the area of the surface is directly as the square of the radius.

60.7 *the resultant intensity ... is proportional to the charge e divided by the square of the distance*: Expressed $R \propto e/r^2$. With the adoption of the electrostatic unit of charge, the force law between charges *e* and *e'* became $F = ee'/r^2$ (41.3). And the resultant electromotive intensity *R* was defined by the equation $F = Re$ (44.3). Combining, $Re = ee'/r^2$ or $R = e'/r^2$; thus the electromotive intensity at all points at distance *r* from charge *e'* is e'/r^2. In the present case Maxwell is representing the central charge as *e* rather than *e'*; thus in the present case $R = e/r^2$.

But notice that Maxwell has offered a *proportion*, not the equation $R = e/r^2$. Nothing would be gained by writing an equation here, since the "whole displacement," which Maxwell will shortly (60.8) denote by *E*, is itself only determinable proportionally, for the same reasons that displacement proper is.

from the center of the sphere: If the surface of a sphere is uniformly charged, it will act in the space surrounding it as though the charge were concentrated at the center. Newton proved this for gravitating matter (*Principia*, Book I, Prop. 74), but the same argument will apply to electricity or to any other instance of inverse-square attraction.

60.8 *the whole displacement ... will be proportional to the resultant intensity multiplied by the area of the spherical surface*: Expressed by the proportion $E \propto RA$. I described Maxwell's "whole displacement" in the comment to (60.3) above. In general, the "whole displacement" will be the surface-integral of displacement proper; and since displacement has been defined as being proportional to electromotive intensity, the "whole displacement" will be proportional to the surface-integral of resultant intensity. In the present case, however, the spherical surface lies at a constant radius, while the electromotive intensity *R* is itself constant for a fixed radius. Thus a simple product of intensity by area, rather than their integral, suffices.

But the resultant intensity is directly as the charge e and inversely as the square of the radius...: This was stated already as the proportion $R \propto e/r^2$.

...while the area of the surface is directly as the square of the radius: Expressed as the proportion $A \propto r^2$. Here again Maxwell has cited a proportion rather than the equation for the surface of a sphere, $A = 4\pi r^2$, since he will be combining it with a representation of *E* that is itself only a proportion (60.7, *comment*).

60.9 Hence the whole displacement, E, is proportional to the charge e, and is independent of the radius.

60.10 To determine the ratio between the charge e, and the quantity of electricity, E, displaced outwards through any one of the spherical surfaces, let us consider the work done upon the medium in the region between two concentric spherical surfaces, while the displacement is increased from E to $E + \delta E$. If V_1 and V_2 denote the potentials at the inner and the outer of these surfaces respectively, the electromotive force by which the additional displacement is produced is $V_1 - V_2$, so that the work spent in augmenting the displacement is $(V_1 - V_2)\delta E$.

60.9 *Hence the whole displacement, E, is proportional to the charge e, and is independent of the radius*: Substituting in $E \propto RA$ above, we have $E \propto \dfrac{e}{r^2} r^2$, which reduces to $E \propto e$.

60.10 *To determine the ratio*: Maxwell has shown *proportionality* between e and E; we now have to determine their ratio; or, as is more commonly expressed today, we have to determine the *proportionality constant* between them.

let us consider the work done upon the medium: Conventional action-at-a-distance theory would insist that *no* work is done on the medium. But Maxwell understands a medium experiencing electric displacement to be in a state of stress (60.), and therefore work must have been done upon it in order to establish that state.

any one of the spherical surfaces: Imagine a nest of concentric spherical surfaces surrounding the charged body e. Maxwell will go on to consider two of these, as indicated in the sketch.

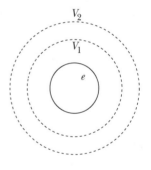

the electromotive force by which the additional displacement is produced is $V_1 - V_2$: Maxwell explained in (49.) that the electromotive force between two points is equal to the difference of potential between those points. But he was not explicit there about the sign convention to be followed; we have to recall from (16.7) that the potential is defined such that the force derived from it has the direction in which the potential *decreases*. In the present configuration V_2 is less than V_1; therefore the electromotive force will be $V_1 - V_2$.

the work spent in augmenting the displacement is $(V_1 - V_2)\delta E$: Since δE is the quantity of electricity that we have supposed displaced through every concentric surface, the situation is equivalent to having transported δE units of electricity from the inner to the outer surface; the energy associated with that electricity thus decreases from $V_1 \delta E$ to $V_2 \delta E$, a loss of $(V_1 - V_2)\delta E$. That energy is of course not "lost"; rather it exists in the medium in the form of stress. The work done upon the medium is therefore $(V_1 - V_2)\delta E$.

60.11 If we now make the inner surface coincide with that of the electrified sphere, and make the radius of the outer infinite, V_1 becomes V, the potential of the sphere, and V_2 becomes zero, so that the whole work done in the surrounding medium is $V\delta E$.

60.12 But by the ordinary theory, the work done in augmenting the charge is $V\delta e$, and if this is spent, as we suppose, in augmenting the displacement, $\delta E = \delta e$, and since E and e vanish together, $E = e$, or—

60.13 *The displacement outwards through any spherical surface concentric with the sphere is equal to the charge on the sphere.*

60.11 V_1 *becomes* V, *the potential of the sphere*: Maxwell here simply denotes the potential of the charged sphere as V.

V_2 *becomes* 0: By definition, the potential at infinite distance from a charged body is zero.

the whole work done in the surrounding medium is $V\delta E$: Obtained by setting $V_1 = V$ and $V_2 = 0$ in $(V_1 - V_2)\delta E$ above.

60.12 *the ordinary theory*: The theory generally held as Maxwell writes.

the work done in augmenting the charge is $V\delta e$: If the sphere carrying charge e has potential V and we add charge δe to it, we shall have to do work equal to the product of the potential and the charge, that is, $V\delta e$.

if this is spent ... in augmenting the displacement: "Ordinary theory" regards this amount of work as having been spent upon the *added charge*; Maxwell attributes it to the *increase of displacement* in the space surrounding the charged body.

$\delta E = \delta e$: If the work traditionally expressed as $V\delta e$ is instead recognized as $V\delta E$, then $\delta E = \delta e$. Thus instead of ascribing that work to an increase δe in the charge supposedly residing on the body, we may choose to ascribe it to an equal increase δE of the displacement in the space surrounding the charged body.

since E *and* e *vanish together,* $E = e$: If there were no charge e on the sphere, there would similarly be no displacement in the region surrounding the sphere; thus "E and e vanish together." And since their increments δE and δe have also been shown equal, the quantities themselves must be equal.

60.13 *The displacement outwards through any spherical surface concentric with the sphere is equal to the charge on the sphere*: It is important to appreciate that this equality is a *conclusion* founded on Maxwell's argument from energy considerations in (60.10–60.12). It is not, as textbook presentations sometimes suggest, a *definition*. Such a definition would brand the concept of electric *displacement* as a merely clever mathematical device, while leaving our inadequate (35., 35.7) ideas of electric *charge* wholly unexamined. In basing the equality on independent energy calculations, Maxwell has given grounds for supplanting our old ideas of electric charge *residing on a body* with the new ideas—or perhaps we should say, the rejuvenated Faradayan ideas—of displacement *in the medium*.

Although Maxwell has derived equality of charge and displacement only for the special case of an electrified sphere within a concentric surface, the conclusion is general, as Maxwell will prove in Arts. 75–76.

60.14 To fix our ideas of electric displacement, let us consider an accumulator formed of two conducting plates *A* and *B*, separated by a stratum of a dielectric *C*. Let *W* be a conducting wire joining *A* and *B*, and let us suppose that by the action of an electromotive force a quantity *Q* of positive electricity is transferred along the wire from *B* to *A*. The positive electrification of *A* and the negative electrification of *B* will produce a certain electromotive force acting from *A* towards *B* in the dielectric stratum, and this will produce an electric displacement from *A* towards *B* within the dielectric. The amount of this displacement, as measured by the quantity of electricity forced across an imaginary section of the dielectric dividing it into two strata, will be, according to our theory, exactly *Q*. See Arts. 75, 76, 111.

60.15 It appears, therefore, that at the same time that a quantity *Q* of electricity is being transferred along the wire by the electromotive force from *B* towards *A*, so as to cross every section of the wire, this same quantity of electricity crosses every section of the dielectric from *A* towards *B* by reason of the electric displacement.

60.16 The displacements of electricity during the discharge of the accumulator will be the reverse of these. In the wire the discharge will be *Q* from *A* to *B*, and in the dielectric the displacement will subside, and a quantity of electricity *Q* will cross every section from *B* towards *A*.

60.17 Every case of charge or discharge may therefore be considered as a motion in a closed circuit, such that at every section of the circuit

60.14 *consider an accumulator formed of two conducting plates A and B, separated by a stratum of a dielectric C*: Maxwell described the class of devices called "accumulator" in (50.1). That venerable name is now inapt for, as we shall see in (60.17), electricity does not accumulate! Alternative terms are "condenser," which Maxwell does not favor, and the modern "capacitor."

suppose that ... a quantity Q of positive electricity is transferred along the wire from B to A. The drawing indicates the result of having transferred charge *Q* along wire *W* from plate *B* to plate *A*; the increased electrification of *A* may alternatively be described as an *electric displacement* from *A* towards *B* within dielectric *C*.

The amount of this displacement ... will be, according to our theory, exactly Q: The equality derived in (60.13) applies here also.

60.17 *Every case of charge or discharge may therefore be considered as a motion in a closed circuit*: There are two regions of the accumulator where this dictum flies in the face of conventional theory. One is in the dielectric, *C*, especially if it be an excellent nonconductor like vacuum or air; for if it is a *nonconductor*, it cannot (conventionally) be part of "a closed circuit"! A second region is at the

the same quantity of electricity crosses in the same time, and this is the case, not only in the voltaic circuit where it has always been recognized, but in those cases in which electricity has been generally supposed to be accumulated in certain places.

61.] We are thus led to a very remarkable consequence of the theory which we are examining, namely, that the motions of electricity are like those of an incompressible fluid, so that the total quantity within an imaginary fixed closed surface remains always the same. This result appears at first sight in direct contradiction to the fact that we can charge a conductor and then introduce it into the closed space, and so alter the quantity of electricity within that space. But we must remember that the ordinary theory takes no account of the electric displacement in the substance of dielectrics which we have been investigating, but confines its attention to the electrification at the bounding surfaces of the conductors and dielectrics. In the case of the charged conductor let us suppose the charge to be positive, then if the surrounding dielectric extends on all sides beyond the closed surface there will be electric polarization,

surfaces of plates *A* and *B*; for if plate *A*, say, is "charged," then (on the conventional view) electricity has *accumulated* on it. But if Maxwell's displacement is equal to the charge that *A* received through the wire, the the plate will have *lost* the same amount that it received!

In fact, that is what Maxwell means when he goes on to say that at every section of the circuit "the same quantity of electricity crosses in the same time," and that this is the case even at places like the plates—places where "electricity has been generally supposed to be accumulated." In the next paragraph Maxwell will develop the image of the "closed circuit" into an even more striking image.

61. *the motions of electricity are like those of an incompressible fluid*: Notice that he is not saying *electricity* is like an incompressible fluid—only that its motions are. But the motions of an incompressible fluid do not include accumulation.

the total quantity within an imaginary fixed closed surface remains always the same: This is another way of saying that electricity does not "accumulate" anywhere. If it did, an imaginary closed surface placed there would encompass an increasing quantity of electricity.

we can charge a conductor and then introduce it into the closed space: Both the ordinary theory and Maxwell's agree that we can do this. The question on which they disagree is, What are the consequences?

and so alter the quantity of electricity within that space: The ordinary theory will claim that we brought electricity into the space while removing none from it; therefore the quantity has increased—in other words, electricity has accumulated—within the space.

But ... ordinary theory takes no account of the electric displacement: Hence ordinary theory fails to acknowledge that electricity exits as well as enters the space.

accompanied with displacement from within outwards all over the closed surface, and the surface-integral of the displacement taken over the surface will be equal to the charge on the conductor within.

61.1 Thus when the charged conductor is introduced into the closed space there is immediately a displacement of a quantity of electricity equal to the charge through the surface from within outwards, and the whole quantity within the surface remains the same.

61.2 The theory of electric polarization will be discussed at greater length in Chapter V, and a mechanical illustration of it will be given in Art. 334, but its importance cannot be fully understood till we arrive at the study of electromagnetic phenomena.

62.] The peculiar features of the theory are:—

62.1 That the energy of electrification resides in the dielectric medium, whether that medium be solid, liquid, or gaseous, dense or rare, or even what is called a vacuum, provided it be still capable of transmitting electrical action.

62.2 That the energy in any part of the medium is stored up in the form of a state of constraint called electric polarization, the amount of which depends on the resultant electromotive intensity at the place.

62.3 That electromotive force acting on a dielectric produces what we have called electric displacement, the relation between the intensity and the displacement being in the most general case of a kind to be afterwards investigated in treating of conduction, but in the most important cases the displacement is in the same direction as the intensity, and is numerically equal to the intensity multiplied by $\frac{1}{4\pi}K$, where K is the specific inductive capacity of the dielectric.

61., continued. *the surface-integral of the displacement taken over the surface will be equal to the charge on the conductor within*: This is a general statement of the conclusion reached in (60.13), and which Maxwell will prove in Arts. 75–76: if a closed surface contains any quantity of electricity, the integral of displacement taken over that surface is equal to the electricity contained. In other words, as much electricity as is brought into a closed space exits from it in the form of displacement.

62. *The peculiar features of the theory*: They are "peculiar" not as being strange or odd but in an older sense of being distinctive, remarkable. The enumeration that follows is to a considerable extent anticipatory; Maxwell has not yet established all the features he is about to mention.

62.1 *the energy of electrification resides in the dielectric medium*: This is in contrast to conventional theory, which ascribes that energy to the electricity itself, or to the electrified body or bodies; compare (60.10, *comment*).

62.3 *the displacement is ... equal to the intensity multiplied by $\frac{1}{4\pi}K$*: Expressed $D = \frac{1}{4\pi}KR$. Textbook presentations sometimes characterize this as a definition,

62.4 That the energy per unit of volume of the dielectric arising from the electric polarization is half the product of the electromotive intensity and the electric displacement, multiplied, if necessary, by the cosine of the angle between their directions.

62.5 That in fluid dielectrics the electric polarization is accompanied by a tension in the direction of the lines of induction, combined with an equal pressure in all directions at right angles to the lines of induction, the tension or pressure per unit of area being numerically equal to the energy per unit of volume at the same place.

62.6 That the surface of any elementary portion into which we may conceive the volume of the dielectric divided must be conceived to be charged so that the surface-density at any point of the surface is equal in magnitude to the displacement through that point of the surface

but it actually follows from earlier results. To begin with, let the dielectric be air or vacuum ($K = 1$). Now (60.7, *comment*) if we adopt the electrostatic unit of charge, then $R = e/r^2$; and the surface-integral of R over a spherical surface of radius r containing charge e will be

$$\frac{e}{r^2} \cdot 4\pi r^2 \quad \text{or} \quad 4\pi e.$$

But the *whole displacement* is the surface-integral of D over the same surface (60.3, *comment*); and by (60.13) it is equal to the enclosed charge, e—that is, to $\frac{1}{4\pi}$ of the surface-integral of R. And since both integrals are taken over the same surface, their integrands are in the same ratio as the integrals; thus $D = \frac{1}{4\pi}R$. Finally, we saw in (52.4, *comment*) that R in a vacuum is equal to K times its value in a dielectric of specific inductive capacity K; in general, then, $D = \frac{1}{4\pi}KR$.

62.4 *the energy per unit of volume ... is half the product of the electromotive intensity and the electric displacement*: Maxwell has stated this result several times already; we will derive it in connection with (84.12).

62.5 *electric polarization is accompanied by a tension ... combined with an equal pressure*: This was Faraday's claim. Maxwell will deduce and calculate it in Arts. 107 and 108 and will recall Faraday's own words in Art. 109.

62.6 *any elementary portion ... of the dielectric*: An infinitesimal volume of dielectric.

the surface-density is equal ... to the displacement...: Consider an element of the dielectric between the charged plates of the accumulator of (60.14). Up to this point Maxwell has maintained that insofar as displacement has occurred within the element, some quantity of electricity q will have moved from its left-hand to its right-hand face; and since the whole displacement is $D\,dy\,dz$, we have $D\,dy\,dz = dq$. Divide through by $dy\,dz$ to obtain displacement D equal to surface-density. *But Maxwell adds a troubling qualification; see on next page.*

reckoned inwards. If the displacement is in the positive direction, the surface of the element will be charged negatively on the positive side of the element, and positively on the negative side. These superficial charges will in general destroy one another when consecutive elements are considered, except where the dielectric has an internal charge, or at the surface of the dielectric.

62.6, continued. *reckoned inwards*: Note Maxwell's use of italics; he realizes that this phrase demands emphatic attention. Indeed, it seems counterintuitive in light of Maxwell's next sentence, as I will discuss in the next two comments.

If the displacement is in the positive direction: That is, in the direction that would be called positive if the displacement were an ordinary conduction current, the direction from a positively-charged electrode towards a negatively-charged one.

the surface of the element will be charged negatively on the positive side of the element, and positively on the negative side: The "positive side" of the element is that which faces the positively-electrified plate, as shown in the sketch. *But then what of Maxwell's earlier claim that the surface-density is equal to the displacement "reckoned inwards"*? As the

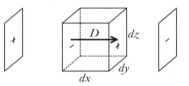

sketch shows, the displacement is directed inwards on the positive side of the element and outwards on the negative side; if we are to "reckon inwards," this would imply *positive* surface-density on the positive face of the element and *negative* surface-density on the negative face. It would ruin Maxwell's interpretation of displacement as motion of positive electricity (60.14), since any such motion would, in the dielectric, have to be in a direction *opposite* to that of the displacement. Finally, it would destroy his dictum that "every case of charge or discharge may therefore be considered as a motion in a closed circuit" (60.17), for in the accumulator of (60.17), the motion of electricity in the dielectric would no longer be a continuation of its motion in the conductor.

Considering that the practice of "reckoning inwards" introduces grave difficulties while offering no obvious advantage, it is tempting to ignore Maxwell's call for it as but a momentary lapse on his part. That is what most textbook accounts of Maxwellian displacement do; it has become commonplace to represent displacement by means of diagrams similar to the sketch above. But the phrase "reckoned inwards" is not just a passing topic of Maxwell's attention. He will employ it again in Art. 111 of Chapter V and there make its contradiction with the dominant ideas of the present paragraph fully explicit. That contradiction has alternately troubled and stimulated a succession of Maxwell's readers representing diverse points of view, beginning with Heinrich Hertz in the nineteenth century. I will review some of Hertz's comments in the editor's introduction to Chapter V.

These superficial charges will in general destroy one another: Adjacent faces of neighboring elements will cancel one another's electrification, except at surfaces adjoining a conductor or a different dielectric.

62.7 That whatsoever electricity may be, and whatever we may understand by the movement of electricity, the phenomenon which we have called electric displacement is a movement of electricity in the same sense as the transference of a definite quantity of electricity through a wire is a movement of electricity, the only difference being that in the dielectric there is a force which we have called electric elasticity which acts against the electric displacement, and forces the electricity back when the electromotive force is removed; whereas in the conducting wire electric elasticity is continually giving way, so that a current of true conduction is set up, and the resistance depends not on the total quantity of electricity displaced from its position of equilibrium, but on the quantity which crosses a section of the conductor in a given time.

62.8 That in every case the motion of electricity is subject to the same condition as that of an incompressible fluid, namely, that at every instant as much must flow out of any given closed surface as flows into it.

62.9 It follows from this that every electric current must form a closed circuit. The importance of this result will be seen when we investigate the laws of electro-magnetism.

62.10 Since, as we have seen, the theory of direct action at a distance is mathematically identical with that of action by means of a medium, the actual phenomena may be explained by the one theory as well as by the other, provided suitable hypotheses be introduced when any difficulty occurs. ...

62.7 *in the conducting wire electric elasticity is continually giving way*: Thus instead of attaining a maximum as in a dielectric, the displacement in a conductor continues to increase so long as an electromotive force is applied—the phenomenon called *conduction current*. Faraday had already interpreted electric current as a continual collapse of tension, rather than as a literal flow of electric charge; see, for example, his *Experimental Researches in Electricity*, Twelfth Series, para. 1370.

the resistance depends ... on the quantity which crosses a section of the conductor in a given time: Since in most conductors resistance is a characteristic property of the material, it would be better to say that the resistance in part *determines* the quantity of electricity which crosses a section in a given time. Maxwell will discuss electric current and resistance more fully in Part II.

62.8 *as much [electricity] must flow out of any given closed surface as flows into it*: Maxwell is freely using the language of "flow," but keep in mind his cautionary words at the opening of the previous paragraph: "whatsoever electricity may be, and whatever we may understand by the movement of electricity...." The mere serviceability of a theory, or of a language, is no guarantee either of the truth of the theory, or of the literal applicability of the language.

62.11 In many parts of physical science, equations of the same form are found applicable to phenomena which are certainly of quite different natures, as, for instance, electric induction through dielectrics, conduction through conductors, and magnetic induction. In all these cases the relation between the intensity and the effect produced is expressed by a set of equations of the same kind, so that when a problem in one of these subjects is solved, the problem and its solution may be translated into the language of the other subjects and the results in their new form will still be true.

62.11 *equations of the same form are found applicable to phenomena ... of quite different natures*: This formal identity is the ground of mathematical *metaphor*; it renders permissible the use of "flow" imagery for both induction through dielectrics and current in conductors.

when a problem in one of these subjects is solved ... its solution may be translated into the language of other subjects: Although Maxwell here emphasizes the practical utility of analogy in problem-solving, it is clear that cultivation of analogy can also be a powerful inspiration to the creative imagination. Maxwell's novel concept of electric displacement, for example, has arisen through his appreciation of the analogy between electric induction and mechanical elasticity. Maxwell will touch on the theme of analogy again in Arts. 243 and 244.

CHAPTER II

ELEMENTARY MATHEMATICAL THEORY OF
STATICAL ELECTRICITY.

* * *

ON ELECTRIC DENSITY.

Distribution in Three Dimensions.

64.] *Definition.* The electric volume-density at a given point in space is the limiting ratio of the quantity of electricity within a sphere whose centre is the given point to the volume of the sphere, when its radius is diminished without limit.

64.1 We shall denote this ratio by the symbol ρ, which may be positive or negative.

Distribution over a Surface.

64.2 It is a result alike of theory and of experiment, that, in certain cases, the charge of a body is entirely on the surface. The density at a point on the surface, if defined according to the method given above, would be infinite. We therefore adopt a different method for the measurement of surface-density.

64.3 *Definition.* The electric density at a given point on a surface is the limiting ratio of the quantity of electricity within a sphere whose centre is the given point to the area of the surface contained within the sphere, when its radius is diminished without limit.

64.4 We shall denote the surface-density by the symbol σ.

* * *

64.9 Similarly, if σ is the surface-density, the whole quantity of electricity on the surface is

$$e = \iint \sigma \, dS,$$

where dS is the element of surface.

64.10 If ρ is the volume-density at any point of space, then the whole electricity within a certain volume is

$$e = \iiint \rho \, dx \, dy \, dz,$$

where $dx \, dy \, dz$ is the element of volume.

$$*\quad *\quad *$$

Law of Force between Charged Bodies.

66.] Coulomb shewed by experiment that the force between charged bodies whose dimensions are small compared with the distance between them, varies inversely as the square of the distance. Hence the repulsion between two such bodies charged with quantities e and e' and placed at a distance r is

$$\frac{ee'}{r^2}.$$

66.1 We shall prove in [Art. 74] that this law is the only one consistent with the observed fact that a conductor, placed in the inside of a closed hollow conductor and in contact with it, is deprived of all electrical charge. Our conviction of the accuracy of the law of the inverse square of the distance may be considered to rest on experiments of this kind, rather than on the direct measurements of Coulomb.

$$*\quad *\quad *$$

Resultant Intensity at a Point.

68.] In order to simplify the mathematical process, it is convenient to consider the action of an electrified body, not on another body of any form, but on an indefinitely small body, charged with an indefinitely small amount of electricity, and placed at any point of the space to which the electrical action extends. By making the charge of this body indefinitely small we render insensible its disturbing action on the charge of the first body.

68.1 Let e be the charge of the small body, and let the force acting on it when placed at the point (x, y, z) be Re, and let the direction-cosines of the force be l, m, n, then we may call R the resultant electric intensity at the point (x, y, z).

68.2 If X, Y, Z denote the components of R, then

$$X = Rl, \qquad Y = Rm, \qquad Z = Rn.$$

66.1 *the observed fact*: It was observed in Experiment VII, Art. 32.

68.1 *the direction-cosines*: Refer to the discussion of direction-cosines in the comment to (21.2) above.

68.3 In speaking of the resultant electric intensity at a point, we do not necessarily imply that any force is actually exerted there, but only that if an electrified body were placed there it would be acted on by a force *Re*, where *e is* the charge of the body.

68.4 *Definition.* The resultant electric intensity at any point is the force which would be exerted on a small body charged with the unit of positive electricity, if it were placed there without disturbing the actual distribution of electricity.

68.5 This force not only tends to move a body charged with electricity, but to move the electricity within the body, so that the positive electricity tends to move in the direction of *R* and the negative electricity in the opposite direction. Hence the quantity *R* is also called the Electromotive Intensity at the point (x, y, z).

68.6 When we wish to express the fact that the resultant intensity is a vector, we shall denote it by the German letter \mathfrak{E}. If the body is a dielectric, then, according to the theory adopted in this treatise, the electricity is displaced within it, so that the quantity of electricity which is forced in the direction of \mathfrak{E} across unit of area fixed perpendicular to \mathfrak{E} is

$$\mathfrak{D} = \frac{1}{4\pi} K \mathfrak{E};$$

where \mathfrak{D} is the displacement, \mathfrak{E} the resultant intensity, and K the specific inductive capacity of the dielectric.

68.7 If the body is a conductor, the state of constraint is continually giving way, so that a current of conduction is produced and maintained as long as \mathfrak{E} acts on the medium.

68.4 *the force … exerted on a small body charged with the unit of positive electricity*: Note therefore that electric intensity is not a measure of force simply, but of force *per unit charge.*

68.5 *This force [tends] to move the electricity within the body*: That is why Maxwell specified a "small" body in the previous paragraph. Such a precaution would not be necessary if, for example, the body were a sphere with its charge symmetrically distributed; for then we could treat the electricity as being concentrated at the sphere's center (60.7, *comment*). But if the electricity migrates within the body, its distribution will be unknown. By making the body small we reduce the uncertainty concerning its effective position.

68.6 *the German letter* \mathfrak{E}: This letter corresponds to the Roman E. Note that Maxwell intends to use German notation "when we wish to emphasize" that a quantity is a vector; a quantity not denoted by a German letter may nevertheless be a vector, though its vectorial character is not there being stressed.

68.7 *If the body is a conductor, the state of constraint is continually giving way*: Maxwell previously characterized conduction in these terms at (62.7).

Line-Integral of Electric Intensity, or Electromotive Force
along an Arc of a Curve.

69.] The Electromotive force along a given arc AP of a curve is numerically measured by the work which would be done by the electric intensity on a unit of positive electricity carried along the curve from A, the beginning, to P, the end of the arc.

69.1 If s is the length of the arc, measured from A, and if the resultant intensity R at any point of the curve makes an angle ε with the tangent drawn in the positive direction, then the work done on unit of electricity in moving along the element of the curve ds will be

$$R \cos \varepsilon \, ds,$$

and the total electromotive force E will be

$$E = \int_0^s R \cos \varepsilon \, ds,$$

the integration being extended from the beginning to the end of the arc.

69.2 If we make use of the components of the intensity, the expression becomes

$$E = \int_0^s \left(X\frac{dx}{ds} + Y\frac{dy}{ds} + Z\frac{dz}{ds} \right) ds.$$

69.3 If X, Y, and Z are such that $X dx + Y dy + Z dz$ is the complete differential of $-V$, a function of x, y, z, then

$$E = \int_A^P \left(X dx + Y dy + Z dz \right) = -\int_A^P dV = V_A - V_P;$$

where the integration is performed in any way from the point A to the point P, whether along the given curve or along any other line between A and P.

69.1 *the work done on unit of electricity in moving along the element of curve ds will be* $R \cos \varepsilon \, ds$: The work done is the product of the resolved part of the intensity, $R \cos \varepsilon$, and the element ds of the curve (14.1, *comment*). It is the work done *on unit of electtricity* because, as noted in the comment to (68.4), electric intensity R measures the force *per unit charge*.

total electromotive force: As defined in Art. 45 above.

69.2 *If we make use of the components of the intensity, the expression becomes...*: A similar expression was derived in the comment to (16.2).

69.3 *complete differential*: Earlier called "exact differential" by Maxwell; see the comment on (16.3).

along the given curve or along any other line between A and P: As Maxwell explained in (16.3), the line-integral of an exact or complete differential, when evaluated between two fixed points, is path-independent.

69.4 In this case V is a scalar function of the position of a point in space, that is, when we know the coordinates of the point, the value of V is determinate, and this value is independent of the position and direction of the axes of reference. See Art. 16.

On Functions of the Position of a Point.

69.5 In what follows, when we describe a quantity as a function of the position of a point, we mean that for every position of the point the function has a determinate value. We do not imply that this value can always be expressed by the same formula for all points of space, for it may be expressed by one formula on one side of a given surface and by another formula on the other side.

On Potential Functions.

70.] The quantity $Xdx + Ydy + Zdz$ is an exact differential whenever the force arises from attractions or repulsions whose intensity is a function of the distances from any number of points. For if r_1 be the distance of one of the points from the point (x, y, z), and if R_1 be the repulsion, then

$$X_1 = R_1 \frac{x - x_1}{r_1} = R_1 \frac{dr_1}{dx},$$

with similar expressions for Y_1 and Z_1, so that

$$X_1 dx + Y_1 dy + Z_1 dz = R_1 dr_1 \, ;$$

and since R_1 is a function of r_1 only, $R_1 dr_1$ is an exact differential of some function of r_1, say $-V_1$.

70.1 Similarly for any other force R_2, acting from a centre at distance r_2,

$$X_2 dx + Y_2 dy + Z_2 dz = R_2 dr_2 = -dV_2 .$$

70. *the distance of one of the points from the point (x, y, z)*: Let P be the center of force, (x, y, z) the point at which unit charge is placed, and r_1 the distance between them. If the charge were moved a distance dx in the x-direction, r_1 would alter by a differential amount dr_1. Then from the similar right triangles formed about the initial point we have $X_1 : R_1 :: dr_1 : dx$, or $X_1 = R_1 \, dr_1 / dx$.

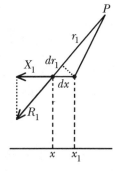

since R_1 is a function of r_1 only, $R_1 dr_1$ is an exact differential: As discussed in the comment to (16.3), the question whether a differential is "exact" or not arises when it is based on a function having multiple variables. Here, since R_1 is a function of the single variable r_1 the question does not pertain; hence $R_1 dr_1$ is an exact differential of some other function of the single variable r_1. We may therefore express $R_1 dr_1$ as the differential $-dV_1$. Maxwell uses this form of expression again in the next paragraph.

But $X = X_1 + X_2 +$ &c., and Y and Z are compounded in the same way, therefore

$$X\,dx + Y\,dy + Z\,dz = -dV_1 - dV_2 - \&c. = -dV.$$

The integral of this quantity, under the condition that it vanishes at an infinite distance, is called the Potential Function.

* * *

70.3 *Definition of Potential.* The Potential at a Point is the work which would be done on a unit of positive electricity by the electric forces if it were placed at that point without disturbing the electric distribution, and carried from that point to an infinite distance: or, what comes to the same thing, the work which must be done by an external agent in order to bring the unit of positive electricity from an infinite distance (or from any place where the potential is zero) to the given point.

Expressions for the Resultant Intensity and its components in terms of the Potential.

71.] Since the total electromotive force along any arc AB is

$$E_{AB} = V_A - V_B,$$

if we put ds for the arc AB we shall have for the intensity resolved in the direction of ds,

$$R\cos\varepsilon = -\frac{dV}{ds};$$

70.1 *it vanishes at an infinite distance*: This requirement lays the foundation for the forthcoming definition in (70.3), which assumes that the potential at an infinite distance is zero.

70.3 *the work which would be done on a unit of positive electricity*: Thus electrical potential is a measure of work *per unit charge*.

what comes to the same thing: Whether the work is done "by the electric forces" in moving from the point in question to an infinite distance, or "by an external agent" in moving over the reverse path, the work will be the same; for by Newton's third law, the force exerted by an external agent will be equal and opposite to the electric force, and the integral is path-independent (16.3). Thus the integral of force over distance is the same in both cases.

71. *the total electromotive force along any arc AB is $E_{AB} = V_A - V_B$*: The total electromotive force between two points is the line integral of the resolved part of the intensity (14.1, 45.); and it is equal to the difference between the potentials at those points (49.); for the order of V_A and V_B see (49., *comment*).

$R\cos\varepsilon = -dV/ds$: Obtained by differentiating both sides of the previous equation.

whence, by assuming ds parallel to each of the axes in succession, we get

$$X = -\frac{dV}{dx}, \quad Y = -\frac{dV}{dy}, \quad Z = -\frac{dV}{dz};$$

$$R = \left[\left(\frac{dV}{dx}\right)^2 + \left(\frac{dV}{dy}\right)^2 + \left(\frac{dV}{dz}\right)^2 \right]^{1/2}.$$

71.1 We shall denote the intensity itself, whose magnitude, or tensor, is R and whose components are X, Y, Z, by the German letter \mathfrak{E}, as in Art. 68.

The Potential at all Points within a Conductor is the same.

72.] A conductor is a body which allows the electricity within it to move from one part of the body to any other when acted on by electromotive force. When the electricity is in equilibrium there can be no electromotive intensity acting within the conductor. Hence $R = 0$ throughout the whole space occupied by the conductor. From this it follows that

$$\frac{dV}{dx} = 0, \quad \frac{dV}{dy} = 0, \quad \frac{dV}{dz} = 0;$$

and therefore for every point of the conductor

$$V = C,$$

where C is a constant quantity.

72.1 Since the potential at all points within the substance of the conductor is C, the quantity C is called the Potential of the conductor. C may be defined as the work which must be done by external agency in order to bring a unit of electricity from an infinite distance to the

assuming ds parallel to each of the axes in succession: That is, let ds equal successively dx, dy, and dz.

71.1 *the intensity itself, whose magnitude, or tensor, is R*: We think of vectors as possessing both magnitude and direction; for quaternions, the corresponding terms are *tensor* and *versor*. Thus R may be referred to as the tensor of \mathfrak{E}. Note: The term "tensor" in modern mathematics has no connection whatever with the tensors of quaternions.

72. *When the electricity is in equilibrium there can be no electromotive intensity acting within the conductor*: For if electromotive intensity were present, the electricity would be in motion and not in equilibrium.

From this it follows that $dV/dx = 0$, $dV/dy = 0$, $dV/dz = 0$: The equation for R in (71.) contains the sum of $(dV/dx)^2$, $(dV/dx)^2$, and $(dV/dx)^2$. Since these quantities are squares they cannot be negative; therefore if their sum is zero each of them individually must be zero.

conductor, the distribution of electricity being supposed not to be disturbed by the presence of the unit.

72.2 It will be shewn at Art. 246 that in general when two bodies of different kinds are in contact, an electromotive force acts from one to the other through the surface of contact, so that when they are in equilibrium the potential of the latter is higher than that of the former. For the present, therefore, we shall suppose all our conductors made of the same metal, and at the same temperature.

72.3 If the potentials of the conductors A and B be V_A and V_B respectively, then the electromotive force along a wire joining A and B will be

$$V_A - V_B$$

in the direction AB, that is, positive electricity will tend to pass from the conductor of higher potential to the other.

72.4 Potential, in electrical science, has the same relation to Electricity the Pressure, in Hydrostatics, has to Fluid, or that Temperature, in Thermodynamics, has to Heat. Electricity, Fluids, and Heat all tend to pass from one place to another, if the Potential, Pressure, or Temperature is greater in the first place than in the second. A fluid is certainly a substance, heat is as certainly not a substance, so that though we may find assistance from analogies of this kind in forming clear ideas of formal relations of electrical quantities, we must be careful not to let the one or the other analogy suggest to us that electricity is either a substance like water, or a state of agitation like heat.

Potential Due to any Electrical System

73.] Let there be a single electrified point charged with a quantity e of electricity, and let r be the distance of the point x', y', z' from it, then

$$V = \int_r^\infty R\,dr = \int_r^\infty \frac{e}{r^2}\,dr = \frac{e}{r}.$$

72.3 *positive electricity will tend to pass from the conductor of higher potential to the other*: Recall that $R\cos\varepsilon = -dV/ds$ (71.). If positive electricity moves in the direction of R, then ds is in the same direction as R, so that $\varepsilon = 0$. Then $-dV/ds$ is positive, which means that V is diminishing. Thus positive electricity tends to pass from higher to lower potential.

72.4 *we must be careful not to let one or the other analogy suggest to us...*: Maxwell is not rescinding his earlier invitation (62.11) to appreciate and even cultivate analogies, but he is reminding us to preserve them *as analogies*.

73. $V = \int_0^\infty R\,dr = \int_0^\infty \frac{e}{r^2}\,dr = \frac{e}{r}$: Maxwell is integrating e/r^2 since that is the expression for the force exerted by a charge e upon a *unit charge* at distance r; compare (70.3).

73.1 Let there be any number of electrified points whose coordinates are (x_1, y_1, z_1), (x_2, y_2, z_2), &c. and their charges e_1, e_2, &c., and let their distances from the point (x', y', z') be r_1, r_2, &c., then the potential of the system at (x', y', z') will be

$$V = \sum \frac{e}{r}.$$

73.2 Let the electric density at any point (x, y, z) within an electrified body be ρ, then the potential due to the body is

$$V = \iiint \frac{\rho}{r} dx\, dy\, dz;$$

where

$$r = \left\{ (x - x')^2 + (y - y')^2 + (z - z')^2 \right\}^{1/2},$$

the integration being extended throughout the body.

On the Proof of the Law of the Inverse Square.

74.] The fact that the force between electrified bodies is inversely as the square of the distance may be considered to be established by direct experiments with the torsion-balance. The results, however, which we derive from such experiments must be regarded as affected by an error depending on the probable error of each experiment, and unless the skill of the operator be very great, the probable error of an experiment with the torsion-balance is considerable. As an argument that the attraction is really, and not merely as a rough approximation, inversely as the square of the distance, Experiment VII (p. [68]) is far more conclusive than any measurements of electrical forces can be.

73.2 *the potential due to the body*: The charge is distributed continuously; therefore instead of summing quotients e/r, as in the preceding equation, we must integrate elements $(\rho/r)\, dx\, dy\, dz$.

74. *On the Proof of the Law of the Inverse Square*: The text of Article 74 reproduced here is that of the first edition (1873). In the second edition Maxwell greatly expanded this topic but omitted the straightforward proof that appears in (74.6–74.14) below.

Experiment VII ... is far more conclusive: Experiment VII directly measures charge rather than force; and Maxwell will explain in (74.3) that the available methods of measuring electrification far exceed the torsion balance in precision. A highly developed version of Experiment VII was performed in 1877–78 at the Cavendish Laboratory under Maxwell's direction; the second and third editions of the Treatise include an extended account of that effort.

74.1 In that experiment a conductor *B*, charged in any manner, was enclosed in a hollow conducting vessel *C*, which completely surrounded it. *C* was also electrified in any manner.

74.2 *B* was then placed in electric communication with *C*, and was then again insulated and removed from *C* without touching it, and examined by means of an electroscope. In this way it was shewn that a conductor, if made to touch the inside of a conducting vessel which completely encloses it, becomes completely discharged, so that no trace of electrification can be discovered by the most delicate electrometer, however strongly the conductor or the vessel has been previously electrified.

74.3 The methods of detecting the electrification of a body are so delicate that a millionth part of the original electrification of *B* could be observed if it existed. No experiments involving the direct measurement of forces can be brought to such a degree of accuracy.

74.4 It follows from this experiment that a non-electrified body in the inside of a hollow conductor is at the same potential as the hollow conductor, in whatever way that conductor is charged. For if it were not at the same potential, then, if it were put in electric connexion with the vessel, either by touching it or by means of a wire, electricity would pass from the one body to the other, and the conductor, when removed from the vessel, would be found to be electrified positively or negatively, which, as we have already stated, is not the case.

74.5 Hence the whole space inside a hollow conductor is at the same potential as the conductor if no electrified body is placed within it. If the law of the inverse square is true, this will be the case whatever be the form of the hollow conductor. Our object at present, however, is to ascertain from this fact the form of the law of attraction.

74.6 For this purpose let us suppose the hollow conductor to be a thin spherical shell. Since everything is symmetrical about its centre, the shell will be uniformly electrified at every point, and we have to enquire what must be the law of attraction of a uniform spherical shell, so as to fulfil the condition that the potential at every point within it shall be the same.

74.1 *In that experiment a conductor B ... was enclosed in a hollow vessel C*: The sketch is an adaptation of Maxwell's Figure 4 from Chapter I. Vessel *C* and its cover are conductive. When the conductive ball *B*, suspended by an insulating thread, is lowered, it hangs within vessel *C* without touching it. The cover is suspended by its own thread, and when it mates with *C*, the ball *B* is completely surrounded.

74.2 *B was then placed in electric communication with C*: One way to do this is to lower the ball further, allowing it to touch the inner surface of *C*.

74.7 Let the force at a distance r from a point at which a quantity e of electricity is concentrated be R, where R is some function of r. All central forces which are functions of the distance admit of a potential, let us write $\dfrac{f(r)}{r}$ for the potential function due to a unit of electricity at a distance r.

74.8 Let the radius of the spherical shell be a, and let the surface density be σ. Let P be any point within the shell at a distance p from the centre. Take the radius through P as the axis of spherical coordinates,

74.7 *central forces ... admit of a potential*: A central force is always directed radially to or from its center. Suppose a body moves along curve AB under the influence of a central force \mathbf{R} with center O. We may approximate the curve by a series of radial increments alternating with circular arcs centered on O. But no work will be done along the arcs, since they are perpendicular to the force. Hence the line-integral from A to B reduces to the sum of integrals along the radial components of the path, that is, to the integral $\int_{r_A}^{r_B} \mathbf{R} \cdot \mathbf{dr}$; and the same will be true for any

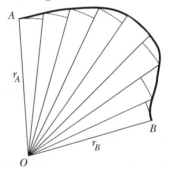

other path joining A and B. Thus if force R is a function of r alone, the line-integral will depends only on the extremities; and this is the condition for existence of a potential (14.1).

write $f(r)/r$ for the potential ... due to a unit electricity at a distance r: Maxwell is adopting an arbitrary form for the potential; but since he has made no assumptions about the function $f(r)$, there is no loss of generality.

74.8 *Take the radius through P as the axis of spherical coordinates*: A typical system of spherical coordinates locates a point P by specifying radius a, azimuth angle φ, and zenith angle θ. Sketch (i), below, shows their relation to rectangular coordinates x, y, z.

This system of spherical coordinates has been imposed on Maxwell's shell in sketch (ii), with the radius through P taken as the zenith axis. The azimuth angle φ is measured in the equatorial plane, shown by the dotted arc.

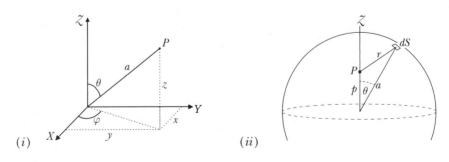

(i) (ii)

and let r be the distance from P to an element dS of the shell. Then the potential at P is

$$V = \iint \sigma \, \frac{f(r)}{r} \, dS, \qquad [i]$$

$$V = \int_0^{2\pi} \int_0^{\pi} \sigma \frac{f(r)}{r} a^2 \sin\theta \, d\theta \, d\varphi. \qquad [ii]$$

74.9 Now
$$r^2 = a^2 - 2ap\cos\theta + p^2,$$
$$r \, dr = ap \sin\theta \, d\theta.$$

74.10 Hence
$$V = 2\pi\sigma \frac{a}{p} \int_{a-p}^{a+p} f(r) \, dr;$$

and V must be constant for all values of p less than a.

74.8, continued. *Then the potential at P is*: Since σ is the surface-density, σdS is the charge of the surface element dS. Then since $f(r)/r$ is the potential due to unit electricity at distance r, its product with σdS will be the potential at P due to the surface element at distance r. Integrate this over the whole surface to find the total potential at P, as in equation $[i]$.

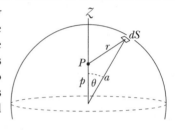

equation $[ii]$: Maxwell has expressed dS as the product of $ad\theta$, the length of its latitudinal arc, and $a\sin\theta d\varphi$, the length of its longitudinal arc. Since Maxwell is expressing angles in radians, arc length is given by the arc's central angle multiplied by the arc's radius. The limits of integration are 0 to π radians (180°) for zenith angle and 0 to 2π radians (360°) for azimuth, thereby encompassing the whole sphere.

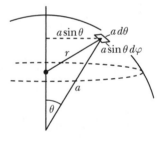

74.9 $r^2 = a^2 - 2ap\cos\theta + p^2$: This is the Law of Cosines written for the triangle having sides r, a, and p. Form the differential of the equation; since a and p are constants this yields $2r \, dr = 2ap \sin\theta \, d\theta$, equivalent to Maxwell's next expression.

74.10 The outer integral of equation $[i]$ is just 2π. And from the previous paragraph we have $\sin\theta d\theta = r \, dr/ap$. Substituting, equation $[i]$ becomes

$$V = 2\pi\sigma \frac{a}{p} \int_{\theta=0}^{\theta=\pi} f(r) \, dr, \quad \text{that is,} \quad V = 2\pi\sigma \frac{a}{p} \int_{a-p}^{a+p} f(r) \, dr.$$

The limits of integration are identical, since $r = a - p$ when $\theta = 0$, and $r = a + p$ when $\theta = \pi$.

V must be constant for all values of p less than a: The equation in (74.10) contains a definite integral, multiplied by constants. Therefore V will be constant under the conditions of the problem, that is, for p less than a.

74.11 Multiplying both sides by p and differentiating with respect to p,

$$V = 2\pi\sigma a\{f(a+p)+f(a-p)\}.$$

74.12 Differentiating again with respect to p,

$$0 = f'(a+p)-f'(a-p).$$

74.13 Since a and p are independent,

$$f'(r)=C, \text{ a constant.}$$

74.14 Hence $\qquad\qquad f(r)=Cr+C',$

and the potential function is

$$\frac{f(r)}{r}=C+\frac{C'}{r}.$$

74.15 The force at distance r is got by differentiating this expression with respect to r, and changing the sign, so that

$$R=\frac{C'}{r^2};$$

or the force is inversely as the square of the distance, and this therefore is the only law of force which satisfies the condition that the potential within a uniform spherical shell is constant.* Now this condition is shewn to be fulfilled by the electric force with the most perfect accuracy. Hence the law of electric force is verified to a corresponding degree of accuracy.

* See Pratt's *Mechanical Philosophy*, p. 144.

74.12 *Differentiating again with respect to p*: On the left hand side we shall have dV/dp, which will equal zero since V has been shown to be a constant. On the right hand side, $f'(r)$ represents $df(r)/dp$.

74.13 *Since a and p are independent, $f'(r)=C$, a constant*: In the preceding paragraph Maxwell deduced that $f'(a+p)-f'(a-p) = 0$. And since a and p are independent, $a+p$ and $a-p$ can each take on arbitrary values. But if a function returns equal values for arbitrary values of its argument, it must be a constant.

74.14 Integrate $f'(r)=C$ to obtain $f(r)=Cr+C'$, where C' is a constant of integration. Divide through by r to obtain $f'(r)/r$, which was the form Maxwell assumed for the potential function in (74.7) above.

75.15 *The force at distance r is got by differentiating this expression and changing the sign*: Recall that when a force R is derivable from a potential V, it has the form $R=-dV/dr$ (16.4, 16.5).

Surface-Integral of Electric Induction, and
Electric Displacement through a surface.

75.] Let R be the resultant intensity at any point of the surface, and ε the angle which R makes with the normal drawn towards the positive side of the surface, then $R \cos \varepsilon$ is the component of the intensity normal to the surface, and if dS is the element of the surface, the electric displacement through dS will be, by Art. 68,

$$\frac{1}{4\pi} KR \cos \varepsilon \, dS.$$

Since we do not at present consider any dielectric except air, $K=1$.

75.1 We may, however, avoid introducing at this stage the theory of electric displacement, by calling $R \cos \varepsilon \, dS$ the Induction through the element dS. This quantity is well known in Mathematical physics, but the name of induction is borrowed from Faraday. The surface-integral of induction is

$$\iint R \cos \varepsilon \, dS,$$

75.] *the positive side of the surface*: "Positive" does not here refer to electric charge. Rather it denotes the outer side of a closed surface, or the convex side of a curved surface, as in the footnote of Art. 21.

75.1 *We may, however, avoid introducing ... the theory of electric displacement*: Note Maxwell's respect for his reader here. He knows that his theory of displacement is novel and must earn the reader's assent. He does not demand that we accept the language of displacement as a condition of continuing the investigation.

the name of induction is borrowed from Faraday: Not that Faraday originated the term; "induction," in its electrical sense, appeared in the *Encyclopædia Britannica* of 1801 and in Humphrey Davy's *Elements of Chemical Philosophy* (1812). But Faraday originated the *conception* of induction with which Maxwell is identifying $R \cos \varepsilon \, dS$. Maxwell characterized that conception as an action by which a dielectric medium transmits electrical effects, and which consists of a polarized state of the particles of the dielectric (29.5, 54.3).

Maxwell's implicit intimation is that besides his own theory of displacement, the only other theory worth considering is Faraday's! But it is evident from the respective mathematical expressions for displacement and induction, above in this paragraph, that the two accounts differ only in the role of the specific inductive capacity K. For Faraday (if he were using mathematical symbolism) the induction depends on R; and R, in turn, varies inversely as the specific inductive capacity of the medium (52.4, *comment*). For Maxwell, on the other hand, the measure of displacement compensates for varying R by multiplying it by K, so that displacement is independent of the medium. Ultimately, Maxwell regards his own theory as essentially identical with Faraday's—or, at least, as identical with what Faraday's thought was striving for.

and it appears by Art. 21, that if X, Y, Z are the components of R, and if these quantities are continuous within a region bounded by a closed surface S, the induction reckoned from within outwards is

$$\iint R \cos \varepsilon \, dS = \iiint \left(\frac{dX}{dx} + \frac{dY}{dy} + \frac{dZ}{dz} \right) dx \, dy \, dz,$$

the integration being extended through the whole space within the surface.

Induction through a Closed Surface due to a single Centre of Force.

76.] Let a quantity e of electricity be supposed to be placed at a point O, and let r be the distance of any point P from O, the intensity at that point is $R = er^{-2}$ in the direction OP.

76.1 Let a line be drawn from O in any direction to an infinite distance. If O is without the closed surface this line will either not cut the surface at all, or it will issue from the surface as many times as it enters. If O is within the surface the line must first issue from the surface, and then it may enter and issue any number of times alternately, ending by issuing from it.

76.2 Let ε be the angle between OP and the normal to the surface drawn outwards where OP cuts it, then where the line issues from the surface, $\cos \varepsilon$ will be positive, and where it enters, $\cos \varepsilon$ will be negative.

76.3 Now let a sphere be described with centre O and radius unity, and let the line OP describe a conical surface of small angular aperture about O as vertex.

76.4 This cone will cut off a small element $d\omega$ from the surface of the sphere, and small elements dS_1, dS_2, &c. from the closed surface at the different places where the line OP intersects it.

76.1 *If O is without the closed surface ... If O is within the surface*: The diagram shows two choices of position for O and several choices of direction for OP.

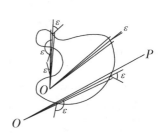

76.2 *where the line issues from the surface, cos ε will be positive*: See the drawing. Since the normal is drawn outward from the surface, a line exiting the surface will make with it an angle ε less than 90°; thus $\cos \varepsilon$ will be positive. Similarly, a line entering the surface will make with the normal an angle ε between 90° and 180°, and $\cos \varepsilon$ will then be negative.

76.5 Then, since any one of these elements dS intersects the cone at a distance r from the vertex and at an obliquity ε,

$$dS = \pm r^2 \sec \varepsilon \, dw;$$

and, since $R = er^{-2}$, we shall have

$$R \cos \varepsilon \, dS = \pm e \, dw;$$

the positive sign being taken when r issues from the surface, and the negative when it enters it. If the point O is without the closed surface, the positive values are equal in number to the negative ones, so that for any direction of r,

$$\sum R \cos \varepsilon \, dS = 0,$$

and therefore

$$\iint R \cos \varepsilon \, dS = 0,$$

the integration being extended over the whole closed surface.

76.6 If the point O is within the closed surface the radius vector OP first issues from the closed surface, giving a positive value of $e \, dw$, and then has an equal number of entrances and issues, so that in this case

$$\sum R \cos \varepsilon \, dS = e \, dw.$$

76.7 Extending the integration over the whole closed surface, we shall include the whole of the spherical surface, the area of which is 4π, so that

$$\iint R \cos \varepsilon \, dS = e \iint dw \ = \ 4\pi e.$$

76.8 Hence we conclude that the total induction outwards through a closed surface due to a centre of force e placed at a point O is zero when O is without the surface, and $4\pi e$ when O is within the surface.

76.5 $dS = \pm r^2 \sec \varepsilon \, dw$: dw is the area cut off on a unit sphere, so $r^2 dw$ is the corresponding area on a sphere of radius r. The oblique projection of $r^2 dw$ corresponds to the hypotenuse of a right triangle having angle ε, while $r^2 dw$ itself corresponds to the adjacent side. Hence the oblique projection is to $r^2 dw$ as 1 is to $\cos \varepsilon$; thus it equals

$$r^2 dw \times 1/\cos \varepsilon \quad \text{or} \quad r^2 dw \sec \varepsilon$$

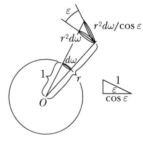

(the secant of ε is defined as $1/\cos \varepsilon$).

$R \cos \varepsilon \, dS = \pm e \, dw$: Since $R \cos \varepsilon \, dS = \frac{e}{r^2} \cos \varepsilon \cdot \frac{r^2 dw}{\cos \varepsilon} = e \, dw$.

76.7 *the spherical surface, the area of which is 4π*: Since the surface area of a sphere of radius r is $4\pi r^2$, the surface of a unit sphere is 4π.

76.8 *the total induction outwards through a closed surface [is] $4\pi e$*: Here e is the quantity of electricity enclosed by the surface. Today the relation between

76.9 Since in air the displacement is equal to the induction divided by 4π, the displacement through a closed surface, reckoned outwards, is equal to the electricity within the surface.

76.10 *Corollary.* It also follows that if the surface is not closed but is bounded by a given closed curve, the total induction through it is ωe, where ω is the solid angle subtended by the closed curve at O. This quantity, therefore, depends only on the closed curve, and the form of the surface of which it is the boundary may be changed in any way, provided it does not pass from one side to the other of the centre of force.

On the Equations of Laplace and Poisson.

77.] Since the value of the total induction of a single centre of force through a closed surface depends only on whether the centre is within the surface or not, and does not depend on its position in my other way, if there are a number of such centres e_1, e_2, &c. within the surface, and e_1', e_2', &c. without the surface, we shall have

$$\iint R\cos\varepsilon\, dS = 4\pi e;$$

induction through a surface and the charge enclosed by that surface is commonly known as Gauss' Theorem; and Maxwell will point out in (76.10) that the relation applies even when the surface is not a closed one. He will restate the closed-surface case for a *distributed* charge in in Art. 77 below, and a related result for a charged surface in (80.4).

76.9 *in air the displacement is equal to the induction divided by* 4π: That is, $D = \frac{1}{4\pi}KR$, as derived in (62.3), and where $K=1$ for air.

the displacement through a closed surface ... is equal to the electricity within the surface: Combine $D = \frac{1}{4\pi}R$ and $\iint R\cos\varepsilon\, dS = 4\pi e$ (76.7) to obtain $\iint D\cos\varepsilon\, dS = e$. Maxwell obtained this result earlier from displacement theory (60.13); by deriving it here on Faraday's terms he shows that the two approaches are mathematically consistent.

76.10 ω *is the solid angle subtended by the closed curve at* O: In the diagram, the solid angle subtended by curve S is measured by the area ω cut off on the unit sphere whose center is O. To compute the induction through the surface bounded by S we integrate $R\cos\varepsilon\, dS$, which equals $ed\omega$ (76.5), over the subtended area ω rather than over the whole sphere.

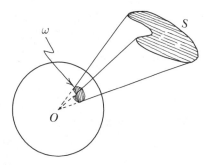

This quantity ... depends only on the closed curve: In his admirable study *Figures of Thought* (Green Lion Press, 2005), Thomas K. Simpson points out that this fact leads us to conceive the surface as a *whole*, characterized by its boundary alone, rather than as a summation of individual area elements.

where e denotes the algebraical sum of the quantities of electricity at all the centres of force within the closed surface, that is, the total electricity within the surface, resinous electricity being reckoned negative.

77.1 If the electricity is so distributed within the surface that the density is nowhere infinite, we shall have by Art. 64,

$$4\pi e = 4\pi \iiint \rho \, dx\, dy\, dz,$$

and by Art. 75,

$$\iint R\cos\varepsilon \, dS = \iiint \left(\frac{dX}{dx} + \frac{dY}{dy} + \frac{dZ}{dz}\right) dx\, dy\, dz.$$

77.2 If we take as the closed surface that of the element of volume $dx\, dy\, dz$, we shall have, by equating these expressions,

$$\frac{dX}{dx} + \frac{dY}{dy} + \frac{dZ}{dz} = 4\pi\rho;$$

and if a potential V exists, we find by Art. 71,

$$\frac{d^2V}{dx^2} + \frac{d^2V}{dy^2} + \frac{d^2V}{dz^2} + 4\pi\rho = 0.$$

77.3 This equation, in the case in which the density is zero, is called Laplace's Equation. In its more general form it was first given by Poisson. It enables us, when we know the potential at every point, to determine the distribution of electricity.

77.1 *If the electricity is ... distributed within the surface*: That is, if it is distributed *throughout the volume* that is bounded by the closed surface. As he did previously in (64.10), Maxwell here expresses the total quantity of electricity e as the volume-integral of its volume-density ρ.

77.2 *If we take as the closed surface that of the element of volume dxdydz*: Since $4\pi e = 4\pi \iiint \rho \, dx\, dy\, dz = \iint r\cos\varepsilon \, dS = \iiint \left(\frac{dX}{dx} + \frac{dY}{dy} + \frac{dZ}{dz}\right) dx\, dy\, dz$, choose the surface of integration in the third expression to be that surface which encloses the volume of integration in the second and last expressions. Then we may equate the two volume-integrals to obtain $4\pi\rho = \frac{dX}{dx} + \frac{dY}{dy} + \frac{dZ}{dz}$. Note: if we take X, Y, Z to be the components of some vector \mathbf{R}, we can also write this equation as div $\mathbf{R} = 4\pi\rho$ or, alternatively, as $\nabla\cdot\mathbf{R} = 4\pi\rho$ (21.1, 25.9, *comments*).

if a potential V exists: Then $X = -dV/dx$ (71.); differentiate again to obtain $dX/dx = -d^2V/dx^2$, and similarly for Y and Z. Substitute in the previous equation and bring all terms to the left to obtain Maxwell's expression.

77.3 *In its more general form*: That is, when $\rho \neq 0$, in which case the equation is called Poisson's.

when we know the potential at every point: That is, when we have V as a function of position. We can then compute the second derivatives and so also ρ.

77.4 We shall denote, as in Art. 26, the quantity

$$\frac{d^2V}{dx^2}+\frac{d^2V}{dy^2}+\frac{d^2V}{dz^2} \quad \text{by} \quad -\nabla^2V,$$

and we may express Poisson's equation in words by saying that the electric density multiplied by 4π is the concentration of the potential. Where there is no electrification, the potential has no concentration, and this is the interpretation of Laplace's equation.

77.5 By Art. 72, V is constant within a conductor. Hence within a conductor the volume-density is zero, and the whole charge must be on the surface.

* * *

Charged Surface of a Conductor.

80.] We have already shewn (Art. 72) that throughout the substance of a conductor in electric equilibrium $X=Y=Z=0$, and therefore V is constant.

Hence
$$\frac{d^2V}{dx^2}+\frac{d^2V}{dy^2}+\frac{d^2V}{dz^2}=4\pi\rho=0,$$

and therefore ρ must be zero throughout the substance of the conductor, or there can be no electricity in the interior of the conductor.

80.1 Hence a superficial distribution of electricity is the only possible distribution in a conductor in equilibrium.

80.2 A distribution throughout the mass of a body can exist only when the body is a non-conductor.

80.3 Since the resultant intensity within the conductor is zero, the resultant intensity just outside the conductor must be in the direction of the normal and equal to $4\pi\sigma$, acting outwards from the conductor.

77.4 *the concentration of the potential*: Maxwell proposed the term *concentration* in (26.3); it has not been generally adopted. Also, as I noted in (26., *comment*), the modern definition omits Maxwell's negative sign, which originated in quaternion multiplication.

80. *V is constant*: From this the next equation follows, since the derivatives of a constant are zero.

therefore ρ must be zero ... or there can be no electricity in the interior: The word "or" is here conjunctive rather than disjunctive; it introduces a consequence, not an alternative.

80.3 *the resultant intensity just outside the conductor must be in the direction of the normal*: If the intensity were in an oblique direction it would have a component parallel to, and therefore within, the conducting surface (46.2, 48.).

and equal to $4\pi\sigma$: This equality follows from Maxwell's earlier result that electrical density σ on an electrified surface is equal to the displacement in the adjacent medium (62.6). For from $D = \sigma$ and $D = R/4\pi$ (in air) we immediately have $R = 4\pi\sigma$. Maxwell will restate this equality in the next paragraph.

80.4 This relation between the surface-density and the resultant intensity close to the surface of a conductor is known as Coulomb's Law, Coulomb having ascertained by experiment that the electromotive intensity near a given point of the surface of a conductor is normal to the surface and proportional to the surface-density at the given point. The numerical relation

$$R = 4\pi\sigma$$

was established by Poisson.

* * *

On Lines of Force.

82.] If a line be drawn whose direction at every point of its course coincides with that of the resultant intensity at that point, the line is called a Line of Force.

82.1 In every part of the course of a line of force, it is proceeding from a place of higher potential to a place of lower potential.

82.2 Hence a line of force cannot return into itself, but must have a beginning and an end. The beginning of a line of force must, by Art. 80, be in a positively charged surface, and the end of a line must be in a negatively charged surface.

82.3 The beginning and the end of the line are called corresponding points on the positive and negative surface respectively.

80.4 *This relation … is known as Coulomb's Law*: Today the title "Coulomb's Law" is reserved for the inverse-square expression of electrostatic or magnetic force between point charges (41.3, 41.4) or magnetic poles (373.3, 373.4). The relation $R = 4\pi\sigma$ for the charged surface of a conductor is rather regarded as a consequence of Gauss's Theorem (76.8, *comment*), although Maxwell obtained it through other arguments (62.6, 76.9). A convenient device for applying Gauss's Theorem to a charged surface is the so-called "Gauss box."

Imagine a mathematical box of infinitesimal thickness, one face of which lies within the charged conductor. If the faces parallel to the conductive surface have area A, the charge enclosed by the box will be σA. By (77.) the total induction through all the faces of the box will be $4\pi\sigma A$; but there is no induction within the body of a conductor, and intensity R must be perpendicular to the surface (80.3). Therefore all the induction must be through the exposed surface A and we shall have $\iint R \, dS = RA = 4\pi\sigma A$, from which $R = 4\pi\sigma$.

82.1 *a line of force [everywhere proceeds] from … higher potential to … lower potential*: This is so because intensity R has the direction from higher to lower potential; see (72.3, *comment*).

82.4 If the line of force moves so that its beginning traces a closed curve on the positive surface, its end will trace a corresponding closed curve on the negative surface, and the line of force itself will generate a tubular surface called a tube of induction. Such a tube is called a Solenoid.*

82.5 Since the force at any point of the tubular surface is in the tangent plane, there is no induction across the surface. Hence if the tube does not contain any electrified matter, by Art. 77 the total induction through the closed surface formed by the tubular surface and the two ends is zero, and the values of $\iint R\cos\varepsilon\, dS$ for the two ends must be equal in magnitude but opposite in sign.

82.6 If these surfaces are surfaces of conductors $\varepsilon = 0$ and $R = -4\pi\sigma$, and $\iint R\cos\varepsilon\, dS$ becomes $-4\pi\iint \sigma\, dS$, or the charge of the surface multiplied by 4π.†

82.7 Hence the positive charge of the surface enclosed within the closed curve at the beginning of the tube is numerically equal to the negative charge enclosed within the corresponding closed curve at the end of the tube.

82.8 Several important results may be deduced from the properties of lines of force.

* From $\sigma\omega\lambda\acute{\eta}\nu$, a tube. Faraday uses (3271) the term 'Sphondyloid' in the same sense.

† {R here is drawn outwards from the *tube*.}

82.4, footnote. Maxwell seems to suggest that Faraday uses the term *sphondyloid* with the sense of a tube with two end surfaces. If so, that is erroneous, since Faraday's paradigmatic sphondyloid image is a figure of rotation (or nest of such figures) that expresses continuity between the internal and external lines of force of a bar magnet, as suggested in this sketch. Unlike the electric lines of force in (82.2), Faraday's magnetic lines *do not terminate*.

82.5 *there is no induction across the surface*: Maxwell gave a general treatment of this problem earlier in Art. 22 and showed a similar result in (22.1). Here the discussion is framed specifically in electrostatic terms.

equal in magnitude but opposite in sign: Maxwell obtained the corresponding general result in his earlier treatment (22.3, *comment*).

82.6 *If these surfaces are surfaces of conductors*: Maxwell refers to the end surfaces of the tube of induction, where the tube runs from one conductor to another.

$\varepsilon = 0$: That is, intensity R is normal to the conducting surface. This result and the relation $R = -4\pi\sigma$ were obtained in (80.3).

$\iint R\cos\varepsilon\, dS$ *becomes* $-4\pi\iint\sigma\, dS$: For in the surface-integral $\iint R\cos\varepsilon\, dS$, ε is 0 (82.6) and R is equal to $-4\pi\sigma$ (80.4).

82.9 The interior surface of a closed conducting vessel is entirely free from charge, and the potential at every point within it is the same as that of the conductor, provided there is no insulated and charged body within the vessel.

82.10 For since a line of force must begin at a positively charged surface and end at a negatively charged surface, and since no charged body is within the vessel, a line of force, if it exists within the vessel, must begin and end on the interior surface of the vessel itself.

82.11 But the potential must be higher at the beginning of a line of force than at the end of the line, whereas we have proved that the potential at all points of a conductor is the same.

82.12 Hence no line of force can exist in the space within a hollow conducting vessel, provided no charged body be placed inside it. If a conductor within a closed hollow conducting vessel is placed in communication with the vessel, its potential becomes the same as that of the vessel, and its surface becomes continuous with the inner surface of the vessel. The conductor is therefore free from charge.

82.13 If we suppose any charged surface divided into elementary portions such that the charge of each element is unity, and if solenoids having these elements for their bases are drawn through the field of force, then the surface-integral for any other surface will be represented by the number of solenoids which it cuts. It is in this sense that Faraday uses his conception of lines of force to indicate not only the direction but the amount of the force at any place in the field.

82.9 *The interior surface of a closed conducting vessel is entirely free from charge*: Experiment VII (32.) yielded this outcome; now Maxwell will explain it in terms of lines of force. The long deductive train that reaches that conclusion began with assuming the inverse-square law in (76.); hence an experiment that verifies the result is itself a confirmation of that law, as Maxwell remarked in (66.1).

82.13 *solenoids having these elements for their bases*: These solenoids are not unit tubes (22.6). Their end surfaces encompass unit *charge*, not unit *induction*, for the induction associated with unit charge is 4π (76.8). But since one tube corresponds to one unit of charge, each of these tubes can serve as a counting unit in the way Faraday intended his lines of force to serve.

not only the direction but the amount of force: For example, in ¶1225 of the *Experimental Researches in Electricity* Faraday describes an exercise in which electric lines of force are diminished in one region through being augmented in another. But it was with respect to magnetic lines that Faraday could demonstrate their quantity most persuasively; through his use of the Moving Wire Faraday could count, in effect, the number of magnetic lines (corresponding to Maxwell's tubes) cut by any surface swept out by the wire. The bulk of his Twenty-eighth Series is devoted to that undertaking.

82.14 We have used the phrase Lines of Force because it has been used by Faraday and others. In strictness, however, these lines should be called Lines of Electric Induction.

82.15 In the ordinary cases the lines of induction indicate the direction and magnitude of the resultant electromotive intensity at every point, because the intensity and the induction are in the same direction and in a constant ratio. There are other cases, however, in which it is important to remember that these lines indicate primarily the induction, and that the intensity is directly indicated by the equipotential surfaces, being normal to these surfaces and inversely proportional to the distances of consecutive surfaces.

On Specific Inductive Capacity.

83a.] In the preceding investigation of surface-integrals we have adopted the ordinary conception of direct action at a distance, and have not taken into consideration any effects depending on the nature of the dielectric medium in which the forces are observed.

82.14 *should be called Lines of Electric Induction*: Lines of *induction* rather than of force because the lines (Maxwell's tubes) relate electrified surfaces through space; and Faraday regards *induction* as the primary relation between electrified bodies. Maxwell's discussion here is confined to the electric field; but he will give a corresponding treatment for magnetic induction in Part III.

82.15 *There are other cases*: As Maxwell noted in (59.4), displacement in some materials is not in the same direction as the electromotive intensity.

The intensity is ... inversely proportional to the distances of consecutive surfaces: One measure of electric intensity is force per unit charge (68.4, *comment*). But if we express force as work per unit distance, as Maxwell chose to do in (12.1) above, then an equivalent measure of electric intensity will be *work per unit distance per unit charge*. Then consider a path that is perpendicular to all the equipotential surfaces it intersects; the number of consecutive equipotential surfaces in unit distance along that path will indicate the intensity.

83a. *we have adopted the ordinary conception of direct action at a distance*: To say that Maxwell has "adopted" the ordinary conception is something of an overstatement. What Maxwell has done throughout the bulk of this chapter is to refrain from using any methods that would be inconsistent with the ordinary theory. He did this in (75.1), for example, when he deliberately employed the formality of induction rather than the language of displacement. It has been possible to exercise such restaint up to now because Maxwell has so far restricted his treatment to the media *air* and *vacuum*. In final few pages of this chapter he revisits the topic of dielectrics in general to give mathematical expression to Faraday's conception of dielectric behavior; sadly, we shall have to pass over that effort.

But what has been Maxwell's purpose in restraining himself in this area? Perhaps to acknowledge that ordinary action-at-a-distance (*continued overleaf*)

83*a*.1 But Faraday has observed that the quantity of electricity induced by a given electromotive force on the surface of a conductor which bounds a dielectric is not the same for all dielectrics. The induced electricity is greater for most solid and liquid dielectrics than for air and gases. Hence these bodies are said to have a greater specific inductive capacity than air, which he adopted as the standard medium.

<p style="text-align:center">*　　*　　*</p>

thinking is formally consistent with the continuous field approach (76.9)—while at the same time bringing out that it is unnecessarily fraught with questionable philosophical conceptions (such as the assumption of isolated objects). Thus moving from it to field theory represents an enrichment rather than an abandonment of the old ways. The same has been said of Einstein's relativity theory in relation to classical mechanics.

CHAPTER III.

ON ELECTRICAL WORK AND ENERGY IN A SYSTEM
OF CONDUCTORS.

84.] *On the Work which must be done by an external agent in order to charge an electrified system in a given manner.*

84.1 The work spent in bringing a quantity of electricity δe from an infinite distance (or from any place where the potential is zero) to a given part of the system where the potential is V, is, by the definition of potential (Art. 70), $V\delta e$.

84.2 The effect of this operation is to increase the charge of the given part of the system by δe, so that if it was e before, it will become $e + \delta e$ after the operation.

84.3 We may therefore express the work done in producing a given alteration in the charges of the system by the integral

$$W = \sum\left(\int V\,\delta e\right); \tag{1}$$

where the summation, (Σ), is to be extended to all parts of the electrified system.

84.4 It appears from the expression for the potential in Art. 73, that the potential at a given point may be considered as the sum of a number of parts, each of these parts being the potential due to a corresponding part of the charge of the system.

84.5 Hence if V is the potential at a given point due to a system of charges which we may call $\Sigma(e)$, and V' the potential at the same point

84.2 *The effect of this operation is to increase the charge*: But not only the charge, since the potential V necessarily increases too. Therefore we will have to integrate $\int V\delta e$ in order to calculate the work done.

84.3 *Equation* (1): The equation is a summation of integrals. First, integral $\int V\delta e$ is performed for each part of the system and represents the work done on or by that part; next, the summation of these for all parts of the system expresses the work done on or by the system as a whole.

141

due to another system of charges which we may call $\Sigma(e')$, the potential at the same point due to both systems of charges existing together would be $V + V'$.

84.6 If, therefore, every one of the charges of the system is altered in the ratio of n to 1, the potential at any given point in the system will also be altered in the ratio of n to 1.

84.7 Let us, therefore, suppose that the operation of charging the system is conducted in the following manner. Let the system be originally free from charge and at potential zero, and let the different portions of the system be charged simultaneously, each at a rate proportional to its final charge.

84.8 Thus if e is the final charge, and V the final potential of any part of the system, then, if at any stage of the operation the charge is ne, the potential will be nV, and we may represent the process of charging by supposing n to increase continuously from 0 to 1.

84.9 While n increases from n to $n + \delta n$, any portion of the system whose final charge is e, and whose final potential is V, receives an increment of charge $e\delta n$, its potential being nV, so that the work done on it during this operation is $eVn\,\delta n$.

84.10 Hence the whole work done in charging the system is

$$\sum(eV)\int_0^1 n\,dn = \tfrac{1}{2}\sum(eV) \tag{2}$$

or half the sum of the products of the charges of the different portions of the system into their respective potentials.

84.11 This is the work which must be done by an external agent in order to charge the system in the manner described, but since the system is a conservative system, the work required to bring the system into the same state by any other process must be the same.

84.8 *e is the final charge, and V the final potential of any part*: Thus e and V are constants, not variables as they were in equation (1) above.

84.10 *Equation* (2): Since e and V are constants they may be written outside the integral $\int n\,dn = \tfrac{1}{2}n^2$. This, evaluated between $n = 0$ and $n = 1$, reduces to $\tfrac{1}{2}$.

84.11 *conservative system*: Maxwell does not define this term in the *Treatise*, but compare Thomson and Tait: "In the particular case of a conservative system, W is a function of the coordinates, irrespectively of the time, or of the paths which have been followed" (*Treatise on Natural Philosophy*, 1867, p. 205).

the work required to bring the system into the same state by any other process must the the same: In Maxwell's example, the work done in charging the system is strictly proportional to the change in potential; it is therefore a function solely of the initial and final states of the system, no matter what particular sequence or process is followed.

84.12 We may therefore call

$$W = \tfrac{1}{2} \sum (eV) \tag{3}$$

the electric energy of the system, expressed in terms of the charges of the different parts of the system and their potentials.

$$* \quad * \quad *$$

84.12 *Equation* (3): This equation is the basis for Maxwell's calculation of *energy density in a dielectric*, the expression for which he has already cited in (59.4), (62.4), and other places. The following derivation of that expression is adapted from one provided by J. J. Thomson for the Third Edition of the *Treatise*, originally as an appendix to Chapter II.

Let there be two parallel plates of area A, oppositely charged, and let the distance s between them be very small compared to the dimensions of the plates. Let V_1, V_2 be the potentials of the plates respectively, and e, $-e$ their respective charges. Then from eq. (3) the total energy of the system will be

$$W = \tfrac{1}{2}(eV_1 - eV_2) = \tfrac{1}{2} eV,$$

where V is the potential difference between the plates. And since the separation distance s is very small, the electric field between the plates will be uniform. Then V is just the product Rs, so that

$$W = \frac{1}{2} e R s. \tag{i}$$

Now displacement is expressed (62.3)

$$D = \frac{1}{4\pi} KR,$$

and the *whole displacement DA* through the dielectric is numerically equal to the charge e (60.13). Thus

$$W = \frac{1}{2} eRs \; = \; \frac{1}{2} \cdot DA \cdot Rs \; = \; \frac{1}{2} \cdot \frac{1}{4\pi} KRA \cdot Rs.$$

Divide each of the last two expressions through by As to obtain the energy per unit volume of dielectric as either

$$\frac{W}{As} = \frac{1}{2} RD \qquad \text{or} \qquad \frac{W}{As} = \frac{1}{8\pi} KR^2. \tag{ii}$$

Remark: We may combine several of the above equations to obtain another useful expression. First eliminate W between equations (i) and (ii) to write

$$\frac{1}{2} eRs \; = \; \frac{1}{8\pi} KR^2 As \qquad \text{or, simplifying,} \qquad es = \frac{1}{4\pi} KRA.$$

But capacity C is defined by the relation $e = CV$ (compare Art. 50) or, for a pair of closely-spaced plates, $e = CRs$. Substituting this, we have

$$C = \frac{1}{4\pi} K \frac{A}{s}.$$

Maxwell had already stated (50.1) that the capacity of a pair of plates is *proportional* to the quotient A/s; now, however, we have an exact expression.

Theory of a system of conductors.

87.] Let A_1, A_2, ... A_n be n conductors of any form; let e_1, e_2, ... e_n be their charges; and V_1, V_2, ... V_n their potentials.

* * *

93.5 ... If the conductors are insulated, they tend to move so that their energy is diminished, and the work done by the electrical forces during the displacement is equal to the diminution of energy.

93.6 If the conductors are connected with batteries, so that their potentials are maintained constant, they tend to move so that the energy of the system is increased, and the work done by the electrical forces

87. *Theory of a system of conductors*: The brief extracts from Arts. 87 and 93 which follow will be useful in connection with Part IV Chapter VII.

93.5 *If the conductors are insulated, they tend to move so that their energy is diminished*: The wording is that of the first edition of the *Treatise*. The argument of Article 93 underwent considerable editorial revision between the first and third editions; the following example incorporates its main points:

Consider a pair of insulated plates bearing opposite charges $+e$ and $-e$. They will possess energy $W = \frac{1}{2}eV$ (84.12, *comment*). But also $e = CV$ (compare Art. 50). if we eliminate V between the two equations, we obtain

$$W = e^2/2C ; \qquad (i)$$

if we eliminate e instead, we shall have

$$W = \tfrac{1}{2}CV^2 . \qquad (ii)$$

In both expressions the energy is exclusively potential, since the plates are stationary.

Nevertheless, being oppositely charged and therefore mutually attracting, the plates tend to approach one another. If they are allowed to do so, their capacity C will increase; but since they are insulated the charge e will be constant. Differentiate equation (i) with respect to C to obtain

$$\frac{dW}{dC} = \frac{e^2}{2}\frac{d}{dC}\left(\frac{1}{C}\right) = -\frac{e^2}{2}\frac{1}{C^2} = -CW\frac{1}{C^2} = -\frac{W}{C} ;$$

then let

$$\Delta W_1 = \int \frac{dW}{dC}dC = -\int \frac{W}{C}dC .$$

As the plates approach, their capacity increases (84.12, *comment*); thus in the integral, W, C, and dC are all positive and hence ΔW_1 will be a negative quantity—the potential energy of the plates will decrease. If the plates are moving freely, that potential energy will have been converted into kinetic energy and the total energy of the system will remain the same. But if the plates are halted at their new position, they will have done mechanical work on the environment and so dissipated their kinetic energy; thus the total energy of the insulated plates will have been diminished by ΔW_1.

93.6 *If the conductors are connected with batteries, ... they tend to move so that the energy of the system is increased*: If the plates are connected to a battery, their

during the displacement is equal to the increment of the energy of the system. The energy spent by the batteries is equal to double of either of these quantities, and is spent half in mechanical, half in electrical work.

* * *

potential difference will be equal to that of the battery, a constant. Differentiate equation (*ii*) in the comment on (93.5), yielding

$$\frac{dW}{dC} = \frac{1}{2}V^2 = \frac{W}{C};$$

interestingly enough, this is just the negative of the previous result. Now if the plates are again allowed to move the same distance as before, the capacity C —which, we recall, is a function of distance (48.12, *comment*)—will again increase by the same amount as before. Integrating,

$$\Delta W_2 = \int \frac{dW}{dC}dC = \int \frac{W}{C}dC.$$

It is therefore clear that $\Delta W_2 = -\Delta W_1$; that is to say, the potential energy of the plates will now increase by the same amount as, previously, it decreased. And this increased energy must be supplied by the battery.

The energy spent by the batteries is equal to double of either of these quantities: In the last example the plates, having moved the same distance as before, will have done the same amount of work on the environment as before. That work too must be, in the present case, ultimately at the expense of the battery, which is thus called upon to supply the sum of ΔW_2 and $(-\Delta W_1)$, that is, $\Delta W_2 + \Delta W_2$, a double portion.

In Part IV, Chapter VII, Maxwell will call attention to a similar result that obtains when parallel current-carrying conductors approach one another.

Editor's Introduction to
Chapter V

In commenting on paragraph 62.6 of Chapter I above, I noted an ambivalence, not to say contradiction, in Maxwell's account of surface charge in relation to electric displacement. One of the earliest of Maxwell's readers to call attention to the difficulty was Heinrich Hertz, who famously declared that "Maxwell's theory *is* Maxwell's system of equations," * thereby relegating Maxwell's physical ideas, his prose narrative, and his rich visual imagery to a merely decorative role.† At first sight, Hertz's remark would appear philosophically gauche, the product of an ostentatious and uncompromising formalism. Yet it is nothing of the sort. Instead, Hertz's introductory chapter reveals that remark to be a cry of despair uttered by a careful and thoughtful reader who has sought, without success, a "consistent conception of Maxwell's ideas."‡ Having found the physical ideas confusing and incoherent, Hertz took refuge in what was to him the only intelligible remnant of Maxwell's work, the equations. Because Hertz was such a careful reader of Maxwell, I hope that reviewing a few stages of his search will clarify the issues—while yet leading to a more satisfactory position than Hertz's own melancholy resignation.

The quest for Maxwell's theory of electric action

Hertz represented diagrammatically what he regarded as the chief candidates for interpretation of electric action, intending to locate Maxwell's standpoint among them. Sketch (*i*) illustrates *action-at-a-distance*, the standard view when Maxwell wrote, but one rejected by him. Here positive and negative electricity has accumulated on the facing surfaces of the "+" and "−" conductors, respectively. Hertz's

positive *Electricität*
negative *Electricität* (*i*)

* Hertz, Heinrich. *Electric Waves*, trans. D. E. Jones (1893; Dover reprint 1962), p. 21 (emphasis added).

† "[W]e should in no wise confuse the simple and homely figure, as it is presented to us by nature, with the gay garment which we use to clothe it." Ibid., p. 28.

‡ Ibid., p. 20.

147

arrows represent the forces exerted by these charges on one another. But although the arrows are drawn in the space between the conductors, *nothing happens in that space*; neither the space nor the aether (if we conceive space as filled with aether) plays any role whatsoever in supporting or transmitting the forces. This inactivity Hertz represents by depicting a region *B* from which we may

positive *Electricität*
negative *Electricität* (*i*)

imagine the aether to have been removed; the persistence of the arrows even within the region *B* is Hertz's way of showing that the forces experienced by the plates would remain wholly unchanged, no matter what changes might occur in the intervening space.

Maxwell, by contrast, posits the phenomenon of *displacement* in the space or other medium between the plates. Hertz's diagram (*ii*) represents one strain of Maxwell's thinking as expressed in various passages of the *Treatise*, some of which we have already encountered in Chapter I. Here, positive and negative electrification appear on the mutually facing surfaces of the conductors; and

positive *Electricität*
negative *Electricität* (*ii*)

there has been a displacement of positive electricity to the right, resulting in negative charge on the surface of the medium facing the "+" conductor, and positive charge on the surface facing the "−" conductor. Within the medium, surface charges on adjacent elements neutralize each other, as Maxwell stated in paragraph 62.6 of Chapter I; indeed, Hertz's diagram is virtually identical to the sketch I supplied in connection with that paragraph. Hertz's arrows, which previously represented forces of action-at-a-distance, he here draws as dotted, indicating that those forces have *vanished*—for this interpretation denies any role for action-at-a-distance forces. This is the configuration Maxwell describes as a "closed circuit"—since any increase or decrease of displacement in the medium simply carries on, in quantity and direction, the charging current that passes (through a wire not shown) from one conductor to the other (60.14, 60.17). Otherwise stated, "the motions of electricity are like those of an incompressible fluid" (61.).

In Hertz's view, this approach treats the electricity on the conductors as primary and causal, and the displacement in the medium as a

secondary effect. In his words, it "explained the nature of the polarisation through the nature of electricity (assumed to be known)..."*

But Hertz also calls attention to an antithetical understanding of displacement found in passages like the following, from paragraph 111.5 of the present chapter:

> In the case of the Leyden jar of which the inner coating is charged positively, any portion of the glass will have its inner side charged positively and its outer side negatively.

Sketch (*iii*) is Hertz's representation of this example. The left-hand conductor corresponds to the Leyden jar's "inner coating" charged positively, while the right-hand conductor represents the (negative) outer coating, the medium being glass; compare the Leyden jar described in (50.1, *comment*). Notice that, just as Maxwell here states, each portion of the glass

positive *Electricität*
negative *Electricität* (*iii*)

has "its inner side charged positively and its outer side negatively." Notice too that this is *precisely the opposite* of the condition illustrated in sketch (*ii*)! On the other hand, it agrees with the statement we found so puzzling in (62.6), Maxwell's claim that "the surface-density ... is equal in magnitude to the displacement through that point of the surface *reckoned inwards*"—a claim, moreover, which Maxwell will repeat nearly verbatim at (111.4) in the present chapter.

In the context of Chapter I, it was hard to see why Maxwell would have introduced a conception so obviously inconsistent with the general tenor of that chapter. Now in Chapter V it becomes evident that this view does honor to Faraday's deep conviction that electric charge is essentially a surface-effect of induction;† the same paragraph 111.5 already quoted goes on to state:

> If this portion be entirely in the interior of the glass, its surface charge will be neutralized by the opposite charge of the parts in contact with it, but if it be in contact with a conducting body, which is incapable of maintaining in itself the inductive state, the surface charge will not be neutralized, but will constitute that apparent charge which is commonly called the Charge of the Conductor. *The charge therefore at the bounding surface of a conductor*

*Ibid., p. 26.

†Faraday, *Experimental Researches in Electricity*, Eleventh Series; see especially paragraph 1299.

and the surrounding dielectric, which on the old theory was called the charge of the conductor, must be called in the theory of induction the surface charge of the surrounding dielectric [emphasis added].

Hertz's sketch (*iii*) depicts this understanding by showing *no electrification* on the conductors; as Maxwell stated, the so-called "charge of the conductor" is really to be understood as a boundary condition *of the medium* insofar as it experiences displacement.

Thus the ambivalence noted in Chapter I will in Chapter V become profound. For the interpretation now being considered does not merely contradict the previous one with respect to the direction of polarization of the medium but completely reverses the causal order: here *displacement* is taken as primary, while *electric charge* is viewed as a secondary effect. In Hertz's words (reprising his earlier quotation), "But whereas [the previous interpretation] explained the nature of the polarisation through the nature of electricity (assumed to be known), we have here to regard the mode of representation as defining the nature of an electric charge through the polarisation of the space (regarded as known)."*

The same contradiction arises again with respect to the natures of displacement and conduction current. Thus we find Maxwell stating in one place that both displacement and conduction represent motion of electricity, at least in a figurative sense (62.7), and in another that neither does, since electricity is a condition found only at the *boundary* between a conductor and a dielectric (111.6, *comment*).

The meaning of Maxwell's dual theories

Maxwell's readers have reacted to his perceived inconsistency in a variety of ways. Hertz believed that the contradiction was for the most part only apparent, that it arose because Maxwell failed to eliminate from his writing a number of expressions that had been derived from his earlier ideas. Pierre Duhem, writing in 1902, was simply horrified: "How," he asked, "could two propositions so manifestly contradictory present themselves at the same moment to Maxwell's mind, both of them securing his adherence?"† More recent writers, while continuing

* Ibid., p. 26.

† Pierre Duhem, *Les thèories èlectriques de J. Clerk Maxwell: ètude historique et critique* (Paris: Hermann, 1902), p. 101. Translation slightly emended from that in P. M. Heimann, "Maxwell, Hertz, and the Nature of Electricity," *Isis*, **62**, 2 (1971), p. 149.

to acknowledge the contradiction, have been willing in varying degrees to pass beyond shocked indignation and investigate the competing motives that shaped Maxwell's thinking.*

The full significance of each of these points of view must await developments that emerge in Part IV of the *Treatise*. What seems clear at this point, at least, is that neither theory, by itself, can be regarded as a viable account. Maxwell's first view, that displacement current simply carries on the conduction current into the dielectric (the "incompressible fluid" view) *explains too much*: by denying accumulation altogether it fails to distinguish what happens at the surface of a conductor from what happens at any imaginary section within the charging current and thus leaves us with no account of what it means for a body to be "charged." His second view, that the surface charge on a conductor is but the manifestation of polarization in the surrounding medium, *explains too little*: in particular it fails to explain the "reckoned inwards" rule (62.6), whereby an electric field in one direction is to be associated with polarization in the reverse direction.

At this point I would only note that a theory which views motion of electricity as formally identical to motion of an incompressible fluid finds a natural place in *electrodynamics*; it offers a clear conception of current flow, and in Part IV of the *Treatise* we shall see that the displacement current carries on the charging current not only in quantity and direction but in magnetic effects too—a continuity central to Maxwell's electromagnetic theory of light. On the other hand, a theory that, in the spirit of Faraday, treats the charge of a "charged body" as a secondary quantity, subordinate to displacement in the medium, has a natural place in *electrostatics*, successfully purging the idea of a "charged body" of any residual elements of both fluid theory and action-at-a-distance thinking. If Maxwell indeed harbors two correspondingly different notions of "displacement," are they mutually consistent? And does his narrative distinguish adequately between them?

In what is perhaps the least contrived among recent efforts to interpret Maxwell's displacement theory, Jed Z. Buchwald answers with a confident *yes* to the first question, a more tentative *yes* to the second.†
The reader may wish to investigate whether Maxwell uses such phrases as "electric displacement," "displacement of electricity," "electric polarization" and "displacement" simply, without qualification, as haphazard

* See, for example, Joan Bromberg, "Maxwell's Electrostatics," *Am. J. Physics*, **36**, 2 (February 1968), p.151; P. M. Heimann, cit., p.157; and Daniel M. Siegel, *Innovation in Maxwell's Electrodynamic Theory*, Cambridge University Press (1991).

† Buchwald, Jed Z. *From Maxwell to Microphysics* (1985).

synonyms; or if, on the contrary, the different phrases appear to distinguish systematically between Maxwell's first characterization of displacement and his second.

CHAPTER V

MECHANICAL ACTION BETWEEN TWO ELECTRICAL SYSTEMS

103.] Let E_1 and E_2 be two electrical systems the mutual action of which we propose to investigate. Let the distribution of electricity in E_1 be defined by the volume-density, ρ_1, of the element whose coordinates are x_1, y_1, z_1. Let ρ_2 be the volume-density of the element of E_2, whose coordinates are x_2, y_2, z_2.

103.1 Then the x-component of the force acting on the element of E_1 on account of the repulsion of the element of E_2 will be

$$\rho_1 \rho_2 \frac{x_1 - x_2}{r^3} \, dx_1 \, dy_1 \, dz_1 \, dx_2 \, dx_2 \, dx_2,$$

where
$$r^2 = \left(x_1 - x_2\right)^2 + \left(y_1 - y_2\right)^2 + \left(z_1 - z_2\right)^2,$$

Arts. 103–106: In these articles Maxwell carries out one of the lengthiest derivations in the *Treatise*, and the most demanding among the present selections. Some readers may wish to pass over it on a first reading. The derivation is of central importance to Maxwell's project, however, for it shows that his theory implies that electric lines of force constitute axes of tension and pressure, just as Faraday had surmised; moreover it establishes the numerical measure of that pressure and tension. Maxwell will announce this result with evident satisfaction in Art. 109: "The distribution of stress considered in this chapter is precisely that to which Faraday was led in his investigation of induction through dielectrics."

103.1 *Then the x-component of the force … will be …*: The equation is just the inverse-square law, multiplied by $(x_1 - x_2)/r$. For $\rho_1 \, dx_1 \, dy_1 \, dz_1$ is the charge contained in the first element, while $\rho_2 \, dx_2 \, dy_2 \, dz_2$ is the charge contained in the second. And by similar triangles, $F_x/F = (x_1 - x_2)/r$; thus component $F_x = F(x_1 - x_2)/r$.

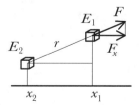

and if A denotes the x-component of the whole force acting on E_1 on account of the presence of E_2

$$A = \iiint\iiint \frac{x_1 - x_2}{r^3} \rho_1\rho_2\, dx_1\, dy_1\, dz_1\, dx_2\, dy_2\, dz_2, \qquad (1)$$

where the integration with respect to x_1, y_1, z_1 is extended throughout the region occupied by E_1 and the integration with respect to x_2, y_2, z_2 is extended throughout the region occupied by E_2.

103.2 Since, however, ρ_1 is zero except in the system E_1 and ρ_2 is zero except in the system E_2, the value of the integral will not be altered by extending the limits of the integrations, so that we may suppose the limits of every integration to be $\pm\infty$.

103.3 This expression for the force is a literal translation into mathematical symbols of the theory which supposes electric forces to act directly between bodies at a distance, no attention being bestowed on the intervening medium.

103.4 If we now define Ψ_2, the potential at the point x_1, y_1, z_1, arising from the presence of system E_2, by the equation

$$\Psi_2 = \iiint \frac{\rho_2}{r}\, dx_2\, dy_2\, dz_2, \qquad (2)$$

Ψ_2 will vanish at an infinite distance, and will everywhere satisfy the equation

$$\nabla^2\Psi_2 = 4\pi\rho_2. \qquad (3)$$

103.5 We may now express A in the form of a triple integral

$$A = -\iiint \frac{d\Psi_2}{dx_1} \rho_1\, dx_1\, dy_1\, dz_1. \qquad (4)$$

103.1, continued. *acting on E_1 on account of the presence of E_2*: Notice that Maxwell has not employed the more conventional locution "exerted on E_1 by E_2; he is preparing to question what is the proper source of action and has therefore offered a more neutral construction.

Equation (1): Maxwell's formalism demands a sixfold integral only because he has specified six variables of integration. Today we would probably define separate volume-elements dV_1 and dV_2 and express the equation as a double integration over the two volumes, respectively.

103.4 *Equation (2)*: This too is a reformulation of an earlier equation—this time the equation for electric potential produced by a charged body (73.2).

Equation (3): This is Poisson's equation, introduced in (77.2ff).

103.5 *Equation (4)*: As Maxwell defined it in (103.1), A is the x-component of the force exerted *on* system E_1. Thus we must multiply the charge of every element of E_1 by the x-component of the intensity subsisting at the location of that element. The x-component of intensity will be $-d\Psi_2/dx_1$ (103.4, 16.5), the charge on an element is $\rho_1\, dx_1\, dy_1\, dz_1$; integrate their product to obtain the total.

103.6 Here the potential Ψ_2 is supposed to have a definite value at every point of the field, and in terms of this, together with the distribution, ρ_1, of electricity in the first system, E_1, the force A is expressed, no explicit mention being made of the distribution of electricity in the second system E_2.

103.7 Now let Ψ_1 be the potential arising from the first system, expressed as a function of x, y, z, and defined by the equation

$$\Psi_1 = \iiint \frac{\rho_1}{r} \, dx_1 \, dy_1 \, dz_1, \qquad (5)$$

Ψ_1 will vanish at an infinite distance, and will everywhere satisfy the equation

$$\nabla^2 \Psi_1 = 4\pi\rho_1. \qquad (6)$$

103.8 We may now eliminate ρ_1 from A and obtain

$$A = -\frac{1}{4\pi} \iiint \frac{d\Psi_2}{dx_1} \nabla^2 \Psi_1 \, dx_1 \, dy_1 \, dz_1. \qquad (7)$$

in which the force is expressed in terms of the two potentials only.

104.] In all the integrations hitherto considered, it is indifferent what limits are prescribed, provided they include the whole of the system E_1. In what follows we shall suppose the systems E_1 and E_2 to be such that a certain closed surface s contains within it the whole of E_1 but no part of E_2.

104.1 Let us also write

$$\rho = \rho_1 + \rho_2, \quad \Psi = \Psi_1 + \Psi_2, \qquad (8)$$

then within s,

$$\rho_2 = 0, \quad \rho = \rho_1,$$

and without s,

$$\rho_1 = 0, \quad \rho = \rho_2. \qquad (9)$$

103.6 *no explicit mention ... of the electricity in the second system E_2*: On the theory of direct action at a distance, the electricity in E_2 is the very source of the action; thus it would seem merely perverse to fail to mention it! But in this splendid example of mathematical rhetoric, Maxwell is discounting the supposed distant source of action, concentrating instead on conditions—in this case the spatial-variation of the potential—that are local to the recipient of the action.

103.8 *We may now eliminate ρ_1 from A*: Having rhetorically dismissed charge as a *source* of action, Maxwell now proceeds to topple it from its supposed role as *recipient*, as well; Poisson's equation (6) permits Maxwell to replace ρ_1, charge density, with $\nabla^2 \Psi$—an expression that refers solely to potential rather than to charge. Equation (7) thus describes a force independently of any supposed source or recipient; it will soon prove to subsist wholly in the medium.

104.2 Now $$A_{11} = -\iiint \frac{d\Psi_1}{dx_1} \rho_1 dx_1\, dy_1\, dz_1 \qquad (10)$$

represents the resultant force, in the direction x, on the system E_1 arising from the electricity in the system itself. But on the theory of direct action this must be zero, for the action of any particle P on another Q is equal and opposite to that of Q on P, and since the components of both actions enter into the integral, they will destroy each other.

104.3 We may therefore write

$$A = -\frac{1}{4\pi} \iiint \frac{d\Psi}{dx} \nabla^2 \Psi\, dx_1\, dy_1\, dz_1, \qquad (11)$$

where Ψ is the potential arising from both systems, the integration being now limited to the space within the closed surface s, which includes the whole of the system E_1 but none of E_2.

105.] If the action of E_2 on E_1 is effected, not by direct action at a distance, but by means of a distribution of stress in a medium extending continuously from E_2 to E_1, it is manifest that if we know the stress at every point of any closed surface s which completely separates E_1 from E_2, we shall be able to determine completely the mechanical action of E_2 on E_1. For if the force on E_1 is not completely accounted for by the stress through s, there must be direct action between something outside of s and something inside of s.

104.2 *Equation* (10): Notice that this equation is identical to equation (4) but with Ψ_2 replaced by Ψ_1. If the earlier equation is taken to represent the force exerted on system 1 by system 2 (as the theory of direct action at a distance does), this equation must represent the force exerted on system 1 *by itself*.

104.3 Ψ *is the potential arising from both systems*: Maxwell defined Ψ as the sum of Ψ_1 and Ψ_2 in equation (8) above.

limited to the space ... which includes the whole of the system E_1 but none of E_2: Since a system does not act on itself, if we integrate only over E_1 then that part of Ψ which is represented by Ψ_1 does not contribute to the integral; thus equation (11) is only a more general form of equation (7).

105. *if we know the stress at every point of any closed surface s which completely separates E_1 from E_2, we shall be able to determine completely the mechanical action*: If the action between systems is sustained by stress, then any surface separating one system from the other must participate in that condition of stress. If, on the contrary, one system could affect another in a way that is *not* reflected in the conditions on a surface separating them, that would appear to exemplify the very meaning of "direct action at a distance"! Maxwell thus has found a definitive formulation for action at a distance, on the one hand, and for its alternative, action through a distribution of stress in a medium, on the other. Both these concepts had suffered from vagueness up to this point.

105.1 Hence if it is possible to account for the action of E_2 on E_1 by means of a distribution of stress in the intervening medium, it must be possible to express this action in the form of a surface-integral extended over any surface s which completely separates E_2 from E_1.

105.2 Let us therefore endeavor to express

$$A = -\frac{1}{4\pi} \iiint \frac{d\Psi}{dx}\left[\frac{d^2\Psi}{dx^2} + \frac{d^2\Psi}{dy^2} + \frac{d^2\Psi}{dz^2}\right] dx\,dy\,dz \qquad (12)$$

in the form of a surface integral.

105.3 By Theorem III, Art. 21, we may do so if we can determine X, Y and Z, so that

$$\frac{d\Psi}{dx}\left(\frac{d^2\Psi}{dx^2} + \frac{d^2\Psi}{dy^2} + \frac{d^2\Psi}{dz^2}\right) = \frac{dX}{dx} + \frac{dY}{dx} + \frac{dZ}{dx}, \qquad (13)$$

105.4 Taking the terms separately,

$$\frac{d\Psi}{dx}\frac{d^2\Psi}{dx^2} = \frac{1}{2}\frac{d}{dx}\left(\frac{d\Psi}{dx}\right)^2,$$

105.1 *it must be possible to express this action in the form of a surface-integral*: The reasoning of the previous paragraph was general; now Maxwell proposes to seek not just any characterization of the bounding surface but specifically a *surface-integral*. This sudden recognition of the significance of the surface-integral marks a peak moment in Maxwell's argument. Notice that he has come upon it not by manipulating symbols but by thinking *physically*. This is characteristic of Maxwell's mathematical physics, as distinguished from mathematics proper: he employs equations to *express* his thinking, but not as the *vehicles* of his thinking.

105.2 *Let us therefore endeavor to express [equation (12)] in the form of a surface integral*: Equation (12) is equation (11), but with $\nabla^2\Psi$ expanded into its individual terms. Equation (11), in turn, was a revised form of equation (7), which expressed the force independently of either source or recipient. Thus if Maxwell can express equation (12) as a surface-integral he will have a complete and coherent account of action subsisting wholly in the medium.

105.3 *we may do so if we can determine X, Y, and Z*: Theorem III—equation (5) in paragraph (21.6)—showed that a volume integral will be equivalent to a surface integral if its integrand has the form $\frac{dX}{dx} + \frac{dY}{dy} + \frac{dZ}{dz}$, the form of a divergence. Accordingly equation (13) attempts to set the integrand of equation (12) equal to a divergence; if Maxwell can find an X, Y, and Z to meet this condition, the attempt will succeed.

105.4 The equation for the first term in equation (13) is an instance of $\frac{d}{dx}\left(u^2\right) = 2u\frac{du}{dx}$, with u taken as $\frac{d\Psi}{dx}$ so that $\frac{du}{dx}$ becomes $\frac{d^2\Psi}{dx^2}$.

$$\frac{d\Psi}{dx}\frac{d^2\Psi}{dy^2} = \frac{d}{dy}\left(\frac{d\Psi}{dx}\frac{d\Psi}{dy}\right) - \frac{d\Psi}{dy}\frac{d^2\Psi}{dxdy},$$

$$= \frac{d}{dy}\left(\frac{d\Psi}{dx}\frac{d\Psi}{dy}\right) - \frac{1}{2}\frac{d}{dx}\left(\frac{d\Psi}{dy}\right)^2.$$

105.5 Similarly $\quad \dfrac{d\Psi}{dx}\dfrac{d^2\Psi}{dz^2} = \dfrac{d}{dz}\left(\dfrac{d\Psi}{dx}\dfrac{d\Psi}{dz}\right) - \dfrac{1}{2}\dfrac{d}{dx}\left(\dfrac{d\Psi}{dz}\right)^2.$

105.6 If, therefore, we write

$$\left.\begin{array}{r}
\left(\dfrac{d\Psi}{dx}\right)^2 - \left(\dfrac{d\Psi}{dy}\right)^2 - \left(\dfrac{d\Psi}{dz}\right)^2 = 8\pi p_{xx}, \\[2ex]
\left(\dfrac{d\Psi}{dy}\right)^2 - \left(\dfrac{d\Psi}{dz}\right)^2 - \left(\dfrac{d\Psi}{dx}\right)^2 = 8\pi p_{yy}, \\[2ex]
\left(\dfrac{d\Psi}{dz}\right)^2 - \left(\dfrac{d\Psi}{dx}\right)^2 - \left(\dfrac{d\Psi}{dy}\right)^2 = 8\pi p_{zz}, \\[2ex]
\dfrac{d\Psi}{dy}\dfrac{d\Psi}{dz} = 4\pi p_{yz} = 4\pi p_{zy}, \\[2ex]
\dfrac{d\Psi}{dz}\dfrac{d\Psi}{dx} = 4\pi p_{zx} = 4\pi p_{xz}, \\[2ex]
\dfrac{d\Psi}{dx}\dfrac{d\Psi}{dy} = 4\pi p_{xy} = 4\pi p_{yz};
\end{array}\right\} \qquad (14)$$

then

$$A = -\frac{1}{4\pi}\iiint\left(\frac{dp_{xx}}{dx} + \frac{dp_{yx}}{dy} + \frac{dp_{zx}}{dz}\right)dxdydz, \qquad (15)$$

105.4, continued. To obtain the equation for the second term of equation (13), apply the product rule $\dfrac{d}{dy}\left(\dfrac{d\Psi}{dx}\dfrac{d\Psi}{dy}\right) = \dfrac{d^2\Psi}{dydx}\dfrac{d\Psi}{dy} + \dfrac{d\Psi}{dx}\dfrac{d^2\Psi}{dy^2}$ and observe that the first term on the right will be canceled in the substitution, while the remaining term is transformed by $\dfrac{d}{dx}\left(u^2\right) = 2u\dfrac{du}{dx}$ as before, but with u now taken as $\dfrac{d\Psi}{dy}$.

105.6 *Equations* (14): Maxwell is in effect defining the quantities p_{xx}, etc., introducing the double indices *xx*, *yx*, *zx*, etc., without explanation. Some of his readers might have recognized them as conforming to the standard notation for *stress*, though Maxwell has not yet identified p_{xx}, p_{yx}, and p_{zx} as stresses. He will do so in (105.9), at which point I will explain the notation system.

the integration being extended throughout the space within s. Transforming the volume-integral by Theorem III, Art. 21,

$$A = \iint \left(l\, p_{xx} + m\, p_{yx} + n\, p_{zx} \right) ds, \tag{16}$$

where ds is an element of any closed surface including the whole of E_1 but none of E_2, and l, m, n are the direction cosines of the normal drawn from ds outwards.

Equation (15): To show that the equation follows from definitions (14), consider its integrand term by term. First,

$$\frac{dp_{xx}}{dx} = \frac{1}{8\pi}\frac{d}{dx}\left(\frac{d\Psi}{dx}\right)^2 - \frac{1}{8\pi}\frac{d}{dx}\left(\frac{d\Psi}{dy}\right)^2 - \frac{1}{8\pi}\frac{d}{dx}\left(\frac{d\Psi}{dz}\right)^2; \tag{a}$$

The first right-hand term of (a) reduces in the way we have seen earlier:

$$\frac{1}{8\pi}\frac{d}{dx}\left(\frac{d\Psi}{dx}\right)^2 = \frac{1}{8\pi}\cdot 2\left(\frac{d\Psi}{dx}\right)\frac{d}{dx}\left(\frac{d\Psi}{dx}\right) = \frac{1}{4\pi}\frac{d\Psi}{dx}\frac{d^2\Psi}{dx^2}.$$

Next,

$$\frac{dp_{yx}}{dy} = \frac{1}{4\pi}\frac{d}{dy}\left(\frac{d\Psi}{dx}\frac{d\Psi}{dy}\right); \tag{b}$$

and finally,

$$\frac{dp_{zx}}{dz} = \frac{1}{4\pi}\frac{d}{dz}\left(\frac{d\Psi}{dz}\frac{d\Psi}{dx}\right); \tag{c}$$

and note that when the three equations are added, the right-hand term of (b) combines with the second right-hand term of (a) to yield

$$\frac{1}{4\pi}\frac{d}{dy}\left(\frac{d\Psi}{dx}\frac{d\Psi}{dy}\right) - \frac{1}{8\pi}\frac{d}{dy}\left(\frac{d\Psi}{dy}\right)^2.$$

This has the same form as the right-hand side of the equation in (105.5) and therefore reduces to

$$\frac{1}{4\pi}\frac{d\Psi}{dx}\frac{d^2\Psi}{dy^2}.$$

Similarly, the right-hand term of (c) combines with the third right-hand term of (a) to yield

$$\frac{1}{4\pi}\frac{d\Psi}{dx}\frac{d^2\Psi}{dz^2}.$$

The overall sum of (a), (b), and (c) is therefore equal to

$$\frac{1}{4\pi}\frac{d\Psi}{dx}\frac{d^2\Psi}{dx^2} + \frac{1}{4\pi}\frac{d\Psi}{dx}\frac{d^2\Psi}{dy^2} + \frac{1}{4\pi}\frac{d\Psi}{dx}\frac{d^2\Psi}{dz^2}, \quad \text{or} \quad \frac{1}{4\pi}\frac{d\Psi}{dx}\left(\frac{d^2\Psi}{dx^2} + \frac{d^2\Psi}{dy^2} + \frac{d^2\Psi}{dz^2}\right).$$

This is the integrand of equation (12); and thus it has been shown to be equivalent to a divergence whose components X, Y, and Z are proportional to p_{xx}, p_{yx}, and p_{zx}, respectively.

Transforming the volume-integral: Compare the transformation between dX/dx, dY/dy, dZ/dz and Xl, Ym, Zn, respectively, in (21.6), discussed in the comments on equations (4), (5), and (6) in that paragraph.

105.7 For the components of the force on E_1 in the directions of y and z we obtain in the same way

$$B = \iint \left(l p_{xy} + m p_{yy} + n p_{zy} \right) ds, \qquad (17)$$

$$C = \iint \left(l p_{xz} + m p_{yz} + n p_{zz} \right) ds. \qquad (18)$$

105.8 If the action of the system E_2 on E_1 does in reality take place by direct action at a distance, without the intervention of any medium, we must consider the quantities p_{xx} &c. as mere abbreviated forms for certain symbolical expressions, and as having no physical significance.

105.9 But if we suppose that the mutual action between E_2 and E_1 is kept up by means of stress in the medium between them, then since the equations (16), (17), (18) give the components of the resultant force arising from the action, on the outside of the surface s, of the stress whose six components are p_{xx} &c., we must consider p_{xx} &c. as the components of a stress actually existing in the medium.

106.] To obtain a clearer view of the nature of this stress let us alter the form of part of the surface s so that the element ds may become part of an equipotential surface. (This alteration of the surface is legitimate provided we do not thereby exclude any part of E_1 or include any part of E_2.)

105.8 *mere abbreviated forms*: With proportionality constants 4π and 8π, the quantities p_{xx}, etc. were contrived to represent the complicated differential functions of Ψ that appear in definitions (14) above. But if the medium plays no role, such simplification is a mere mathematical convenience.

105.9 *components of the resultant force*: The quantities A, B, C in equations (16), (17), and (18) were defined as components of force in (103.1) and (105.7).

we must consider p_{xx} &c. as the components of a stress: For each of these quantities, multiplied by the appropriate direction-cosine l, m, or n, yields the resolved part of itself in the direction of the normal to surface-element ds. If, then, the integral of products like $p_{xy} ds$ yields a *force*, then p_{xy} must have dimensions of *force per unit area*; these are the dimensions of *pressure* or *stress*.

If the quantities p are interpreted as stresses, the standard notation system works this way: p_{xx} denotes a stress which, referred to a plane perpendicular to the x-direction, also acts in the x-direction. It is called a "normal" stress since its direction is normal to its reference plane. Similarly p_{yx} denotes a stress which, while referred to a plane perpendicular to the y-*direction*, acts in the the x-*direction*. Finally, p_{zx} denotes a stress which, while referred to a plane perpendicular to the z-*direction*, acts in the x-*direction*. These last two, acting parallel to their respective reference planes, are called "shear" stresses.

106.1 Let ν be a normal to ds drawn outwards.

106.2 Let $R = -d\Psi/d\nu$ be the electromotive intensity in the direction of ν, then

$$\frac{d\Psi}{dx} = -Rl, \quad \frac{d\Psi}{dy} = -Rm, \quad \frac{d\Psi}{dz} = -Rn.$$

106.3 Hence the six components of the stress are

$$p_{xx} = \frac{1}{8\pi} R^2 \left(l^2 - m^2 - n^2 \right), \quad p_{yz} = \frac{1}{4\pi} R^2 mn,$$

$$p_{yy} = \frac{1}{8\pi} R^2 \left(m^2 - n^2 - l^2 \right), \quad p_{zx} = \frac{1}{4\pi} R^2 nl,$$

$$p_{zz} = \frac{1}{8\pi} R^2 \left(n^2 - l^2 - m^2 \right), \quad p_{xy} = \frac{1}{4\pi} R^2 lm.$$

106. *element ds may become part of an equipotential surface*: Surface s was arbitrary at first. If part of it now coincides with an equipotential surface, the normal stresses on that part will be in the direction of the electromotive intensity.

106.2 *Rl, Rm, Rn*: Recall that l, m, n are the direction-cosines of the normal to surface ds (105.6). Since we are now considering an equipotential surface, R is in the direction of that normal; therefore Rl, Rm, and Rn are the x-, y-, and z-components of R respectively, and thus they equal $-d\Psi/dx$, $-d\Psi/dy$, and $-d\Psi/dz$.

106.3 *the six components of the stress*: Formally, the stress applied to a cubical solid will have 18 components, three for each of the six faces of the cube. But if the stress is free of *thrust*—that is, if it has no tendency to cause translational motion—then (*i*) the normal components acting on opposite faces of the cube must balance one another pairwise, reducing the number of independent components by three; and (*ii*) pairs of shear components acting on opposite faces but in the same direction must balance one another likewise, further reducing the number of independent components by six. Thus we shall be left with *nine* independent components, three normal and six shear.

Now if in addition the stress in question has no *rotational* tendency, it is evident that (*iii*) pairs of shear components acting on adjacent faces and perpendicular to the same axis must equal one another, or there would be rotation about that axis. In the drawing on the facing page, for example, $p_{xz} = p_{zx}$; similarly $p_{xy} = p_{yx}$, and $p_{zy} = p_{zx}$, reducing the number of independent shear components to three and the total number of independent components to *six*. Maxwell has assumed these equalities in his definitions (14).

You can find this argument in the same place Maxwell himself probably found it, Thomson and Tait's *Treatise on Natural Philosophy* (1867)—the book that contained pretty much everything a mid-nineteenth century mathematical physicist would be expected to know; see Arts. 661ff. in that work.

p_{xx}, etc. Obtain Maxwell's expressions for these quantities by substituting the expressions set out in (106.2) into definitions (14).

106.4 If a, b, c are the components of the force on ds per unit of area,

$$a = l\,p_{xx} + m\,p_{yx} + n\,p_{zx} = \frac{1}{8\pi}R^2 l,$$

$$b = \frac{1}{8\pi}R^2 m,$$

$$c = \frac{1}{8\pi}R^2 n.$$

106.5 Hence the force exerted by the part of the medium outside ds on the part of the medium inside ds is normal to the element and directed outwards, that is to say, it is a tension like that of a rope, and

106.4 *a, b, and c are the components of the force on ds per unit of area*: As before, let the direction of ν, the normal to area ds, be specified by direction-cosines l, m, n. Then the projections of ds onto the three reference planes will have areas

 $l\,ds$ (in the plane normal to the x-axis)

 $m\,ds$ (in the plane normal to the y-axis)

 $n\,ds$ (in the plane normal to the x-axis)

Now let a, b, and c be respectively the x-, y-, and z-components of the stress exerted on surface ds. Since stress is reckoned as force per unit area, the product $a\,ds$ will be a force in the x-direction. It must then equal the sum of forces acting in the x-direction on the projections of ds, so that

$$a \cdot ds = p_{xx} \cdot l\,ds + p_{yx} \cdot m\,ds + p_{zx} \cdot n\,ds.$$

Dividing by ds, $a = l p_{xx} + m p_{yx} + n p_{zx}.$

Similarly we obtain $b = l p_{xy} + m p_{yy} + n p_{zy}$

and $c = l p_{xz} + m p_{yz} + n p_{zz}.$

Obtain Maxwell's totals by substituting the expressions for p_{xx}, etc. from (106.3), and employing the identity $l^2 + m^2 + n^2 = 1$.

106.5 *the force exerted ... is normal to the element and directed outwards*: For a, b, and c are the x-, y-, and z-coordinates of the stress. And they are to one another as l, m, n, the direction-cosines of the normal (106.4); therefore the stress of which they are the components must be parallel to the normal. Note that this is also the direction of the resultant intensity R, since ds was taken as part of an equipotential surface (106.).

It is also evident from (106.4) that a, b, and c must be positive, since R^2 is positive; and so their sum, the total stress on element ds, will also be positive. But the positive direction of the normal was taken outwards (106.1); hence the stress on element ds is outwards.

it is a tension like that of a rope: Consider a section of a *tube of induction* (82.4). By definition its cylindrical walls coincide with lines of force and its two faces, being perpendicular to those lines, are equipotential surfaces; thus each face will experience stress T in the outward direction. The drawing should suggest the similarity to a rope.

its value per unit of area is $\dfrac{1}{8\pi}R^2$.

106.6 Let us next suppose that the element ds is at right angles to the equipotential surfaces which cut it, in which case

$$l\frac{d\Psi}{dx} + m\frac{d\Psi}{dy} + n\frac{d\Psi}{dz} = 0.\tag{19}$$

106.7 Now

$$8\pi\left(lp_{xx} + mp_{yx} + np_{zx}\right) = l\left[\left(\frac{d\Psi}{dx}\right)^2 - \left(\frac{d\Psi}{dy}\right)^2 - \left(\frac{d\Psi}{dz}\right)^2\right]$$

$$+ 2m\frac{d\Psi}{dx}\frac{d\Psi}{dy} + 2n\frac{d\Psi}{dx}\frac{d\Psi}{dz}.\tag{20}$$

106.8 Multiplying (19) by $2\,d\Psi/dx$ and subtracting from (20), we find

$$8\pi\left(lp_{xx} + mp_{yx} + np_{zx}\right) = -l\left[\left(\frac{d\Psi}{dx}\right)^2 - \left(\frac{d\Psi}{dy}\right)^2 - \left(\frac{d\Psi}{dz}\right)^2\right]$$

$$= -lR^2.\tag{21}$$

its value per unit of area: If the total stress is T, then $T^2 = a^2 + b^2 + c^2$ or $\left(\frac{1}{8\pi}R^2\right)^2\left(l^2 + m^2 + n^2\right)$. But $l^2 + m^2 + n^2 = 1$ (106.4, *comment*); hence $T = \frac{1}{8\pi}R^2$.

106.6 *Let us next suppose that the element ds is at right angles to the equipotential surfaces*: In other words, let ds coincide with a portion of the cylindrical wall of a tube of induction such as the one depicted on the facing page.

Equation (19): This is the mathematical statement of perpendicularity between the normal ν to the surface and the resultant intensity R. For the unit vector normal to ds is

$$\nu = \mathbf{i}l + \mathbf{j}m + \mathbf{k}n,$$

while the intensity is

$$R = \mathbf{i}\frac{d\Psi}{dx} + \mathbf{j}\frac{d\Psi}{dy} + \mathbf{k}\frac{d\Psi}{dz}.$$

Now form the dot product $\nu\cdot R$ (see pages 49–51, comments on quaternions); when multiplied out, it is identical to the lefthand side of equation (19). But $\nu\cdot R$ is also equal to the product νR, multiplied by the cosine of the angle between them. Then if $\nu\cdot R$ is set equal to zero, and neither ν nor R is zero, the cosine of the angle between them must be zero, and that angle will be 90°.

106.7 *Equation* (20): Form the quantity $8\pi(lp_{xx} + mp_{yx} + np_{zx})$ from the expressions for p_{xx}, etc. that appear in definitions (14).

106.9 Hence the components of the tension per unit area of *ds* are

$$a = -\frac{1}{8\pi}R^2 l, \quad b = -\frac{1}{8\pi}R^2 m, \quad c = -\frac{1}{8\pi}R^2 n.$$

106.10 Hence if the element *ds* is at right angles to an equipotential surface, the force which acts on it is normal to the surface, and the numerical value per unit of area is the same as in the former case, but the direction of the force is different, for it is a pressure instead of a tension.

106.11 We have thus completely determined the type of the stress at any given point in the medium.

106.12 The direction of the electromotive intensity at the point is a principal axis of stress, and the stress along such an axis is a tension whose numerical magnitude is

$$p = \frac{1}{8\pi}R^2, \tag{22}$$

where *R* is the electromotive intensity.

106.13 Any direction at right angles to this is also a principal axis of stress, and the stress along such an axis is a pressure whose numerical magnitude is also *p*.

106.14 The stress as thus defined is not of the most general type, for it has two of its principal stresses equal to each other, and the third has the same value with the sign reversed.

106.9 *the components of the tension per unit area of ds*: Maxwell obtains the values of components *a*, *b*, and *c* in essentially the same way as he obtained the values in (106.4) above. But bear in mind that the orientation of *ds* is now perpendicular to its earlier orientation.

106.10 *the force which acts on it is normal to the surface*: The reasoning is the same as that offered previously in (106.5, *comment*).

and the numerical value ... is the same: As before, take the square root of $a^2 + b^2 + c^2$, employing the identity for direction-cosines, to obtain $\frac{1}{8\pi}R^2$, the same value as for the first case.

for it is a pressure instead of a tension: Maxwell referred to "components of the tension" in (106.9) because he had adopted *outwards from the surface* as the positive direction and the presumptive direction of the stress; an outward stress is a tension. But notice that he has obtained *negative* values for those components, showing that they are actually directed inwards and are therefore to be counted as components of pressure rather than of tension.

106.12 *principal axis of stress*: One of three axes such that the normal stresses along them, without any shear components, suffice to represent the stress completely. A set of three principal axes can be found for every stress.

106.14 *not of the most general type*: In general the principal stresses are independent; but in the present case the two transverse pressures are equal, while the axial tension is the same in magnitude though opposite in sign.

106.15 These conditions reduce the number of independent variables which determine the stress from six to three, accordingly it is completely determined by the three components of the electromotive intensity

$$-\frac{d\Psi}{dx}, \quad -\frac{d\Psi}{dy}, \quad -\frac{d\Psi}{dz}.$$

The three relations between the six components of stress are

$$\left.\begin{aligned}
p_{yz}^2 &= \left(p_{xx} + p_{yy}\right)\left(p_{zz} + p_{xx}\right), \\
p_{zx}^2 &= \left(p_{yy} + p_{zz}\right)\left(p_{xx} + p_{yy}\right), \\
p_{xy}^2 &= \left(p_{zz} + p_{xx}\right)\left(p_{yy} + p_{zz}\right).
\end{aligned}\right\} \tag{23}$$

106.15 *These conditions reduce the number of independent variables ... from six to three*: Equations (23) show that the three shear components previously supposed independent (106.3, *comment*), are actually determined by the three normal components—reducing the number of independent components to three. To see how these equations arise, suppose the reference cube (105.9, *comment*) is oriented so that the axial stress **A** bears on the *z*-face, not necessarily normal to it. Then transverse stresses **B** and **C** will bear on the *x*- and *y*-faces, respectively. Each will be the (vector) sum of the stress components on its corresponding face, as

$$\mathbf{A} = \mathbf{i}p_{zx} + \mathbf{j}p_{zy} + \mathbf{k}p_{zz}$$
$$\mathbf{B} = \mathbf{i}p_{xx} + \mathbf{j}p_{xy} + \mathbf{k}p_{xz}$$
$$\mathbf{C} = \mathbf{i}p_{yx} + \mathbf{j}p_{yy} + \mathbf{k}p_{yz}.$$

Now axial stress **A** being perpendicular to transverse stresses **B** and **C**, their dot products (106.6, *comment*) are zero, so that

$$\mathbf{A}\cdot\mathbf{B} = p_{zx}p_{xx} + p_{zy}p_{xy} + p_{zz}p_{xz} = 0$$

and

$$\mathbf{A}\cdot\mathbf{C} = p_{zx}p_{yx} + p_{zy}p_{yy} + p_{zz}p_{yz} = 0.$$

Based on the equalities discussed in (106.3, *comment*), set

$$p_{xy} = p_{yx} = a, \quad p_{xz} = p_{zx} = b, \quad p_{zy} = p_{yz} = c.$$

Substituting in the above yields

$$b(p_{xx} + p_{zz}) + ac = 0 \quad \text{and} \quad c(p_{yy} + p_{zz}) + ab = 0;$$

Multiply these two equations to get

$$(p_{xx} + p_{zz})(p_{yy} + p_{zz}) = a = p_{xy}^2,$$

the third of equations (23). Obtain the first two equations in similar fashion.

it is completely determined by the three components of the electromotive intensity: Orient the reference cube so that electromotive intensity *R* acts normal to one face; then all normal components are known in terms of *R*, as expressed in equation (22). Thus intensity *R* completely determines the stress.

This result is very important to Maxwell, since it replaces the conventional understanding of electromotive intensity in terms of *force at a distance* with its relation to *stress in the medium*—as Maxwell states in (110.2) below.

* * *

109.] The distribution of stress considered in this chapter is precisely that to which Faraday was led in his investigation of induction through dielectrics. He sums up in the following words:—

"(1297) The direct inductive force, which may be conceived to be exerted in lines between the two limiting and charged conducting surfaces, is accompanied by a lateral or transverse force equivalent to a dilatation or repulsion of these representative lines (1224); or the attractive force which exists amongst the particles of the dielectric in the direction of the induction is accompanied by a repulsive or a diverging force in the transverse direction.

"(1298) Induction appears to consist in a certain polarized state of the particles, into which they are thrown by the electrified body sustaining the action, the particles assuming positive and negative points or parts, which are symmetrically arranged with respect to each other and the inducting surfaces or particles. The state must be a forced one, for it is originated and sustained only by force, and sinks to the normal or quiescent state when that force is removed. It can be *continued* only in insulators by the same portion of electricity, because they only can retain this state of the particles."

109.1 This is an exact account of the conclusions to which we have been conducted by our mathematical investigation. At every point of the medium there is a state of stress such that there is tension along the lines of force and pressure in all directions at right angles to these lines, the numerical magnitude of the pressure being equal to that of the tension, and both varying as the square of the resultant force at the point.

109.2 The expression 'electric tension' has been used in various senses by different writers. I shall always use it to denote the tension along the lines of force, which, as we have seen, varies from point to point, and is always proportional to the square of the resultant force at the point.

110.] The hypothesis that a state of stress of this kind exists in a fluid dielectric, such as air or turpentine, may at first sight appear at variance

109.1 *an exact account of the conclusions to which we have been conducted by our mathematical investigation*: The force of this remark is not that the mathematical treatment confirms Faraday's conclusions, as though Faraday's thought stood in need of mathematical foundation. On the contrary, it is the *mathematics* that has been put to the test, as Maxwell intimated in his Preface (pages 4–5): whether formal mathematics is *capable* of expressing what Faraday's prose could express.

109.2 *proportional to the square of the resultant force*: Better, proportional to the square of the resultant *intensity*, R, as in (106.12), above.

with the established principle that at any point in a fluid the pressures in all directions are equal. But in the deduction of this principle from a consideration of the mobility and equilibrium of the parts of the fluid it is taken for granted that no action such as that which we here suppose to take place along the lines of force exists in the fluid. The state of stress which we have been studying is perfectly consistent with the mobility and equilibrium of the fluid, for we have seen that, if any portion of the fluid is devoid of electric charge, it experiences no resultant force from the stresses on its surface, however intense these may be. It is only when a portion of the fluid becomes charged that its equilibrium is disturbed by the stresses on its surface, and we know that in this case it actually tends to move. Hence the supposed state of stress is not inconsistent with the equilibrium of a fluid dielectric.

110.1 The quantity W, which was investigated in Chapter IV, Art. 99a, may be interpreted as the energy in the medium due to the distribution of stress. It appears from the theorems of that chapter that the distribution of stress which satisfies the conditions there given also makes W an absolute minimum. Now when the energy is a minimum for any configuration, that configuration is one of equilibrium, and the equilibrium is stable. Hence the dielectric, when subjected to the inductive action of electrified bodies, will of itself take up a state of stress distributed in the way we have described.

110.2 It must be carefully borne in mind that we have made only one step in the theory of the action of the medium. We have supposed it to be in a state of stress, but we have not in any way accounted for this stress, or explained how it is maintained. This step, however, seems to me to be an important one, as it explains, by the action of the consecutive parts of the medium, phenomena which were formerly supposed to be explicable only by direct action at a distance.

110. *in a fluid the pressures in all directions are equal*: Pascal formulated this principle in his *Treatise on the Equilibrium of Liquids* (1653); trans. I. H. B. and A. G. H. Spiers in *The Physical Treatises of Pascal*, Columbia University Press (1937).

110.1 *The quantity W*: Maxwell introduced W in a general form in Art. 99a, not included in our selections. But we saw a special form of it in the comment to (84.12), as $W = \frac{1}{8\pi} KR^2 As$, where W is the energy contained in a volume As of dielectric bordered by parallel charged plates.

110.2 *we have not ... accounted for this stress*: This bears further on my remark in (109.1, *comment*). Although the mathematical expression of Faraday's theory of stress in the medium is a magnificent feat, it is essentially an achievement of translation and enunciation; in itself it possesses no greater explanatory power than did Faraday's narrative.

111.] I have not been able to make the next step, namely, to account by mechanical considerations for these stresses in the dielectric. I therefore leave the theory at this point, merely stating what are the other parts of the phenomenon of induction in dielectrics.

111.1 I. *Electric Displacement.* When induction is transmitted through a dielectric, there is in the first place a displacement of electricity in the direction of the induction. For instance, in a Leyden jar, of which the inner coating is charged positively and the outer coating negatively, the direction of the displacement of positive electricity in the substance of the glass is from within outwards.

111.2 Any increase of this displacement is equivalent, during the time of increase, to a current of positive electricity from within outwards, and any diminution of the displacement is equivalent to a current in the opposite direction.

111.3 The whole quantity of electricity displaced through any area of a surface fixed in the dielectric is measured by the quantity which we have already investigated (Art. 75) as the surface integral of induction through that area, multiplied by $K/4\pi$, where K is the specific inductive capacity of the dielectric.

111.4 II. *Surface charge of the particles of the dielectric.* Conceive any portion of the dielectric, large or small, to be separated (in imagination)

111. *to account by mechanical considerations*: What does Maxwell mean by a "mechanical" account? Surely not one based on things like wheels and gears, despite his habit of imagining mechanical models like this fanciful rendition of current induction in "On Physical Lines of Force" (1861–1862). On the whole, Maxwell's mechanisms are *metaphors*, visible structures intended to lead the mind to discern the intelligibility inherent in natural systems. As he stated in "A Dynamical Theory of the Electromagnetic Field" (1864), "I wish merely to direct the mind of the reader to mechanical phenomena which will assist him in understanding the electrical ones." Nevertheless, in Part IV of the present Treatise Maxwell will adopt a highly abstract conception of mechanism in which visibility is abandoned in favor of generality.

111.1 *Leyden jar*: The Leyden jar was illustrated and discussed in (50.1, *comment*); see also the sketch accompanying (111.5, *comment*) on the opposite page.

within outwards: Here, these terms make reference to the jar as a whole. But in (111.4) below, Maxwell will use the term "inwards" in relation to *a portion of the dielectric*, not to the whole jar.

from the rest by a closed surface, then we must suppose that on every elementary portion of this surface there is a charge measured by the total displacement of electricity through that element of surface *reckoned inwards*.

111.5 In the case of the Leyden jar of which the inner coating is charged positively, any portion of the glass will have its inner side charged positively and its outer side negatively. If this portion be entirely in the interior of the glass, its surface charge will be neutralized by the opposite charge of the parts in contact with it, but if it be in contact with a conducting body, which is incapable of maintaining in itself the inductive state, the surface charge will not be neutralized, but will constitute that apparent charge which is commonly called the Charge of the Conductor.

111.4 *there is a charge measured by the total displacement...*: Maxwell showed that charge is equal to, and therefore measured by, total displacement in the paragraphs leading up to (60.13). Although his proof there is specific to a spherical configuration, he also obtained the general result in Arts. 75–76.

reckoned inwards: "Reckoning inwards" means taking the inward direction (with respect to the element) as positive. Thus if displacement *D* has the direction sketched here, it will be *positive* at face *a* and *negative* at face *b*; face *a* will exhibit a positive charge and face *b* a negative charge.

111.5 *In the case of the Leyden jar of which the inner coating is charged positively...*: The inner and outer coatings are represented in this sketch of a "dissected" Leyden jar. The central post is connected to the inner coating and provides the means for charging it.

... any portion of the glass will have its inner side charged positively and its outer side negatively: If the drawing above (111.4, *comment*) represents a portion of the glass, face *a* will be the inner side—facing the inner coating—and will exhibit positive charge, while face *b* will be the outer side and will exhibit negative charge. Notice that this assignment of positive and negative charge is the reverse of what Maxwell asserted in the penultimate sentence of (62.6); I discuss this evident contradiction in the editor's introduction to the present chapter.

incapable of maintaining ... the inductive state: We saw (72.) that there is no electromotive intensity, and therefore no induction, within the body of a conductor.

the surface charge will not be neutralized, but will constitute that apparent charge which is commonly called the Charge of the Conductor: The surface charge, not being countered by an opposing power, will be "apparent," that is, it will manifest its presence. But although it is commonly called "charge on the conductor," it more truly reflects a condition in the *dielectric*, as Maxwell is about to state explicitly.

The charge therefore at the bounding surface of a conductor and the surrounding dielectric, which on the old theory was called the charge of the conductor, must be called in the theory of induction the surface charge of the surrounding dielectric.

111.6 According to this theory, all charge is the residual effect of the polarization of the dielectric. The polarization exists throughout the interior of the substance, but it is there neutralized by the juxtaposition of oppositely charged parts, so that it is only at the surface of the dielectric that the effects of the charge become apparent.

111.7 The theory completely accounts for the theorem of Art. 77, that the total induction through a closed surface is equal to the total quantity of electricity within the surface multiplied by 4π. For what we have called the induction through the surface is simply the electric displacement multiplied by 4π, and the total displacement outwards is necessarily equal to the total charge within the surface.

111.8 The theory also accounts for the impossibility of communicating an 'absolute charge' to matter. For every particle of the dielectric has

111.5, continued. *the charge of the conductor must be called ... the surface charge of the surrounding dielectric*: Charge does not reside on, adhere to, or in any sense belong to, the "charged" conductor; it is rather a characteristic of the *dielectric* to which the so-called charged body is adjacent.

111.6 *All charge is the residual effect of the polarization of the dielectric*: If charge is a "residual effect" it surely can enjoy only a derivative existence. And if so, it would seem we can no longer regard displacement as being constituted by movement of charge—despite Maxwell's several locutions to that effect in Art. 60, the term "displacement" itself, and especially Maxwell's oft-repeated maxim that "the motions of electricity are like those of an incompressible fluid" (61.). But then, what *is* displacement?

111.7 *The theory completely accounts for the theorem of Art. 77*: That is, from the new point of view (that charge is merely a manifestation of displacement), the theorem becomes almost trivial, as Maxwell goes on to explain.

what we have called the induction ... is simply the electric displacement multiplied by 4π: In (75.) Maxwell cited $\frac{1}{4\pi}KR\cos\varepsilon\,dS$ as the expression for displacement, and in (75.1) he defined induction as $R\cos\varepsilon\,dS$; thus induction is $4\pi/K$ times the displacement. (Maxwell has ignored the specific inductive capacity K, perhaps because he is here thinking only of air or vacuum.)

and the total displacement outwards is necessarily equal to the total charge within the surface: Maxwell *deduced* that displacement is equal to charge in (60.13). Here he reasons that if charge is but the manifestation of displacement, it must *necessarily* equal the displacement. Then if the induction is 4π times the displacement (in air or vacuum), while the displacement equals the charge, it immediately follows that the induction equals 4π times the charge. See also (76.9).

equal and opposite charges on its opposite sides, if it would not be more correct to say that these charges are only the manifestations of a single phenomenon, which we may call Electric Polarization.

111.9 A dielectric medium, when thus polarized, is the seat of electric energy, and the energy in unit of volume of the medium is numerically equal to the electric tension on unit of area, both quantities being equal to half the product of the displacement and the resultant electromotive intensity, or

$$p = \tfrac{1}{2}\mathfrak{D}\mathfrak{E} = \frac{1}{8\pi}K\mathfrak{E}^2 = \frac{2\pi}{K}\mathfrak{D}^2,$$

where p is the electric tension, \mathfrak{D} the displacement, \mathfrak{E} the electromotive intensity, and K the specific inductive capacity.

111.10 If the medium is not a perfect insulator, the state of constraint, which we call electric polarization, is continually giving way. The medium yields to the electromotive force, the electric stress is relaxed, and the potential energy of the state of contraint is converted into heat. The rate at which this decay of the state of polarization takes place depends on

111.9 $p = \tfrac{1}{2}\mathfrak{D}\mathfrak{E}$, *etc.*: We derived $p = \tfrac{1}{8\pi}K\mathfrak{E}^2$ in the comment to (84.12). Obtain the alternate forms given here by substituting $\mathfrak{D} = \tfrac{1}{4\pi}K\mathfrak{E}$ (62.3, 68.6).

111.10 *the state of constraint ... is continually giving way*: Maxwell introduced this conception of current conduction earlier at (62.7).

the potential energy of the state of constraint is converted into heat: No doubt readers will at this point wish to ask, "But *how* is it so converted?" On the fluid-flow conception of current, heat production is typically attributed to loss of kinetic energy by friction or collision between particles of the supposed fluid and the material of the conductor. In contrast, Maxwell here makes no use of mechanical hypotheses; his characterization operates on the assumption of energy conservation and conversion, simply.

Today we regard a theory as defective if it is unable to specify the microscopic processes that underlie the behavior of observable quantities. But as Jed Z. Buchwald perceptively explains (*From Maxwell to Microphysics*, 1985), Maxwell and his adherents believed profoundly in the *continuum*—whether under the name of "aether" or of space itself. To them, atoms and molecules were not the fundamental realities we tend to suppose today but somewhat artificial constructs, while the electron was simply unheard of. Thus the drive to exhibit theories in terms of the constitution of matter would have seemed as inconsequntial to Maxwell as it is thought orthodox today.

In Chapter V of Part IV Maxwell will explore a method of theorizing based on Lagrange's equations, an activity that reveals what Maxwell himself expects of a theory. In a word, that is *flexibility*; he does not value a system that articulates all that we may think is true today, but rather one that says no more than we really know, but which can develop to accommodate what we hope to learn tomorrow.

the nature of the medium. In some kinds of glass, days or years may elapse before the polarization sinks to half its original value. In copper, a similar change is effected in less than the billionth of a second.

111.11 We have supposed the medium after being polarized to be simply left to itself. In the phenomenon called the electric current the constant passage of electricity through the medium tends to restore the state of polarization as fast as the conductivity of the medium allows it to decay. Thus the external agency which maintains the current is always doing work in restoring the polarization of the medium, which is continually becoming relaxed, and the potential energy of this polarization is continually becoming transformed into heat, so that the final result of the energy expended in maintaining the current is to gradually raise the temperature of the conductor, till as much heat is lost by conduction and radiation from its surface as is generated in the same time by the electric current.

111.11 *We have supposed the medium after being polarized to be simply left to itself*: For example, remove the glass or copper of the previous paragraph from the electromotive force which had polarized it. While the glass will maintain its polarized condition for days or years, polarization in the copper disappears almost immediately.

the electric current ... tends to restore the state of polarization as fast as the conductivity of the medium allows it to decay: As a given material undergoes displacement in proportion to its *specific inductive capacity*, K (111.3), so too that displacement decays at a rate proportional to the material's *conductivity*, C. We will find a quantitative definition of C in (609.).

Notice that Maxwell is following Faraday (109.) in making no sharp distinction between insulators and conductors. *All* materials exhibit both inductive capacity (12.10) and conductivity; whether a given material is ranked as an insulator or a conductor depends on whether K or C predominates.

CHAPTER VII

FORMS OF THE EQUIPOTENTIAL SURFACES AND LINES OF INDUCTION IN SIMPLE CASES.

117.] WE have seen that the determination of the distribution of electricity on the surface of conductors may be made to depend on the solution of Laplace's equation

$$\frac{d^2V}{dx^2} + \frac{d^2V}{dy^2} + \frac{d^2V}{dz^2} = 0,$$

V being a function of x, y, and z, which is always finite and continuous, which vanishes at an infinite distance, and which has a given constant value at the surface of each conductor.

117.1 It is not in general possible by known mathematical methods to solve this equation so as to fulfil arbitrarily given conditions, but it is easy to write down any number of expressions for the function V which shall satisfy the equation, and to determine in each case the forms of the conducting surfaces, so that the function V shall be the true solution.

117.2 It appears, therefore, that what we should naturally call the inverse problem of determining the forms of the conductors when the expression for the potential is given is more manageable than the direct problem of determining the potential when the form of the conductors is given.

117.1 *the distribution of electricity ... the solution of Laplace's equation*: For a given shape of conductor, a solution to Laplace's equation will yield V as a function of position throughout the space surrounding that conductor; from this we can calculate the intensity R at points near its surface and thence find the surface-density σ by means of the relation $R = 4\pi\sigma$ (80.4).

117.1 *It is not in general possible by known mathematical methods to solve this equation...*: What Maxwell points out here is true of any second-order (or higher) differential equation. There are very few methods for solving such equations explicitly. Instead we resort in most cases to a trial-and-error procedure; but with experience one becomes familiar with the likely candidates for solving equations of various forms; and then, as Maxwell goes on to state, "it is easy to write down any number of expressions ... which shall satisfy the equation."

117.3 In fact, every electrical problem of which we know the solution has been constructed by this inverse process. It is therefore of great importance to the electrician that he should know what results have been obtained in this way, since the only method by which he can expect to solve a new problem is by reducing it to one of the cases in which a similar problem has been constructed by the inverse process.

117.4 This historical knowledge of results can be turned to account in two ways. If we are required to devise an instrument for making electrical measurements with the greatest accuracy, we may select those forms for the electrified surfaces which correspond to cases of which we know the accurate solution. If, on the other hand, we are required to estimate what will be the electrification of bodies whose forms are given, we may begin with some case in which one of the equipotential surfaces takes a form somewhat resembling the given form, and then by a tentative method we may modify the problem till it more nearly corresponds to the given case. This method is evidently very imperfect considered from a mathematical point of view, but it is the only one we have, and if we are not allowed to choose our conditions, we can make only an approximate calculation of the electrification. It appears, therefore, that what we want is a knowledge of the forms of equipotential surfaces and lines of induction in as many different cases as we can collect together and remember. In certain classes of cases, such as those relating to spheres, there are known mathematical methods by which we may proceed. In other cases we cannot afford to despise the humbler method of actually drawing tentative figures on paper, and selecting that which appears least unlike the figure we require.

117.5 This latter method I think may be of some use, even in cases in which the exact solution has been obtained, for I find that an eye-knowledge of the forms of the equipotential surfaces often leads to a right selection of a mathematical method of solution.

117.6 I have therefore drawn several diagrams of systems of equipotential surfaces and lines of induction, so that the student may make himself familiar with the forms of the lines. The methods by which such diagrams may be drawn will be explained in Art. 123.

118.] In the first figure [on the page opposite] we have the sections of the equipotential surfaces surrounding two points charged with quantities of electricity of the same kind and in the ratio of 20 to 5.

118.1 Here each point is surrounded by a system of equipotential surfaces which become more nearly spheres as they become smaller, though none of them are accurately spheres. If two of these surfaces, one surrounding each point, be taken to represent the surfaces of two conducting bodies, nearly but not quite spherical, and if these bodies be

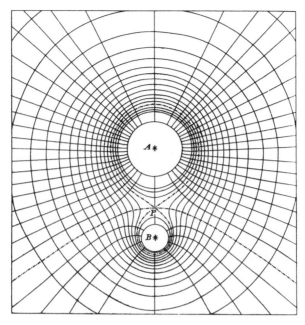

Lines of Force and Equipotential Surfaces.

A = 20 B = 5 . P, Point of Equilibrium. AP = $\frac{2}{3}$AB

charged with the same kind of electricity, the charges being as 4 to 1, then the diagram will represent the equipotential surfaces, provided we expunge all those which are drawn inside the two bodies. It appears from the diagram that the action between the bodies will be the same as that between two points having the same charges, these points being not exactly in the middle of the axis of each body, but each somewhat more remote than the middle point from the other body.

118.2 The same diagram enables us to see what will be the distribution of electricity on one of the oval figures, larger at one end than the other, which surround both centres. Such a body, if charged with 25 units of electricity and free from external influence, will have the surface-density greatest at the small end, less at the large end, and least in a circle somewhat nearer the smaller than the larger end.*

> *{This can be seen by comparing the distances between the equipotential surfaces in various parts of the field.}

118.1 *provided we expunge all those which are drawn inside the two bodies*: There can be no difference of potential either within a solid conductor (51., 72.) or inside a closed conducting surface (46.9); thus neither lines of induction nor equipotentials ought to be drawn there.

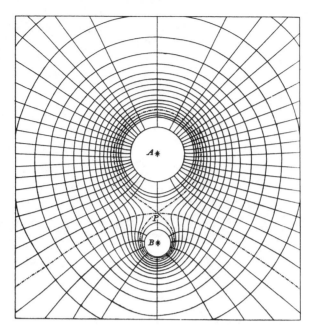

Lines of Force and Equipotential Surfaces.

$A = 20$ $B = 5$. P, *Point of Equilibrium*. $AP = \frac{2}{3}AB$

118.3 There is one equipotential surface, indicated by a dotted line, which consists of two lobes meeting at the conical point P. That point is a point of equilibrium, and the surface-density on a body of the form of this surface would be zero at this point.

118.4 The lines of force in this case form two distinct systems, divided from one another by a surface of the sixth degree. indicated by a dotted line, passing through the point of equilibrium, and somewhat resembling one sheet of the hyperboloid of two sheets.

118.3 *point of equilibrium*: A point charge placed at P would experience no force. For if the potential V is continuous, the point where two equipotential lines intersect must lie on an equipotential surface, even if only an infinitesimal one. For a gravitational analogy, think of Figure I as a topographic map of peaks A and B. The region surrounding point P is a *saddle surface* and point P is a saddle point, which means that the infinitesimal area around point P is *flat*. A ball placed there would have no tendency to roll in any direction; it would be in equilibrium. But the slightest departure from point P—except along the equipotential line Maxwell mentions—would constitute either an ascent or a descent.

Electrically, the infinitesimal "flat" region about P is a region where potential V is constant; it is therefore a region where the electromotive intensity R is zero; thus from the relation $R = 4\pi\sigma$ (80.4) it follows that, as Maxwell goes on to state, the surface-density σ is zero at point P.

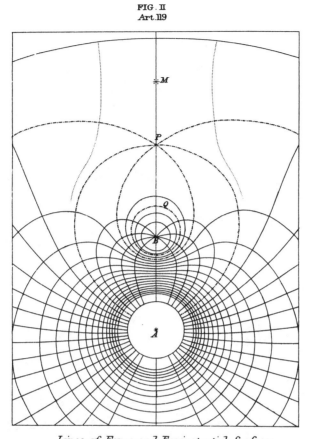

FIG . II
Art.119

Lines of Force and Equipotential Surfaces.

$A = 20$ $B = -5$ *P.Point of Equilibrium .* $AP = 2AB$
Q ,Spherical surface of Zero potential .
M.Point of Maximum Force along the axis .
The dotted line is the Line of Force $\Psi = 0.1$. *thus* _._._._

118.5 This diagram may also be taken to represent the lines of force and equipotential surfaces belonging to two spheres of gravitating matter whose masses are as 4 to 1.

119.] In [Figure II] we have again two points whose charges are as 20 to 5, but the one positive and the other negative. In this case one of the equipotential surfaces, that, namely, corresponding to potential zero, is

118.5 *This diagram may also be taken to represent the lines of force and equipotential surfaces belonging to two spheres of gravitating matter*: Not so, since Figure I is a case of repulsion, and gravitating bodies only attract. Maxwell's remark should properly refer to Figure II, which does diagram a case of attraction. The electrical drawing is capable of representing the gravitational case as well because the force laws for gravitational and electrical attraction, being both inverse-square, are formally identical.

FIG. II
Art. 119

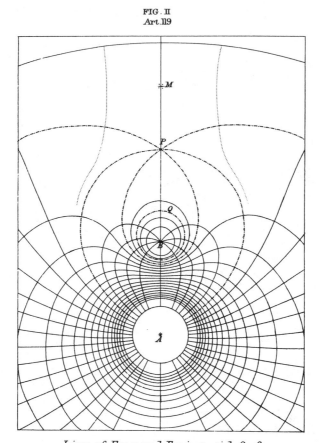

Lines of Force and Equipotential Surfaces.

a sphere. It is marked in the diagram by the dotted circle Q. The importance of this spherical surface will be seen when we come to the theory of Electrical Images.

119.1 We may see from this diagram that if two round bodies are charged with opposite kinds of electricity they will attract each other as much as two points having the same charges but placed somewhat nearer together than the middle points of the round bodies.

119.2 Here, again, one of the equipotential surfaces, indicated by a dotted line, has two lobes, an inner one surrounding the point whose charge is 5 and an outer one surrounding both bodies, the two lobes meeting in a conical point P which is a point of equilibrium.

119.1 *they will attract each other as much as two points having the same charges but placed somewhat nearer together than the middle points*: Maxwell alluded to this in (43.1) as a difficulty that plagues measurements with the torsion balance. But the more distant the spherical bodies are from one another, the more nearly can their charges be treated as concentrated at the centers.

FIG . III.
Art .120.

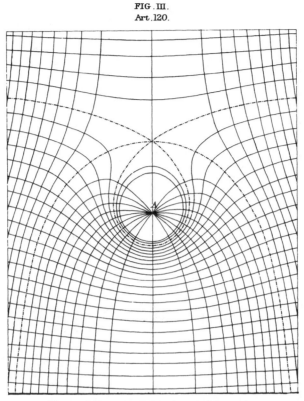

Lines of Force and Equipotential surfaces

119.3 If the surface of a conductor is of the form of the outer lobe, a roundish body having, like an apple, a conical dimple at one end of its axis, then, if this conductor be electrified, we shall be able to determine the surface-density at any point. That at the bottom of the dimple will be zero.

119.4 Surrounding this surface we have others having a rounded dimple which flattens and finally disappears in the equipotential surface passing through the point marked *M*.

119.5 The lines of force in this diagram form two systems divided by a surface which passes through the point of equilibrium.

119.6 If we consider points on the axis on the further side of the point *B*, we find that the resultant force diminishes to the double point *P*, where it vanishes. It then changes sign, and reaches a maximum at *M*, after which it continually diminishes.

119.7 This maximum, however, is only a maximum relatively to other points on the axis, for if we consider a surface through *M* perpendicular to the axis, *M* is a point of minimum force relatively to neighbouring points on that surface.

120.] Figure III represents the equipotential surfaces and lines of induction due to a point whose charge is 10 placed at *A*, and surrounded

FIG . III.
Art.120.

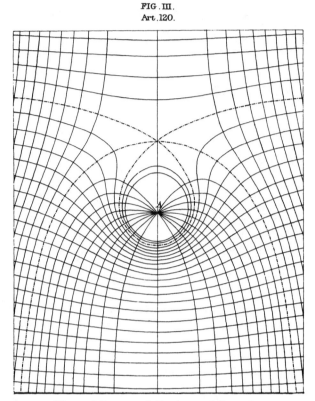

Lines of Force and Equipotential surfaces

by a field of force, which, before the introduction of the charged point, was uniform in direction and magnitude at every part.*

120.1 The equipotential surfaces have each of them an asymptotic plane. One of them, indicated by a dotted line, has a conical point and a lobe surrounding the point A. Those below this surface have one sheet with a depression near the axis. Those above have a closed portion surrounding A and a separate sheet with a slight depression near the axis.

120.2 If we take one of the surfaces below A as the surface of a conductor, and another a long way below A as the surface of another conductor at a different potential, the system of lines and surfaces between the two conductors will indicate the distribution of electric force. If the lower conductor is very far from A its surface will be very nearly plane, so that

*{Maxwell does not give the strength of the field. M. Cornu however has calculated the strength of the uniform field from the diagram of the lines of force, and finds that its electromotive intensity before the introduction of the charged body was 1.5.}

we have here the solution of the distribution of electricity on two sur-
faces, both of them nearly plane and parallel to each other, except that
the upper one has a protuberance near its middle point, which is more
or less prominent according to the particular equipotential surface we
choose.

* * *

122.] These diagrams should be studied as illustrations of the
language of Faraday in speaking of 'lines of force,' the 'forces of an elec-
trified body,' &c.

122.1 The word Force denotes a restricted aspect of that action
between two material bodies by which their motions are rendered differ-
ent from what they would have been in the absence of that action. The
whole phenomenon, when both bodies are contemplated at once, is
called Stress, and may be described as a transference of momentum
from one body to the other. When we restrict our attention to the first of
the two bodies, we call the stress acting on it the Moving Force, or simply
the Force on that body, and it is measured by the momentum which that
body is receiving per unit of time.

122.2 The mechanical action between two charged bodies is a stress,
and that on one of them is a force. The force on a small charged body
is proportional to its own charge, and the force per unit of charge is
called the Intensity of the force.

122.3 The word Induction was employed by Faraday to denote the
mode in which the charges of electrified bodies are related to each other,
every unit of positive charge being connected with a unit of negative

122.1 *Force denotes a restricted aspect of that action between two material bodies*:
Notice how Maxwell subordinates *force*—which is conventionally treated as an
elementary quantity—to *mutual action* between bodies.

122.2 *The mechanical action between two charged bodies is a stress, and on one of
them is a force*: This interesting distinction between force and stress explicitly
emphasizes point of view. In contrast, modern nomenclature distinguishes
between *force*, simply, and *force distributed over a cross-sectional area*, the latter
being called stress. In both locutions, however, the *measure* of stress is that of
force per unit area.

The force on a small charged body is proportional to its own charge: Maxwell is of
course speaking specifically of the *electric* force—the force that a charged body
experiences insofar as it *is* charged.

122.3 *The word Induction was employed by Faraday to denote the mode in which the
charges of electrified bodies are related to each other*: In Maxwell's view, then,
Faraday's own concept of induction presupposes mutual action.

charge by a line, the direction of which, in fluid dielectrics, coincides at every part of its course with that of the electric intensity. Such a line is often called a line of Force, but it is more correct to call it a line of Induction.

122.4 Now the quantity of electricity in a body is measured, according to Faraday's ideas, by the number of lines of force, or rather of induction, which proceed from it. These lines of force must all terminate somewhere, either on bodies in the neighbourhood, or on the walls and roof of the room, or on the earth, or on the heavenly bodies, and wherever they terminate there is a quantity of electricity exactly equal and opposite to that on the part of the body from which they proceeded. By examining the diagrams this will be seen to be the case. There is therefore no contradiction between Faraday's views and the mathematical results of the old theory, but, on the contrary, the idea of lines of force throws great light on these results, and seems to afford the means of rising by a continuous process from the somewhat rigid conceptions of the old theory to notions which may be capable of greater expansion, so as to provide room for the increase of our knowledge by further researches.

123.] These diagrams are constructed in the following manner: First, take the case of a single centre of force, a small electrified body with a charge e. The potential at a distance r is $V = e/r$; hence, if we make $r = e/V$, we shall find r, the radius of the sphere for which the potential

122.3, continued. *a line, the direction of which, in fluid dielectrics, coincides ... with that of the electric intensity*: Maxwell mentioned in (59.4) that in some solid dielectrics, the direction of polarization is not everywhere the same as the direction of electric intensity.

a line of Force, but it is more correct to call it a line of Induction: Notwithstanding that "line of force" is Faraday's own term, on the basis of Maxwell's characterization of Force in (122.1) that term fails to express the mutuality of action he regards as essential to Faraday's idea of induction.

122.4 *the quantity of electricity in a body is measured, according to Faraday's ideas, by the number of lines of force, or rather of induction, which proceed from it*: Since Faraday's lines of force are infinitely dense in their distribution, their "number" can only be relative. Maxwell has made their measure strictly quantitative by defining a tube of induction that terminates in *unit charge* (82.13). Thus the number of these tubes issuing from the surface of a charged body will be numerically equal to the charge of the body; and they will perform exactly the role that Faraday intended for his lines of force.

But for Maxwell, unit charge corresponds to 4π units of induction (76.8). Thus Maxwell will represent induction in terms of the *unit tube* (22.6), rather than the tubes that correspond to Faraday's lines. Maxwell has not abandoned Faraday's principle of "counting" lines; he has only changed the counting unit.

FIG.6.

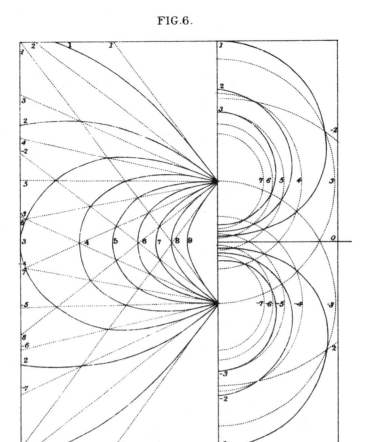

Lines of Force. | *Equipotential Surfaces.*

Method of drawing
Lines of Force and Equipotential Surfaces.

is V. If we now give to V the values I, 2, 3, &c., and draw the correspond-ing spheres, we shall obtain a series of equipotential surfaces, the potentials corresponding to which are measured by the natural numbers. The sections of these spheres by a plane passing through their common centre will be circles, each of which we may mark with the number denoting its potential. These are indicated by the dotted semicircles on the right hand of Fig. 6.

123.1 If there be another centre of force, we may in the same way draw the equipotential surfaces belonging to it, and if we now wish to find the form of the equipotential surfaces due to both centres together, we must remember that if V_1 be the potential due to one centre, and V_2 that due to the other, the potential due to both will be $V_1 + V_2 = V$. Hence, since at every intersection of the equipotential surfaces belonging to the two series we know both V_1 and V_2, we also know the value of V. If therefore

FIG.6.

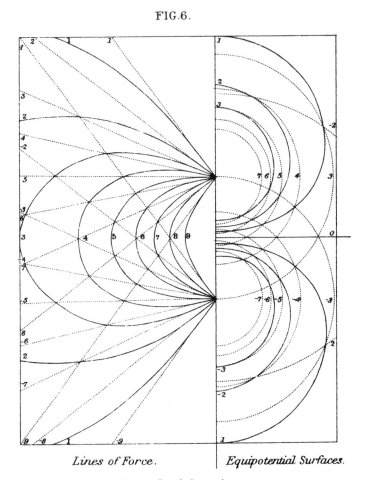

Lines of Force. | *Equipotential Surfaces.*

Method of drawing
Lines of Force and Equipotential Surfaces.

we draw a surface which passes through all those intersections for which the value of V is the same, this surface will coincide with a true equipotential surface at all these intersections, and if the original systems of surfaces are drawn sufficiently close, the new surface may be drawn with any required degree of accuracy. The equipotential surfaces due to two points whose charges are equal and opposite are represented by the continuous lines on the right hand side of Fig. 6.

123.2 This method may be applied to the drawing of any system of equipotential surfaces when the potential is the sum of two potentials, for which we have already drawn the equipotential surfaces.

123.3 The lines of force due to a single centre of force are straight lines radiating from that centre. If we wish to indicate by these lines the intensity as well as the direction of the force at any point, we must draw them so that they mark out on the equipotential surfaces

portions over which the surface-integral of induction has definite values. The best way of doing this is to suppose our plane figure to be the section of a figure in space formed by the revolution of the plane figure about an axis passing through the centre of force. Any straight line radiating from the centre and making an angle θ with the axis will then trace out a cone, and the surface-integral of the induction through that part of any surface which is cut off by this cone on the side next the positive direction of the axis is $2\pi e\,(1-\cos\theta)$.

123.4 If we further suppose this surface to be bounded by its intersection with two planes passing through the axis, and inclined at the angle whose arc is equal to half the radius, then the induction through the surface so bounded is $\frac{1}{2}e(1-\cos\theta)=\Phi$, say;

123.3 *they mark out on the equipotential surfaces portions over which the surface-integral of induction has definite values*: If they do this, then portions of the successive equipotentials that have equal surface-integrals will be crossed by the same number of unit tubes (122.4, *comment*). And since induction is the surface-integral of electric intensity (75.1), the areas of the respective portions will be inversely as the intensities. The spacing of the lines so constructed will thus indicate the intensity at various regions.

the induction through that part of any surface which is cut off by this cone: Recall that the equipotential surfaces about a center of force are spherical, and consider a surface of radius r. A line of force proceeding from center O and making angle θ with the vertical axis will cut off a cap like the one drawn here. Now the area of the dotted ring having width $r\,d\theta$ and radius s is

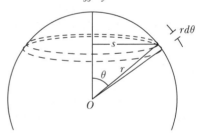

$dA=2\pi s\cdot r\,d\theta,$

where $s=r\sin\theta;$

hence $dA=2\pi r^2\sin\theta\,d\theta$

and the area of the cap is then $A=\int dA=2\pi r^2\int\sin\theta\,d\theta$, integrated between 0 and θ. Thus $A=-2\pi r^2(\cos\theta-\cos0)=2\pi r^2(1-\cos\theta).$

Now if the charge at O equals e, the total induction through the spherical surface (area $4\pi r^2$) is $4\pi e$ (76.8); so the induction through the cap will be $A/4\pi r^2$ of the total, or $4\pi e A/4\pi r^2$, which reduces to $2\pi e\,(1-\cos\theta)$.

123.4 *two planes ... inclined at the angle whose arc is equal to half the radius*: That is, an angle of *one-half radian*. Then since 2π radians make a full circle, the bounded portion will represent $1/4\pi$ (about one twelfth) of the cap just calculated—PQR in this diagram. Since the induction through the whole cap is $2\pi e(1-\cos\theta)$, that through sector PQR will be $1/4\pi$ of this, or $\frac{1}{2}e(1-\cos\theta)$.

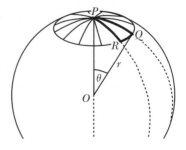

and
$$\theta = \cos^{-1}\left(1 - 2\frac{\Phi}{e}\right).$$

123.5 If we now give to Φ a series of values 1, 2, 3 ... e, we shall find a corresponding series of values of θ, and if e be an integer, the number of corresponding lines of force, including the axis, will be equal to e.

123.6 We have thus a method of drawing lines of force so that the charge of any centre is indicated by the number of lines which diverge from it, and the induction through any surface cut off in the way described is measured by the number of lines of force which pass through it. The dotted straight lines on the lefthand side of Fig. 6 represent the lines of force due to each of two electrified points whose charges are 10 and –10 respectively.

123.7 If there are two centres of force on the axis of the figure we may draw the lines of force for each axis corresponding to values of Φ_1 and Φ_2, and then, by drawing lines through the consecutive intersections of these lines for which the value of $\Phi_1 + \Phi_2$ is the same, we may find the

123.4, continued. Obtain Maxwell's expression by solving the previous equation, $\frac{1}{2}e(1-\cos\theta) = \Phi$, for θ.

123.5 *If we now give to Φ a series of values 1, 2, 3 ... e, we shall find a corresponding series of values of θ:* For example, let $\Phi = 1$. Then the induction through sector PQR will be 1 unit, and that through the whole cap will be 4π. Solve for θ to find the cone (better, the *solid angle*) which encompasses 4π units of induction due to the charge at O, and similarly for successive values. Finally when $\Phi = e$, the total charge at O, the induction through the whole cap will be $4\pi e$. Solving for θ in this case yields $\theta = \cos^{-1}(-1)$, that is, $\theta = \pi$ radians or 180°; the containing figure is not a cone but a sphere. This agrees with what Maxwell has already established (76.8), that the induction through a closed surface surrounding a charge e will be $4\pi e$.

123.6 *We thus have a method of drawing lines of force*: With this method our drawings will accurately represent the lines of force due to any number of units of charge, and the number of lines crossing any surface will accurately represent the induction through that surface.

The dotted straight lines ... represent the lines of force due to ... points whose charges are 10 and –10 respectively: Maxwell has labeled the lines from each center of force; you can see that there are 10 positive and 10 negative lines; they are drawn at angles representing the cones that encompass induction due to units of charge from 1 to 10. Notice that the line that would be labeled "10" corresponds with the vertical axis; that is, it lies (as it must) at the angle of 180°. I have corrected a few erroneous labels in Maxwell's published diagram.

FIG.6.

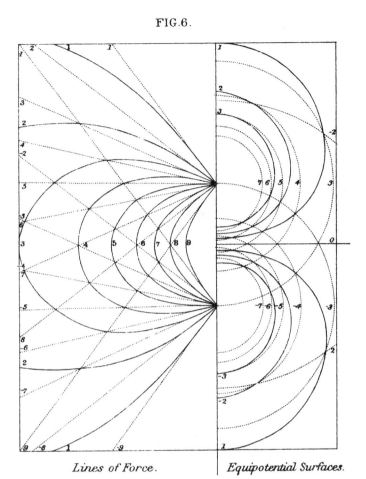

Lines of Force. | Equipotential Surfaces.

Method of drawing
Lines of Force and Equipotential Surfaces.

lines of force due to both centres, and in the same way we may combine any two systems of lines of force which are symmetrically situated about the same axis. The continuous curves on the left-hand side of Fig. 6 represent the lines of force due to the two charged points acting at once.

123.8 After the equipotential surfaces and lines of force have been constructed by this method, the accuracy of the drawing may be tested by observing whether the two systems of lines are everywhere orthogonal...

* * *

123.8 *whether the two systems of lines are everywhere orthogonal*: Equipotential lines ought to be perpendicular to lines of force, since any inclination of the equipotential lines towards the lines of force would imply a change of potential along the equipotential line—which is absurd.

PART II

ELECTROKINEMATICS.

CHAPTER I.

THE ELECTRIC CURRENT.

230.] WE have seen, in Art. 45, that when a conductor is in electrical equilibrium the potential at every point of the conductor must be the same.

230.1 If two conductors A and B are charged with electricity so that the potential of A is higher than that of B, then, if they are put in communication by means of a metallic wire C touching both of them, part of the charge of A will be transferred to B, and the potentials of A and B will become in a very short time equalized.

231.] During this process certain phenomena are observed in the wire C, which are called the phenomena of the electric conflict or current.

231. *electric conflict or current*: The term "electric conflict" is Oersted's (see Art. 475 below) and has been the source of much puzzlement. Despite Maxwell's phraseology here it is not really a synonym for "current"; for Maxwell is about to characterize current as, essentially, the transference of a state of electrification from one place to another, along a certain path (231.1, 231.2). Oersted's term, however, was more inclusive and denoted phenomena that other investigators would regard as ancillary to the current proper, such as *heat, light,* and Oersted's own celebrated discovery of *magnetic action.* Maxwell will discuss the current's magnetic action at length in Part IV, Chapter II.

But to a modern reader it is probably no more evident why "conflict" is any more suitable a term for heat, light, and magnetic action than it is for the transference of electrification. It is therefore useful to know that Oersted was very likely being influenced by a basic doctrine in the tradition of *Naturphilosophie.* F. W. J. Schelling, the tradition's most illustrious champion, had taught that all natural phenomena are founded upon the "polar opposition" of fundamental forces; from that point of view, then, the phenomena of what we now call electric current are to be ranked as a distinctive species of opposition or conflict. The story of Oersted's relation to the nature-philosophical tradition is engagingly told by L. Pearce Williams in *The Origins of Field Theory*, Random House (1966), Chapter II.

231.1 The first of these phenomena is the transference of positive electrification from *A* to *B* and of negative electrification from *B* to *A*. This transference may be also effected in a slower manner by bringing a small insulated body into contact with *A* and *B* alternately. By this process, which we may call electrical convection, successive small portions of the electrification of each body are transferred to the other. In either case a certain quantity of electricity, or of the state of electrification, passes from one place to another along a certain path in the space between the bodies.

231.2 Whatever therefore may be our opinion of the nature of electricity, we must admit that the process which we have described constitutes a current of electricity.

<div align="center">* * *</div>

On Steady Currents.

232.] In the case of the current between two insulated conductors at different potentials the operation is soon brought to an end by the equalization of the potentials of the two bodies, and the current is therefore essentially a Transient Current.

232.1 But there are methods by which the difference of potentials of the conductors may be maintained constant, in which case the current will continue to flow with uniform strength as a Steady Current.

The Voltaic Battery.

232.2 The most convenient method of producing a steady current is by means of the Voltaic Battery.

231.1 *convection*: the action of carrying, from L. *convehere*. The name is appropriate whenever an actual *carrier* can be identified. For example, in (55.8) Maxwell cited "the motion of charged particles" as a case of convection, the particles serving as the carriers. In his present example, the "small insulated body" provides that function. Nor is the example merely hypothetical; in Art. 242 below Maxwell will allude to a device that utilizes the very mechanism here mentioned in order to ring a pair of bells.

231.2 *the process we have described constitutes a current*: More precisely, the process of convection just discussed enables us to grasp the essential nature of current as being *transference of electrification along a path*. But the "carrier" element in convection is not essential to current: there is no compelling reason to assume that an electric current requires a material carrier. We will have understood that point already from Maxwell's discussion of the various fluid theories in Arts. 36–37; he reminds us of that discussion with his proviso, "Whatever therefore may be our opinion of the nature of electricity..."

232. *Transient*: brief, momentary, fleeting; from L. *trans* (across) + *ire* (to go).

232.1 *the current will continue to flow with uniform strength*: Here Maxwell is tacitly taking the *strength* of a current to be its *time rate of transfer of electricity*. He will speak more explicitly in Art. 237 below.

232.3 For the sake of distinctness we shall describe Daniell's Constant Battery:—

232.4 A solution of sulphate of zinc is placed in a cell of porous earthenware, and this cell is placed in a vessel containing a saturated solution of sulphate of copper. A piece of zinc is dipped into the sulphate of zinc, and a piece of copper is dipped into the sulphate of copper. Wires are soldered to the zinc and to the copper above the surfaces of the liquids. This combination is called a cell or element of Daniell's battery.

* * *

234.] Let us now consider a voltaic battery with its ends insulated from each other. The copper end will be positively or vitreously electrified, and the zinc end will be negatively or resinously electrified.

234.1 Let the two ends of the battery be now connected by means of a wire. An electric current will commence, and will in a very short time attain a constant value. It is then said to be a Steady Current.

Properties of the Current.

235.] The current forms a closed circuit in the direction from copper to zinc through the wires, and from zinc to copper through the solutions.

232.3 *Daniell's Constant Battery*: As Maxwell describes it in the following paragraph, zinc *Z* and the zinc sulphate solution are separated from copper *C* and the copper sulphate solution by the porous pot *P*.

232.4 *a cell or element of Daniell's battery*: When several of these *cells* are connected in series (the copper of one to the zinc of the next), the array constitutes a *battery*.

234. *positively or vitreously … negatively or resinously electrified*: Vitreous and resinous electrification were identified with positive and negative electrification respectively in (27.5) and (27.6).

235. *the current forms a closed circuit in the direction from copper to zinc in the wires…*: The direction of a current is conventionally taken to be the direction in which positive electrification would move. And since in a conductor electricity tends to move from a higher to a lower potential (45.3), that direction in a wire joining the battery terminals would be from the positive (copper) to the negative (zinc)—and similarly in those conductors within the battery that join individual cells to one another.

…and from zinc to copper through the solutions: It is important to understand that the convention regarding current direction applies *only to a conductor*; it does not apply to a voltaic cell or other source of electromotive force. For if positive electrification moves in, as Maxwell says, "a closed circuit," then *within each cell* it must move from zinc to copper, that is, from the lower to the higher potential. The voltaic cell is ordained, Sisyphus-like, to raise quantities of electricity to a place of high potential in order that the electricity may return, in the form of current in a conductor, to a place of lower potential.

235.1 If the circuit be broken by cutting any of the wires which connect the copper of one cell with the zinc of the next in order, the current will be stopped, and the potential of the end of the wire in connexion with the copper will be found to exceed that of the end of the wire in connexion with the zinc by a constant quantity, namely, the total electromotive force of the circuit.

* * *

Electrolytic Action of the Current.

236.] As long as the circuit is broken no chemical action goes on in the cells, but as soon as the circuit is completed, zinc is dissolved from the zinc in each of the Daniell's cells, and copper is deposited on the copper:

236.1 The quantity of sulphate of zinc increases, and the quantity of sulphate of copper diminishes unless more is constantly supplied.

236.2 The quantity of zinc dissolved, and also that of copper deposited, is the same in each of the Daniell's cells throughout the circuit, whatever the size of the plates of the cell, and if any one of the cells be of a different construction, the amount of chemical action in it bears a constant proportion to the action in the Daniell's cell. For instance, if one of the cells consists of two platinum plates dipped into sulphuric acid diluted with water, oxygen will be given off at the surface of the plate where the current enters the liquid, namely, the plate in metallic connexion with the copper of Daniell's cell, and hydrogen at the surface of the plate where the current leaves the liquid, namely, the plate connected with the zinc of Daniell's cell.

236.3 The volume of the hydrogen is exactly twice the volume of the oxygen given off in the same time, and the weight of the oxygen is exactly eight times the weight of the hydrogen.

236.4 In every cell of the circuit the weight of each substance dissolved, deposited, or decomposed is equal to a certain quantity called

235.1 *the total electomotive force of the circuit*: Maxwell defined the total electromotive force over a path, and its relation to potential, in Art. 45. As he there explained, in the electrostatic case the total electromotive force about a closed conducting path must vanish; but if a current flows the total electromotive force about the path does not vanish.

236.3 *The volume of the hydrogen is exactly twice the volume of the oxygen...*: In electrolysis—Maxwell will so name this process in (237.)—substances are liberated (or deposited, or dissolved, or decomposed) in the same ratios as their combining weights and, if gaseous, in the ratios of their combining volumes. These are their *chemically equivalent* ratios.

the electro-chemical equivalent of that substance, multiplied by the strength of the current and by the time during which it has been flowing.

* * *

237.] Substances which are decomposed in this way are called Electrolytes. The process is called Electrolysis. The places where the current enters and leaves the electrolyte are called Electrodes. Of these the electrode by which the current enters is called the Anode, and that by which it leaves the electrolyte is called the Cathode. The components into which the electrolyte is resolved are called Ions: that which appears at the anode is called Anion, and that which appears at the cathode is called the Cation.

237.2 If a homogeneous electrolyte is placed in a tube of variable section, and if the electrodes are placed at the ends of this tube, it is found that when the current passes, the anion appears at the anode and the cation at the cathode, the quantities of these ions being electrochemically equivalent, and such as to be together equivalent to a certain quantity of the electrolyte. In the other parts of the tube, whether the section be large or small, uniform or varying, the composition of the electrolyte remains unaltered. Hence the amount of electrolysis which takes place across every section of the tube is the same. Where the section is small the action must therefore be more intense than where the section is large, but the total amount of each ion which crosses any complete section of the electrolyte in a given time is the same for all sections.

236.4 *the electrochemical equivalent of that substance*: Specifically, the amount of that substance dissolved, deposited, or decomposed by current of unit strength passing for unit time. But we have yet to define *unit strength* for a current.

237. *the electrode by which the current enters is called the Anode*: From ἄνω, up, and ὅδος, a way. *Cathode* is similarly derived from κατά, down, and ὅδος. Maxwell is following Faraday's definitions (*Experimental Researches in Electricity*, para. 663), which differ significantly from modern usage; see Fisher, Howard J., *Faraday's Experimental Researches in Electricity: Guide to a First Reading* (2001), pp. 197–198.

Ions: from ιόν, neuter participle of ιέναι, go.

237.2 *the quantities ... being electrochemically equivalent*: That is, they are liberated by the same current in the same time period and therefore represent the same amount of electrolytic action. They are also *chemically equivalent* (236.3, *comment*) to "a certain quantity of the electrolyte," namely, that quantity which is decomposed by the same current in the same time period.

Hence the amount of electrolysis ... across every section of the tube is the same: If more electrolyte were decomposed in one region than in another, or—were it possible—decomposed in different proportions, either the composition or the concentration of electrolyte, or both, would vary from region to region.

237.3 The strength of the current may ... be measured by the amount of electrolysis in a given time. An instrument by which the quantity of the electrolytic products can be readily measured is called a Voltameter.

237.4 The strength of the current, as thus measured, is the same at every part of the circuit, and the total quantity of the electrolytic products in the voltameter after any given time is proportional to the amount of electricity which passes any section in the same time.

Magnetic Action of the Current.

239.] Oersted discovered that a magnet placed near a straight electric current tends to place itself at right angles to the plane passing through the magnet and the current. See Art. 475.

239.1 If a man were to place his body in the line of the current that the current from copper through the wire to zinc should flow from his head to his feet, and if he were to direct his face towards the centre of the magnet, then that end of the magnet which tends to point to the north would, when the current flows, tend to point towards the man's right hand.

239.2 The nature and laws of this electromagnetic action will be discussed when we come to the fourth part of this treatise. What we are concerned with at present is the fact that the electric current has a magnetic action which is exerted outside the current, and by which its

237.4 *the total quantity of electrolytic products ... is proportional to the amount of electricity which passes*: Then since current strength is measured by the time rate of electrolysis (237.3), it is also measured by the time rate of transfer of electricity, that is, *the quantity of electricity transferred per unit time*.

239. *a magnet ... tends to place itself at right angles to the plane passing through the magnet and the current*: that is, to the plane passing through the *center* of the magnet, as sketched here; the current is in the direction of the arrow, while the magnet is shown in two different positions. You can see that if the magnet remains a constant distance from the wire, its successive positions will be as the tangents to a circle. Maxwell adopts a more colorful image in the next paragraph; he will give a fuller account of the Oersted experiments in Art. 475.

239.1 *If a man were to place his body in the line of the current...*: This body-centered image has the advantage of expressing direction, which the bald geometrical statement of (239.) could not do. It is also an intriguingly personal image, inviting us to occupy the very territory of the action. I will offer some additional remarks about this image in the editor's introduction to Part IV, Chapter I.

239.2 *magnetic action ... exerted outside the current*: In the present context, Maxwell calls attention to the "outside" character of this action primarily because, being "outside," it provides a practical means of measuring the current without interrupting it. But in Part IV of the Treatise Maxwell will question the

existence can be ascertained and its intensity measured without breaking the circuit or introducing anything into the current itself.

239.3 The amount of the magnetic action has been ascertained to be strictly proportional to the strength of the current as measured by the products of electrolysis in the voltameter, and to be quite independent of the nature of the conductor in which the current is flowing, whether it be a metal or an electrolyte,

240.] An instrument which indicates the strength of an electric current by its magnetic effects is called a Galvanometer.

240.1 Galvanometers in general consist of one or more coils of silk-covered wire within which a magnet is suspended with its axis horizontal. When a current is passed through the wire the magnet tends to set itself with its axis perpendicular to the plane of the coils. If we suppose the plane of the coils to be placed parallel to the plane of the earth's equator, and the current to flow round the coil from east to west in the direction of the apparent motion of the sun, then the magnet within will tend to set itself with its magnetization in the same direction as that of the earth considered as a great magnet, the north pole of the earth being similar to that end of the compass needle which points south.

extent to which we must regard "the current" as being confined to the wire. In Art. 549 he will find at least one attribute of the current that is clearly not localized within the conductor.

239.3 *the magnetic action has been ascertained to be strictly proportional to the strength of the current as measured by the products of electrolysis*: Maxwell noted the measurement of current by its electrolytic action in (237.3). Electrolytic measurements are favored for their great precision; but unlike a current's magnetic action, electrolysis requires substantial interference with the very current being measured. Thus it is of tremendous practical importance that the measure of current by its magnetic effect has been shown to be "strictly proportional" to the measure according to its electrolytic effect.

240.1 *Galvanometers in general consist of one or more coils ... within which a magnet is suspended*: A simplified movement is shown. The magnetized needle normally points north and south; the coil axis (dotted line) runs east and west. The coils are wired in series; when a current passes through them the poles of the needle experience attractive or repulsive forces parallel to the axis, causing the needle to turn from its north-south orientation; with suitable calibration the degree of deflection indicates the strength of the current. In a practical instrument the coils would be much closer together, and the needle's position would be indicated on an angular scale.

240.2 The galvanometer is the most convenient instrument for measuring the strength of electric currents. We shall therefore assume the possibility of constructing such an instrument in studying the laws of these currents, reserving the discussion of the principles of the instrument for our fourth part. When therefore we say that an electric current is of a certain strength we suppose that the measurement is effected by the galvanometer.

<div align="center">* * *</div>

240.2 *When ... we say that an electric current is of a certain strength we suppose that the measurement is effected by the galvanometer*: Maxwell styles the galvanometer the "most convenient" instrument for measuring current; but at this stage *time rate of electrolysis* remains the primary measure of current. The galvanometer, however convenient, provides a measure only because of the demonstrated proportionality of a current's magnetic effects to its electrolytic effects (239.3, *comment*). Nevertheless, the galvanometer will assume primacy when Maxwell adopts an *electromagnetic unit of current* in Art. 479 of Part IV.

CHAPTER II.

CONDUCTION AND RESISTANCE.

241.] IF by means of an electrometer we determine the electric potential at different points of a circuit in which a constant electric current is maintained, we shall find that in any portion of the circuit consisting of a single metal of uniform temperature throughout, the potential at any point exceeds that at any other point farther on in the direction of the current by a quantity depending on the strength of the current and on the nature and dimensions of the intervening portion of the circuit. The difference of the potentials at the extremities of this portion of the circuit is called the External electromotive force acting on it. If the portion of the circuit under consideration is not homogeneous, but contains transitions from one substance to another, from metals to electrolytes, or from hotter to colder parts, there may be, besides the external electromotive force, Internal electromotive forces which must be taken into account.

241.1 The relations between Electromotive Force, Current, and Resistance were first investigated by Dr. G. S. Ohm, in a work published in 1827, entitled *Die Galvanische Kette Mathematisch Bearbeitet*, translated in Taylor's *Scientific Memoirs*. The result of these investigations in the case of homogeneous conductors is commonly called 'Ohm's Law.'

Ohm's Law.

241.2 The electromotive force acting between the extremities of any part of a circuit is the product of the strength of the current and the resistance of that part of the circuit.

241.2 *The electromotive force ... is the product of the strength of the current and the resistance*: This is a statement of Ohm's Law. Put in equation form, $E = CR$, where E is the electromotive force, C the current, and R the resistance of the conductor. Maxwell defines *resistance* in the next paragraph and will introduce the symbols C and R in (242.2) below. Later he will offer an alternative formulation of Ohm's Law as $C = cE$, where c is the reciprocal of the resistance, named *conductance*.

241.3 Here a new term is introduced, the Resistance of a conductor, which is defined to be the ratio of the electromotive force to the strength of the current which it produces. The introduction of this term would have been of no scientific value unless Ohm had shewn, as he did experimentally, that it corresponds to a real physical quantity, that is, that it has a definite value which is altered only when the nature of the conductor is altered.

241.4 In the first place, then, the resistance of a conductor is independent of the strength of the current flowing through it.

241.5 In the second place the resistance is independent of the electric potential at which the conductor is maintained, and of the density of the distribution of electricity on the surface of the conductor.

241.6 It depends entirely on the nature of the material of which the conductor is composed, the state of aggregation of its parts, and its temperature.

241.3 *Resistance ... is defined to be the ratio of the electromotive force to the strength of the current*: Stated in equation form, $R = E/C$. But this is algebraically identical to the statement of Ohm's Law itself (241.2, *comment*). If it is a definition, then Ohm's Law becomes nothing but a tautology. That is what Maxwell means when he goes on to state that the mere introduction of the term "resistance" would in itself have been of no scientific value.

Ohm [*showed*] *that it corresponds to a real physical quantity, that is, ... it has a definite value which is altered only when the nature of the conductor is altered*: The real content of Ohm's Law, then, in Maxwell's view, is to reveal the quotient E/C—the resistance—as a *characteristic* of every conductor. Ohm's Law is a *physical* statement and not (as so often taught, even today) a mere formal relation among mathematical quantities.

Maxwell's examples in the paragraphs following all indicate that resistance indeed changes "only when the nature of the conductor is altered." But we should be aware that, experimentally, resistance also changes when the *dimensions* of the conductor are altered; Maxwell alluded to that dependence in (241.) and it is curious that he neglects to mention it again both here and in (241.6).

241.4 *the resistance ... is independent of the strength of the current*: The current does not affect the conductor's nature and therefore should not affect its resistance—and indeed (for most materials) it does not.

241.5 *the resistance is independent of the electric potential*: Again, the potential does not affect the nature of the conductor and therefore should not affect the resistance either.

241.6 *It depends entirely on ... the material ... the state of aggregation of its parts, and its temperature*: Evidently, then, the "nature" of the conductor encompasses more than its material. Is our reasoning circular if we keep redefining a conductor's "nature" so as to include all the factors (other than physical dimensions) that prove to affect its resistance?

241.7 The resistance of a conductor may be measured to within one ten thousandth or even one hundred thousandth part of its value, and so many conductors have been tested that our assurance of the truth of Ohm's Law is now very high. In the sixth chapter we shall trace its applications and consequences.

* * *

Generation of Heat by the Current.

242.] We have seen that when an electromotive force causes a current to flow through a conductor, electricity is transferred from a place of higher to a place of lower potential. If the transfer had been made by convection, that is, by carrying successive charges on a ball from the one place to the other, work would have been done by the electrical forces on the ball, and this might have been turned to account. It is actually turned to account in a partial manner in those dry pile circuits where the electrodes have the form of bells, and the carrier ball is made to swing like a pendulum between the two bells and strike them alternately. In this way the electrical action is made to keep up the swinging of the pendulum and to propagate the sound of the bells to a distance. In the case of the conducting wire we have the same transfer of electricity from a place of high to a place of low potential without any external work being done. The principle of the Conservation of Energy therefore leads us to look for internal work in the conductor. In an electrolyte this

241.7 *resistance ... may be measured to within one ten thousandth or even one hundred thousandth part*: We measure electromotive force with the electrometer as outlined in (241.) and current with the galvanometer (240.) The impressive precision Maxwell cites for the quotient of these quantities indicates the high levels of performance to which those instruments had attained.

242. *convection*: Maxwell discussed electrical convection earlier in (231.1).

242. *dry pile circuits where ... the carrier ball is made to swing like a pendulum between the two bells*: The "dry pile" (also known as the Zamboni pile) is a voltaic battery comprising thousands of alternate layers of metal foil and manganese dioxide-coated paper. With an output of thousands of volts but very tiny currents, its capabilities resemble those of typical electrostatic generators. For this reason it was sometimes called an "electrostatic battery"; nevertheless it is a voltaic device.

The photograph shows the famous Clarendon Bell, housed in the Clarendon Laboratory of Oxford University. The two large cylinders are the dry piles, connected in series to the bell-shaped electrodes. The device has been ringing without interruption since 1840.

internal work consists partly of the separation of its components. In other conductors it is entirely converted into heat.

242.1 The energy converted into heat is in this case the product of the electromotive force into the quantity of electricity which passes. But the electromotive force is the product of the current into the resistance, and the quantity of electricity is the product of the current into the time. Hence the quantity of heat multiplied by the mechanical equivalent of unit of heat is equal to the square of the strength of the current multiplied into the resistance and into the time.

242.2 The heat developed by electric currents in overcoming the resistance of conductors has been determined by Dr. Joule, who first established that the heat produced in a given time is proportional to the square of the current, and afterwards by careful absolute measurements of all the quantities concerned, verified the equation

$$JH = C^2 Rt,$$

where J is Joule's dynamical equivalent of heat, H the number of units of heat, C the strength of the current, R the resistance of the conductor, and t the time during which the current flows. These relations between electromotive force, work, and heat, were first fully explained by Sir W. Thomson in a paper on the application of the principle of mechanical effect to the measurement of electromotive forces.*

* Phil. Mag., Dec. 1851.

242.1 *the energy* [*is*] *the product of the electromotive force into the quantity of electricity*: The dimensions of electromotive force are those of work per unit charge (45., *comment*). Its product with the quantity (that is, number of units) of electricity will therefore yield work or energy. Expressed as an equation, $W = Eq$, where W is the energy, E the electromotive force, and q the quantity of electricity transferred.

But the electromotive force is the product of the current into the resistance: Expressed as an equation (241.2, *comment*), $E = CR$; Maxwell is about to introduce C to represent the current and R the resistance (242.2).

and the quantity of electricity is the product of the current into the time: The measure of current is the quantity of electricity transferred per unit time (237.4, *comment*). Expressed as an equation, $C = q/t$ or $q = Ct$, where t is the time during which the current flows.

the quantity of heat multiplied by the mechanical equivalent of unit of heat is equal to the square of the strength of the current multiplied into the resistance and into the time: Combining the three foregoing equations, $W = Eq = CR \cdot Ct = C^2 Rt$. This expresses the *energy* that is converted to heat by a current C passing through resistance R. But if one unit of heat represents, say, J units of mechanical work, then J will be "the mechanical equivalent of unit of heat" and the energy represented by H units of heat will be HJ; thus $HJ = C^2 Rt$. Maxwell will cite this equation in the next paragraph.

243.] The analogy between the theory of the conduction of electricity and that of the conduction of heat is at first sight almost complete. If we take two systems geometrically similar, and such that the conductivity for heat at any part of the first is proportional to the conductivity for electricity at the corresponding part of the second, and if we also make the temperature at any part of the first proportional to the electric potential at the corresponding point of the second, then the flow of heat across any area of the first will be proportional to the flow of electricity across the corresponding area of the second.

243.1 Thus, in the illustration we have given, in which flow of electricity corresponds to flow of heat, and electric potential to temperature, electricity tends to flow from places of high to places of low potential, exactly as heat tends to flow from places of high to places of low temperature.

244.] The theory of electric potential and that of temperature may therefore be made to illustrate one another; there is, however, one remarkable difference between the phenomena of electricity and those of heat.

244.1 Suspend a conducting body within a closed conducting vessel by a silk thread, and charge the vessel with electricity: The potential of the vessel and of all within it will be instantly raised, but however long and however powerfully the vessel be electrified, and whether the body within be allowed to come in contact with the vessel or not, no signs of electrification will appear within the vessel, nor will the body within shew any electrical effect when taken out.

244.2 But if the vessel is raised to a high temperature, the body within will rise to the same temperature, but only after a considerable time, and if it is then taken out it will be found hot, and will remain so till it has continued to emit heat for some time.

244.3 The difference between the phenomena consists in the fact that bodies are capable of absorbing and emitting heat, whereas they have no corresponding property with respect to electricity. A body cannot be made hot without a certain amount of heat being supplied to it, depending on the mass and specific heat of the body, but the electric potential of a body may be raised to any extent in the way already described without communicating any electricity to the body.

244. *The theory of electric potential and that of temperature may therefore be made to illustrate one another*: Such mutual illustration is the express mark, as well as the fruit, of analogy; compare (62.11). Maxwell called attention to a limited thermal analogy with electric potential in (53.4) and to an analogy between electric displacement and the deformation of an elastic body in (60.4).

245.] Again, suppose a body first heated and then placed inside the closed vessel. The outside of the vessel will be at first at the temperature of surrounding bodies, but it will soon get hot, and will remain hot till the heat of the interior body has escaped.

245.1 It is impossible to perform a corresponding electrical experiment. It is impossible so to electrify a body, and so to place it in a hollow vessel, that the outside of the vessel shall at first shew no signs of electrification but shall afterwards become electrified. It was for some phenomenon of this kind that Faraday sought in vain under the name of an absolute charge of electricity.

245.2 Heat may be hidden in the interior of a body so as to have no external action, but it is impossible to isolate a quantity of electricity so as to prevent it from being constantly in inductive relation with an equal quantity of electricity of the opposite kind.

245.3 There is nothing therefore among electric phenomena which corresponds to the capacity of a body for heat. This follows at once from the doctrine which is asserted in this treatise, that electricity obeys the same condition of continuity as an incompressible fluid. It is therefore impossible to give a bodily charge of electricity to any substance by forcing an additional quantity of electricity into it. See Arts. 61, 111, 329, 334.

<div style="text-align:center">* * *</div>

245. *the outside of the vessel will be at first at the temperature of the surrounding bodies, but it will soon get hot*: Absorption of heat is implicated because the temperature of the outer surface reflects the interior heat content only belatedly. I discussed another example of heat absorption in (53.4, *comment*).

245.1 *electrify a body, and ... place it in a hollow vessel, that the outside of the vessel shall at first shew no signs of electrification but shall afterwards become electrified*: Such an outcome would contradict Experiment II in Art. 28.

absolute charge: That is, a charge not related to any other charge. If the delay in electrification here described were to occur, then during that delay some or all of the charge within the vessel would be *unrelated* to the charge of the outer surface. In fact, as Maxwell expresses it in the next paragraph, any quantity of electricity must be "constantly in inductive relation with an equal quantity of electricity of the opposite kind." See also Art. 54.

245.3 *electricity obeys the same condition of continuity as an incompressible fluid*: Unfortunately, it is not clear that Maxwell's doctrine of continuity is consistent with his treatment of displacement. I discussed this difficulty in the editor's introduction to Part I Chapter V.

Editor's Introduction to
Part III Chapter I

Parts I and II of the *Treatise* concerned themselves with electricity; now Parts III and IV will study magnetism. In several respects, Maxwell's treatment of elementary magnetism in Part III parallels his treatment of electrostatics in Part I. To begin with, Maxwell founded his electrical theory on a careful definition of electricity as a mathematical quantity; in the opening chapter of Part III he will establish magnetism as a mathematical quantity in an equally scrupulous way. Secondly, Maxwell favored an agnostic position as to whether or not the quantity called "electricity" is to be regarded as a fluid or other material substance; similarly, we shall see, he exhibits an almost ponderous reserve with respect to the various theories of "magnetic matter."

Lastly, Maxwell's electrical résumé embodied a characteristic ambivalence. Consider, for example, an accumulator consisting of two parallel metal plates separated by a sheet of dielectric (60.14). When the accumulator is electrified, we commonly say that the plates are *charged*. At the same time, as Maxwell urged, we understand the dielectric to be in a state of electrical tension—a *polarized* condition (109., 111.6) in which every elementary portion manifests opposite electrical charges on opposite sides. Within the body of the dielectric, as Maxwell explained, these opposites nullify one another; but at its surfaces they remain uncanceled and manifest themselves freely. Ambivalence arises when we ask whether *charge* or *polarization* is to be taken as fundamental. Is the condition of the glass simply its response to the influence of the charged plates? Or are the charges on the plates themselves only the superficial manifestation of a more fundamental interior condition of the dielectric? I noted Maxwell's ambivalent treatment of this matter when introducing Part I, Chapter V.

As we shall see, a similar ambivalence holds in the case of magnets. One way of looking at, say, a bar magnet is to think of it as harboring, near its two ends, two quantities of magnetism (called "free magnetism"

by Maxwell), equal in amount but opposite in kind. Although these quantities are conceived as distributed throughout both end regions, the action of an ordinary bar magnet at large distances is just about the same as it would be if the free magnetism were concentrated at two points near its ends. These points are called the *poles* of the magnet; but Maxwell will make it clear in the present Chapter that this concept of a magnetic pole is highly idealized.

From another point of view, however, magnetization is not something found exclusively at a magnet's two ends but a condition extending throughout its whole substance. Maxwell will represent the body of a magnet as being polarized the way a dielectric under tension is. In the magnetic case, this means that every elementary portion of the magnet is itself a separate magnet, as suggested in this sketch. Again, just as in a polarized dielectric, the adjacent faces of these elementary portions nullify one another; thus the central regions of a magnet show little or no activity, but at the ends the opposite conditions of the elementary portions manifest freely. Here, then, is a magnetic ambivalence paralleling the electrical one: Are the *poles themselves* to be regarded as fundamental, or is it rather the *polarization of the whole substance* that is prior?

Although this question will appear most explicitly in Chapter II, two articles in the present chapter touch upon it. For if magnetic poles are indeed independent and fundamental beings, we would expect to be able to exhibit them separately. Yet in Art. 378 Maxwell notes that if you break a bar magnet in two, you do not end up with a north magnet and a south magnet but with two smaller magnets, each of which exhibits both a "north" and a "south" end. It is, in other words, impossible to separate a magnet's north and south poles physically from one another. Such impossibility would seem to resolve the question in the direction *against* primacy of magnetic poles; yet by continuing to employ freely the language of "surface magnetism" (equivalent to "free magnetism" distributed over a surface) Maxwell effectively maintains an ambivalence instead of resolving it.

Later in the chapter the same ambivalence appears again, this time dictating alternate forms for one of Maxwell's definitions. In Art. 384, Maxwell will define Intensity of Magnetization both as the *magnetic moment per unit volume*, and alternatively as the *pole strength per unit cross-section* of a magnet. The two forms are mathematically equivalent; but notice that the first expresses magnetization as extending throughout a

material volume, while the second characterizes it as a surface phenomenon. By recognizing both forms of the definition, Maxwell is again implicitly preserving the ambivalent status of *poles* and *polarization*.

Why does Maxwell tolerate ambivalence in this area? Why does he not regard it as *demanding* resolution—as we might expect him to do if he were concerned primarily with the *physical existence* of magnetic poles (or electric charges)? We are seeing here, I think, an early sign of what will become a highly explicit theme in Part IV of the *Treatise*: Maxwell's determination to distinguish between mathematical quantities and physical ones. In the present chapter, Maxwell indeed defines mathematical measures for magnetic pole strength and intensity of polarization. But these measures by no means guarantee that anything like either poles or the polarized state itself has a physical existence; and whether poles are the *cause* or the *effect* of polarization is a question that can properly be asked only of physical entities.

Maxwell will do all he can to shed experimental light on such questions—in part through the great "electromagnetic top" experiment we shall encounter in Part IV, Chapter VI. But—if I may anticipate remarks made in the introduction to Part IV Chapter XX—we must understand that Maxwell is still in the initial phase of his work, surveying both the phenomena themselves and the competing languages that aspire to represent them. It will be opportune to question the fate of this and other ambivalences in connection with the avowedly theoretical portion of his *Treatise*, which begins with Part IV Chapter VI.

PART III.

MAGNETISM.

CHAPTER I.

ELEMENTARY THEORY OF MAGNETISM.

371.] CERTAIN bodies, as, for instance, the iron ore called loadstone, the earth itself, and pieces of steel which have been subjected to certain treatment, are found to possess the following properties, and are called Magnets.

371.1 If, near any part of the earth's surface except the Magnetic Poles, a magnet be suspended so as to turn freely about a vertical axis, it will in general tend to set itself in a certain azimuth, and if disturbed from this position it will oscillate about it. An unmagnetized body has no such tendency, but is in equilibrium in all azimuths alike.

372.] It is found that the force which acts on the body tends to cause a certain line in the body, called the Axis of the Magnet, to become parallel to a certain line in space, called the Direction of the Magnetic Force.

372.1 Let us suppose the magnet suspended so as to be free to turn in all directions about a fixed point. To eliminate the action of its weight we may suppose this point to be its centre of gravity. Let it come to a position of equilibrium. Mark two points on the magnet, and note their positions in space. Then let the magnet be placed in a new position of equilibrium, and note the positions in space of the two marked points on the magnet.

371.1 *azimuth*: horizontal angular direction, which Maxwell measures counter-clockwise from north; see (372.6).

372.1 *We may suppose this point to be its centre of gravity*: This is a bit oversimplified; the methods for finding the center of gravity of a body experimentally become quite complicated if the body is already magnetized (compare 377.3). For this thought experiment, we may suppose the center of gravity to have been determined prior to magnetization; or, if the body is a homogeneous rectilinear solid, to have been calculated geometrically.

Let it come to a position of equilibrium: that is, equilibrium with respect to the magnetic force, since by suspending it at its center of gravity we will have eliminated the action of its weight.

a new position of equilibrium: but keeping the point of suspension fixed in space.

372.2 Since the axis of the magnet coincides with the direction of magnetic force in both positions, we have to find that line in the magnet which occupies the same position in space before and after the motion. It appears, from the theory of the motion of bodies of invariable form, that such a line always exists, and that a motion equivalent to the actual motion might have taken place by simple rotation round this line.

372.3 To find the line, join the first and last positions of each of the marked points, and draw planes bisecting these lines at right angles. The intersection of these planes will be the line required, which indicates the direction of the axis of the magnet and the direction of the magnetic force in space.

372.4 The method just described is not convenient for the practical determination of these directions. We shall return to this subject when we treat of Magnetic Measurements.

372.5 The direction of the magnetic force is found to be different at different parts of the earth's surface. If the end of the axis of the magnet which points in a northerly direction be marked, it has been found that the direction in which it sets itself in general deviates from the true meridian to a considerable extent, and that the marked end points on the whole downwards in the northern hemisphere and upwards in the southern.

372.6 The azimuth of the direction of the magnetic force, measured from the true north in a westerly direction, is called the Variation, or the Magnetic Declination. The angle between the direction of the magnetic force and the horizontal plane is called the Magnetic Dip. These two angles determine the direction of the magnetic force, and, when the magnetic intensity is also known, the magnetic force is completely determined. The determination of the values of these three elements at different parts of the earth's surface, the discussion of the manner in which they vary according to the place and time of observation, and the investigation of the causes of the magnetic force and its variations, constitute the science of Terrestrial Magnetism.

372.3 *to find the line*...: The procedure is designed for bodies of irregular shape, but it is easier to visualize for the rectangular bar drawn here. Let the magnet, suspended at O, find two successive positions of equilibrium; and suppose points a, b in the first position occupy positions a', b' in the second. Maxwell pointed out in the previous paragraph that the two equilibrium positions are related by a rotation about the magnetic axis (in this example a 90° rotation is illustrated); the sketch suggests how planes bisecting lines aa' and bb' at right angles intersect along the axis of rotation AA, the magnetic axis.

373.] Let us now suppose that the axes of several magnets have been determined, and the end of each which points north marked. Then, if one of these magnets be freely suspended and another brought near it, it is found that two marked ends repel each other, that a marked and an unmarked end attract each other, and that two unmarked ends repel each other.

373.1 If the magnets are in the form of long rods or wires, uniformly and longitudinally magnetized, (see below, Art. 384,) it is found that the greatest manifestation of force occurs when the end of one magnet is held near the end of the other, and that the phenomena can be accounted for by supposing that like ends of the magnets repel each other, that unlike ends attract each other, and that the intermediate parts of the magnets have no sensible mutual action.

373.2 The ends of a long thin magnet are commonly called its Poles. In the case of an indefinitely thin magnet, uniformly magnetized throughout its length, the extremities act as centres of force, and the rest of the magnet appears devoid of magnetic action. In all actual magnets the magnetization deviates from uniformity, so that no single points can be taken as the poles. Coulomb, however, by using long thin rods magnetized with care, succeeded in establishing the law of force between two like magnetic poles* {the medium between them being air}.

373.3 The repulsion between two like magnetic poles is in the straight line joining them, and is numerically equal to the product of the strengths of the poles divided by the square of the distance between them.

374.] This law, of course, assumes that the strength of each pole is measured in terms of a certain unit, the magnitude of which may be deduced from the terms of the law.

374.1 The unit-pole is a pole which points north, and is such that, when placed at unit distance in air from another unit-pole, it repels it with unit of force, the unit of force being defined as in Art. 6. A pole which points south is reckoned negative.

374.2 If m_1 and m_2 are the strengths of two magnetic poles, l the distance between them, and f the force of repulsion, all expressed numerically, then

$$f = \frac{m_1 m_2}{l^2}.$$

374.3 But if $[m]$, $[L]$ and $[F]$ be the concrete units of magnetic pole, length and force, then

$$f[F] = \left[\frac{m}{L}\right]^2 \frac{m_1 m_2}{l^2},$$

* Coulomb, *Mém. de l'Acad.* 1785, p. 603, and in Biot's *Traité de Physique*, tome iii.

whence it follows that

$$\left[m^2\right]=\left[L^2F\right]=\left[L^2\frac{ML}{T^2}\right],$$

or
$$[m]=\left[L^{\frac{3}{2}}T^{-1}M^{\frac{1}{2}}\right].$$

The dimensions of the unit-pole are therefore $\frac{3}{2}$ as regards length, (-1) as regards time, and $\frac{1}{2}$ as regards mass. These dimensions are the same as those of the electrostatic unit of electricity, which is specified in exactly the same way in Arts. 41, 42.

375.] The accuracy of this law may be considered to have been established by the experiments of Coulomb with the Torsion Balance, and confirmed by the experiments of Gauss and Weber, and of all observers in magnetic observatories, who are every day making measurements of magnetic quantities, and who obtain results which would be inconsistent with each other if the law of force had been erroneously assumed. It derives additional support from its consistency with the laws of electro-magnetic phenomena.

376.] The quantity which we have hitherto called the strength of a pole may also be called a quantity of 'Magnetism,' provided we attribute no properties to 'Magnetism' except those observed in the poles of magnets.

376.1 Since the expression of the law of force between given quantities of 'Magnetism' has exactly the same mathematical form as the law of force between quantities of 'Electricity' of equal numerical value, much of the mathematical treatment of magnetism must be similar to that of electricity. There are, however, other properties of magnets which must be borne in mind, and which may throw some light on the electrical properties of bodies.

Relation between the Poles of a Magnet.

377.] The quantity of magnetism at one pole of a magnet is always equal and opposite to that at the other, or more generally thus:—

377.1 *In every Magnet the total quantity of Magnetism* (reckoned algebraically) *is zero.*

376. *we attribute no properties to 'Magnetism' except those observed in the poles of magnets*: In particular, recall the caution Maxwell urged in Art. 35 regarding electricity: "we must not too hastily assume that it is, or is not, a substance...."

377.1 *reckoned algebraically*: that is, such that addition of a negative quantity is equivalent to subtraction of a positive quantity.

377.2 Hence in a field of force which is uniform and parallel through-out the space occupied by the magnet, the force acting on the marked end of the magnet is exactly equal, opposite and parallel to that on the unmarked end, so that the resultant of the forces is a statical couple, tending to place the axis of the magnet in a determinate direction, but not to move the magnet as a whole in any direction.

377.3 This may be easily proved by putting the magnet into a small vessel and floating it in water. The vessel will turn in a certain direction, so as to bring the axis of the magnet as near as possible to the direction of the earth's magnetic force, but there will be no motion of the vessel as a whole in any direction; so that there can be no excess of the force towards the north over that towards the south, or the reverse. It may also be shewn from the fact that magnetizing a piece of steel does not alter its weight. It does alter the apparent position of its centre of gravity, caus-ing it in these latitudes to shift along the axis towards the north. The centre of inertia, as determined by the phenomena of rotation, remains unaltered.

378.] If the middle of a long thin magnet be examined, it is found to possess no magnetic properties, but if the magnet be broken at that point, each of the pieces is found to have a magnetic pole at the place of fracture, and this new pole is exactly equal and opposite to the other pole belonging to that piece. It is impossible, either by magnetization, or by breaking magnets, or by any other means, to procure a magnet whose poles are unequal.

378.1 If we break the long thin magnet into a number of short pieces we shall obtain a series of short magnets, each of which has poles of nearly the same strength as those of the original long magnet. This mul-tiplication of poles is not necessarily a creation of energy, for we must remember that after breaking the magnet we have to do work to sepa-rate the parts, in consequence of their attraction for one another.

379.] Let us now put all the pieces of the magnet together as at first. At each point of junction there will be two poles exactly equal and of oppo-site kinds, placed in contact, so that their united action on any other pole

377.2 *statical couple*: A pair of equal forces acting in parallel but opposite directions (sketch *a*). They tend to turn a body so that the line between their application points coincides with both forces, as in sketch (*b*).

377.3 *It does alter the apparent position of its centre of gravity*: It is this alteration that complicates practical determination of the center of gravity of a magne-tized body, as noted in a comment to (372.1).

will be null. The magnet, thus rebuilt, has therefore the same properties as at first, namely two poles, one at each end, equal and opposite to each other, and the part between these poles exhibits no magnetic action.

379.1 Since, in this case, we know the long magnet to be made up of little short magnets, and since the phenomena are the same as in the case of the unbroken magnet, we may regard the magnet, even before being broken, as made up of small particles, each of which has two equal and opposite poles. If we suppose all magnets to be made up of such particles, it is evident that since the algebraical quantity of magnetism in each particle is zero, the quantity in the whole magnet will also be zero, or in other words, its poles will be of equal strength but of opposite kind.

Theory of 'Magnetic Matter.'

380.] Since the form of the law of magnetic action is identical with that of electric action, the same reasons which can be given for attributing electric phenomena to the action of one 'fluid' or two 'fluids' can also be used in favour of the existence of a magnetic matter, or of two kinds of magnetic matter, fluid or otherwise. In fact, a theory of magnetic matter, if used in a purely mathematical sense, cannot fail to explain the phenomena, provided new laws are freely introduced to account for the actual facts.

380.1 One of these new laws must be that the magnetic fluids cannot pass from one molecule or particle of the magnet to another, but that the process of magnetization consists in separating to a certain extent the two fluids within each particle, and causing the one fluid to be more concentrated at one end, and the other fluid to be more concentrated at the other end of the particle. This is the theory of Poisson.

380.2 A particle of a magnetizable body is, on this theory, analogous to a small insulated conductor without charge, which on the two-fluid theory contains indefinitely large but exactly equal quantities of the two electricities. When an electromotive force acts on the conductor, it separates the electricities, causing them to become manifest at opposite sides

380. *one fluid or two fluids*: Maxwell discussed "fluid" theories of electricity in Arts. 36 and 37.

a theory of magnetic matter, if used in a purely mathematical sense: that is, in a sense that attributes "no properties to 'Magnetism' except those observed in the poles of magnets" (376.)—the properties of *magnitude* and *sign* (+ or −).

380.2 *A particle of a magnetizable body is, on this theory, analogous to a small insulated conductor...*: Maxwell refers to Aepinus's (1759) and subsequent magnetic theories, which were more or less deliberately modeled on fluid theories of electricity.

of the conductor. In a similar manner, according to this theory, the magnetizing force causes the two kinds of magnetism, which were originally in a neutralized state, to be separated, and to appear at opposite sides of the magnetized particle.

380.3 In certain substances, such as soft iron and those magnetic substances which cannot be permanently magnetized, this magnetic condition, like the electrification of the conductor, disappears when the inducing force is removed. In other substances, such as hard steel, the magnetic condition is produced with difficulty, and, when produced, remains after the removal of the inducing force.

380.4 This is expressed by saying that in the latter case there is a Coercive Force, tending to prevent alteration in the magnetization, which must be overcome before the power of a magnet can be either increased or diminished. In the case of the electrified body this would correspond to a kind of electric resistance, which, unlike the resistance observed in metals, would be equivalent to complete insulation for electromotive forces below a certain value.

380.5 This theory of magnetism, like the corresponding theory of electricity, is evidently too large for the facts, and requires to be restricted by artificial conditions. For it not only gives no reason why one body may not differ from another on account of having more of both fluids, but it enables us to say what would be the properties of a body containing an excess of one magnetic fluid. It is true that a reason is given why such a body cannot exist, but this reason is only introduced as an after-thought to explain this particular fact. It does not grow out of the theory.

381.] We must therefore seek for a mode of expression which shall not be capable of expressing too much, and which shall leave room for the introduction of new ideas as these are developed from new facts.

380.4 *Coercive Force ... would correspond to a kind of electric resistance*: This is an extension of the magnetic/electric parallel sought by fluid theorists (380.2, *comment*). As electrical resistance tends to prevent change in the distribution of electric charge (by resisting current), the coercive force is conceived by its adherents as opposing any change in the accumulation of magnetic matter. Note that Maxwell is characterizing, not asserting, this analogy.

below a certain value: The phenomena typically ascribed to "coercive force" exhibit a threshold below which a material's magnetic response is hardly detectable.

380.5 *but this reason is introduced only as an afterthought to explain this particular fact*: The "reason" is the hypothesis, indicated in (380.1), that the austral and boreal magnetic fluids cannot flow independently of one another, but can only undergo symmetrical separation. There is no independent foundation for such a hypothesis; nor, does it (as Maxwell puts it) "grow out of the theory." On the contrary, it seems to be tacked on for the sole reason of *saving* the theory.

This, I think, we shall obtain if we begin by saying that the particles of a magnet are Polarized.

Meaning of the term 'Polarization.'

381.1 When a particle of a body possesses properties related to a certain line or direction in the body, and when the body, retaining these properties, is turned so that this direction is reversed, then if as regards other bodies these properties of the particle are reversed, the particle, in reference to these properties, is said to be polarized, and the properties are said to constitute a particular kind of polarization.

381.2 Thus we may say that the rotation of a body about an axis constitutes a kind of polarization, because if, while the rotation continues, the direction of the axis is turned end for end, the body will be rotating in the opposite direction as regards space.

381.3 A conducting particle through which there is a current of electricity may be said to be polarized, because if it were turned round, and if the current continued to flow in the same direction as regards the particle, its direction in space would be reversed.

381.4 In short, if any mathematical or physical quantity is of the nature of a vector, as defined in Art. 11, then any body or particle to which this directed quantity or vector belongs may be said to be Polarized,* because it has opposite properties in the two opposite directions or poles of the directed quantity. The poles of the earth, for example, have reference to its rotation, and have accordingly different names.

Meaning of the term 'Magnetic Polarization.'

382.] In speaking of the state of the particles of a magnet as magnetic polarization, we imply that each of the smallest parts into which a magnet may be divided has certain properties related to a definite direction

* The word Polarization has been used in a sense not consistent with this in Optics, where a ray of light is said to be polarized when it has properties relating to its sides, which are identical on opposite sides of the ray. This kind of polarization refers to another kind of Directed Quantity, which may be called a Dipolar Quantity, in opposition to the former kind, which may be called Unipolar.

When a dipolar quantity is turned end for end it remains the same as before. Tensions and Pressures in solid bodies, Extensions, Compressions, and Distortions and most of the optical, electrical, and magnetic properties of crystallized bodies are dipolar quantities.

The property produced by magnetism in transparent bodies of twisting the plane of polarization of the incident light, is, like magnetism itself, a unipolar property....

through the particle, called its Axis of Magnetization, and that the properties related to one end of this axis are opposite to the properties related to the other end.

382.1 The properties which we attribute to the particle are of the same kind as those which we observe in the complete magnet, and in assuming that the particles possess these properties, we only assert what we can prove by breaking the magnet up into small pieces, for each of these is found to be a magnet.

Properties of a Magnetized Particle.

383.] Let the element $dx\,dy\,dz$ be a particle of a magnet, and let us assume that its magnetic properties are those of a magnet the strength of whose positive pole is m, and whose length is ds. Then if P is any point in space distant r from the positive pole and r' from the negative pole, the magnetic potential at P will be m/r due to the positive pole, and $-m/r'$ due to the negative pole, or

$$V = \frac{m}{rr'}(r' - r). \qquad (1)$$

383.1 If ds, the distance between the poles, is very small, we may put

$$r' - r = ds\cos\varepsilon, \qquad (2)$$

where ε is the angle between the vector drawn from the magnet to P and the axis of the magnet,* or in the limit

$$V = \frac{mds}{r^2}\cos\varepsilon. \qquad (3)$$

*{The positive direction of the axis is from the negative to the positive pole.}

383. *its magnetic properties are those of a magnet the strength of whose positive pole is m, and whose length is ds*: The positive pole is north-seeking (374.1); the arrangement is as sketched here.

the magnetic potential at P will be m/r due to the positive pole: Since the force law for magnetic poles is the same as for electric charges (376.1), the expression for potential will have the same form as that derived for charges; see (73., *comment*).

383.1 *If ds ... is very small, we may put r′−r = ds cosε*: In the diagram above, note that the smaller ds is, the more nearly will r' and r coincide, thus the more nearly will $ds\cos\varepsilon$ be their difference.

in the limit: Substitute equation (2) into equation (1). Then as ds becomes vanishingly small, r' becomes equal to r, yielding equation (3).

Magnetic Moment.

384.] The product of the length of a uniformly and longitudinally magnetized bar magnet into the strength of its positive pole is called its Magnetic Moment.

Intensity of Magnetization.

384.1 The intensity of magnetization of a magnetic particle is the ratio of its magnetic moment to its volume. We shall denote it by I.

384.2 The magnetization at any point of a magnet may be defined by its intensity and its direction. Its direction may be defined by its direction-cosines λ, μ, ν.

Components of Magnetization.

384.3 The magnetization at a point of a magnet (being a vector or directed quantity) may be expressed in terms of its three components referred to the axes of coordinates. Calling these A, B, C,

$$A=I\lambda,\ B=I\mu,\ C=I\nu,\tag{4}$$

and the numerical value of I is given by the equation

$$I^2 = A^2+B^2+C^2\tag{5}$$

384. *Magnetic Moment*: Consider a narrow bar magnetized along its length l. If each of its poles has strength m, its magnetic moment is the product ml. Notice that since the force experienced by a magnetic pole is proportional to m, the product ml resembles the product of force and distance that measures the *turning moment* (torque) of a weighted horizontal lever. Thus the magnetic moment of a magnet indicates the tendency it would have to turn under magnetic influence directed perpendicular to its magnetic axis.

384.1 *The intensity of magnetization of a magnetic particle is the ratio of its magnetic moment to its volume*: Maxwell will apply this definition not only to "particles" but to magnets of any size. Consider a magnetized bar having length l and cross-sectional area A, and let each of its poles have strength m. Then its magnetic moment will be the product ml, and its intensity of magnetization I will be ml/lA or m/A, the quotient of the pole strength to the surface area of the end of the bar. Later, Maxwell will refer to this quotient as the "surface-density of magnetism." Thus intensity of magnetization I has the same dimensions as does the surface-density of magnetism on a polar face.

384.2 *direction-cosines*: defined in (21.2, comment).

384.3 Equations (4) and (5) state, in different ways, that I is the diagonal of a rectangular solid having edges A, B, C.

385.] If the portion of the magnet which we consider is the differential element of volume $dx\,dy\,dz$, and if I denotes the intensity of magnetization of this element, its magnetic moment is $I\,dx\,dy\,dz$. Substituting this for $m\,ds$ in equation (3), and remembering that

$$r\cos\varepsilon = \lambda(\xi-x)+\mu(\eta-y)+\nu(\zeta-z),\tag{6}$$

where ξ, η, ζ are the coordinates of the extremity of the vector r drawn from the point (x, y, z), we find for the potential at the point (ξ, η, ζ) due to the magnetized element at (x, y, z),

$$\left\{A(\xi-x)+B(\eta-y)+C(\zeta-z)\right\}\frac{1}{r^3}dx\,dy\,dz,\tag{7}$$

385.1 To obtain the potential at the point (ξ, η, ζ) due to a magnet of finite dimensions, we must find the integral of this expression for every element of volume included within the space occupied by the magnet, or

$$V = \iiint\left\{A(\xi-x)+B(\eta-y)+C(\zeta-z)\right\}\frac{1}{r^3}dx\,dy\,dz,\tag{8}$$

385. *its magnetic moment is* $I\,dx\,dy\,dz$: Since intensity of magnetization was defined as the ratio of magnetic moment to volume, the product of the intensity and the volume of the portion being considered will equal the magnetic moment, which is also equal to $m\,ds$ according to the definition in Art. 384.

Equation (6): Maxwell is relying on a general theorem for resolving one line in the direction of another, but modern readers can obtain this result using the vector dot product. Let **r** be a vector with components r_x, r_y, r_z and let **u** be a unit vector having components u_x, u_y, u_z; let the angle between them be ε. (For simplicity, the diagram shows two dimensions only.) Now the dot product **r·u** $= ru\cos\varepsilon = r_xu_x+r_yu_y+r_zu_z$ is the resolved part of **r** in the direction of **u**. And Maxwell makes clear that $r_x = \xi-x$, $r_y = \eta-y$, and $r_z = \zeta-z$. Moreover the direction-cosines of a line are simply the x-, y-, and z-components of a unit vector in the direction of the line. Thus from the dot product above,

$$ru\cos\varepsilon = r_xu_x+r_yu_y+r_zu_z$$

or
$$r\cos\varepsilon = (\xi-x)\lambda+(\eta-y)\mu+(\zeta-z)\nu.$$

More generally, *the projection of one line onto another equals the sum of products of the rectangular components of the projected line with the corresponding direction-cosines of the line targeted.*

Equation (7): As Maxwell directed earlier in the paragraph, substitute $I\,dx\,dy\,dz$ for the magnetic moment $m\,ds$ in equation (3):

$$V = \frac{m\,ds}{r^2}\cos\varepsilon = \frac{I\,dx\,dy\,dz}{r^3}r\cos\varepsilon = \frac{I\,dx\,dy\,dz}{r^3}\left[\lambda(\xi-x)+\mu(\eta-y)+\nu(\zeta-z)\right].$$

But $I\lambda = A$, $I\mu = B$, $I\nu = C$, as in equation (4). Substitution yields the equation.

385.2 Integrated by parts, this becomes

$$V = \iint A \frac{1}{r} dy\,dz + \iint B \frac{1}{r} dz\,dx + \iint C \frac{1}{r} dx\,dy$$

$$- \iiint \frac{1}{r} \left(\frac{dA}{dx} + \frac{dB}{dy} + \frac{dC}{dz} \right) dx\,dy\,dz,$$

where the double integration in the first three terms refers to the surface of the magnet, and the triple integration in the fourth to the space within it.

385.3 If l, m, n denote the direction-cosines of the normal drawn outwards from the element of surface dS, we may write, as in Art. 21, for the sum of the first three terms

$$\iint (lA + mB + nC) \frac{1}{r} dS,$$

where the integration is to be extended over the whole surface of the magnet.

385.4 If we now introduce two new symbols σ and ρ, defined by the equations

385.2 *Integrated by parts*: Unlike his earlier use of this term in (24.9), Maxwell is here integrating "by parts" in the conventional sense, that is by making use of the identity $\int v\,du = uv - \int u\,dv$. We begin by rewriting equation (8),

$$V = \iiint [A(\xi - x) + B(\eta - y) + C(\zeta - z)] \frac{1}{r^3} dx\,dy\,dz, \quad \text{as}$$

$$V = \iint \left[\int \frac{A(\xi - x)}{r^3} dx \right] dy\,dz + \iint \left[\int \frac{B(\eta - y)}{r^3} dy \right] dx\,dz + \iint \left[\int \frac{C(\zeta - z)}{r^3} dz \right] dy\,dz.$$

If we now choose $v = A$ and $u = 1/r$, then $du = -r^{-2} dr$. But $r^2 = (\xi - x)^2 + (\eta - y)^2 + (\xi - z)^2$; hence $2r\,dr = -2(\xi - x)\,dx$ (holding y and z constant) and thus $dr = -(\xi - x)r^{-1} dx$. Then du becomes $(\xi - x)r^{-3} dx$. Now write the identity

$$\int v\,du = uv - \int u\,dv.$$

Substituting, $\quad \int \frac{A(\xi - x)}{r^3} dx = \frac{A}{r} - \int \frac{1}{r} dA = \frac{A}{r} - \int \frac{1}{r} \frac{dA}{dx} dx.$

Perform a similar procedure on each of the bracketed integrals above to obtain

$$\int \frac{B(\eta - y)}{r^3} dy = \frac{B}{r} - \int \frac{1}{r} \frac{dB}{dy} dy \quad \text{and} \quad \int \frac{C(\zeta - z)}{r^3} dz = \frac{C}{r} - \int \frac{1}{r} \frac{dC}{dz} dz.$$

When these are substituted into the equation for V we have Maxwell's results.

385.3 *We may write, as in Art. 21, for the sum of the first three terms*: In particular, compare equation (1) in paragraph 21.2.

$$\sigma = lA + mB + nC,$$

$$\rho = -\left(\frac{dA}{dx} + \frac{dB}{dy} + \frac{dC}{dz}\right),$$

the expression for the potential may be written

$$V = \iint \frac{\sigma}{r}\,dS + \iiint \frac{\rho}{r}\,dx\,dy\,dz.$$

386.] This expression is identical with that for the electric potential due to a body on the surface of which there is an electrification whose surface-density is σ, while throughout its substance there is a bodily electrification whose volume-density is ρ. Hence, if we assume σ and ρ to be the surface- and volume-densities of the distribution of an imaginary substance, which we have called 'magnetic matter,' the potential due to this imaginary distribution will be identical with that due to the actual magnetization of every element of the magnet.

386.1 The surface-density σ is the resolved part of the intensity of magnetization I in the direction of the normal to the surface drawn outwards, and the volume-density ρ is the 'convergence' (see Art. 25) of the magnetization at a given point in the magnet.

386.2 This method of representing the action of a magnet as due to a distribution of 'magnetic matter' is very convenient, but we must always remember that it is only an artificial method of representing the action of a system of polarized particles.

* * *

386.1 *the resolved part of the intensity of magnetization I in the direction of the normal to the surface*: Since A, B, and C are the rectangular components of the intensity of magnetization and l, m, n are the direction-cosines of the normal to the surface (385.3), then in accordance with the theorem stated in (385., *comment on equation* 6), the sum of products $lA + mB + nC$ is the resolved part of the intensity in the direction of the normal.

the 'convergence' (see Art. 25) of the magnetization at a given point: Although Maxwell indeed offers a full discussion of convergence in Art. 25, the remarks offered in (21.1, *comment*) may be more applicable here. Note again that modern nomenclature prefers "divergence" and omits the negative sign, writing

$$\text{div } I = \frac{dA}{dx} + \frac{dB}{dy} + \frac{dC}{dz}.$$

386.2 *an artificial method of representing the action of polarized particles*: As with the analogous fluid theories of electricity, the "polarized particles" Maxwell speaks of are particles of actual materials like iron and steel—materials from which magnets may be fabricated. The theoretical "magnetic matter" or "magnetic fluid" is not itself to be conceived as consisting of polarized particles.

On the Potential Energy of a Magnet placed in a Magnetic Field.

389.] Let V be the magnetic potential due to any system of magnets acting on the magnet under consideration. We shall call V the potential of the external magnetic force.

389.1 If a small magnet whose strength is m, and whose length is ds, be placed so that its positive pole is at a point where the potential is V, and its negative pole at a point where the potential is V', the potential energy of this magnet will be $m(V-V')$, or, if ds is measured from the negative pole to the positive,

$$m\frac{dV}{ds}ds. \tag{1}$$

389.2 If I is the intensity of the magnetization, and λ, μ, ν its direction-cosines, we may write,

$$mds = I\,dx\,dy\,dz,$$

and
$$\frac{dV}{ds} = \lambda\frac{dV}{dx} + \mu\frac{dV}{dy} + \nu\frac{dV}{dz}$$

* * *

393.] At all parts of the earth's surface, except some parts of the Polar regions, one end of a magnet points towards the north, or at least in a northerly direction, and the other in a southerly direction. In speaking of the ends of a magnet we shall adopt the popular method of calling the end which points to the north the north end of the magnet. When, however, we speak in the language of the theory of magnetic fluids we shall use the words Boreal and Austral. Boreal magnetism is an imaginary

389.1 *the potential energy of this magnet will be* $m(V-V')$: In the expression that follows, since ds is presumed very small, $(V-V')$ is likewise presumed very small; hence Maxwell denotes it as dV.

389.2 λ, μ, ν *its direction-cosines*: That is, the direction-cosines of the magnetization I. But since the direction of magnetization coincides with the line joining the magnet's poles, λ, μ, ν are also the direction-cosines of the axis (372.) of the magnet.

$mds = I\,dx\,dy\,dz$: Since mds is the magnetic moment (384.), this equation just restates the definition of intensity of magnetization given in (384.1).

$\frac{dV}{ds} = \lambda\frac{dV}{dx} + \mu\frac{dV}{dy} + \nu\frac{dV}{dz}$: Since dV/dx, dV/dy, dV/dz are the components of ∇V (Art. 17) and λ, μ, ν are the direction-cosines of the length of the magnet, then by the theorem noted in my comment to (385.), the right hand side of this equation represents the resolved part of ∇V in the direction of orientation of the magnet, that is, dV/ds.

kind of matter supposed to be most abundant in the northern parts of the earth, and Austral magnetism is the imaginary magnetic matter which prevails in the southern regions of the earth. The magnetism of the north end of a magnet is Austral, and that of the south end is Boreal. When therefore we speak of the north and south ends of a magnet we do not compare the magnet with the earth as the great magnet, but merely express the position which the magnet endeavours to take up when free to move. When, on the other hand, we wish to compare the distribution of imaginary magnetic fluid in the magnet with that in the earth we shall use the more grandiloquent words Boreal and Austral magnetism.

394.] In speaking of a field of magnetic force we shall use the phrase Magnetic North to indicate the direction in which the north end of a compass needle would point if placed in the field of force.

394.1 In speaking of a line of magnetic force we shall always suppose it to be traced from magnetic south to magnetic north, and shall call this direction positive. In the same way the direction of magnetization of a magnet is indicated by a line drawn from the south end of the magnet towards the north end, and the end of the magnet which points north is reckoned the positive end.

394.2 We shall consider Austral magnetism, that is, the magnetism of that end of a magnet which points north, as positive. If we denote its numerical value by m, then the magnetic potential

$$V = \sum \frac{m}{r},$$

and the positive direction of a line of force is that in which V diminishes.

394. *we shall use the phrase Magnetic North to indicate the direction in which the north end of a compass needle would point*: Note therefore that "magnetic north" denotes a *direction*, not a species of magnetism.

394.1 *In speaking of a line of magnetic force we shall always suppose it to be traced from magnetic south to magnetic north, and shall call this direction positive*: Thus the positive direction of the earth's magnetic lines of force is from geographic south to geographic north; and the positive direction of any magnetic line of force is from a north-seeking to a south-seeking pole—or, as Maxwell will make clear in the next paragraph, from a positive pole to a negative one.

394.2 *the magnetic potential*: For the quotient m/r see (383., *comment*). Maxwell uses a summation notation (\sum) to allow for the possibility of multiple poles m, each at its corresponding distance r from the point at which the potential is measured.

Editor's Introduction to
Chapter II

The magnetic condition within a magnetized medium

In Chapter I Maxwell studied the forces between magnets and developed measures for pole strength and for intensity of magnetization. These are properties that can be studied "from the outside," so to speak. But what about the conditions in the *interior* of a magnet? In order to make measurements inside a mass of iron or steel, we would have to hollow out within it a cavity capable of receiving the test element—in this case a hypothetical *unit magnetic pole*.

A hollow space creates a problem, however: just as Maxwell represents dielectric materials as polarizing under the action of electromotive intensity, so he conceives magnetic materials like iron and steel as undergoing a comparable polarization (defined in Art. 382) under magnetic action. But, as Maxwell will explain in the present chapter, such polarized material must necessarily develop magnetization on at least some of its surfaces; and this "surface magnetism" will affect our test pole—the effect evidently depending on the shape and proportions of the cavity. What shape of cavity, then, will best reveal the condition of magnetic force within the body of the magnet?

Magnetic force and magnetic induction

As Maxwell will explain, a cylindrical cavity, suitably oriented within the magnetized substance, will develop surface magnetism on its circular faces only, none on its cylindrical walls. In Arts. 396–400 Maxwell will consider such a cavity, illustrated here in section. First (sketch *a*) he lets the circular faces of the cylinder shrink to negligible area while its length remains fixed; thus the surface-magnetism on those faces will

(a)

also become negligible, and Maxwell concludes that the force on the test pole will be the same as it would be in the absence of any polarization— it will consist of the force exerted by the magnet's own poles, along with any external magnetic sources that may reside in the vicinity. Maxwell will call this force the *magnetic force* proper, symbolized \mathfrak{H}.

Thus we find that a long narrow crevasse actually *conceals* the very condition that distinguishes a magnet's interior, that is, the polarization

of the material surrounding the cavity. To detect that condition, Maxwell (sketch *b*) lets the length shrink to neglible size while the circular faces remain fixed. This circumstance permits the surface-magnetism on the faces to contribute to the force on the unit pole. Maxwell does not

(*b*)

propose a name for this contribution, but he shows that it is proportional to the intensity of magnetization \mathfrak{J}, being equal to $4\pi\mathfrak{J}$.

Note that each cavity shape brings out an aspect of force that is conceived to exist in the material even when there is no cavity there. Thus the contribution $4\pi\mathfrak{J}$ to the force within the cavity does not replace \mathfrak{H} but adds to it; Maxwell will name the sum of both quantities the *magnetic induction*, \mathfrak{B}, so that $\mathfrak{B} = \mathfrak{H} + 4\pi\mathfrak{J}$. It will be evident that in air or vacuum (neither of which undergoes magnetization) \mathfrak{B} reduces to \mathfrak{H}. In any material capable of magnetization, however, the distinction between \mathfrak{B} and \mathfrak{H} is consequential—not only numerically but conceptually.

In Chapter V Maxwell will introduce another locution respecting \mathfrak{B} and \mathfrak{H}, speaking of \mathfrak{H} *causally* as the "magnetizing force," which acts to magnetize a material "by induction." This way of speaking is particularly apt when \mathfrak{H} arises from an *external source*. It gives additional resonance to the name of "magnetic induction" for the quantity \mathfrak{B}, which is thus viewed as having been induced by action of \mathfrak{H}. Although this form of expression makes a rather inconspicuous appearance in Maxwell's narrative, it has become the dominant idiom in modern treatments of magnetism.

CHAPTER II.

MAGNETIC FORCE AND MAGNETIC INDUCTION.

395.] WE have already (Art. 385) determined the magnetic potential at a given point due to a magnet, the magnetization of which is given at every point of its substance, and we have shewn that the mathematical result may be expressed either in terms of the actual magnetization of every element of the magnet, or in terms of an imaginary distribution of 'magnetic matter,' partly condensed on the surface of the magnet and partly diffused throughout its substance.

395.1 The magnetic potential, as thus defined, is found by the same mathematical process, whether the given point is outside the magnet or within it. The force exerted on a unit magnetic pole placed at any point outside the magnet is deduced from the potential by the same process of differentiation as in the corresponding electrical problem. If the components of this force are α, β, γ,

$$\alpha = -\frac{dV}{dx}, \quad \beta = -\frac{dV}{dy}, \quad \gamma = -\frac{dV}{dz}. \tag{1}$$

395.2 To determine by experiment the magnetic force at a point within the magnet we must begin by removing part of the magnetized substance, so as to form a cavity within which we are to place the magnetic pole. The force acting on the pole will depend, in general, on the form of this cavity, and on the inclination of the walls of the cavity to the direction of magnetization. Hence it is necessary, in order to avoid ambiguity in speaking of the magnetic force within a magnet, to specify the form and position of the cavity within which the force is to be measured. It is manifest that when the form and position of the cavity is specified, the point within it at which the magnetic pole is placed must be regarded as

395.1 *unit magnetic pole*: defined in (374.1). Maxwell here imagines an isolated pole, one that is not part of an actual magnet. Although such a "monopole" cannot actually exist, we may calculate the force that would be exerted on it if it did exist.

in the corresponding electrical problem: or in any case—not necessarily an electrical one—in which a force is derived from a potential, as in (16.4).

no longer within the substance of the magnet, and therefore the ordinary methods of determining the force become at once applicable.

396.] Let us now consider a portion of a magnet in which the direction and intensity of the magnetization are uniform. Within this portion let a cavity be hollowed out in the form of a cylinder, the axis of which is parallel to the direction of magnetization, and let a magnetic pole of unit strength be placed at the middle point of the axis.

396.1 Since the generating lines of this cylinder are in the direction of magnetization, there will be no superficial distribution of magnetism on the curved surface, and since the circular ends of the cylinder are perpendicular to the direction of magnetization, there will be a uniform superficial distribution, of which the surface-density is I for the negative end, and $-I$ for the positive end.

396.2 Let the length of the axis of the cylinder be $2b$, and its radius a. Then the force arising from this superficial distribution on a magnetic pole placed at the middle point of the axis is that due to the attraction of the disk on the positive side, and the repulsion of the disk on the negative side. These two forces are equal and in the same direction, and their sum is

$$R = 4\pi I\left(1 - \frac{b}{\sqrt{a^2 + b^2}}\right). \tag{2}$$

396.1 *the surface-density is I for the negative end, and $-I$ for the positive end*: The negative and positive ends of the cavity are the ones nearest to the negative and positive poles of the magnet, respectively; thus the negative end of the cavity sketched here is on the *left*. And we saw in (386.1) that the surface-density σ is the resolved part of the intensity I in the direction of the normal to the surface drawn outwards. At the left (negative) end of the cylindrical cavity, then, the normal drawn *outwards* (that is, away from the *magnetized substance*) has the same direction as I, and thus σ is equal to I and has the same positive sign. The sign will be opposite—negative—at the positive end of the cavity.

396.2 *Equation* (2): I will first derive this equation, then offer a simpler alternative that does not involve complicated integration.

The inverse-square force law for a surface element takes the form $dR = dm/r^2$, where dm is the pole strength of the element and r is the distance from that element. But $dm = \sigma dA = I dA$ while $r^2 = b^2 + y^2$. Now a ring-shaped element of radius y and width dy will have area $dA = 2\pi y\, dy$, and will therefore manifest pole strength $dm = 2\pi I y\, dy$ and contribute $dR = 2\pi I y\, dy / r^2$ to the magnetic force R at the center of the cylinder. But R is directed along r, and its transverse components

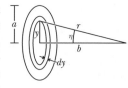

396.3 From this expression it appears that the force depends, not on the absolute dimensions of the cavity, but on the ratio of the length to the diameter of the cylinder. Hence, however small we make the cavity, the force arising from the surface distribution on its walls will remain, in general, finite.

* * *

397.1 If, therefore, we assume the dimensions of the cylinder so small that the magnetization of the part removed may be regarded as everywhere parallel to the axis of the cylinder, and of constant magnitude I, the force on a magnetic pole placed at the middle point of the axis of the cylindrical hollow will be compounded of two forces. The first of these is that due to the distribution of magnetic matter on the outer surface of the magnet, and throughout its interior, exclusive of the portion hollowed out. The components of this force are α, β and γ, derived from the potential by equations (1). The second is the force R, acting along the axis of the cylinder in the direction of magnetization. The value of this force depends on the ratio of the length to the diameter of the cylindric cavity.

will cancel one another; thus we need integrate only the axial components $dR\cos\eta$, that is, $dR\,b/r$. Integrating from $y = 0$ to $y = a$, the radius of a face, we have

$$R = \int_0^a \frac{2\pi I y}{b^2 + y^2} \frac{b}{\sqrt{b^2 + y^2}}\, dy = 2\pi I b \int_0^a \frac{y}{\left(b^2 + y^2\right)^{\frac{3}{2}}}\, dy.$$

Evaluation of this integral yields the contribution to R by a single face; Maxwell doubles the result since the two faces supplement each other, one repelling, one attracting.

There is a simpler alternative to this treatment. Shortly, in Art. 398, Maxwell will consider a case in which a becomes much smaller than b. If we make this assumption at the start, no integration is necessary. Instead we argue this way:

Consider each disk-shaped face of the cylinder as representing pole strength equal to the product of its area and the surface-density, that is, $\pi a^2 \sigma$ or $\pi a^2 I$ (since we showed that intensity of magnetization is equal to surface-density). Then if radius a is very much smaller than b, all portions of the face may be considered as being distance b from the test pole at the center of the cylinder. Then by the inverse-square law, each disk exerts magnetic force $\pi a^2 I / b^2$ upon the unit pole, for a total force of $2\pi I a^2 / b^2$. This expression will suffice for Maxwell's discussion in Art. 398.

397.1 *the force on a magnetic pole placed at the middle point ... will be compounded of two forces*: As Maxwell goes on to state, the first of these forces is that described by equation (1) above and is due to the magnet's own poles, plus any external magnetic sources; the second is described by equation (2) and depends on the shape of the cavity. In subsequent Articles Maxwell will consider two paradigm cases: first when the cavity shrinks to a *line*, second when it shrinks to a *surface*.

398.] Case I. Let this ratio be very great, or let the diameter of the cylinder be small compared with its length. Expanding the expression for R in powers of a/b, we find

$$R = 4\pi I\left\{\frac{1}{2}\frac{a^2}{b^2} - \frac{3}{8}\frac{a^4}{b^4} + \&c.\right\},\qquad(3)$$

a quantity which vanishes when the ratio of b to a is made infinite. Hence, when the cavity is a very narrow cylinder with its axis parallel to the direction of magnetization, the magnetic force within the cavity is not affected by the surface distribution on the ends of the cylinder, and the components of this force are simply α, β and γ, where

$$\alpha = -\frac{dV}{dx},\quad \beta = -\frac{dV}{dy},\quad \gamma = -\frac{dV}{dz}.\qquad(4)$$

398.1 We shall define the force within a cavity of this form as the magnetic force within the magnet. Sir William Thomson has called this the Polar definition of magnetic force. When we have occasion to consider this force as a vector we shall denote it by \mathfrak{H}.

398. *Expanding ... in powers of a/b*: In the expression $R = 4\pi I\left(1 - \dfrac{b}{\sqrt{a^2+b^2}}\right)$

divide numerator and denominator of the fraction by b to obtain

$$R = 4\pi I\left(1 - \frac{1}{\sqrt{a^2/b^2+1}}\right),\text{ that is, } 4\pi I\left[1 - \left(1+a^2/b^2\right)^{-\frac{1}{2}}\right].$$

The binomial expansion $(1+x)^{-\frac{1}{2}}$ is $1 - \frac{1}{2}x + \frac{3}{8}x^2\cdots$; therefore let $x = a^2/b^2$ and the expression for R becomes

$$R = 4\pi I\left[1 - (1+x)^{-\frac{1}{2}}\right] = 4\pi I\left[1 - \left(1 - \tfrac{1}{2}x + \tfrac{3}{8}x^2\cdots\right)\right] = 4\pi I\left(\tfrac{1}{2}x - \tfrac{3}{8}x^2\cdots\right);$$

and since we set $x = a^2/b^2$ this is identical to equation (3). But recall the alternative derivation $R = 2\pi I a^2/b^2$, obtained in (396.2, *comment*). You can see that it is equivalent to the first term of equation (3) and vanishes under the same conditions. Some readers will prefer this simpler approach.

Equations (4): Maxwell is repeating equations (1) from (395.1). He is able to do this because, as he here states, "the magnetic force within the cavity is not affected by surface distribution" on its boundaries and is therefore calculated in the same way as it would be for points outside the magnet—that is, from the gradient of the potential (17.5, *comment*). Remember that the potential in question reflects both the magnet's own poles and any external magnetic sources (397.1, *comment*).

398.1 *Polar definition of magnetic force*: See Thomson's *Reprint of Papers on Electrostatics and Magnetism* (1872), p.397. Thomson's nomenclature was never

399.] Case II. Let the length of the cylinder be very small compared with its diameter, so that the cylinder becomes a thin disk. Expanding the expression for R in powers of b/a, it becomes

$$R = 4\pi I\left\{1 - \frac{b}{a} + \frac{1}{2}\frac{b^3}{a^3} - \&c.\right\},\tag{5}$$

the ultimate value of which, when the ratio of a to b is made infinite, is $4\pi I$.

399.1 Hence, when the cavity is in the form of a thin disk, whose plane is normal to the direction of magnetization, a unit magnetic pole placed at the middle of the axis experiences a force $4\pi I$ in the direction of magnetization, arising from the superficial magnetism on the circular surfaces of the disk.

399.2 Since the components of I are A, B and C, the components of this force are $4\pi A$, $4\pi B$, and $4\pi C$. This must be compounded with the force whose components are α, β, γ.

widely adopted; most writers follow Maxwell in naming the quantity defined by equations (4) the "magnetic force," simply.

When we have occasion to consider this force as a vector we shall denote it by \mathfrak{H}. It is interesting that Maxwell does not assign a special symbol to the force \mathfrak{H} when it is *not* considered as a vector, restricting himself instead to circumlocutions like "the force whose components are α, β, γ," as in (399.2) below. In Part IV of the *Treatise* (Arts 478 and 479) he will introduce T for that purpose; modern practice generally prefers H for the magnitude of Maxwell's vector \mathfrak{H}.

Notice that inasmuch as H is defined by equations (4), it is derived from a potential, V, in the same way as resultant intensity R was derived from a potential (16.4–16.5). Thus, just as the resultant intensity of an electric field may be represented by a system of lines of electric force (82.), so too H may be represented by the *lines of magnetic force* Maxwell discussed in Art. 394.

399. *Expanding ... in powers of* b/a: This time in the expression for R divide numerator and denominator of the fraction by a to obtain

$$R = 4\pi I\left(1 - \frac{b/a}{\sqrt{1+b^2/a^2}}\right) \quad \text{or} \quad R = 4\pi I\left[1 - \frac{b}{a}\left(1+b^2/a^2\right)^{-\frac{1}{2}}\right].$$

Using the same binomial expansion as in (398.), let $x = b^2/a^2$ to obtain eq. (5).

the ultimate value of which, when the ratio of a to b is made infinite, is $4\pi I$: For a simpler alternative, go back to equation (2) in (396.2), namely

$$R = 4\pi I\left(1 - \frac{b}{\sqrt{a^2+b^2}}\right),$$

and note that when b approaches zero, the expression reduces to $R = 4\pi I$.

400.] Let the actual force on the unit pole be denoted by the vector \mathfrak{B}, and its components by a, b and c, then

$$\left.\begin{array}{l} a = \alpha + 4\pi A, \\ b = \beta + 4\pi B, \\ c = \gamma + 4\pi C. \end{array}\right\} \qquad (6)$$

400.1 We shall define the force within a hollow disk, whose plane sides are normal to the direction of magnetization, as the Magnetic Induction within the magnet. Sir William Thomson has called this the Electromagnetic definition of magnetic force.

400.2 The three vectors, the magnetization \mathfrak{J}, the magnetic force \mathfrak{H} and the magnetic induction \mathfrak{B}, are connected by the vector equation

$$\mathfrak{B} = \mathfrak{H} + 4\pi\mathfrak{J}. \qquad (7)$$

Line-Integral of Magnetic Force.

401.] Since the magnetic force, as defined in Art. 398, is that due to the distribution of free magnetism on the surface and through the interior of the magnet, and is not affected by the surface-magnetism of the cavity, it may be derived directly from the general expression for the potential of the magnet, and the line-integral of the magnetic force taken along any curve from the point A to the point B is

$$\int_A^B \left(\alpha \frac{dx}{ds} + \beta \frac{dy}{ds} + \gamma \frac{dz}{ds} \right) ds = V_A - V_B, \qquad (8)$$

where V_A and V_B denote the potentials at A and B respectively.

400. *the actual force on the unit pole*: that is, the sum of the two forces mentioned in (399.2)—the force due to magnetization I (eq. 5) and the force derived from potential V (eq. 4). Equations (6) express the summation for each component.

400.1 *the Electromagnetic definition of magnetic force*: Again compare Thomson, *loc. cit.* (398.1, *comment*). The name may seem inapt since there was no reference whatever to electromagnetism in \mathfrak{B}'s derivation; but Maxwell will point out in Art. 404 that the lines disclosed by Faraday's Moving Wire (which *does* involve electric current) correspond to lines of \mathfrak{B}. We shall see, too, that the quantity \mathfrak{B} arises when electric currents—or even changing electric fields—are present, even in the absence of material magnets. Nevertheless, writers have generally preferred Maxwell's term, Magnetic Induction.

400.2 *the magnetization \mathfrak{J}*: the magnetization I, now considered as a vector.

$\mathfrak{B} = \mathfrak{H} + 4\pi\mathfrak{J}$: Notice that since both \mathfrak{H} and $4\pi\mathfrak{J}$ are vector quantities, and \mathfrak{H} is represented by a system of lines of force (398.1, *comment*), then their sum, the magnetic induction \mathfrak{B}, may similarly be represented by a system of *lines of induction*; Maxwell will introduce that phrase in (404.) below.

Surface-Integral of Magnetic Induction.

402.] The magnetic induction through the surface S is defined as the value of the integral

$$Q = \iint \mathfrak{B} \cos \varepsilon \, dS, \tag{9}$$

where \mathfrak{B} denotes the magnitude of the magnetic induction at the element of surface dS, and ε the angle between the direction of the induction and the normal to the element of surface, and the integration is to be extended over the whole surface, which may be either closed or bounded by a closed curve.

402.1 If a, b, c denote the components of the magnetic induction, and l, m, n the direction-cosines of the normal, the surface-integral may be written

$$Q = \iint (la + mb + nc) \, dS. \tag{10}$$

402.2 If we substitute for the components of the magnetic induction their values in terms of those of the magnetic force, and the magnetization as given in Art. 400, we find

$$Q = \iint (l\alpha + m\beta + n\gamma) \, dS + 4\pi \iint (lA + mB + nC) \, dS. \tag{11}$$

402.3 We shall now suppose that the surface over which the integration extends is a closed one, and we shall investigate the value of the two terms on the right-hand side of this equation.

402.4 Since the mathematical form of the relation between magnetic force and free magnetism is the same as that between electric force and

401. *Equation* (8): Recall that α, β, γ, the components of magnetic force \mathfrak{H}, are equal to $-dV/dx$, $-dV/dy$, $-dV/dz$, respectively (398.). Thus the integrand in the equation is

$$\left(-\frac{dV}{dx}\frac{dx}{ds} - \frac{dV}{dy}\frac{dy}{ds} - \frac{dV}{dz}\frac{dz}{ds} \right) ds;$$

and this, integrated from A to B, yields $(-V_B) - (-V_A)$ or $V_A - V_B$.

402. *The magnetic induction through the surface S*: We must always be careful to distinguish the magnetic induction *through a surface*, which is measured by the integral of \mathfrak{B}, from the magnetic induction *at a point*, which is \mathfrak{B} simply. Maxwell will sometimes write "magnetic induction" without explicitly supplying the prepositional qualifier, trusting the alert reader to gather whether the quantity \mathfrak{B} or its integral is to be understood. Modern readers find this shared terminology awkward, but evidently Maxwell did not. Note that in the language of *electrical* induction (which was already more firmly established by the time of Maxwell's writing) the quantity R does *not* share a name with its own surface-integral (75.1), and no comparable ambiguity arises.

402.2 *Equation* (11): Combine equations (6) and equation (10).

free electricity, we may apply the result given in Art. 77 to the first term in the value of Q by substituting α, β, γ, the components of magnetic force, for X, Y, Z, the components of electric force in Art. 77, and M, the algebraic sum of the free magnetism within the closed surface, for e, the algebraic sum of the free electricity.

402.5 We thus obtain the equation

$$\iint (l\alpha + m\beta + n\gamma)\,dS = 4\pi M. \tag{12}$$

402.6 Since every magnetic particle has two poles, which are equal in numerical magnitude but of opposite signs, the algebraic sum of the magnetism of the particle is zero. Hence, those particles which are entirely within the closed surface S can contribute nothing to the algebraic sum of the magnetism within S. The value of M must therefore depend only on those magnetic particles which are cut by the surface S.

402.7 Consider a small element of the magnet of length s and transverse section k^2, magnetized in the direction of its length, so that the strength of its poles is m. The moment of this small magnet will be ms, and the intensity of its magnetization, being the ratio of the magnetic moment to the volume, will be

$$I = \frac{m}{k^2}. \tag{13}$$

402.8 Let this small magnet be cut by the surface S, so that the direction of magnetization makes an angle ε' with the normal drawn outwards from the surface, then if dS denotes the area of the section,

$$k^2 = dS \cos\varepsilon'. \tag{14}$$

The negative pole $-m$ of this magnet lies within the surface S. Hence, if we denote by dM the part of the free magnetism within S which is contributed by this little magnet,

$$dM = -m = -Ik^2 = -I\cos\varepsilon'\,dS. \tag{15}$$

402.9 To find M, the algebraic sum of the free magnetism within the closed surface S, we must integrate this expression over the closed

402.4 *the result given in Art. 77*: that is, the equation $\iint R\cos\varepsilon\,dS = 4\pi e$ in the first paragraph of that Article. Note that $R\cos\varepsilon\,dS$ may also be written as $(lX + mY + nZ)\,dS$, since X, Y, Z are the components of R and l, m, n are the direction-cosines of the normal to dS (385, *comment on equation* 6). Then, as Maxwell directs, replace the electrical quantities X, Y, Z and e with their magnetic counterparts α, β, γ and M.

402.5 *Equation* (12): Note that this equation pertains solely to magnetic force \mathfrak{H} and corresponds to the first term of equation (11).

402.7 *Equation* (13): This is the same relation we noted earlier in a comment to (384.1).

surface, so that

$$M = -\iint I \cos \varepsilon' \, dS,$$

or writing A, B, C for the components of magnetization, and l, m, n for the direction-cosines of the normal drawn outwards,

$$M = -\iint (lA + mB + nC) \, dS. \tag{16}$$

402.10 This gives us the value of the integral in the second term on the right-hand side of equation (11). The value of Q in that equation may therefore be found from equations (12) and (16),

$$Q = 4\pi M - 4\pi M = 0, \tag{17}$$

or, *the surface-integral of the magnetic induction through any closed surface is zero.*

403.] If we assume as the closed surface that of the differential element of volume $dx\,dy\,dz$, we obtain the equation

$$\frac{da}{dx} + \frac{db}{dy} + \frac{dc}{dz} = 0. \tag{18}$$

403.1 This is the solenoidal condition, which is always satisfied by the components of the magnetic induction.

402.9 *the direction-cosines of the normal*: Here again Maxwell uses the theorem proved in (385., *comment on equation* 6).

402.10 *Equation* (17): Note the reasoning implicit in this equation. Even if there exists an isolated magnetic pole M (first term), the effect of magnetization ℑ (second term) undoes its contribution to the magnetic induction 𝕭, so that 𝕭 = 0 overall. Thus 𝕭 has nothing to do with magnetic poles or other points of termination (404.6). Perhaps its independence of hypothetical poles is one reason why in (404.7) below, Maxwell will name 𝕭 rather than 𝕳 as the physically significant magnetic quantity.

403. *Equation* (18): Maxwell obtains this equation for magnetism just as he previously derived $\dfrac{dX}{dx} + \dfrac{dY}{dy} + \dfrac{dZ}{dz} = 4\pi\rho$ for electricity in (77.2). But since he here begins with equation (17) in which Q equals zero, the magnetic case will correspond to $\rho = 0$.

403.1 *the solenoidal condition...*: The solenoidal condition denotes the nonexistence of sources or termini of a quantity. Since the right-hand sides of equations (17) and (18) equal zero they state respectively that no spatial volume, finite or infinitesimal, can act as a terminus for the magnetic induction 𝕭; see also (404.6, *comment*). By contrast, equation (12) shows that in any region containing a magnetic pole $M \neq 0$ the magnetic force 𝕳 will *not* obey the solenoidal condition.

... is always satisfied by the components of the magnetic induction: It is "always satisfied" by 𝕭 because we derived equation (18) from the integral of magnetic induction 𝕭 in equation (9) making no special assumptions. In contrast, the solenoidal condition is not always satisfied by electric intensity X, Y, Z since ρ need not be zero in the equation cited in (403., *comment*), above.

403.2 Since the distribution of magnetic induction is solenoidal, the induction through any surface bounded by a closed curve depends only on the form and position of the closed curve, and not on that of the surface itself.

404.] Surfaces at every point of which

$$la + mb + nc = 0 \tag{19}$$

are called surfaces of no induction, and the intersection of two such surfaces is called a line of induction. The conditions that a curve, s, may be a line of induction are

$$\frac{1}{a}\frac{dx}{ds} = \frac{1}{b}\frac{dy}{ds} = \frac{1}{c}\frac{dz}{ds}. \tag{20}$$

404.1 A system of lines of induction drawn through every point of a closed curve forms a tubular surface called a Tube of induction.

404.2 The induction across any section of such a tube is the same. If the induction is unity the tube is called a Unit tube of induction.

404.3 All that Faraday* says about lines of magnetic force and magnetic sphondyloids is mathematically true, if understood of the lines and tubes of magnetic induction.

404.4 The magnetic force and the magnetic induction are identical outside the magnet, but within the substance of the magnet they must be carefully distinguished.

* *Exp. Res.*, series xxviii.

403.2 *the induction through any surface bounded by a closed curve depends only on the form and position of the closed curve, and not on that of the surface itself*: Maxwell continues to rehearse results obtained earlier for electricity and which can be reasserted for magnetism. He drew this conclusion for electric induction in (76.10).

404. *Surfaces at every point of which* $la + mb + nc = 0$ *are called surfaces of no induction*: Since a, b, c are the components of the magnetic induction \mathfrak{B} and l, m, n are the direction-cosines of the normal to to the surface, the sum $la + mb + nc$ represents the resolved part of \mathfrak{B} with respect to the normal (385, *comment on equation* 6). Since it totals zero, there is no induction at any point for which the equation holds. If it holds at every point of a surface, that surface is a *surface of no induction*.

The conditions that a curve, s, may be a line of induction: Any differential portion ds of the curve must be parallel to \mathfrak{B}; hence dx, dy, dz (the components of ds) must be in the same ratio as a, b, c (the components of \mathfrak{B}) respectively. Write the proportions $dx/dy = a/b$ and $dy/dz = b/c$; then divide dx, dy, and dz by ds to obtain equations (20).

404.1 *Tube of induction*: that is, of magnetic induction. Maxwell defined the tube of electrical induction in (84.3).

404.3 *if understood of the lines and tubes of magnetic induction*: Still pursuing the translation of Faraday's prose into mathematical symbolism, Maxwell corrects (but very lovingly) Faraday's terminology. Faraday did not distinguish between magnetic force (\mathfrak{H}) and magnetic induction (\mathfrak{B}); but doing so mathematically

404.5 In a straight uniformly magnetized bar the magnetic force due to the magnet itself is from the end which points north, which we call the positive pole, towards the south end or negative pole, both within the magnet and in the space without.

404.6 The magnetic induction, on the other hand, is from the positive pole to the negative outside the magnet, and from the negative pole to the positive within the magnet, so that the lines and tubes of induction are re-entering or cyclic figures.

404.7 The importance of the magnetic induction as a physical quantity will be more clearly seen when we study electromagnetic phenomena. When the magnetic field is explored by a moving wire, as in Faraday's *Exp. Res.* 3076, it is the magnetic induction and not the magnetic force which is directly measured.

The Vector-Potential of Magnetic Induction.

405.] Since, as we have shewn in Art. 403, the magnetic induction through a surface bounded by a closed curve depends on the closed

has revealed (403.1) that the continuity and physical significance which Faraday ascribed to *lines of force* are more correctly asserted of *lines of induction*. Maxwell will state explicitly that the lines of induction are continuous in (404.6).

404.5 *the magnetic force due to the magnet itself is from … the positive pole, towards the … negative pole, both within the magnet and the space without*: The sketch shows lines of magnetic force \mathfrak{H}. Their direction is everywhere from the positive (north) pole to the negative (south) pole and they must therefore *terminate* at those polar regions, each of which is conceived as possessing a quantity of free magnetism M in accordance with equation (12).

404.6 *the lines and tubes of induction are re-entering or cyclic figures*: The quantity Q is the integral, over a closed surface, of magnetic induction (402.1). And Q was found to be zero for every closed surface (402.10). Therefore the lines and tubes of induction *cannot terminate*; for if they did, a closed surface drawn about the region of their terminus would sustain induction Q not equal to zero. For the same reason, the lines of induction \mathfrak{B} must be continuous— *even when they pass from one material to another*. The continuity of \mathfrak{B} (in contrast to \mathfrak{H}) is perhaps the most compelling reason to regard \mathfrak{B}, not \mathfrak{H}, as the physically significant quantity.

404.7 *it is the magnetic induction and not the magnetic force which is directly measured*: Faraday characterized the Moving Wire as summing up the lines of force which it cut in its motion, but it actually measures the surface-integral of \mathfrak{B}. Maxwell will offer an example in Art. 532. Notice that Maxwell's "translation" of Faraday does not move exclusively from Faraday's prose to mathematical terminology; sometimes, as here, mathematical considerations prompt an emendation of Faraday's terms.

curve, and not on the form of the surface which is bounded by it, it must be possible to determine the induction through a closed curve by a process depending only on the nature of that curve, and not involving the construction of a surface forming a diaphragm of the curve.

405.1 This may be done by finding a vector \mathfrak{A} related to \mathfrak{B}, the magnetic induction, in such a way that the line-integral of \mathfrak{A}, extended round the closed curve, is equal to the surface-integral of \mathfrak{B}, extended over a surface bounded by the closed curve.

405.2 If, in Art. 24, we write F, G, H for the components of \mathfrak{A}, and a, b, c for the components of \mathfrak{B}, we find for the relation between these components

$$a = \frac{dH}{dy} - \frac{dG}{dz}, \quad b = \frac{dF}{dz} - \frac{dH}{dx}, \quad c = \frac{dG}{dx} - \frac{dF}{dy}. \tag{21}$$

405.3 The vector \mathfrak{A}, whose components are F, G, H, is called the vector-potential of magnetic induction.

405.4 If a magnetic molecule whose moment is m and the direction of whose axis of magnetization is (λ, μ, ν) be at the origin of coordinates, the potential at a point (x, y, z) distance r from the origin is, by Art. 387,

405. *it must be possible to determine the induction through a closed curve by a process ... not involving the construction of a surface*: Since induction involves a surface-integral *by definition*, this striking insight of Maxwell's might, upon first consideration, appear highly improbable! Only the reason Maxwell cites—that the induction depends upon the boundary of the surface, and not upon the surface itself—could have suggested it.

405.1 *This may be done by finding a vector \mathfrak{A} ... [such] that the line-integral of \mathfrak{A}, extended round the closed curve, is equal to the surface-integral of \mathfrak{B}*: Once stated, this seems obvious: if the induction through a surface indeed depends only on the boundary of the surface (405.), there must be some integral along the boundary that will equal the induction through the surface. But the clarity with which Maxwell seizes upon this insight marks an especially beautiful moment in the argument.

405.2 *Equation* (21): Maxwell is employing Theorem IV, Stokes' Theorem, in Art. 24. Compare equation (1) in (24.2), which states the conditions under which the surface-integral of a vector having components ξ, η, ζ may be expressed as the line-integral of a vector whose components are X, Y, Z. Substitute F, G, H for X, Y, Z and a, b, c for ξ, η, ζ to obtain the equation.

Equation (21) is the definition of vector \mathfrak{A}. Textbook presentations often try to characterize the definition as essentially strategic, motivated mainly by the convenient results it effectuates. I hope it is evident that, on the contrary, the definition arises naturally out of the insight expressed in (405.1). Moreover, the vector \mathfrak{A} will prove to possess an unexpected *physical* significance, as we shall see in Part IV, Chapter IX, below.

405.4 *a magnetic molecule whose moment is m*: Confusingly, Maxwell has decided to change his notation for magnetic moment. Previously, m represented pole

$$-m\left(\lambda\frac{d}{dx}+\mu\frac{d}{dy}+\nu\frac{d}{dz}\right)\frac{1}{r};$$ [21a]

$$\therefore c = m\left(\lambda\frac{d^2}{dx\,dz}+\mu\frac{d^2}{dy\,dz}+\nu\frac{d^2}{dz^2}\right)\frac{1}{r},$$ [21b]

strength, so that the magnetic moment of a particle with length ds was $m\,ds$ (384.). On that basis he derived the potential produced by a magnetic particle as

$$V=\frac{m\,ds}{r^2}\cos\varepsilon,$$

where r is the distance from the particle and ε the angle with the particle's axis of magnetization (383.1). With Maxwell's present notation that potential is expressed

$$V=\frac{m}{r^2}\cos\varepsilon. \qquad (i)$$

Equation [21a]: The equation numbers in square brackets that appear in this paragraph and the next are not Maxwell's; I have introduced them to facilitate commentary. Although Maxwell traces expression [21a] to omitted Art. 387, it is also equivalent to equation (i), which was derived, albeit with different notation, in (383.1). We can show this by proving that

$$-m\left(\lambda\frac{d}{dx}+\mu\frac{d}{dy}+\nu\frac{d}{dz}\right)\frac{1}{r}=\frac{m}{r^2}\cos\varepsilon,$$

or, equivalently, that

$$\lambda\frac{d}{dx}\left(\frac{1}{r}\right)+\mu\frac{d}{dy}\left(\frac{1}{r}\right)+\nu\frac{d}{dz}\left(\frac{1}{r}\right)=-\frac{1}{r^2}\cos\varepsilon. \qquad (ii)$$

Now $\dfrac{d}{dx}\left(\dfrac{1}{r}\right)$, $\dfrac{d}{dy}\left(\dfrac{1}{r}\right)$, $\dfrac{d}{dz}\left(\dfrac{1}{r}\right)$ are the rectangular components of $\dfrac{d}{dr}\left(\dfrac{1}{r}\right)$; while λ,μ,ν are the direction-cosines of ds, the particle's axis of magnetization. Then by the theorem in (385., *comment*) the sum of products on the left-hand side of eq. (ii) represents the resolved part of $\dfrac{d}{dr}\left(\dfrac{1}{r}\right)$ in the direction of ds. But $\dfrac{d}{dr}\left(\dfrac{1}{r}\right)=-\dfrac{1}{r^2}$, and to obtain the resolved part of that quantity in the direction of ds we multiply by the cosine of the angle between dr and ds, that is, by $\cos\varepsilon$. Thus expression [21a] for the potential of a magnetic particle is equivalent to equation (i), already derived.

Equation [21b]: This is the derivative, with respect to z, of equation [21a] for potential V; thus it basically states that $c=-dV/dz$. Maxwell can make this assertion because, first, c is the z-component of the magnetic induction \mathfrak{B}; and second, since the point (x,y,z) is taken in empty space, $\mathfrak{B}=\mathfrak{H}$ (404.4). Therefore c will also be numerically equal to the z-component of \mathfrak{H}, which in turn is $-dV/dz$ (398., equation 4); thus $c=-dV/dz$. The symbol \therefore means "therefore."

which, by Laplace's equation, may be thrown into the form

$$c = m\frac{d}{dx}\left(\lambda\frac{d}{dz} - \nu\frac{d}{dx}\right)\frac{1}{r} - m\frac{d}{dy}\left(\nu\frac{d}{dy} - \mu\frac{d}{dz}\right)\frac{1}{r}. \qquad [21c]$$

405.5 The quantities a, b may be dealt with in a similar manner.

405.4, continued. *Laplace's equation*: In Art. 77 Maxwell obtained Poisson's equation,

$$\frac{d^2V}{dx^2} + \frac{d^2V}{dy^2} + \frac{d^2V}{dz^2} = -4\pi\rho,$$

where V represents electrostatic potential and ρ the volume-density of electric charge. For points at which there is *no* electric charge we have $\rho = 0$, and in this special case the equation is called Laplace's equation. Applying the argument of Art. 77 to the magnetic case, for points at which there is no magnetic pole, we can write Laplace's equation as

$$\frac{d^2V}{dx^2} + \frac{d^2V}{dy^2} + \frac{d^2V}{dz^2} = 0$$

where, in this case, V represents the magnetic potential. Now we know (383.) that the potential produced by a magnetic pole is proportional to $1/r$. Thus at any point where there is no pole we may write

$$\frac{d^2}{dx^2}\left(\frac{1}{r}\right) + \frac{d^2}{dy^2}\left(\frac{1}{r}\right) + \frac{d^2}{dz^2}\left(\frac{1}{r}\right) = 0.$$

which ... may be thrown into the form...: By comparing equations [21b] and [21c] term-by-term, you can verify that the equations are indeed identical, provided that

$$\frac{d^2}{dx^2}\left(\frac{1}{r}\right) + \frac{d^2}{dy^2}\left(\frac{1}{r}\right) + \frac{d^2}{dz^2}\left(\frac{1}{r}\right) = 0. \qquad (iii)$$

And this relation, an instance of Laplace's equation, is true at every point where there is no magnetic pole, as we just acknowledged. Hence equation [21b] may be transformed to equation [21c], as Maxwell states. Note: in the original text, Maxwell did not write out the left-hand member of equation [21c]; I have supplied it for clarity.

405.5 *quantities a, b may be dealt with in a similar manner*: Doing so, we obtain

$$a = m\frac{d}{dy}\left(\mu\frac{d}{dx} - \lambda\frac{d}{dy}\right)\frac{1}{r} - m\frac{d}{dz}\left(\lambda\frac{d}{dz} - \nu\frac{d}{dx}\right)\frac{1}{r}$$

and $$b = m\frac{d}{dz}\left(\nu\frac{d}{dy} - \mu\frac{d}{dz}\right)\frac{1}{r} - m\frac{d}{dx}\left(\mu\frac{d}{dx} - \nu\frac{d}{dy}\right)\frac{1}{r},$$

and we note that the expressions for a, b, and c have the forms

$$a = \frac{d[\]}{dy} - \frac{d[\]}{dz}, \quad b = \frac{d[\]}{dz} - \frac{d[\]}{dx}, \quad c = \frac{d[\]}{dx} - \frac{d[\]}{dy}.$$

Hence

$$F = m\left(\nu\frac{d}{dy} - \mu\frac{d}{dz}\right)\frac{1}{r},$$

$$= \frac{m(\mu z - \nu y)}{r^3}.$$

[21d]

405.6 From this expression G and H may be found by symmetry. We thus see that the vector-potential at a given point, due to a magnetized particle placed at the origin, is numerically equal to the magnetic moment of the particle divided by the square of the radius vector and multiplied by the sine of the angle between the axis of magnetization and the radius vector, and the direction of the vector-potential is perpendicular to the plane of the axis of magnetization and the radius vector,

Equations [21d]: Comparing the forms just noted with equations (21), it is clear that F must be the quantity that in [21b] is differentiated with respect to z and in [21c] is differentiated with respect to y, that is,

$$F = m\left(\nu\frac{d}{dy} - \mu\frac{d}{dz}\right)\frac{1}{r},$$

as Maxwell states. In the same way we obtain these expressions for G and H, as Maxwell will go on to state in the next paragraph:

$$G = m\left(\lambda\frac{d}{dz} - \nu\frac{d}{dx}\right)\frac{1}{r}, \quad H = m\left(\mu\frac{d}{dx} - \lambda\frac{d}{dy}\right)\frac{1}{r}.$$

Procure the second of equations [21d] by noting that

$$\frac{d}{dy}\left(\frac{1}{r}\right) = \frac{d}{dy}(x^2 + y^2 + z^2)^{-\frac{1}{2}} = -\frac{1}{2}(x^2 + y^2 + z^2)^{-\frac{3}{2}} \cdot 2y = -y(r^2)^{-\frac{3}{2}} = -\frac{y}{r^3}$$

and similarly that

$$\frac{d}{dz}\left(\frac{1}{r}\right) = -\frac{z}{r^3}.$$

Substitution yields the desired result, and we can obtain analogous expressions for G and H in a corresponding way.

405.6 *the magnetic moment of the particle divided by the square of the radius vector and multiplied by the sine of the angle...*: For example, consider a point in the yz-plane, and suppose the particle is magnetized exclusively in the z-direction; then λ and μ will each equal zero and the magnetic moment of the particle will be $m\nu$ (405.4 and *comment*). Then the second of equations [21d] will reduce to $F = -\frac{m\nu}{r^2} \cdot \frac{y}{r}$. This agrees with Maxwell's statement, since the quotient y/r is the sine of the angle between the direction of magnetization and the radius (angle θ in the sketch). Note that F, the x-component of the vector-potential, is in the negative x-direction, and thus perpendicular to the yz-plane.

and is such that to an eye looking in the positive direction along the axis of magnetization the vector-potential is drawn in the direction of rotation of the hands of a watch.

405.7 Hence, for a magnet of any form in which A, B, C are the components of magnetization at the point (x, y, z), the components of the vector-potential at the point (ξ, η, ζ) are

$$F = \iiint \left(B\frac{dp}{dz} - C\frac{dp}{dy} \right) dx\,dy\,dz,$$

$$G = \iiint \left(C\frac{dp}{dx} - A\frac{dp}{dz} \right) dx\,dy\,dz, \qquad (22)$$

$$H = \iiint \left(A\frac{dp}{dy} - B\frac{dp}{dx} \right) dx\,dy\,dz;$$

where p is put, for conciseness, for the reciprocal of the distance between the points (ξ, η, ζ) and (x, y, z), and the integrations are extended over the space occupied by the magnet.

405.6, continued. *the vector-potential is drawn in the direction of rotation of the hands of a watch*: The sketch represents F at a point in the yz-plane. You can see that, sighting from the origin in the positive z-direction, F does indeed point clockwise about the z-axis, while the expressions obtained for G and H reveal that both are zero for the point in question. Similarly we can confirm that for a point in the xz-plane, G will point in the clockwise direction while F and H are zero.

405.7 *A, B, C are the components of magnetization at the point (x, y, z)*: Thus what Maxwell has been writing as $m\lambda$ he now writes as A. Similarly, $m\mu = B$ and $m\nu = C$.

the components of the vector-potential at the point (ξ, η, ζ) are: With the substitutions just noted, the first of equations [21d] will read

$$F = C\frac{d}{dy}\left(\frac{1}{r}\right) - B\frac{d}{dz}\left(\frac{1}{r}\right);$$

but this expresses the component F due to an infinitesimal magnetic particle (405.4). Therefore F itself is an infinitesimal quantity. (For that reason, modern notation would demand that it be designated dF rather than F.) To calculate the value of F arising from a finite magnet at (x, y, z) we integrate the above expression over the whole volume occupied by the magnet, that is,

$$F = \iiint dF = \iiint \left[C\frac{d}{dy}\left(\frac{1}{r}\right) - B\frac{d}{dz}\left(\frac{1}{r}\right) \right] dx\,dy\,dz = \iiint \left[C\frac{dp}{dy} - B\frac{dp}{dz} \right] dx\,dy\,dz.$$

Note that Maxwell substitutes p in place of $1/r$, "for conciseness."

Maxwell's expression in equations (22), however, is the *negative* of this; that is because up to now he has been considering a magnetic particle located at the

406.] The scalar, or ordinary, potential of magnetic force, Art. 385, becomes when expressed in the same notation,

$$V = \iiint \left(A\frac{dp}{dx} + B\frac{dp}{dy} + C\frac{dp}{dx} \right) dx\,dy\,dz. \tag{23}$$

406.1 Remembering that $\dfrac{dp}{dx} = -\dfrac{dp}{d\xi}$, and that the integral

$$\iiint A\left(\frac{d^2p}{dx^2} + \frac{d^2p}{dy^2} + \frac{d^2p}{dz^2} \right) dx\,dy\,dz$$

has the value $-4\pi(A)$ when the point (ξ, η, ζ) is included within the limits of integration, and is zero when it is not so included, (A) being the value of A at the point (ξ, η, ζ), we find for the value of the x-component of the magnetic induction,

origin of coordinates $(0,0,0)$, while calculating the vector-potential at point (x, y, z). But from here on—rather inconveniently for the reader—he places the *magnet* at points (x, y, z) and calculates the vector-potential at some point (ξ, η, ζ), which can always be taken as the origin, though it need not be. Since he has effectively reversed his point of view, the signs of F, G, and H also reverse.

406. *The scalar, or ordinary, potential of magnetic force, Art. 385*: Maxwell refers to equation (8) in (385.1)*, $V = \iiint \left[A(\xi-x) + B(\eta-y) + C(\zeta-z) \right] \dfrac{1}{r^3} dx\,dy\,dz.$

Let (ξ, η, ζ) be the origin; then the equation becomes

$$V = \iiint (-Ax - By - Cz)\frac{1}{r^3} dx\,dy\,dz,$$

the first term of whose integrand is $-Ax/r^3$. But it is identical to the first term of the integrand of equation (23), namely

$$A\frac{dp}{dx} = A\frac{d}{dx}\left(\frac{1}{r}\right) = -A\frac{1}{r^2}\frac{dr}{dx} = -A\frac{1}{r^2}\frac{x}{r} = -A\frac{x}{r^3},$$

and similarly for the remaining terms. Thus eq. (8) in (385.1), representing the scalar potential, can also be expressed as eq. (23) of the present paragraph.

406.1 *Remembering that $dp/dx = -dp/d\xi$*: Let r be drawn from the point whose coordinate is ξ to the point whose coordinate is x. If then ξ increases by $d\xi$, r will decrease by an infinitesimal amount dr. But if then x increases by dx equal to $d\xi$, it is clear that r will be restored to its original length. Thus equal increases in $d\xi$ and dx produce mutu-ally cancelling changes in r, and therefore also in $1/r$, that is, in p. Hence $dp/dx + dp/d\xi = 0$, or $dp/dx = -dp/d\xi$.

the integral ... has the value $4\pi(A)$...: When the limits of integration do not include any part of the magnet, the integrand is A times Laplace's equation (405.4, *comment*) and equals zero. Otherwise it has the nonzero value styled $4\pi(A)$ by Maxwell. Today we would more typically use a subscript notation, as $4\pi A_0$.

$$a = \frac{dH}{d\eta} - \frac{dG}{d\zeta}$$

$$= \iiint \left\{ A\left(\frac{d^2p}{dy\,d\eta} + \frac{d^2p}{dz\,d\zeta}\right) - B\frac{d^2p}{dx\,d\eta} - C\frac{d^2p}{dx\,d\zeta} \right\} dx\,dy\,dz$$

$$= -\frac{d}{d\xi} \iiint \left\{ A\frac{dp}{dx} + B\frac{dp}{dy} + C\frac{dp}{dz} \right\} dx\,dy\,dz$$

$$- \iiint A\left(\frac{d^2p}{dx^2} + \frac{d^2p}{dy^2} + \frac{d^2p}{dz^2}\right) dx\,dy\,dz. \tag{24}$$

406.2 The first term of this expression is evidently $-dV/d\xi$, or, α the component of the magnetic force.

406.3 The quantity under the integral sign in the second term is zero for every element of volume except that in which the point (ξ, η, ζ) is included. If the value of A at the point (ξ, η, ζ) is (A), the value of the second term is easily proved to be $4\pi(A)$, where (A) is evidently zero at all points outside the magnet.

406.4 We may now write the value of the x-component of the magnetic induction

$$a = \alpha + 4\pi(A), \tag{25}$$

an equation which is identical with the first of those given in Art. 400. The equations for b and c will also agree with those of Art. 400.

406.5 We have already seen that the magnetic force \mathfrak{H} is derived from the scalar magnetic potential V by the application of Hamilton's operator ∇ so that we may write, as in Art. 17,

$$\mathfrak{H} = -\nabla V, \tag{26}$$

406.1, continued. *Equation* (24): The expression for a—the x-component of magnetic induction—is that of equations (21), but with η and ζ replacing y and z, as noted in the comment to (405.7).

406.2 *The first term ... is evidently* $-dV/d\xi$: For it is the negative derivative of equation (23), which expresses the potential V (406.). As such, it expresses "the component of the magnetic force," as Maxwell will remind us again in (406.5).

406.3 *The quantity under the integral sign in the second term is zero for every element of volume except...*: Maxwell is repeating the observation he made in (406.1).

If the value of A at the point (ξ, η, ζ) is (A): Maxwell introduced this notation in (406.1) to designate a particular value of the variable A. Today we would probably write "Let the value of A at the point (ξ, η, ζ) be A_0."

406.5 *We have already seen...*: In (395.1), and again in (398.). Compare equations (1–4) in Art. 17.

$\mathfrak{H} = -\nabla V$: This quaternion equation is expressed no differently in modern

and that this equation is true both without and within the magnet.

406.6 It appears from the present investigation that the magnetic induction \mathfrak{B} is derived from the vector-potential \mathfrak{A} by the application of the same operator, and that the result is true within the magnet as well as without it.

406.7 The application of this operator to a vector-function produces, in general, a scalar quantity as well as a vector. The scalar part, however, which we have called the convergence of the vector-function, vanishes when the vector-function satisfies the solenoidal condition

$$\frac{dF}{d\xi} + \frac{dG}{d\eta} + \frac{dH}{d\zeta} = 0. \tag{27}$$

By differentiating the expressions for F, G, H in equations (22), we find that this equation is satisfied by these quantities.

vector notation. Its individual components, $\alpha = -dV/dx$, $\beta = -dV/dy$, and $\gamma = -dV/dz$, appeared in (395.1) as equations (1) and in (398.) as equations (4).

this equation is true both without and within the magnet: Recall that \mathfrak{H} "within" a magnet was found to obey the same equations as for outside the magnet. See the comment on equations (4) in paragraph 398.

406.6 *the magnetic induction \mathfrak{B} is derived from the vector potential \mathfrak{A} by the application of the same operator*: that is, operator ∇. But while application of ∇ to a scalar produces a vector, as in equation (26), application of ∇ to vector \mathfrak{A} is more complicated, as Maxwell will note in the next paragraph.

406.7 *The application of this operator to a vector-function produces … a scalar quantity as well as a vector*: In my earlier note on quaternions (25., *comment*), I explained why it is that, in quaternion algebra, the product of two vectors has both a scalar and a vector part, which for the modern reader's convenience I expressed in terms of vector algebra. Thus the scalar part of the quaternion product **ab** may be written as $-\mathbf{a} \cdot \mathbf{b}$, and the vector part as $\mathbf{a} \times \mathbf{b}$. When the product is $\nabla \mathfrak{A}$, as here, the scalar part will be $-\nabla \cdot \mathfrak{A}$ and the vector part $\nabla \times \mathfrak{A}$.

the scalar part … vanishes: Writing out $-\nabla \cdot \mathfrak{A} = -\left(\dfrac{dF}{d\xi} + \dfrac{dG}{d\eta} + \dfrac{dH}{d\zeta} \right)$, where ξ, η, ζ are the spatial coordinates, we recognize the right-hand side as the expression which determines whether the vector having components F, G, H satisfies the solenoidal condition. If the vector \mathfrak{A} is solenoidal, the expression will equal zero, and thus the scalar part of $\nabla \mathfrak{A}$ will vanish.

Differentiating the expressions for F, G, H in equations (22): For example, the first of equations (22) will become $\dfrac{dF}{dx} = \iint B \, dp \, dy - \iint C \, dp \, dz$. When dG/dy and dH/dz are added to it, the terms of all three will be found to cancel pairwise. Note again Maxwell's unannounced shift of notation from x, y, z to ξ, η, ζ.

406.8 We may therefore write the relation between the magnetic induction and its vector-potential

$$\mathfrak{B} = \nabla\mathfrak{A},$$

which may be expressed in words by saying that the magnetic induction is the curl of its vector-potential. See Art. 25.

406.8 *We may therefore write* $\mathfrak{B} = \nabla\mathfrak{A}$: Once having shown that the scalar part of $\nabla\mathfrak{A}$ is zero (406.7 and *comment*), the expression reduces (in modern notation) to $\mathfrak{B} = \nabla\times\mathfrak{A}$. Now Maxwell has defined the *x*-, *y*- and *z*-components of \mathfrak{B} in equations (21); thus

$$\mathfrak{B} = a\mathbf{i} + b\mathbf{j} + c\mathbf{k} = \left(\frac{dH}{dy} - \frac{dG}{dz}\right)\mathbf{i} + \left(\frac{dF}{dz} - \frac{dH}{dx}\right)\mathbf{j} + \left(\frac{dG}{dx} - \frac{dF}{dy}\right)\mathbf{k},$$

while

$$\nabla\times\mathfrak{A} = \left(\frac{d}{dx}\mathbf{i} + \frac{d}{dy}\mathbf{j} + \frac{d}{dz}\mathbf{k}\right)\times\left(F\mathbf{i} + G\mathbf{j} + H\mathbf{k}\right).$$

Multiply out the second expression term-by-term, and recall the rules for vector multiplication of **i**, **j**, **k** (25.3, *comment*). Since products involving $\mathbf{i}\times\mathbf{i}$, $\mathbf{j}\times\mathbf{j}$, $\mathbf{k}\times\mathbf{k}$ vanish, we shall be left with six terms which, as you can readily verify, are respectively equal to the six terms in the expression for \mathfrak{B}; thus $\mathfrak{B} = \nabla\mathfrak{A}$ or, in vector notation, $\mathfrak{B} = \nabla\times\mathfrak{A}$.

the curl of its vector-potential: Earlier (25.11) Maxwell had proposed, "with great diffidence," the name "rotation" for the result of applying operator ∇ to a vector quantity. His present term "curl," however, has become standard. In modern vector algebra the expression $\nabla\times\mathbf{E}$ is regularly read "curl of **E**."

In present-day mechanics, gravitational theory, and electrostatics, "potential" is exclusively a *scalar*; thus the idea of a *vector* potential may appear bizarre to modern readers. Notice, though, that in quaternion notation the application of operator ∇ to scalar potential *V* and to vector potential \mathfrak{A}—that is, ∇V and $\nabla\mathfrak{A}$—have the same form, and so suggest a parallel between the scalar and the vector potentials. Not so in modern vector algebra, in which the operations ∇V and $\nabla\times\mathfrak{A}$ are so dissimilar in notation as to obscure any rationale for calling both *V* and \mathfrak{A} "potentials."

CHAPTER III.

MAGNETIC SOLENOIDS AND SHELLS.*

On Particular Forms of Magnets.

407.] IF a long narrow filament of magnetic matter like a wire is magnetized everywhere in a longitudinal direction, then the product of any transverse section of the filament into the mean intensity of the magnetization across it is called the strength of the magnet at that section. If the filament were cut in two at the section without altering the magnetization, the two surfaces, when separated, would be found to have equal and opposite quantities of superficial magnetization, each of which is numerically equal to the strength of the magnet at the section. A filament of magnetic matter, so magnetized that its strength is the same at every section, at whatever part of its length the section be made, is called a Magnetic Solenoid. If m is the strength of the solenoid, ds an element of its length, s being measured from the negative to the positive pole of the magnet, r the distance of that element from a given point, and ε the angle which r makes with the axis of magnetization of the element, the potential at the given point due to the element is

$$\frac{m\,ds\cos\varepsilon}{r^2} = -\frac{m}{r^2}\frac{dr}{ds}\,ds.$$

* See Sir W. Thomson's 'Mathematical Theory of Magnetism,' *Phil. Trans.*, June 1849 and June 1850, or *Reprint* of Papers on Electrostatics and Magnetism, p. 340.

407. *a long narrow filament of magnetic matter*: Better, *magnetizable* matter, like iron or steel. Maxwell does not here mean to invoke the hypothetical magnetic matter he introduced in Art. 380.

the strength of the magnet at that section: We noted earlier (384.1, *comment*) that intensity of magnetization I has dimensions of pole-strength per unit area; thus the strength, being the product of I with the area of a transverse section, has the dimensions of pole strength.

Magnetic Solenoid: Since it is defined as having uniform magnetization I, the strength of a magnetic solenoid is equal to the strength at any of its cross-sections.

the potential at the given point due to the element: The left hand side repeats equation (3) in paragraph (383.1) and is derived in the same way. But in the present example, r is being measured from the element to the given point— opposite to the earlier case. Here, therefore, $dr/ds = -\cos\varepsilon$.

245

407.1 Integrating this expression with respect to s, so as to take into account all the elements of the solenoid, the potential is found to be

$$V = m\left(\frac{1}{r_1} - \frac{1}{r_2}\right),$$

r_1 being the distance of the positive end of the solenoid, and r_2 that of the negative end from the point where V is measured.

407.2 Hence the potential due to a solenoid, and consequently all its magnetic effects, depend only on its strength and the position of its ends, and not at all on its form, whether straight or curved, between these points.

407.3 Hence the ends of a solenoid may be called in a strict sense its poles.

407.4 If a solenoid forms a closed curve the potential due to it is zero at every point, so that such a solenoid can exert no magnetic action, nor can its magnetization be discovered without breaking it at some point and separating the ends.

407.5 If a magnet can be divided into solenoids, all of which either form closed curves or have their extremities in the outer surface of the magnet, the magnetization is said to be solenoidal, and, since the action of the magnet depends entirely upon that of the ends of the solenoids, the distribution of imaginary magnetic matter will be entirely superficial.

407.1 The third edition of the *Treatise* has, erroneously, r^2 instead of r_2 in the equation for V; I have corrected it here.

407.3 *Hence the ends of a solenoid may be called in a strict sense its poles*: The sense is a "strict" one because the ends act as though they were the only magnetically active regions; or, if we adopt the language of "magnetic matter," they act as though they are the only places where magnetic matter resides.

407.4 *If a solenoid forms a closed curve*: Imagine, for example, an iron ring or "doughnut"—mathematicians tend to favor the more formal term *torus*—magnetized throughout in the direction of its circumference.

nor can its magnetization be discovered without breaking it at some point: This interesting instance of a magnet whose action is wholly internal (so long as it remains intact) is, like another example to be offered later, a device to help us develop what we may call *magnetic literacy*. Like Faraday's sphondyloid, it can serve as a powerful interpretive image. For example, we can understand a horseshoe magnet as a bent sphondyloid; but we can also usefully think of it as a broken torus.

407.5 *the distribution of imaginary magnetic matter will be entirely superficial*: that is, entirely on the end surfaces.

407.6 Hence the condition of the magnetization being solenoidal is

$$\frac{dA}{dx} + \frac{dB}{dy} + \frac{dC}{dz} = 0,$$

where A, B, C are the components of the magnetization at any point of the magnet.

* * *

409.] If a thin shell of magnetic matter is magnetized in a direction everywhere normal to its surface, the intensity of the magnetization at any place multiplied by the thickness of the shell at that place is called the Strength of the magnetic shell at that place.

409.1 If the strength of a shell is everywhere the same, it is called a Simple magnetic shell; if it varies from point to point it may be conceived to be made up of a number of simple shells superposed and overlapping each other. It is therefore called a Complex magnetic shell.

409.2 Let dS be an element of the surface of the shell at Q, and Φ the strength of the shell, then the potential at any point, P, due to the element of the shell, is

$$dV = \Phi \frac{1}{r^2} dS \cos \varepsilon,$$

where ε is the angle between the vector QP, or r, and the normal drawn outwards from the positive side of the shell.

407.6 *the condition of the magnetization being solenoidal is…*: Since A, B, C are the components of magnetization **I**, the equation states that $\nabla \cdot \mathbf{I} = 0$; or, in other terms, that the divergence of **I** is zero (25.9, *comment*).

409. *the Strength of the magnetic shell at that place*: Notice that Maxwell defines the strength of a shell differently from that of a magnet or magnetic element. While the strength of a magnet or element was defined as I times the transverse *area* (407.), he here defines the strength of a shell as I times the *thickness* of the shell. A motivation for this definition will appear in (409.2).

409.2 *the potential at any point, P, due to the element of the shell*: In (407.) we obtained the potential due to a magnetized element having pole strength m as $\frac{m ds \cos \varepsilon}{r^2}$. For a shell composed of many such elements, the pole strength m of a single element must be replaced by the product of magnetic surface-density and shell area dS, that is $I dS$ (since magnetization I was shown to equal the surface-density). Similarly the shell's thickness t takes the place of element length ds; we obtain $\frac{I dS t \cos \varepsilon}{r^2}$. Finally, in accordance with Maxwell's definition, replace the product It with shell strength Φ.

ε *is the angle between the vector QP … and the normal*: In the drawing, QN is the normal, thus $\varepsilon = \angle NQP$.

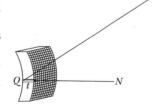

409.3 But if $d\omega$ is the solid angle subtended by dS at the point P

$$r^2 d\omega = dS \cos\varepsilon,$$

whence $$dV = \Phi\,d\omega,$$

and therefore in the case of a simple magnetic shell

$$V = \Phi\omega,$$

or, *the potential due to a magnetic shell at any point is the product of its strength into the solid angle subtended by its edge at the given point.**

410.] The same result may be obtained in a different way by supposing the magnetic shell placed in any field of magnetic force, and determining the potential energy due to the position of the shell.

*This theorem is due to Gauss, *General Theory of Terrestrial Magnetism*, §38.

409.3 *if $d\omega$ is the solid angle subtended by dS at the point P*: Maxwell will define solid angle measure in (417.1). Briefly, the surface area of a unit sphere (sphere of unit radius) is 4π, and is said to subtend a solid angle of 4π steradians at its center. Similarly, the surface area of a hemisphere with unit radius is 2π, and is said to subtend a solid angle of 2π steradians. By extension, any area $d\omega$ marked out on the surface of a unit sphere may be said to subtend a solid angle equal to $d\omega$ steradians. Finally, any *radial projection* of area $d\omega$—such as area dS in the diagram—is likewise said to subtend a solid angle equal to $d\omega$.

$r^2 d\omega = dS \cos\varepsilon$: Now let an element of the shell have surface area dS. Since it subtends solid angle $d\omega$ at the point P, the radii from P to its boundaries will mark out an area $d\omega$ on the surface of a unit sphere centered on P; hence they will also delineate an area $r^2 d\omega$ on the surface of a sphere of radius r.

If the shell were perpendicular to radius QP, its area dS and its projection $r^2 d\omega$ would coincide and be equal. But since the shell departs from perpendicularity by angle $NQP = \varepsilon$, its effective cross-section is diminished by $\cos\varepsilon$; thus $r^2 d\omega = dS \cos\varepsilon$.

a simple magnetic shell: a shell of uniform strength, as defined in (409.1), so that when integrating $dV = \Phi\,d\omega$, Φ is constant.

the potential due to a magnetic shell at any point is the product of its strength into the solid angle subtended by its edge at the given point: This beautiful theorem has no generally accepted name. It deserves to be called Gauss's Theorem, but many writers reserve that title for Theorem III (21.1), which, however, has more recently been named the Divergence Theorem.

Notice that since the solid angle ω is wholly determined by the *boundaries* of the shell, it is the form of the edge, not that of the shell itself, that determines the potential. Maxwell will obtain this result explicitly in (410.3).

410.1 If V is the potential at the element dS, then the energy due to this element is

$$\Phi\left(l\frac{dV}{dx}+m\frac{dV}{dy}+n\frac{dV}{dz}\right)dS,$$

or, *the product of the strength of the shell into the part of the surface-integral of $dV/d\nu$ due to the element dS of the shell.*

410.2 Hence, integrating with respect to all such elements, the energy due to the position of the shell in the field is equal to the product of the strength of the shell and the surface-integral of the magnetic induction taken over the surface of the shell.

410.1 $\Phi\left(l\dfrac{dV}{dx}+m\dfrac{dV}{dy}+n\dfrac{dV}{dz}\right)dS$: Maxwell stated $\dfrac{dV}{ds}=\lambda\dfrac{dV}{dx}+\mu\dfrac{dV}{dy}+\nu\dfrac{dV}{dz}$ in

(389.2), representing the spatial rate of change of potential in the direction of orientation of a magnetic element having length ds. (Note that Maxwell has shifted his notation for the direction-cosines of the axis of the magnetic element from λ,μ,ν in (389.2) to l,m,n in the present paragraph.) And from (389.1), if that element has pole strength m its energy will be

$$m\frac{dV}{ds}ds.$$

Then we may express the energy of the element as

$$m\frac{dV}{ds}ds=m\left(\lambda\frac{dV}{dx}+\mu\frac{dV}{dy}+\nu\frac{dV}{dz}\right)ds. \qquad (i)$$

But the strength Φ of an element of magnetic shell was defined in (409.) as the product of the magnetization I and the shell thickness; thus, recalling the definition of I (384.1), this becomes

$$\Phi=I\,ds=\frac{mds}{ds\,dS}ds=\frac{mds}{dS}, \quad \text{that is,} \quad mds=\Phi\,dS.$$

Substitute for mds in the right hand side of (i) above to obtain Maxwell's expression.

410.2 *the surface-integral of the magnetic induction*: Integrating the expression in (410.1), we shall have

$$\Phi\iint\left(l\frac{dV}{dx}+m\frac{dV}{dy}+n\frac{dV}{dz}\right)dS.$$

The integrand of this expression is written in terms of differences of scalar potential, which properly yield *magnetic force* \mathfrak{H} rather than *magnetic induction* \mathfrak{B}; compare (401. and *comment*). But as Maxwell will point out in a similar discussion later (489.3), "the shell does not belong to the magnetic system to which the potential V is due." Hence we are concerned only with points not occupied by any magnetic material. At such points \mathfrak{H} and \mathfrak{B} are equal (404.1), and thus the claim of this paragraph is true.

410.3 Since this surface-integral is the same for any two surfaces which have the same bounding edge and do not include between them any centre of force, the action of the magnetic shell depends only on the form of its edge.

411.] If a magnetic pole m starts from a point on the negative surface of a magnetic shell, and travels along any path in space so as to come round the edge to a point close to where it started but on the positive side of the shell, the solid angle will vary continuously, and will increase by 4π during the process. The work done by the pole will be $4\pi\Phi m$, and

410.3 *this surface-integral is the same for any two surfaces which have the same bounding edge*: Maxwell showed in (76.7) that if a closed surface contains no electric charge, the total induction over that surface must be zero. And since the respective force laws for electric charges and for magnetic poles are formally identical (376.1), a similar theorem must apply to magnetic force \mathfrak{H}. Now two surfaces that, without coinciding, share a common boundary form a single closed surface; the same will be true of two magnetic shells. And the combined shells will contain "no centre of force" (no magnetic pole) between them, for "in every magnet the total quantity of magnetism is zero" (377.1). Thus the surface-integral of magnetic force \mathfrak{H} over the combined surfaces will be zero, just as in the electrical case. Hence the respective integrals over the separate portions must be equal but opposite—one directed inward, the other outward. Considering each surface by itself, then, the surface-integral of \mathfrak{H} over one will equal the surface-integral of \mathfrak{H} over the other.

For magnetic induction \mathfrak{B} we can dispense with this argument, as I will show in the next comment.

the action of a magnetic shell depends only on the form of its edge: Since all shells having a common boundary exhibit the same surface-integral of \mathfrak{H}, the value of that integral (and hence the energy of the shell) can depend only on the shape of the boundary, not on the shape of the shell.

For magnetic induction \mathfrak{B}, the conclusion is almost immediate. Since we know that magnetic induction *always* has the solenoidal character, the desired conclusion follows directly, as stated in (403.2).

Recall that Maxwell already obtained this result implicitly (409.3, *comment*).

411. *the solid angle ... will increase by 4π during the process*: Imagine that the earth is a flat disk whose surfaces are distinguished as positive and negative. If you are living at some point A on the negative surface, the disk will obscure one celestial hemisphere; but if you could travel to some point B above the surface, the disk would obscure less than before; and at point C (in the plane of the disk), it will obscure nothing at all. Thus the solid angle subtended by the negative side would decrease from 2π steradians 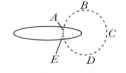 (a hemisphere) to zero. Similarly, if you continued on to point E on the positive side, the solid angle subtended by that side would increase from zero to 2π. If we now adopt the convention that solid angles are to be measured always in the direction approaching the positive side, then the first journey's decrease of 2π in the negative direction counts as an increase of 2π in the positive

the potential at any point on the positive side of the shell will exceed that at the neighbouring point on the negative side by $4\pi\Phi$.

411.1 If a magnetic shell forms a closed surface, the potential outside the shell is everywhere zero, and that in the space within is everywhere $4\pi\Phi$, being positive when the positive side of the shell is inward. Hence such a shell exerts no action on any magnet placed either outside or inside the shell.

412.] If a magnet can be divided into simple magnetic shells, either closed or having their edges on the surface of the magnet, the distribution of magnetism is called Lamellar. If φ is the sum of the strengths of all the shells traversed by a point in passing from a given point to a point (x, y, z) by a line drawn within the magnet, then the conditions of lamellar magnetization are

$$A = \frac{d\varphi}{dx}, \quad B = \frac{d\varphi}{dy}, \quad C = \frac{d\varphi}{dz}.$$

The quantity, φ, which thus completely determines the magnetization at any point may be called the Potential of Magnetization. It must be carefully distinguished from the Magnetic Potential.

* * *

direction. So the overall journey from A to E results in an overall increase of 4π steradians—a whole sphere.

The work done by the pole will be $4\pi\Phi\omega m$: Since the potential of the shell is $\Phi\omega$ (409.3), the potential will have increased by $4\pi\Phi\omega$, and the work done by the pole will be the product of this increase and m.

411.1 *If a magnetic shell forms a closed surface, the potential outside the shell is everywhere zero*: The sketch shows a shell in the final stages of closing. Its *boundary* (marked with an arrow) is in the process of shrinking to a point. When it does so, it will subtend a solid angle of *zero* with respect to any exterior point; and the potential of the shell will likewise be zero.

and [the potential] in the space within is everywhere $4\pi\Phi$: Since a closed surface completely surrounds any point within it, it subtends a solid angle of 4π with respect to each such point; thus the potential everywhere within the closed shell is $4\pi\Phi$.

Hence such a shell exerts no action...: Although the potential has different values inside and outside the shell, in both regions it is *constant*; no force is exerted in a region of constant potential. Like the earlier magnetized torus (407.4), this fanciful example helps build intuition. In part it does so by presenting puzzles; for example: Can such a shell sustain magnetic lines of force or of induction (Arts. 394, 404)? For there is no continuous path joining its inside and its outside that does not penetrate the shell. On the other hand, if such lines do not exist within the shell how can it be, as Maxwell has just claimed, that the potential inside the shell differs from that outside by a *determinate* amount?

417.] We have already proved that at any point P the potential due to a magnetic shell is equal to the solid angle subtended by the edge of the shell multiplied by the strength of the shell. As we shall have occasion to refer to solid angles in the theory of electric currents, we shall now explain how they may be measured.

417.1 *Definition.* The solid angle subtended at a given point by a closed curve is measured by the area of a spherical surface whose centre is the given point and whose radius is unity, the outline of which is traced by the intersection of the radius vector with the sphere as it traces the closed curve. This area is to be reckoned positive or negative according as it lies on the left- or the right-hand of the path of the radius vector as seen from the given point.

<p style="text-align:center">* * *</p>

Potential Energy of a Magnetic Shell placed in a Magnetic Field.

423.] We have already shewn, in Art. 410, that the potential energy of a shell of strength φ placed in a magnetic field whose potential is V, is

$$M = \varphi \iint \left(l \frac{dV}{dx} + m \frac{dV}{dy} + n \frac{dV}{dz} \right) dS, \tag{12}$$

where l, m, n are the direction-cosines of the normal to the shell drawn outwards from the positive side, and the surface-integral is extended over the shell.

423.1 Now this surface-integral may be transformed into a line-integral by means of the vector-potential of the magnetic field, and we may write

$$M = -\varphi \int \left(F \frac{dx}{ds} + G \frac{dy}{ds} + H \frac{dz}{ds} \right) ds, \tag{13}$$

417. *We have already proved...*: in (409.3) above.

417.1 *This area is to be reckoned positive or negative...*: Seen from the center, if the radius vector traces a clockwise path around the curve, then with respect to a creature moving forward in that same direction around the curve, the area enclosed by the path will lie to the right; and it will be positive. If the radius vector traces a counterclockwise path, the enclosed area will be negative.

423.1 *this surface-integral may be transformed into a line-integral by means of the vector-potential of the magnetic field*: Recall that in (405.1) Maxwell *defined* the vector-potential \mathfrak{A} such that its line-integral over a closed curve equals the integral of magnetic induction \mathfrak{B} over the surface bounded by that curve. Now the components of magnetic force \mathfrak{H} are respectively equal to $-dV/dx$, $-dV/dy$, $-dV/dz$ (398.); thus the double integral in equation (12) properly represents the surface-integral of $-\mathfrak{H}$. But except for places occupied by magnetic material, \mathfrak{H} and \mathfrak{B} are equal (404.1). Thus the integral may be taken as the surface-integral

where the integration is extended once round the closed curve s which forms the edge of the magnetic shell, the direction of ds being opposite to that of the hands of a watch when viewed from the positive side of the shell.

423.2 If we now suppose that the magnetic field is that due to a second magnetic shell whose strength is φ', we may determine the value of F directly from the results of Art. 416 or from Art. 405. If l', m', n' be the direction-cosines of the normal to the element dS' of the second shell, we have

$$F = \varphi' \iint \left(m' \frac{d}{dz'} \frac{1}{r} - n' \frac{d}{dy'} \frac{1}{r} \right) dS',$$

where r is the distance between the element dS' and a point on the boundary of the first shell.

of $-\mathfrak{B}$ and hence set equal to the line-integral of $-\mathfrak{A}$, that is, to

$$\int -\mathfrak{A} \cdot \mathbf{ds} = -\int (\mathbf{F} \cdot \mathbf{ds} + \mathbf{G} \cdot \mathbf{ds} + \mathbf{H} \cdot \mathbf{ds}) = -\int \left(F \frac{dx}{ds} + G \frac{dy}{ds} + H \frac{dz}{ds} \right) ds.$$

Compare equation (11) of Art. 24 (Stokes' Theorem). Substitute this expression in place of the double integral in equation (12) to obtain equation (13.)

the direction of ds being opposite to the hands of a watch...: Maxwell outlined the sign convention for rotation under the right-hand system in Art. 23, especially (23.2 and *comment*). The direction Maxwell here describes as *counter-clockwise* if looking onto the shell from its positive surface would be *clockwise* if looking outward from the shell, that is, looking in the positive direction of its magnetization.

423.2 *we may determine the value of F ... from Art. 405*: See (405.7), where Maxwell

writes $F = \iiint \left(B \frac{d}{dz} \frac{1}{r} - C \frac{d}{dy} \frac{1}{r} \right) dx\,dy\,dz$. This expresses the x-component of the

vector potential produced at some point P by a magnet having magnetization B in the y-direction and C in the z-direction, where r is the distance between the point P and an element $dx\,dy\,dz$ of the magnet. If the magnet is a shell of strength φ', then B and C become $d\varphi'/dy$ and $d\varphi'/dz$, respectively (412.). Let the elements of the shell have coordinates x', y', z'; the expression becomes

$$F = \iiint \frac{d\varphi'}{dy'} \frac{d}{dz'} \frac{1}{r} dx'\,dy'\,dz' - \iiint \frac{d\varphi'}{dz'} \frac{d}{dy'} \frac{1}{r} dx'\,dy'\,dz'$$

$$= \iint \varphi' \frac{d}{dz'} \frac{1}{r} dx'\,dz' - \iint \varphi' \frac{d}{dy'} \frac{1}{r} dx'\,dy'.$$

Finally, $dx'\,dz'$, being the y-component of the surface element dS', can be written as $m'dS'$; in the same way we have $dx'\,dy' = n'\,dS'$. Substitute in the above to obtain Maxwell's expression. Note that he is taking the shell strength φ' as constant.

423.3 Now this surface-integral may be converted into a line-integral round the boundary of the second shell; viz. it is

$$\varphi' \int \frac{1}{r} \frac{dx'}{ds'} ds'. \tag{14}$$

423.4 In like manner

$$G = \varphi' \int \frac{1}{r} \frac{dy'}{ds'} ds',$$

$$H = \varphi' \int \frac{1}{r} \frac{dz'}{ds'} ds'.$$

423.5 Substituting these values in the expression for M we find

$$M = -\varphi \varphi' \iint \frac{1}{r} \left(\frac{dx}{ds} \frac{dx'}{ds'} + \frac{dy}{ds} \frac{dy'}{ds'} + \frac{dz}{ds} \frac{dz'}{ds'} \right) ds\, ds', \tag{15}$$

where the integration is extended once round s and once round s'. This expression gives the potential energy due to the mutual action of the two shells, and is, as it ought to be, the same when s and s' are interchanged. This expression with its sign reversed, when the strength of each shell is unity, is called the potential of the two closed curves s and s'. It is a quantity of great importance in the theory of electric currents. If we write ε for the angle between the directions of the elements ds and ds', the potential of s and s' may be written

$$\iint \frac{\cos\varepsilon}{r} ds\, ds'. \tag{16}$$

It is evidently a quantity of the dimensions of a line.

423.3 *this surface-integral may be converted into a line-integral*: by using Stokes' Theorem, discussed in Art. 24. There Maxwell showed that a surface-integral of form $\iint \left(\frac{dX}{d\alpha} \frac{dx}{d\beta} - \frac{dX}{d\beta} \frac{dx}{d\alpha} \right) d\beta\, d\alpha$ (24.6, equation 8) can also be expressed as the line-integral $\int X \frac{dx}{ds} ds$ (24.11, equation 10). Set $X = 1/r$, $z = \alpha$, $y = \beta$, supply the appropriate primes, and multiply through by φ' to obtain eq. (14).

423.5 *write ε for the angle between ... elements ds and ds'*: For example, suppose the elements lie in the xy-plane; then $dx/ds = \cos\alpha$ and $dx'/ds' = \cos\beta$, while $dy/ds = \sin\alpha$ and $dy'/ds' = \sin\beta$. Thus in equation (15) the quantity in parentheses becomes

$$\cos\alpha\cos\beta + \sin\alpha\sin\beta,$$

that is, $\cos(\beta-\alpha)$ or $\cos\varepsilon$, yielding equation (16) when φ and φ', the respective shell strengths, are each made equal to unity.

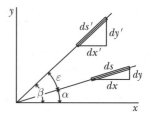

CHAPTER IV.

INDUCED MAGNETIZATION.

424.] WE have hitherto considered the actual distribution of magnetization in a magnet as given explicitly among the data of the investigation. We have not made any assumption as to whether this magnetization is permanent or temporary, except in those parts of our reasoning in which we have supposed the magnet broken up into small portions, or small portions removed from the magnet in such a way as not to alter the magnetization of any part.

424.1 We have now to consider the magnetization of bodies with respect to the mode in which it may be produced and changed. A bar of iron held parallel to the direction of the earth's magnetic force is found to become magnetic, with its poles turned the opposite way from those of the earth, or the same way as those of a compass needle in stable equilibrium.

424.2 Any piece of soft iron placed in a magnetic field is found to exhibit magnetic properties. If it be placed in a part of the field where the magnetic force is great, as between the poles of a horseshoe magnet, the magnetism of the iron becomes intense. If the iron is removed from the magnetic field, its magnetic properties are greatly weakened or disappear entirely. If the magnetic properties of the iron depend entirely on the magnetic force of the field in which it is placed, and vanish when it is removed from the field, it is called Soft iron. Iron which is soft in the magnetic sense is also soft in the literal sense. It is easy to bend it and give it a permanent set, and difficult to break it.

424.3 Iron which retains its magnetic properties when removed from the magnetic field is called Hard iron. Such iron does not take up the magnetic state so readily as soft iron. The operation of hammering, or any other kind of vibration, allows hard iron under the influence of magnetic force to assume the magnetic state more readily, and to part with it more readily when the magnetizing force is removed. Iron which is magnetically hard is also more stiff to bend and more apt to break.

424.4 The processes of hammering, rolling, wire-drawing, and sudden cooling tend to harden iron, and that of annealing tends to soften it.

424.4 *annealing*: The process of heating followed by slow cooling, in order to strengthen a material, typically metal or glass, and reduce brittleness.

424.5 The magnetic as well as the mechanical differences between steel of hard and soft temper are much greater than those between hard and soft iron. Soft steel is almost as easily magnetized and demagnetized as iron, while the hardest steel is the best material for magnets which we wish to be permanent.

424.6 Cast iron, though it contains more carbon than steel, is not so retentive of magnetization.

424.7 If a magnet could be constructed so that the distribution of its magnetization is not altered by any magnetic force brought to act upon it, it might be called a rigidly magnetized body. The only known body which fulfils this condition is a conducting circuit round which a constant electric current is made to flow.

424.8 Such a circuit exhibits magnetic properties, and may therefore be called an electromagnet, but these magnetic properties are not affected by the other magnetic forces in the field. We shall return to this subject in Part IV.

424.9 All actual magnets, whether made of hardened steel or of loadstone, are found to be affected by any magnetic force which is brought to bear upon them.

424.10 It is convenient, for scientific purposes, to make a distinction between the permanent and the temporary magnetization, defining the permanent magnetization as that which exists independently of the magnetic force, and the temporary magnetization as that which depends on this force. We must observe, however, that this distinction is not founded on a knowledge of the intimate nature of the magnetizable substances: it is only the expression of an hypothesis introduced for the sake of bringing calculation to bear on the phenomena. We shall return to the physical theory of magnetization in Chapter VI.

425.] At present we shall investigate the temporary magnetization on the assumption that the magnetization of any particle of the substance depends solely on the magnetic force acting on that particle. This magnetic force may arise partly from external causes, and partly from the temporary magnetization of neighboring particles.

425.1 A body thus magnetized in virtue of the action of magnetic force is said to be magnetized by induction, and the magnetization is said to be induced by the magnetizing force.

425.1 *the magnetization is said to be induced by the magnetizing force*: Maxwell is beginning to introduce the idiom expressing magnetizing force (\mathfrak{H}) as cause, and induced magnetization (\mathfrak{I}) and, consequently, magnetic induction (\mathfrak{B}) as effects, as I noted when introducing Chapter II. He will employ this locution more prominently in Art. 428, below, especially (428.18).

425.2 The magnetization induced by a given magnetizing force differs for different substances. It is greatest in the purest and softest iron, in which the ratio of the magnetization to the magnetic force may reach the value 32, or even 45.* Other substances, such as the metals nickel and cobalt, are capable of an inferior degree of magnetization, and all substances when subjected to a sufficiently strong magnetic force are found to give indications of polarity. When the magnetization is in the same direction as the magnetic force, as in iron, nickel, cobalt, &c., the substance is called Paramagnetic, Ferromagnetic, or more simply Magnetic. When induced magnetization is in the direction opposite to the magnetic force, as in bismuth, &c., the substance is said to be Diamagnetic.

425.3 In all these diamagnetic substances the ratio of the magnetization to the magnetic force which produces it is exceedingly small, being only about $-1/400,000$ in the case of bismuth, which is the most highly diamagnetic substance known. In crystallized, strained, and organized substances the direction of the magnetization does not always coincide with that of the magnetic force which produces it. ...

425.4 In soft iron, the direction of the magnetization coincides with that of the magnetic force at the point, and for small values of the magnetic force the magnetization is nearly proportional to it. As the magnetic force increases, however, the magnetization increases more slowly, and it would appear from experiments described in Chap. VI, that there is a limiting value of the magnetization, beyond which it cannot pass, whatever be the value of the magnetic force.

425.5 In the following outline of the theory of induced magnetism, we shall begin by supposing the magnetization proportional to the magnetic force, and in the same line with it.

Definition of the Coefficient of Induced Magnetization.

426.] Let \mathfrak{H} be the magnetic force, defined as in Art. 398, at any point of the body, and let \mathfrak{I} be the magnetization at that point, then the ratio of \mathfrak{I} to \mathfrak{H} is called the Coefficient of Induced Magnetization.

*Thalén, *Nova Acta, Reg. Soc. Sc.*, Upsal, 1863.

425.3 $-1/400,000$: The ratio of magnetization to magnetic force is negative because in diamagnetic materials the induced magnetization is in the direction opposite to the magnetic force (425.2).

organized substances: Here, organic matter, as vegetable and animal substances. We do not ordinarily think of such materials as magnetic, but Faraday studied the magnetic behavior of olive oil, turpentine, sugar, starch, wood, ivory, spermaceti, mutton, beef, blood, leather, apple, and bread (*Experimental Researches in Electricity*, ¶2280).

426.1 Denoting this coefficient by κ, the fundamental equation of induced magnetism is

$$\mathfrak{I} = \kappa\mathfrak{H} \tag{1}$$

426.2 The coefficient κ is positive for iron and paramagnetic substances, and negative for bismuth and diamagnetic substances. It reaches the value {1600} in iron, and it is said to be large in the case of nickel and cobalt, but in all other cases it is a very small quantity, not greater than 0.00001.

426.3 The force \mathfrak{H} arises partly from the action of magnets external to the body magnetized by induction, and partly from the induced magnetization of the body itself. Both parts satisfy the condition of having a potential.

<div align="center">* * *</div>

428.] The problem of induced magnetism may be treated in a different manner by introducing the quantity which we have called, with Faraday, the Magnetic Induction.

428.1 The relation between \mathfrak{B}, the magnetic induction, \mathfrak{H}, the magnetic force, and \mathfrak{I}, the magnetization, is expressed by the equation

$$\mathfrak{B} = \mathfrak{H} + 4\pi\mathfrak{I}. \tag{12}$$

428.2 The equation which expresses the induced magnetization in terms of the magnetic force is

$$\mathfrak{I} = \kappa\mathfrak{H}. \tag{13}$$

428.3 Hence, eliminating \mathfrak{I}, we find

$$\mathfrak{B} = (1 + 4\pi\kappa)\mathfrak{H} \tag{14}$$

as the relation between the magnetic induction and the magnetic force in substances whose magnetization is induced by magnetic force.

428.4 In the most general case κ may be a function, not only of the position of the point in the substance, but of the direction of the vector \mathfrak{H}, but in the case which we are now considering κ is a numerical quantity.

428.1 *Equation* (12): obtained earlier in (400.2).

428.2 *Equation* (13): obtained in (426.1).

428.4 *a numerical quantity*: a pure number (without dimensions); see (428.21, *comment*). By contrasting the present circumstance to "the most general case," in which κ may vary, Maxwell may seem to be implying that a numerical quantity is a *constant*. That, however, is not so; there have been any number of passages throughout the *Treatise* in which he described a variable quantity as "numerical."

Note also that κ is dimensionless only in the *electromagnetic system* of units. See the comment on (428.21).

428.5 If we next write

$$\mu = 1 + 4\pi\kappa, \tag{15}$$

we may define μ as the ratio of the magnetic induction to the magnetic force, and we may call this ratio the magnetic inductive capacity of the substance, thus distinguishing it from κ, the coefficient of induced magnetization.

* * *

428.13 The magnetic force is derived from the magnetic potential, precisely as the electric force is derived from the electric potential.

428.14 The magnetic induction is a quantity of the nature of a flux, and satisfies the same conditions of continuity as the electric current does.

428.15 In isotropic media the magnetic induction depends on the magnetic force in a manner which exactly corresponds with that in which the electric current depends on the electromotive force.

428.16 The specific magnetic inductive capacity in the one problem corresponds to the specific conductivity of the other. Hence Thomson,

428.5 *Equation* (15): This equation may seem to imply that μ is dimensionless—a pure number. For example, someone might try to argue that since $1 + 4\pi\kappa = \mu$, both 1 and $4\pi\kappa$ must have the same dimensions as μ; and that since 1 and 4π are both pure numbers, κ and μ must be pure numbers too—"numerical quantities," as Maxwell wrote in 428.4.

But we cannot infer that a quantity appearing in an equation is dimensionless just because it is expressed as a number rather than a letter! Rather, we have to examine it in the context of the entire equation. In the present example, go back to the (true) observation that 1 and $4\pi\kappa$ must have the same dimensions as μ. We cannot assume that 1 is a pure number here; but π is a pure number by definition; therefore 4π is a pure number too. Then 1, μ, and κ all have the same dimensions; and since $\mathfrak{J}=\kappa\mathfrak{H}$, their dimensions must be those of $\mathfrak{J}/\mathfrak{H}$.

Now \mathfrak{J} was defined as the magnetic moment per unit volume (384.1); so its dimensions are $[mL/L^3]$ or $[m/L^2]$, where $[m]$ denotes the dimensions of the unit magnetic pole. And \mathfrak{H} was defined as the force per unit pole (395.1); its dimensions are therefore $[ML/T^2m]$. Thus the dimensions of $\mathfrak{J}/\mathfrak{H}$, and hence of κ, are $[m/L^2][T^2m/ML]$ or $[T^2m^2/ML^3]$.

Now in the electromagnetic system $[m]$ is defined by the force law, so that $[m^2]=[L^3M/T^2]$ (374.3). Thus in the electromagnetic system the dimensions $[T^2m^2/ML^3]$ reduce to $[T^2/ML^3][L^3M/T^2]$, a dimensionless quantity, just as Maxwell stated—but *only* in the electromagnetic system. In the electrostatic system, on the other hand, $[m]$ will prove to have the dimensions $[L^{1/2}M^{1/2}]$ (625.); thus in that system $[T^2m^2/ML^3]$, the dimensions of κ, reduce to $[T^2/L^2]$ —*not* dimensionless! Maxwell will offer a résumé of the chief systems of units in Part IV, Chapter X.

we may define μ as the ratio of the magnetic induction to the magnetic force: Combine equation (14) and equation (15) to obtain $\mathfrak{B}=\mu\mathfrak{H}$, in which μ is indeed the ratio of \mathfrak{B} to \mathfrak{H}. This expression will prove important later. See also (428.21, *comment*).

in his *Theory of Induced Magnetism* (*Reprint*, 1872, p. 484) has called this quantity the *permeability* of the medium.

428.17 We are now prepared to consider the theory of induced magnetism from what I conceive to be Faraday's point of view.

428.18 When magnetic force acts on any medium, whether magnetic or diamagnetic, or neutral, it produces within it a phenomenon called Magnetic Induction.

428.19 Magnetic induction is a directed quantity of the nature of a flux, and it satisfies the same conditions of continuity as electric currents and other fluxes do.

428.20 In isotropic media the magnetic force and the magnetic induction are in the same direction, and the magnetic induction is the product of the magnetic force into a quantity called the coefficient of induction, which we have expressed by μ.

428.21 In empty space the coefficient of induction is unity. In bodies capable of induced magnetization the coefficient of induction is $1 + 4\pi\kappa = \mu$, where κ is the quantity already defined as the coefficient of induced magnetization.

<p style="text-align:center">* * *</p>

428.21 *In empty space the coefficient of induction is unity*: This is true *by definition* in the electromagnetic system of units (628.7); but not in the electrostatic system, as I will explain shortly.

In bodies capable of induced magnetization the coefficient of induction is $1 + 4\pi\kappa = \mu$: Substituting this definition of μ into equation (14) yields $\mathfrak{B} = \mu\mathfrak{H}$. This expression of proportionality between magnetic force and magnetic induction is very convenient. Strangely, Maxwell will not write out it explicitly until Art. 614.

Like equation (15), which I discussed in a comment to (428.5), the relation $\mathfrak{B} = \mu\mathfrak{H}$ must be interpreted with care. For example, it might seem that we *must* take μ to be unity in air and vacuum, since in (404.4) and elsewhere, Maxwell points out that "the magnetic force and the magnetic induction are identical outside the magnet." If $\mathfrak{B} = \mathfrak{H}$ outside the magnet (that is, in air or vacuum), does it not follow that $\mu = 1$ no matter what system of units we may employ?

No, it does not. Maxwell determined \mathfrak{H} and \mathfrak{B} by measuring the force on a unit magnetic pole (396.). But we defined the unit pole based on the magnetic force law (374.3)—and that definition is the foundation of the *electromagnetic* system of units! (Maxwell will discuss this fully in Part IV, Chapter X; see paragraphs 620.1 and 625.1). Thus the whole procedure described in Chapter II above is restricted to the electromagnetic system. In particular, we cannot assume that $\mathfrak{B} = \mathfrak{H}$ for empty space if we are working in the electrostatic system.

The electrostatic system is founded on a definition of *unit quantity of electricity*, not unit magnetic pole; as a result, it imposes dimensionality on κ and μ (428.5, *comment*). That, in turn, means that \mathfrak{H} and \mathfrak{B} differ in dimensions from one another—they cannot, in the electrostatic system, simply be set *equal*. As we shall see in Part IV, Chapter X, in the electrostatic system not only is the value of μ for empty space *not unity*, it will take on a highly significant value.

Editor's Introduction to
Part IV, Chapter I

With this chapter we begin the fourth and final Part of the Treatise, titled *Electromagnetism*. Maxwell's topic for the chapter is *electromagnetic force*, but this is a force that shows itself in several guises. One is the force experienced by a magnet when brought near a current-carrying wire, an effect discovered by Oersted. By means of an elegant experimental device (Art. 478), Maxwell readily determines the magnitude of this force. Articulating its direction, however, is a challenge; for unlike the central forces of Newton, this force is *circular* in its effect. It therefore involves three directional relations—two within the plane of its circle, and a third specifying the circle's axis.

Electromagnetic geometry and body-centered experience

When it comes to expressing directional relations in space, Maxwell seems to nourish a certain fondness for images that invoke bodily experience. In a footnote to Art. 23 he drew upon the kinesthetics of the right arm for that purpose, and in Art. 239 he employed not just the arm but the whole body as a frame of reference. In the present chapter he will again utilize an arresting bodily image:

> 477. ... If a man were to place himself in imagination in the position of the wire, so that the current should flow from his head to his feet, then a magnet suspended freely before him would set itself so that the end which points north would, under the action of the current, point to his right hand.

Notice the striking immediacy that characterizes this image. Maxwell is not imagining detached observation but personal occupation of territory. Modern readers are apt to be somewhat startled when such "whole body" images appear in physics; we are more at ease with mnemonic aids that focus on the hand, such as the various "right-hand rules" that many of us learned in school. Why so? Have we been taught to associate the sciences with *manipulation* and *control* (by means of the hand), rather than with *inhabiting* (with the body) a larger natural order?

(a) (b)

Notwithstanding their interesting rhetorical differences, the two images are fully consistent in their content. In the drawing above, Maxwell's "man" is represented in (a); the direction of current is from his head to his feet (downwards) and the north-seeking end of the compass needle before him points to his right. The "right-hand grip rule" is depicted in (b); the thumb points in the direction of current (upwards) and the curled fingers indicate the pointing of the compass. The two representations are consistent because even though their compasses point in opposite directions, their currents are also reversed, thus preserving the direction relations between electric current and magnetic force.

You might try to unify the two images in a single view, but that will be awkward: If you stand upright, current passing from your head to your feet (as Maxwell specifies) obviously has the *downward* direction. But if you then try to point your right thumb downward too, you will find that either you cannot place the hand directly in front of you, or you cannot see the fingertips when you do—try it.

Unity of the electric circuit

The strange circularity of the electromagnetic force is not its only mystery. Another is the identity of the agent—what, exactly, *exerts* the force? Ampère had constructed a brilliant analysis in which *each elementary length* of current-carrying wire directs its own share of force upon the magnet—just as, in gravitational theory, each element of the earth's mass exerts gravitational force on a falling body. This diagram represents a "current element" idl in rela-

tion to a magnet at P. But surely such an independent element can only be a mathematical fiction. Real electric currents form *closed circuits*, so no isolated piece of wire can carry a current, nor can it be counted as a physical agent.

Maxwell will offer a very different understanding by showing that the whole electric circuit is equivalent to a *magnetic shell*, which in turn applies force to a magnet at *P*. A requirement for such equivalence is that the current path coincide with the border of the shell, as shown in this diagram. And Maxwell showed in Art. 490 that the magnetic potential established at any point *P* can be expressed in terms of the solid angle ω subtended at *P* by the shell.

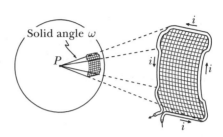

As Thomas Simpson has pointed out,* this device of the magnetic shell with its solid angle measure preserves the unity of the circuit and avoids the Amperian fiction of atomistic elements of electromagnetic action. At issue here is the sense of wholeness, not merely of an electric circuit but, more widely, of science's representation of nature itself. A vision of the unity of natural powers is what Maxwell admired in Faraday, and he continues to strive for mathematical methods that reflect and do justice to that unity.

The infinite straight conductor

Maxwell's current/shell equivalence will prove a powerful conceptual tool throughout this chapter. Remarkably, however, the very first example Maxwell chooses to discuss is one that would seem to defy it: he considers an *infinite straight current-carrying conductor*—but how can a conductor that is both infinite and straight serve as the border of a magnetic shell, as required for equivalence? Maxwell's imaginative solution, offered in Art. 481, is to take the infinite straight conductor in question as forming *one side* of an infinite closed rectangular loop, then to fill that infinite loop with an infinite magnetic shell. In this sketch, *ad* represents a finite length of the conductor and *abcd* represents a finite section of the infinite shell. In my comments on Art. 485 I will show how solid angle measure can be applied to the infinite shell. Under this interpretation, then, even infinitude is no barrier to the unity of a current!

The forces on electricity and on matter

The force that an electric current exerts upon a magnet is one manifestation of electromagnetic force. But if Newton's Third Law applies to electromagnetic forces, that action of a current on a magnet should lead

*Simpson, Thomas King. *Figures of Thought* (2005), pp. 13–14, and *Maxwell's Mathematical Rhetoric* (2010), p. 115, both published by Green Lion Press.

us to seek a second, reciprocal case—action by the magnet upon the electric circuit. As Maxwell will explain, there are in fact two such effects. One is the generation of electromotive force (and consequent current) in a circuit that is subjected to a magnetic field of varying intensity; this is the induced current discovered by Faraday. A second effect is the force experienced by a current-carrying conductor when it resides in a magnetic field—even if the intensity of that field is *not* changing. As Maxwell will state at (490.1) below, "Every portion of the circuit ... is acted on by a force urging it across the lines of magnetic induction so as to include a greater number of these lines within the embrace of the circuit."

Although both these instances represent a magnetic system acting upon an electric circuit, Maxwell wishes to draw a sharp distinction between them. In Art. 501 he will write:

> It must be carefully remembered, that the mechanical force which urges a conductor carrying a current across the lines of magnetic force, acts, not on the electric current, but on the conductor which carries it. ... The only force which acts on electric currents is electromotive force, which must be distinguished from the mechanical force which is the subject of this chapter.

Maxwell is distinguishing between the *mechanical force* which acts on ponderable mass as such and the *electromotive force* which acts on electricity to produce displacement in dielectrics and continuous current in conductors. Such a distinction is likely to seem opaque to modern readers. We habitually think in terms of "elementary" charged particles like the electron, whose mass in relation to its charge was established in 1897.* For us, accordingly, any agency that acts upon *charge* necessarily acts upon *charged particles*—and therefore acts upon mass too.

But Maxwell had no evidence of any such *de facto* relation between charge and mass; and while he did perform experiments to determine whether charge is *inherently* inertial, the answer seemed to be that it is not.† Regarding, then, charge and mass as essentially independent quantities, it was natural and even responsible for him to distinguish between forces that act on one and forces that act on the other.

Nevertheless, several of Maxwell's readers found the distinction less than persuasive. Edwin H. Hall, then a student at the Johns Hopkins

*The ratio was measured by J. J. Thomson, who had served as editor for the Third Edition of Maxwell's *Treatise* in 1891. The discovery was of course unknown to Maxwell, who died in 1879.

†See Arts. 574–577. I suspect that Maxwell would not have accepted as readily as we do that a certain quantity of charge "just happens" to accompany a certain quantity of matter in the electron. He was interested in essential relations, and therefore did not frame his theories to accommodate incidental ones.

University, noted (among other observations) that when a current-carrying wire is affected by a magnet, the force it experiences depends only on the current, not on the physical properties of the conductor; is it not then more natural to regard the force as acting directly on the current and only indirectly on the conductor?*

The Hall Effect

Hall found, in fact, that when a current-carrying conductor is fixed in place, a magnet acting upon it tends to *displace the current* within the body of the conductor. The effect appears most clearly in a later form of his apparatus, shown here in a drawing adapted from Hall's own.

Hall clamps a small sheet of gold leaf *G* to a piece of board, and passes a current through it in the direction from *a* to *b*. A galvanometer connected at the terminals shown reads zero, which means that the arms at *c* and *d* are at equal potential. But when he applies a magnetic field perpendicularly to the sheet *G*, the galvanometer deflects; its reading increases with increasing current, increasing magnetic field, or both—indicating that under magnetic action, the current tends to deflect either upward or downward from its horizontal path. Thus magnetic action evidently, and contrary to Maxwell, affects a current directly; and according to both Maxwell's thinking and Hall's, it must do so by developing an *electromotive* force.

The question here is not merely one of names. If the force acts on the current directly, how then does the current in turn shift the conductor (rather than simply escaping from it) in a case like Maxwell's at (490.1), where the conductor is free to move? The current must apply a *mechanical force* to the conductor, but how can it do that unless the *current itself* has ponderable mass? Thus the Hall Effect disturbingly suggests that current is not pure electricity but represents a transport of matter as well. And this would seem to bring the old electric fluid ideas, with all their difficulties, back into play.

Maxwell did not learn of Hall's discovery before his death in November 1879. J. J. Thomson, editor for the Third (1891) Edition of

*See E. H. Hall, "On a New Action of the Magnet on Electric Currents" in *American Journal of Mathematics*, vol. 2 (1879); republished in *Phil. Mag.*, March 1880.

the *Treatise*, remarked on the Effect in the form of a footnote to Art. 501. I have included that note, along with some comments, in the text of the present chapter.

PART IV.

ELECTROMAGNETISM.

CHAPTER I.

ELECTROMAGNETIC FORCE.

475.] IT HAD BEEN noticed by many different observers that in certain cases magnetism is produced or destroyed in needles by electric discharges through them or near them, and conjectures of various kinds had been made as to the relation between magnetism and electricity, but the laws of these phenomena, and the form of these relations, remained entirely unknown till Hans Christian Örsted,* at a private lecture to a few advanced students at Copenhagen, observed that a wire connecting the ends of a voltaic battery affected a magnet in its vicinity. This discovery he published in a tract entitled *Experimenta circa effectum Conflictûs Electrici in Acum Magneticam*, dated July 21, 1820.

475.1 Experiments on the relation of the magnet to bodies charged with electricity had been tried without any result till Örsted endeavoured to ascertain the effect of a wire heated by an electric current. He discovered, however, that the current itself, and not the heat of the wire, was the cause of the action, and that the 'electric conflict acts in a revolving manner,' that is, that a magnet placed near a wire transmitting an electric current tends to set itself perpendicular to the wire, and with the same end always pointing forwards as the magnet is moved round the wire.

476.] It appears therefore that in the space surrounding a wire transmitting an electric current a magnet is acted on by forces dependent on the position of the wire and on the strength of the current. The space in which these forces act may therefore be considered as a magnetic field, and we may study it in the same way as we have already studied the field in the neighbourhood of ordinary magnets, by tracing the course of the lines of magnetic force, and measuring the intensity of the force at every point.

* See another account of Örsted's discovery in a letter from Professor Hansteen in the *Life of Faraday* by Dr. Bence Jones, vol. ii. p. 395.

475.1 *electric conflict*: The term "conflict" was discussed in (231., *comment*).

477.] Let us begin with the case of an indefinitely long straight wire carrying an electric current. If a man were to place himself in imagination in the position of the wire, so that the current should flow from his head to his feet, then a magnet suspended freely before him would set itself so that the end which points north would, under the action of the current, point to his right hand.

477.1 The lines of magnetic force are everywhere at right angles to planes drawn through the wire, and are therefore circles each in a plane perpendicular to the wire, which passes through its centre. The pole of a magnet which points north, if carried round one of these circles from left to right, would experience a force acting always in the direction of its motion. The other pole of the same magnet would experience a force in the opposite direction.

478.] To compare these forces let the wire be supposed vertical, and the current a descending one, and let a magnet be placed on an apparatus which is free to rotate about a vertical axis coinciding with the wire. It is found that under these circumstances the current has no effect in causing the rotation of the apparatus as a whole about itself as an axis. Hence the action of the vertical current on the two poles of the magnet is such that the statical moments of the two forces about the current as an axis are equal and opposite. Let m_1 [Fig. 21] and m_2 be the strengths of the two poles, r_1 and r_2 their distances from the axis of the wire, T_1 and T_2, the intensities of the magnetic force due to the current at the

477. *If a man were to place himself in imagination in the position of the wire, so that the current should flow from his head to his feet…*: Maxwell's image alters an earlier mnemonic device of Ampère, who in 1820 called upon his readers to imagine themselves swimming in the direction of the current. As noted in the editor's introduction to this chapter, the image is consistent with the "right-hand grip rule" more frequently encountered today.

477.1 *The lines of magnetic force are everywhere at right angles to planes drawn through the wire*: Maxwell is reasoning from the image just introduced, for as the man "in the position of the wire" looks upon the suspended magnet, the medial plane of his body is itself a plane through the wire. And since the north pole of the magnet points to his right, the magnetic lines of force also run from his left to his right—that is, perpendicular to the plane.

478. *the … moments of the two forces*: In this and similar phrases, the word "moment" retains an early (16th century) connotation of *significance* or *importance*—in this case significance with respect to producing *rotation*. The turning moment of a force is the product of the force and its perpendicular distance from the axis; the modern term is *torque*.

intensities of the magnetic force: Intensity of magnetic force (Maxwell will often write *magnetic force*, simply) denotes the force on a *unit* magnetic pole; it also qualifies as an "intensity" in the sense discussed in Art. 12.

two poles respectively, then the force on m_1 is $m_1 T_1$, and since it is at right angles to the axis its moment is $m_1 T_1 r_1$. Similarly that of the force on the other pole is $m_2 T_2 r_2$, and since there is no motion observed,

$$m_1 T_1 r_1 + m_2 T_2 r_2 = 0.$$

478.1 But we know that in all magnets

$$m_1 + m_2 = 0.$$

478.2 Hence $\quad T_1 r_1 = T_2 r_2,$

or the electromagnetic force due to a straight current of infinite length is perpendicular to the current, and varies inversely as the distance from it.

479.] Since the product Tr depends on the strength of the current it may be employed as a measure of the current. This method of measure-

Fig. 21.

ment is different from that founded upon electrostatic phenomena, and as it depends on the magnetic phenomena produced by electric currents it is called the Electromagnetic system of measurement. In the electromagnetic system if i is the current,

$$Tr = 2i.$$

480.] If the wire be taken for the axis of z, then the rectangular components of T are $X = -2i\dfrac{y}{r^2},\quad Y = 2i\dfrac{x}{r^2},\quad Z = 0.$

the force on m_1 is $m_1 T_1$: Thus it is clear that T represents the force per unit pole, that is, the *magnetic force* \mathfrak{H}, which Maxwell defined in (398.1) as distinguished from *magnetic induction* \mathfrak{B}. There he employed \mathfrak{H} to represent magnetic force insofar as it was considered as a vector; here he introduces T to represent the same quantity, *not* considered as a vector. As I mentioned earlier, modern practice generally writes H in preference to Maxwell's T.

478.1 *in all magnets $m_1 + m_2 = 0$*: the poles are equal and opposite, so their sum is zero (Art. 377).

479. *In the electromagnetic system ... $Tr = 2i$*: Exasperatingly, Maxwell leaves the value of the proportionality constant unexplained. In fact the value "2" results from the application of magnetic shell theory (Part III, Chapter III) to electric currents; we will derive it in connection with (485.3), below.

480. *the rectangular components of T*: that is, the x-, y-, and z-components. Since the wire coincides with the z-axis, magnetic force T is in the xy-plane. It has magnitude $2i/r$ in the tangential direction; and by similar triangles, $X/T = y/r$ and $Y/T = x/r$, as shown in the diagram; the expressions for X and Y follow from these. Since the current is in the $+z$ direction, T is directed counter-clockwise; so for the position shown, X is negative and Y is positive.

480.1 Here $Xdx + Ydy + Zdz$ is a complete differential, being that of

$$2i\tan^{-1}\frac{y}{x} + C.$$

480.2 Hence the magnetic force in the field can be deduced from a potential function, as in several former instances, but the potential is in this case a function having an infinite series of values whose common difference is $4\pi i$. The differential coefficients of the potential with respect to the coordinates have, however, definite and single values at every point.

480.1 *Here $Xdx + Ydy + Zdz$ is a complete differential...*: For with X, Y, Z defined as above in (480.), we have

$$Xdx + Ydy + Zdz = -2i\frac{ydx}{r^2} + 2i\frac{xdy}{r^2} = \frac{2i}{r^2}(-ydx + xdy). \tag{i}$$

But the complete differential of $2i\tan^{-1}\frac{y}{x} + C$ (where $C =$ constant) is

$$2i\frac{d}{dx}\left(\tan^{-1}\frac{y}{x}\right)dx + 2i\frac{d}{dy}\left(\tan^{-1}\frac{y}{x}\right)dy = \frac{2i}{1+\left(\frac{y}{x}\right)^2}\left[\frac{d}{dx}\left(\frac{y}{x}\right)dx + \frac{d}{dy}\left(\frac{y}{x}\right)dy\right]$$

$$= \frac{2ix^2}{x^2+y^2}\left[\frac{d}{dx}\left(\frac{y}{x}\right)dx + \frac{d}{dy}\left(\frac{y}{x}\right)dy\right] = \frac{2ix^2}{r^2}\left(-\frac{y}{x^2}dx + \frac{1}{x}dy\right) = \frac{2i}{r^2}(-ydx + xdy),,$$

agreeing with (i) above.

$2i\tan^{-1}\frac{y}{x} + C$: A function of the form $\tan^{-1}a$ denotes *inverse tangent*—an angle whose tangent is a. Also known as *arc tangent*, written arc tan a.

480.2 *Hence the magnetic force ... can be deduced from a potential function*: that is, its components X, Y, Z can be obtained as the spatial derivatives of some function V, as Maxwell showed in Art. 16. In the present example that function, the potential, is $2i\tan^{-1}\frac{y}{x} + C$.

a function having an infinite series of values whose common difference is $4\pi i$: As was noted above, the inverse tangent function specifies an *angle*. But angles that differ by 2π radians ($= 360°$) are equivalent. Hence if the function $2i\tan^{-1}\frac{y}{x} + C$ has some particular value $2i\theta + C$, it will also have the values $2i(\theta + 2\pi) + C$, $2i(\theta + 4\pi) + C$, $2i(\theta + 6\pi) + C$, etc., with the common difference $2i(2\pi)$ or $4\pi i$.

But this cannot be satisfactory. By definition, the relative potential between two points must be independent of the path between those points (16.3). Such is clearly not the case here, since paths whose endpoints differ by θ and $\theta + 2\pi$ respectively will be associated with relative potentials that differ by $4\pi i$. Maxwell will address this difficulty in Art. 481.

differential coefficients: that is, derivatives. The derivatives of the potential with respect to the coordinates are dV/dx, dV/dy, dV/dz. It is important that these derivatives have "definite and single values at every point" since their negatives are the components of the magnetic force, in accordance with (16.4). Thus the force will have, as it must, a definite value at every point.

480.3 The existence of a potential function in the field near an electric current is not a self-evident result of the principle of the conservation of energy, for in all actual currents there is a continual expenditure of the electric energy of the battery in overcoming the resistance of the wire, so that unless the amount of this expenditure were accurately known, it might be suspected that part of the energy of the battery was employed in causing work to be done on a magnet moving in a cycle. In fact, if a magnetic pole, m, moves round a closed curve which embraces the wire, work is actually done to the amount of $4\pi mi$. It is only for closed paths which do not embrace the wire that the line-integral of the force vanishes. We must therefore for the present consider the law of force and the existence of a potential as resting on the evidence of the experiment already described.

481.] If we consider the space surrounding an infinite straight line we shall see that it is a cyclic space, because it returns into itself. If we now conceive a plane, or any other surface, commencing at the straight line and extending on one side of it to infinity, this surface may be regarded as a diaphragm which reduces the cyclic space to an acyclic one. If from any fixed point lines be drawn to any other point without cutting the diaphragm, and the potential be defined as the line-integral of the force taken along one of these lines, the potential at any point will then have a single definite value.

480.3 *if a magnetic pole, m, moves around a closed curve which embraces the wire, work is actually done to the amount of $4\pi mi$*: Suppose the pole follows a circular path of radius r. The force on the pole will be $mT = 2im/r$ (479.); then the work done on the pole will be the line-integral of mT, that is, $2\pi r \cdot 2im/r$ or $4\pi mi$.

481. *it is a cyclic space, because it returns into itself*: Maxwell offered an example of cyclic space in an omitted portion (Art. 18) of his Preliminaries: "If the region has one bounding surface, we may suppose that surface to contract inwards without breaking its continuity or cutting itself. If the region is one of simple continuity, such as a sphere, this process may be continued till it is reduced to a point; but if the region is like a ring, the result will be a closed curve..." The *closed curve* and the *point* respectively illustrate the difference between a cyclic space that "returns into itself" and an acyclic space that does not.

In our present example, the space surrounding an infinite straight wire is like a cylindrical ring whose inner surface coincides with the outer surface of the wire.

a diaphragm which reduces a cyclic space to an acyclic one: We can eliminate the cyclic character of the cylindrical ring sketched here by passing diaphragm PP' through it as shown. To correspond to our present case, let the diaphragm be infinite, one edge coinciding with the infinite wire. This effectively restricts the expression $\tan^{-1}\frac{y}{x}$ to angles between 0 and 2π, eliminating the difficulty noted in (480.2, *comment*). For more on Maxwell's diaphragm, see Thomas K. Simpson, *Maxwell's Mathematical Rhetoric*, pp. 118–141.

481.1 The magnetic field is now identical in all respects with that due to a magnetic shell coinciding with this surface, the strength of the shell being *i*. This shell is bounded on one edge by the infinite straight line. The other parts of its boundary are at an infinite distance from the part of the field under consideration.

482.] In all actual experiments the current forms a closed circuit of finite dimensions. We shall therefore compare the magnetic action of a finite circuit with that of a magnetic shell of which the circuit is the bounding edge.

482.1 It has been shewn by numerous experiments, of which the earliest are those of Ampère, and the most accurate those of Weber, that the magnetic action of a small plane circuit at distances which are great compared with the dimensions of the circuit is the same as that of a magnet whose axis is normal to the plane of the circuit, and whose magnetic moment is equal to the area of the circuit multiplied by the strength of the current.*

482.2 If the circuit be supposed to be filled up by a surface bounded by the circuit and thus forming a diaphragm, and if a magnetic shell of strength *i* coinciding with this surface be substituted for the electric

*{Ampère, *Théorie des phénomènes électrodynamiques*, 1826; Weber, *Elektro-dynamische Maasbestimmungen* (*Abhandlungen der königlich Sachs. Gesellschaft zu Leipzig*, 1850–1852.)}

481.1 *The magnetic field...*: Maxwell has replaced in thought the "infinite straight line" of his preceding paragraph with an infinite straight current-carrying conductor like the one discussed in Arts. 477–479. Such a conductor will be surrounded by a magnetic field, as shown by the experiment there described.

...is now identical in all respects with that due to a magnetic shell coinciding with this surface: This equation of an infinite conductor with a half-infinite magnetic shell is rather abrupt, but Maxwell is anticipating a principle that he will cite in the next paragraph. As first stated, the principle applies only to a finite conductor circling a finite magnetic shell, but Maxwell will develop it in the next three Articles, after which he will be able to apply it to the case of an infinite straight conductor in (485.2).

482.1 *magnetic moment ... equal to the area of the circuit multiplied by the strength of the current*: In Art. 384 Maxwell defined magnetic moment as the length of a small bar magnet multiplied by the strength of its positive, that is, north-seeking (394.2), pole. Then a magnet having pole strength m and length l will be equivalent to a loop carrying current i and enclosing area A, provided $ml = iA$. Bear in mind that A is the area of the *circuit*, and that it can have any value provided the current is adjusted so as to preserve the relation $ml = iA$.

482.2 *a magnetic shell of strength i*: Consider the magnet of (482.1), and suppose that its pole face has area a. Now intensity of magnetization I was defined in (384.1) as the ratio of magnetic moment to volume; thus for each magnet,

current, then the magnetic action of the shell on all distant points will be identical with that of the current.

483.] Hitherto we have supposed the dimensions of the circuit to be small compared with the distance of any part of it from the part of the field examined. We shall now suppose the circuit to be of any form and size whatever, and examine its action at any point P not in the conducting wire itself. The following method, which has important geometrical applications, was introduced by Ampère for this purpose.

483.1 Conceive any surface S bounded by the circuit and not passing through the point P. On this surface draw two series of lines crossing each other so as to divide it into elementary portions, the dimensions of which are small compared with their distance from P, and with the radii of curvature of the surface.

483.2 Round each of these elements conceive a current of strength i to flow, the direction of circulation being the same in all the elements as it is in the original circuit.

483.3 Along every line forming the division between two contiguous elements two equal currents of strength i flow in opposite directions.

483.4 The effect of two equal and opposite currents in the same place is absolutely zero, in whatever aspect we consider the currents. Hence

$I = ml/al$ or m/a. And since one such magnet was shown to be equivalent to a current i enclosing area A, n of them will be equivalent to current ni enclosing the same area.

Then let a parallel array of n magnets fill up area A, so that $na = A$. It will constitute a *shell* of thickness l. The strength of this shell will be the product of its intensity of magnetization and its thickness (409.); and its intensity at each point is simply that of each of its constituent magnets. Thus the magnetic shell that is equivalent to a current ni must have strength $\Phi = Il = ml/a$. But in (482.1) we set $ml = iA$, thus we have $\Phi = iA/a = ni$; or *a magnetic shell is equivalent to a current loop if it occupies the same area as the loop and has strength numerically equal to the current*. This is the principle referred to in (481.1, *comment*), but as yet derived only for small circuits. In the next two Articles Maxwell will extend it to circuits of any size and shape.

483.1 *On this surface draw two series of lines ... so as to divide it into elementary portions*: The sketch shows the elementary portions; also shown are currents circulating about each of them; Maxwell will introduce these currents in the next paragraph.

483.3 *two equal currents of strength i flow in opposite directions*: The sketch shows how the currents encounter one another in opposite directions at the boundaries of adjacent elements.

their magnetic effect is zero. The only portions of the elementary circuits which are not neutralized in this way are those which coincide with the original circuit. The total effect of the elementary circuits is therefore equivalent to that of the original circuit.

484.] Now since each of the elementary circuits may be considered as a small plane circuit whose distance from P is great compared with its dimensions, we may substitute for it an elementary magnetic shell of strength i whose bounding edge coincides with the elementary circuit. The magnetic effect of the elementary shell on P is equivalent to that of the elementary circuit. The whole of the elementary shells constitute a magnetic shell of strength i, coinciding with the surface S and bounded by the original circuit, and the magnetic action of the whole shell on P is equivalent to that of the circuit.

484.1 It is manifest that the action of the circuit is independent of the form of the surface S, which was drawn in a perfectly arbitrary manner so as to fill it up. We see from this that the action of a magnetic shell depends only on the form of its edge and not on the form of the shell itself. This result we obtained before, in Art. 410, but it is instructive to see how it may be deduced from electromagnetic considerations.

483.4 *The only portions of the elementary circuits which are not neutralized in this way are those which coincide with the original circuit*: As the sketch shows, currents along the outer segments of the elementary portions are unhindered; together they constitute a single current circulating about the boundary of the array.

484. *since each of the elementary circuits may be considered as a small plane circuit ... we may substitute for it an elementary magnetic shell of strength i*: As before, the relation $ml = iA$ must apply; but now the area A of each elementary current must equal the area of the corresponding element of shell. Then as in (482.2, *comment*), the strength of each shell element will be $\Phi = Il = ml/A = iA/A = i$.

the whole of the elementary shells constitute a magnetic shell of strength i: Since the shell elements have identical strength i, they make up a simple magnetic shell (409.1) of strength i.

the magnetic action of the whole shell on P is equivalent to that of the circuit: Maxwell garnered this principle from experiments on "small" circuits (482.2, *comment*); now he has shown that it may be applied to circuits "of any form and size whatever," as he set out to do in (483.) above.

484.1 *the action of a magnetic shell depends only on the form of its edge and not on the form of the shell itself*: Previous grounds for this conclusion were the geometry of solid angles (409.3) and the mathematical properties of the surface-integral (410.3). The present argument is more physical, in that a current would appear to consist solely of a linear path, having nothing to do with surfaces. On

484.2 The magnetic force due to the circuit at any point is therefore identical in magnitude and direction with that due to a magnetic shell bounded by the circuit and not passing through the point, the strength of the shell being numerically equal to that of the current. The direction of the current in the circuit is related to the direction of magnetization of the shell, so that if a man were to stand with his feet on that side of the shell which we call the positive side, and which tends to point to the north, the current in front of him would be from right to left.

485.] The magnetic potential of the circuit, however, differs from that of the magnetic shell for those points which are in the substance of the magnetic shell.

485.1 If ω is the solid angle subtended at the point P by the magnetic shell, reckoned positive when the positive or austral side of the shell is next to P, then the magnetic potential at any point not in the shell itself is $\omega\varphi$, where φ is the strength of the shell. ...

485.2 ... In the case of the electric circuit, on the other hand, the magnetic potential at every point not in the conducting wire itself is equal to $i\omega$, where i is the strength of the current, and ω is the solid angle subtended by a circuit at the point, and is reckoned positive when the current, as seen from P, circulates in the direction opposite to that of the hands of a watch.

the other hand, experiments to be described in Chapter IV suggest that "current" may not in all respects be confined to its conductor; see especially (549., 552.2).

484.2 *a magnetic shell bounded by the circuit ..., the strength of the shell being numerically equal to that of the current*: The equivalence principle of (482.2) has now been established for circuits of any size and shape, so Maxwell will be justified in applying it to the infinite straight wire of (481.1).

if a man were to stand with his feet on that side of the shell...: This image is different from the one in (477.). Here the *magnetic force* runs from feet to head, while earlier the *current* ran from head to feet. Nevertheless the images agree in their content. How could you add a second human figure to sketch (a) (page 262) so as to demonstrate that consistency? (Hint: let the added figure take the place of one of the compass needles.)

485.1 *the positive or austral side of the shell*: that is, the north-seeking side, as in (484.2) above.

the magnetic potential ... is $\omega\varphi$: Maxwell deduced this in (409.3).

485.2 *the magnetic potential ... is equal to $i\omega$*: The product $\omega\varphi$ in the preceding paragraph may be set equal to $i\omega$, since a current must be numerically equal to the strength of its equivalent magnetic shell (484.2).

485.3 The quantity $i\omega$ is a function having an infinite series of values whose common difference is $4\pi i$. The differential coefficients of $i\omega$ with respect to the coordinates have, however, single and determinate values for every point of space.

485.3 *infinite series of values whose common difference is* $4\pi i$: Maxwell obtained this result in (411.)

The differential coefficients: that is, the derivatives. The derivatives of $i\omega$ are the derivatives of the potential (485.2), and the negatives of these equal the components of magnetic force \mathfrak{H} in the corresponding directions (395.1, 398.1). It is important, then, that they indeed have "single and determinate values for every point of space," since the magnetic force \mathfrak{H} must be single-valued.

We are ready now to return to the infinite straight conductor and account for the proportionality constant "2" in Maxwell's expression for magnetic force T (479.). As noted in the editor's introduction to this chapter, the equivalent magnetic shell will be a half-infinite shell of strength $\varphi = i$, bounded at one edge by the infinite straight wire. The potential produced at some point P by this shell will be $i\omega$, where ω is the solid angle which the shell subtends at P.

To visualize ω, first consider the solid angle that would be subtended at P by the finite area $abcd$; it would be the apex angle of the pyramid having faces Pda, Pab, Pbc, and Pcd. Then imagine line ab displaced infinitely far along the conductor in direction oa and line dc displaced infinitely along the conductor in the opposite direction; you can see that lines drawn from P to a and to d will become parallel to line da, thus the finite face Pda will ultimately become the infinite plane Poa. Similarly, the finite face Pbc will become more and more nearly parallel to plane $abcd$ and will ultimately become the infinite plane ePq. Finally, the former apex of the finite pyramid ultimately becomes the ridge oPq. Angle oPq. then, determines ω; the next step is to calculate it.

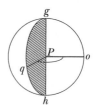

In a unit sphere, draw equatorial angle oPq with P at the center. Solid angle ω is the portion of the sphere's surface area that is subtended by oPq; and since a whole sphere corresponds to a solid angle of 4π steradians, we have

$$\omega = \frac{\angle oPq}{2\pi} \cdot 4\pi \quad \text{or} \quad \omega = 2\angle oPq.$$

Now let point P be at distance r from the conductor, and draw arc $s = r\theta$. Note too that $\angle oPq = \pi - \theta$, so that

$$\omega = 2(\pi - \theta) = 2\left(\pi - \frac{s}{r}\right).$$

And since Maxwell has already established that the magnetic force T is disposed circularly about the conductor (477.1), the force at P will be directed along arc s; then

$$T = -\frac{dV}{ds} = -\frac{d(i\omega)}{ds} = -2i\frac{d}{ds}\left(\pi - \frac{s}{r}\right) = \frac{2i}{r}. \quad \text{Q.E.D.}$$

486.] If a long thin flexible solenoidal magnet were placed in the neighbourhood of an electric circuit, the north and south ends of the solenoid would tend to move in opposite directions round the wire, and if they were free to obey the magnetic force the magnet would finally become wound round the wire in a closed coil. If it were possible to obtain a magnet having only one pole, or poles of unequal strength, such a magnet would be moved round and round the wire continually in one direction, but since the poles of every magnet are equal and opposite, this result can never occur. Faraday, however, has shewn how to produce the continuous rotation of one pole of a magnet round an electric current by making it possible for one pole to go round and round the current while the other pole does not. That this process may be repeated indefinitely, the body of the magnet must be transferred from one side of the current to the other once in each revolution. To do this without interrupting the flow of electricity, the current is split into two branches, so that when one branch is opened to let the magnet pass the current continues to flow through the other. Faraday used for this purpose a circular trough of mercury, as shewn in Fig. 23. The current enters the trough through the wire *AB*, it is divided at *B*, and after flowing through the arcs *BQP* and *BRP* it unites at *P*, and leaves the trough through the wire *PO*, the cup of mercury *O*, and a vertical wire beneath *O*, down which the current flows.

Figure 23

486. *solenoidal magnet*: for example, a cylindrical or bar magnet.

Faraday used ... a circular trough of mercury, as shewn in Fig. 23: The figure is a little misleading as to the design of Faraday's device. Faraday's actual rotator is shown here in a detail taken from Plate IV of his *Experimental Researches in Electricity*, vol. 2. The device shown in Figure 23, however, works on the same principle and makes the "splitting" technique more evident than Faraday's does.

As originally drawn, Figure 23 was also misleading in another respect. Maxwell means the wire *OP* to revolve about *O*; but that was not evident in the published figure, in which wires *OP* and *AB* were drawn too long and would have collided; I have altered the diagram to shorten both of them.

486.1 The magnet (not shewn in the figure) is mounted so as to be capable of revolving about a vertical axis through O, and the wire OP revolves with it. The body of the magnet passes through the aperture of the trough, one pole, say the north pole, being beneath the plane of the trough, and the other above it. As the magnet and the wire OP revolve about the vertical axis, the current is gradually transferred from the branch of the trough which lies in front of the magnet to that which lies behind it, so that in every complete revolution the magnet passes from one side of the current to the other. The north pole of the magnet revolves about the descending current in the direction N.E.S.W. ...

487.] Let us now endeavour to form a notion of the state of the magnetic field near a linear electric circuit.

487.1 Let the value of ω, the solid angle subtended by the circuit, be found for every point of space, and let the surfaces for which ω is constant be described. These surfaces will be the equipotential surfaces. ...

487.2 Figure XVIII, [on the following page], represents a section of the equipotential surfaces due to a circular current. The small circle represents a section of the conducting wire, and the horizontal line at the bottom of the figure is the perpendicular to the plane of the circular current through its centre. The equipotential surfaces, 24 of which are drawn corresponding to a series of values of ω differing by $\pi/6$, are

486.1 *The magnet* (*not shown in the figure*): Since the position of the magnet is essential, I have supplied a side view. In the sketch, the magnet and the wire OP are understood to be joined so that they both revolve together about axis OO'.

The north pole of the magnet revolves about the descending current in the direction N.E.S.W.: that is, clockwise, looking downward—in accordance with either of the directional rules discussed in the introduction to this chapter.

487. *the state of the magnetic field near a linear electric circuit*: By a "linear" circuit Maxwell does not here mean a *straight conductor* as previously discussed in Art. 477; in fact he is about to illustrate the case of a circular current. Rather, a linear current is one that exists in a *wire* of any shape—as opposed to a current-sheet, which is a current that is distributed over a surface.

487.1 *These surfaces will be the equipotential surfaces*: since, by the theorem of (409.3), the potential is proportional to ω, which is constant for each of the surfaces in question. Although the theorem was stated for a magnetic shell, in (485.3) we found the shell that is equivalent to an infinite straight current-carrying conductor.

FIG .XVIII.

Art. 487, 702

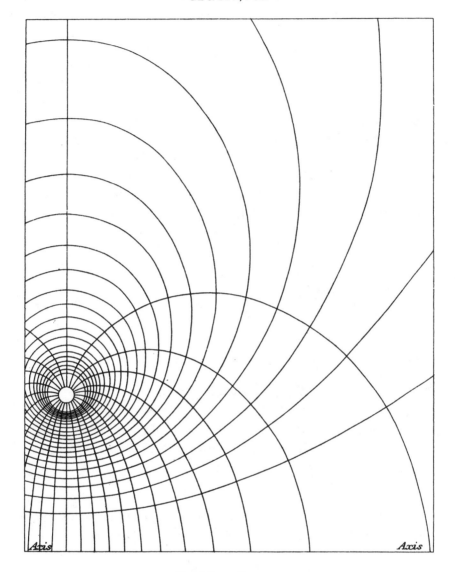

Circular Current

surfaces of revolution, having this line for their common axis. They are evidently oblate figures, being flattened in the direction of the axis. They meet each other in the line of the circuit at angles of 15°.

487.2 *a section of the equipotential surfaces*: In Figure XVIII, the lines which terminate at the conductor are equipotential lines. The lines which encircle the conductor are lines of induction.

oblate figures: Figures of rotation, compressed or flattened at the poles.

487.3 The force acting on a magnetic pole placed at any point of an equipotential surface is perpendicular to this surface, and varies inversely as the distance between consecutive equipotential surfaces. The closed curves surrounding the section of the wire in Fig. XVIII are the lines of force. They are copied from Sir W. Thomson's Paper on 'Vortex Motion.'* ...

Action of an Electric Circuit on any Magnetic System.

488.] We are now able to deduce the action of an electric circuit on any magnetic system in its neighbourhood from the theory of magnetic shells. For if we construct a magnetic shell, whose strength is numerically equal to the strength of the current, and whose edge coincides in position with the circuit, while the shell itself does not pass through any part of the magnetic system, the action of the shell on the magnetic system will be identical with that of the electric current.

Reaction of the Magnetic System on the Electric Circuit.

489.] From this, applying the principle that action and reaction are equal and opposite, we conclude that the mechanical action of the magnetic system on the electric circuit is identical with its action on a magnetic shell having the circuit for its edge.

489.1 The potential energy of a magnetic shell of strength φ placed in a field of magnetic force of which the potential is V, is, by Art. 410,

$$= \varphi \iint \left(l\frac{dV}{dx} + m\frac{dV}{dy} + n\frac{dV}{dz} \right) dS,$$

where l, m, n are the direction-cosines of the normal drawn from the positive side of the element dS of the shell, and the integration is extended over the surface of the shell.

489.2 Now the surface-integral $N = \iint (la + mb + nc)dS$, where a, b, c are the components of the magnetic induction, represents the quantity of magnetic induction through the shell, or, in the language of Faraday,

*Trans. R. S. Edin., vol. xxv. p. 217, (1869).

489.1 *the potential energy of a magnetic shell*: The equation in this paragraph is the integral of an expression in (410.1) for the energy of a magnetic shell element; but Maxwell stated it explicitly for an integral shell in (423.).

489.2 *the quantity of magnetic induction through the shell*: More strictly, the quantity through the *area* that happens to be occupied by a shell. This distinction will become important in (490.3), where Maxwell uses the same equation, but where no magnetic shell is present.

in the language of Faraday: Compare Art. 404.

the number of lines of magnetic induction, reckoned algebraically, which pass through the shell from the negative to the positive side, lines which pass through the shell in the opposite direction being reckoned negative.

489.3 Remembering that the shell does not belong to the magnetic system to which the potential V is due, and that the magnetic force is therefore equal to the magnetic induction, we have

$$a = -\frac{dV}{dx}, \quad b = -\frac{dV}{dy}, \quad c = -\frac{dV}{dz},$$

and we may write the value of M,

$$M = -\varphi N.$$

489.4 If δx_1 represents any displacement of the shell, and X_1 the force acting on the shell so as to aid the displacement, then by the principle of conservation of energy,

$$X_1 \delta x_1 + \delta M = 0,$$

or
$$X_1 = \varphi \frac{dN}{dx_1}.$$

489.5 We have now determined the nature of the force which corresponds to any given displacement of the shell. It aids or resists that displacement accordingly as the displacement increases or diminishes N, the number of lines of induction which pass through the shell.

489.6 The same is true of the equivalent electric circuit. Any displacement of the circuit will be aided or resisted according as it increases or diminishes the number of lines of induction which pass through the circuit in the positive direction.

489.3 *the magnetic force is therefore equal to the magnetic induction*: As in (410.2) earlier, since the system responsible for the potential does not include the shell itself, the derivatives of potential V are being taken in empty space, where magnetic force and magnetic induction are equal (400.2).

the value of M: Maxwell introduced the quantity M in Art. 423, where it denoted the energy of a magnetic shell placed in a region of potential V; thus it equals the expression in (489.1), while N was defined in (489.2). Then since

$$M = \varphi \iint \left(l \frac{dV}{dx} + m \frac{dV}{dy} + n \frac{dV}{dz} \right) dS,$$

when we substitute as directed for the spatial derivatives of V, we shall have $M = -\varphi N$.

489.4 *by the principle of conservation of energy*: Better, by the work-energy equivalence. The term $X_1 \delta x_1$, being the product of force and distance, represents work, while δM represents change in the energy of the magnetic shell.

489.7 We must remember that the positive direction of a line of magnetic induction is the direction in which the pole of a magnet which points north tends to move along the line, and that a line of induction passes through the circuit in the positive direction, when the direction of the line of induction is related to the direction of the current of vitreous electricity in the circuit as the longitudinal to the rotational motion of a right-handed screw. See Art. 23.

490.] It is manifest that the force corresponding to any displacement of the circuit as a whole may be deduced at once from the theory of the magnetic shell. But this is not all. If a portion of the circuit is flexible, so that it may be displaced independently of the rest, we may make the edge of the shell capable of the same kind of displacement by cutting up the surface of the shell into a sufficient number of portions connected by flexible joints. Hence we conclude that if by the displacement of any portion of the circuit in a given direction the number of lines of induction which pass through the circuit can be increased, this displacement will be aided by the electromagnetic force acting on the circuit.

490.1 Every portion of the circuit therefore is acted on by a force urging it across the lines of magnetic induction so as to include a greater number of these lines within the embrace of the circuit, and the work done by the force during this displacement is numerically equal to the number of the additional lines of induction multiplied by the strength of the current.

490.2 Let the element ds of a circuit, in which a current of strength i is flowing, be moved parallel to itself through a space δx, it will sweep out an area in the form of a parallelogram whose sides are parallel and equal to ds and δx respectively.

489.7 *vitreous electricity*: that is, positive electricity; the direction of a current is reckoned from positive to negative.

the direction of the line of induction is related to the direction of the current ... as the longitudinal to the rotational motion of a right-handed screw: the direction relations are sketched here.

490.1 *the work done ... is numerically equal to the number of the additional lines of induction multiplied by the strength of the current*: From $M = -\varphi N$ (489.3) it follows that any change ΔN in the number of lines embraced by the circuit will be associated with a change in energy $\Delta M = -\varphi \Delta N$. Then since the work done by the circuit is the negative of the change in its energy (489.4 and *comment*), the circuit will do work equal to $\varphi \Delta N$. Finally, since a circuit carrying current i is equivalent to a magnetic shell of strength φ whose boundary coincides with the circuit (494.), we may also express this work as $i \Delta N$.

490.3 If the magnetic induction is denoted by \mathfrak{B}, and if its direction makes an angle ε with the normal to the parallelogram, the value of the increment of N corresponding to the displacement is found by multiplying the area of the parallelogram by $\mathfrak{B} \cos \varepsilon$. The result of this operation is represented geometrically by the volume of a parallelopiped whose edges represent in magnitude and direction δx, ds, and \mathfrak{B}, and it is to be reckoned positive if when we point in these three directions in the order here given the pointer moves round the diagonal of the parallelopiped in the direction of the hands of a watch.* The volume of this parallelopiped is equal to $X\delta x$.

490.4 If θ is the angle between ds and \mathfrak{B}, the area of the parallelogram whose sides are ds and \mathfrak{B} is $ds\cdot\mathfrak{B}\sin\theta$, and if η is the angle which the displacement δx makes with the normal to this parallelogram, the volume of the parallelopiped is $ds.\mathfrak{B}\sin\theta.\delta x\cos\eta = \delta N$.

490.5 Now $$X\delta x = i\delta N = i\,ds.\mathfrak{B}\sin\theta\,\delta x\cos\eta,$$

and $$X = ids.\mathfrak{B}\sin\theta\cos\eta$$

is the force which urges ds, resolved in the direction δx.

*{In this rule ds is drawn in the direction of i and the observer is supposed to be at that corner of the parallelopiped from which dx, ds and \mathfrak{B} are drawn.}

490.3 *multiplying the area of the parallelogram by* \mathfrak{B} *cos* ε: "The parallelogram" has sides δx, ds (490.2). In (489.2) Maxwell wrote $N = \iint (la + mb + nc)dS$ for the number of lines of magnetic induction that penetrate the area over which the integration is performed. Since a, b, c are the components of magnetic induction \mathfrak{B}, while l, m, n are the direction-cosines of the normal to area dS, the integrand represents the resolved part of \mathfrak{B} in the direction of the normal, or \mathfrak{B} *cos* ε. Since \mathfrak{B} *cos* ε is constant it can be removed from the integrand, so as to multiply $\iint dS$, which is "the area of the parallelogram."

the volume of a parallelopiped whose edges represent ... δx, ds, *and* \mathfrak{B}: A parallelopiped (more commonly spelled *parallelepiped*) is a solid with six faces, each a parallelogram and each parallel to its opposite face. The figure Maxwell describes is suggested by edges δx, ds, *and* \mathfrak{B} in the drawing.

490.4 $ds.\mathfrak{B}\sin\theta.dx\cos\eta = dN$: In this and several equations to follow, the "dots" represent ordinary multiplication. Today we would write them (if at all) above the baseline, as $ds\cdot\mathfrak{B}\sin\theta\cdot dx\cos\eta$.

490.5 $X\delta x = idN$: Eliminating M from the equations in (489.3, 489.4) yields $X\delta x = \varphi\,dN$ or $X\delta x = i\,dN$. Substitute for dN to obtain the expression for X.

resolved in the direction δx: X was defined as "the force acting on the shell so as to aid the displacement" (489.4)—that is, the component (of the total force) parallel to displacement δx.

490.6 The direction of this force is therefore perpendicular to the parallelogram, and its magnitude is equal to $i.ds.\mathfrak{B}\sin\theta$.

490.7 This is the area of a parallelogram whose sides represent in magnitude and direction ids and \mathfrak{B}. The force acting on ds is therefore represented in magnitude by the area of this parallelogram, and in direction by a normal to its plane drawn in the direction of the longitudinal motion of a right-handed screw, the handle of which is turned from the direction of the current ids to that of the magnetic induction \mathfrak{B}.

490.8 We may express in the language of Quaternions, both the direction and the magnitude of this force by saying that it is the vector part of the result of multiplying the vector ids, the element of the current, by the vector \mathfrak{B}, the magnetic induction.

491.] We have thus completely determined the force which acts on any portion of an electric circuit placed in a magnetic field. If the circuit is moved in any way so that, after assuming various forms and positions, it returns to its original place, the strength of the current remaining constant during the motion, the whole amount of work done by the electromagnetic forces will be zero. Since this is true of any cycle of motions of the circuit, it follows that it is impossible to maintain by electromagnetic forces a motion of continuous rotation in any part of a linear circuit of constant strength against the resistance of friction, &c.

490.6 *The direction of this force is … perpendicular to the parallelogram, and its magnitude is equal to* $ids.\mathfrak{B}\sin\theta$: The drawing has been repeated from the preceding page; this time "the parallelogram" is the one with sides \mathfrak{B} and ds (490.4). Since the component of force parallel to δx is $ids\cdot\mathfrak{B}\sin\theta\cdot\cos\eta$ (490.5) drop the

factor $\cos\eta$ to obtain the force itself, which will make angle η with δx and will be parallel to n, that is, perpendicular to both \mathfrak{B} and ds.

490.7 *a right-handed screw, the handle of which is turned from the direction of the current ids to that of the magnetic induction* \mathfrak{B}: Although we tend to think of screws as having heads rather than handles, the right-handed screw in this bookbinder's press is what Maxwell has in mind. Turning its handle clockwise from the direction of ids to the direction of \mathfrak{B} advances the screw downward, showing the direction of the force.

490.8 *the vector part of the result of multiplying the vector ids … by the vector* \mathfrak{B}: "The vector part" of this quaternion multiplication corresponds to the modern vector (cross) product (23.11, *comment*); thus $\mathbf{F}=i\mathbf{ds}\times\mathfrak{B}$, where \mathbf{F}, $i\mathbf{ds}$, and \mathfrak{B} have the directions indicated in the drawing.

491.1 It is possible, however, to produce continuous rotation provided that at some part of the course of the electric current the current passes from one conductor which slides or glides over another.

<p style="text-align:center">* * *</p>

492.] We are now able to pass from the mutual action of magnets and currents to the action of one circuit on another. For we know that the magnetic properties of an electric circuit C_1, with respect to any magnetic system M_2, are identical with those of a magnetic shell S_1, whose edge coincides with the circuit, and whose strength is numerically equal to that of the electric current. Let the magnetic system M_2 be a magnetic shell S_2, then the mutual action between S_1 and S_2 is identical with that between S_1 and a circuit C_2, coinciding with the edge of S_2 and equal in numerical strength, and this latter action is identical with that between C_1 and C_2.

492.1 Hence the mutual action between two circuits C_1 and C_2 is identical with that between the corresponding magnetic shells S_1 and S_2.

492.2 We have already investigated, in Art. 423, the mutual action of two magnetic shells whose edges are the closed curves s_1 and s_2. If we make $M = \int_0^{s_2} \int_0^{s_1} \frac{\cos\varepsilon}{r}\, ds_1\, ds_2$, where ε is the angle between the directions of the elements ds_1 and ds_2, and r is the distance between them, the integrations being extended one round s_2 and one round s_1, and if we call M the potential of the two closed curves s_1 and s_2, then the potential energy due to the mutual action of two magnetic shells whose strengths are i_1 and i_2 bounded by the two circuits is
$$-i_1 i_2 M,$$
and the force X, which aids any displacement δx, is
$$-i_1 i_2 \frac{dM}{dx}.$$

492.3 The whole theory of the force acting on any portion of an electric circuit due to the action of another electric circuit may be deduced from this result.

491.1 *the current passes from one conductor which slides or glides over another*: As in Faraday's rotator, a version of which Maxwell discussed in Art 486 and diagrammed in Figure 23.

492.2 *the force X, which aids any displacement δx*: That is, the force whose direction is that of δx. We can see from the expression that when X and δx are both positive, $dM/\delta x$ will be positive. In other words, we must apply a force in order to effect a displacement that increases the potential energy M.

493.] The method which we have followed in this chapter is that of Faraday. Instead of beginning, as we shall do, following Ampère, in the next chapter, with the direct action of a portion of one circuit on a portion of another, we shew, first, that a circuit produces the same effect on a magnet as a magnetic shell, or, in other words, we determine the nature of the magnetic field due to the circuit. We shew, secondly, that a circuit when placed in any magnetic field experiences the same force as a magnetic shell. We thus determine the force acting on the circuit placed in any magnetic field. Lastly, by supposing the magnetic field to be due to a second electric circuit we determine the action of one circuit on the whole or any portion of the other.

494.] Let us apply this method to the case of a straight current of infinite length acting on a portion of a parallel straight conductor.

494.1 Let us suppose that a current i in the first conductor is flowing vertically downwards. In this case the end of a magnet which points north will point to the right-hand of a man (with his feet downwards) looking at it from the axis of the current.

494.2 The lines of magnetic induction are therefore horizontal circles, having their centres in the axis of the current, and their positive direction is north, east, south, west.

493. *as we shall do, following Ampère, in the next chapter*: The "next chapter" (Chapter II of Part IV) is omitted from these selections. I explain why in the editor's introduction to Chapter III.

Instead of beginning ... with the direct action of a portion of one circuit on a portion of another: Such was Ampère's method, as noted in the editor's introduction to this chapter. "Direct action" denotes action-at-a-distance with no intermediary.

we shew, first, ... secondly, ... Lastly ...: Maxwell's use of the simple present "we shew" points neither to what he has already done earlier in this chapter nor to what he is about to do in the sequel. Rather, it identifies the definitive steps of what he calls Faraday's method. To review these steps: (1) we show that a given circuit is equivalent to a magnetic shell in its ability to produce magnetic effects; then (2) that a second circuit is equivalent to a second shell in its susceptibility to magnetic action. Finally (3) we equate the action between circuits with the action between their equivalent shells.

Maxwell contrasts this three-step analysis with Ampère's method and, to be sure, representing the circuit as an integral magnetic shell does preserve the wholeness of the circuit, at least in thought. But why is the interaction between even whole circuits any less "direct" than that between Ampère's "portions"? And why does Maxwell think the method he has outlined is Faraday's? We shall have to keep these questions in mind as Maxwell proceeds.

494. *Let us apply this method to ... a straight current of infinite length*: Of course Maxwell has already treated the infinite straight conductor in Arts. 477–479. But while he there expressed the force it exerts on a magnetic pole, he now undertakes to determine the force it exerts on a portion of another conductor.

494.3 Let another descending vertical current be placed due west of the first. The lines of magnetic induction due to the first current are here directed towards the north. The direction of the force acting on the second circuit is to be determined by turning the handle of a right-handed screw from the nadir, the direction of the current, to the north, the direction of the magnetic induction. The screw will then move towards the east, that is, the force acting on the second circuit is directed towards the first current, or, in general, since the phenomenon depends only on the relative position of the currents, two parallel circuits conveying currents in the same direction attract each other.

494.4 In the same way we may shew that two parallel circuits conveying currents in opposite directions repel one another.

495.] The intensity of the magnetic induction at a distance r from a straight current of strength i is, as we have shewn in Art. 479, $2i/r$.

495.1 Hence, a portion of a second conductor parallel to the first, and carrying a current i' in the same direction, will be attracted towards the first with a force

$$F = 2ii'\frac{a}{r},$$

where a is the length of the portion considered, and r is its distance from the first conductor.

494.1 *the end of a magnet which points north will point to the right-hand*: For this and the other geometrical relations noted by Maxwell, see the editor's introduction to this chapter.

494.3 *The lines of magnetic induction due to the first current are here directed towards the north*: "Here" refers to the position of the second current, located due west of the first. Since the lines of induction surrounding the first current have clockwise direction (looking downward), the portions lying west of the current will be directed towards the north.

The direction of the force ... is to be determined by turning the hand of a right-handed screw...: Maxwell is applying the right-hand rule he introduced in (490.7) above. See the comment to that paragraph for a diagram.

495. *The intensity of the magnetic induction*: The expression Maxwell is about to give is properly that for magnetic force \mathfrak{H}. In air, however, the magnetic force \mathfrak{H} and magnetic induction \mathfrak{B} are not distinguished (404.4).

as we have shewn in Art. 479: Maxwell asserted the expression in Art. 479 but did not really show it. See rather (485.3, *comment*).

495.1 *a portion ... carrying current i' ... will be attracted with a force F*: Apply $\mathbf{F} = i\mathbf{ds} \times \mathfrak{B}$ (490.8, *comment*) to the current i'. Here, $\mathfrak{B} = 2i/r$ and is perpendicular to $i'\mathbf{ds}$. Thus the force on the element $i'ds$ is $(2i/r) \cdot i'ds$ and is perpendicular to both of them. Since every element $i'ds$ experiences the same force, to obtain the total force on a length a we need only substitute $i'a$ for $i'ds$ in the previous product to obtain Maxwell's expression for F.

495.2 Since the ratio of a to r is a numerical quantity independent of the absolute value of either of these lines, the product of two currents measured in the electromagnetic system must be of the dimensions of a force, hence the dimensions of the unit current are

$$[i] = \left[F^{\frac{1}{2}} \right] = \left[M^{\frac{1}{2}} L^{\frac{1}{2}} T^{-1} \right].$$

496.] Another method of determining the direction of the force which acts on a circuit is to consider the relation of the magnetic action of the current to that of other currents and magnets.

496.1 If on one side of the wire which carries the current the magnetic action due to the current is in the same or nearly the same direction as that due to other currents, then, on the other side of the wire, these forces will be in opposite or nearly opposite directions, and the force acting on the wire will be from the side on which the forces strengthen each other to the side on which they oppose each other.

496.2 Thus, if a descending current is placed in a field of magnetic force directed towards the north, its magnetic action will be to the north on the west side, and to the south on the east side. Hence the forces strengthen each other on the west side and oppose each other on the east side, and the circuit will therefore be acted on by a force from west to east. See Fig. 22.

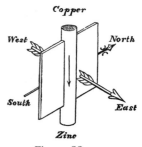

Figure 22

496.3 In Fig. XVII [on the opposite page] the small circle represents a section of the wire carrying a descending current, and placed in a uniform field of magnetic force acting towards the left-hand of the figure.

496.2 *Figure 22*: The figure depicts the facts but not their causes. The current is represented as passing from the copper to the zinc terminal of a voltaic battery (not pictured). The compass rose depicts magnetic north, while the ornate arrows from west to east indicate the force experienced by the conductor. The actual terms of Maxwell's explanation are better illustrated in Figure XVII, which he discusses in the next paragraph.

496.3 *placed in a uniform field … acting towards the left-hand*: The lines of force or induction surrounding the descending current are directed clockwise. If the uniform field acts towards the left, its lines of force or induction will agree with those of the current in the lower half of the drawing but will oppose those in the upper half. In any region where lines of force have opposing directions they cannot coalesce, so their numbers there will be fewer. On Faraday's principle that parallel lines of force exert lateral pressure (compare Art. 109), a conductor (and its associated pattern of lines of force) will be urged away from higher concentrations, and toward lower concentrations, of lines of force.

FIG.XVII

Art .496

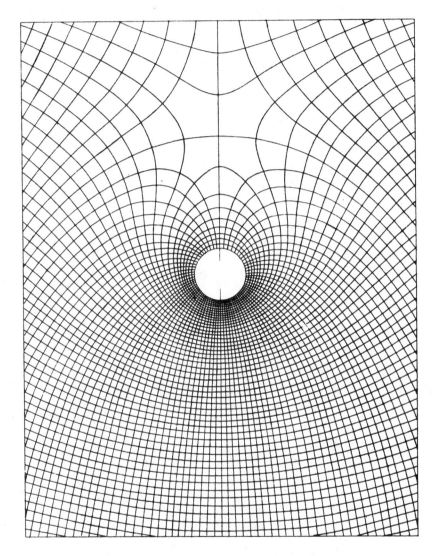

Uniform magnetic field disturbed by an Electric
Current in a straight conductor.

The magnetic force is greater below the wire than above it. It will there-
fore be urged from the bottom towards the top of the figure.

* * *

497.2 In considering the mutual action of two currents it is not necessary to bear in mind the relations between electricity and magnetism which we have endeavoured to illustrate by means of a right-handed screw. Even if we have forgotten these relations we shall arrive at correct results, provided we adhere consistently to one of the two possible forms of the relation.

Fig. 24. Relation between the electric current and the lines of magnetic induction indicated by a right-handed screw.

498.] Let us now bring together the magnetic phenomena of the electric circuit so far as we have investigated them.

498.1 We may conceive the electric circuit to consist of a voltaic battery, and a wire connecting its extremities, or of a thermoelectric arrangement, or of a charged Leyden jar with a wire connecting its positive and negative coatings, or of any other arrangement for producing an electric current along a definite path.

498.2 The current produces magnetic phenomena in its neighbourhood.

497.2 *the relations between electricity and magnetism which we have endeavoured to illustrate by means of a right-handed screw*: In Figure 24, unadorned arrows indicate the direction of electric current, which proceeds from the copper (C) to the zinc (Z) plate of the voltaic cell. The ornate arrows indicate the direction of magnetic induction. The right-handed screws (or helices) drawn around the conductor embody the relation between a current and its lines of magnetic induction. Similarly, the right-handed screw drawn about a line of magnetic induction epitomizes the relation between a line of induction and the current that produces it.

one of the two possible forms of the relation: That is, either a right-hand or a left-hand rule. Even though a left-hand rule would violate standard conventions for the relation between current and magnetic induction, if we apply that rule consistently we shall obtain correct results for the interaction of two currents.

498.3 If any closed curve be drawn, and the line-integral of the magnetic force taken completely round it, then, if the closed curve is not linked with the circuit, the line-integral is zero, but if it is linked with the circuit, so that the current i flows through the closed curve, the line-integral is $4\pi i$, and is positive if the direction of integration round the closed curve would coincide with that of the hands of a watch as seen by a person passing through it in the direction in which the electric current flows. To a person moving along the closed curve in the direction of integration, and passing through the electric circuit, the direction of the current would appear to be that of the hands of a watch. We may express this in another way by saying that the relation between the directions of the two closed curves may be expressed by describing a right-handed screw round the electric circuit and a right-handed screw round the closed curve. If the direction of rotation of the thread of either, as we pass along it, coincides with the positive direction in the other, then the line-integral will be positive, and in the opposite case it will be negative.

499.] *Note.*—The line-integral $4\pi i$ depends solely on the quantity of the current, and not on any other thing whatever. It does not depend on the nature of the conductor through which the current is passing, as, for instance, whether it be a metal or an electrolyte, or an imperfect conductor. We have reason for believing that even when there is no proper conduction, but merely a variation of electric displacement, as in the glass of a Leyden jar during charge or discharge, the magnetic effect of the electric movement is precisely the same.

498.3 *If any closed curve be drawn, and the line-integral of the magnetic force taken completely round it...*: Note that here through (499.1) Maxwell discusses the line-integral of *magnetic force* \mathfrak{H} or T (478., *comment*); but in (500.) he will abruptly shift his attention to the surface-integral of *magnetic induction* \mathfrak{B}.

if the closed curve ... linked with the circuit: A closed curve is said to be linked with a circuit (which is itself a closed curve) if the two pass through one another like links of a chain.

if ... current i flows through the closed curve, the line-integral is $4\pi i$: Compare (480.3).

499. *We have reason for believing that even when there is no proper conduction, but merely a variation of electric displacement ... the magnetic effect of the electric movement is precisely the same*: The casual manner in which Maxwell offers this remark disguises its importance—and its boldness. Maxwell initially identified a varying electric displacement as a *current* (60.5) in the sense that it represented *electricity in motion*. Now he claims that varying displacement constitutes current in the additional sense of *originating a magnetic field*.

What is Maxwell's new claim based on? It does not follow from any theoretical principle but is instead a brilliant hypothesis, one that will serve as a cornerstone of his electromagnetic theory of light in Chapter XX. We will see one "reason for believing" it in (607.9, *comment*).

499.1 Again, the value of the line-integral $4\pi i$ does not depend on the nature of the medium in which the closed curve is drawn. It is the same whether the closed curve is drawn entirely through air, or passes through a magnet, or soft iron, or any other substance, whether paramagnetic or diamagnetic.

500.] When a circuit is placed in a magnetic field the mutual action between the current and the other constituents of the field depends on the surface-integral of the magnetic induction through any surface bounded by that circuit. If by any given motion of the circuit, or of part of it, this surface-integral can be increased, there will be a mechanical force tending to move the conductor or the portion of the conductor in the given manner.

499.1 *the value of the line-integral $4\pi i$ does not depend on the nature of the medium in which the closed curve is drawn*: Recall that the magnetic field H within a magnetizable substance is that measured within a longitudinal cavity, so it will have essentially the same value as if the material were not present (398.1); since H does not depend on the material, neither will its line-integral. Mathematically, too, we saw that the value $4\pi i$ represents a difference of *potential* (401.) and must therefore have the same value for all paths sharing the same endpoints. If the integral depended on the material, it might be contrived to depend on the path, which is impossible.

paramagnetic or diamagnetic: These terms were introduced in (425.2).

500. *If by any given motion of the circuit ... this surface-integral can be increased, there will be a mechanical force tending to move the conductor ... in the given manner*: Maxwell is repeating a point he made earlier at (490.) where he stated that "if by the displacement of any portion of the circuit in a given direction the number of lines of induction which pass through the circuit can be increased, this displacement will be aided by the electromagnetic force acting on the circuit."

For example, suppose a rigid conducting loop free to rotate about axis AA. By the right-hand grip rule (page 262), a current i about the loop will produce magnetic induction B_1 within the loop in the direction perpendicular to it. Now apply an external, uniform, magnetic field B_2 in any direction perpendicular to axis AA. If B_2 is in the same general direction as B_1 so as to augment it (the case shown), forces F will act on the arms of the loop indicated; the loop will rotate in the direction shown, bringing B_1 more nearly into coincidence with B_2—thereby increasing the total magnetic induction through the loop. Conversely, if B_2 is in a direction opposing B_1, the forces F will be

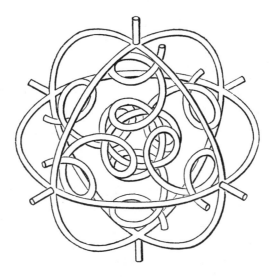

Fig. 25. Relations between the positive directions of motion and of rotation indicated by three right-handed screws.

500.1 The kind of motion of the conductor which increases the surface integral is motion of the conductor perpendicular to the direction of the current and across the lines of induction.

reversed and the loop will rotate in the reverse direction, reducing the opposing component—and, again, increasing the total induction through the loop.

But why does Maxwell specify that the force acts on the *circuit* rather than on the current itself? He will fortify that claim in Art. 501 below.

500.1 *The kind of motion of the conductor which increases the surface integral is ... across the lines of induction*: Faraday had characterized a conductor that moves in this way as "cutting" the lines of force" (*Exp. Res.* ¶¶217ff.). Maxwell's locution emphasizes the conductor's role as enclosing, rather than cutting, lines of induction. Here again we find Maxwell distinctively treating the circuit as a whole—for while short sections of a circuit can "cut" lines, only the whole circuit can enclose or corral them.

Modern textbooks generally neglect Maxwell's rhetoric of the *whole* circuit. The common right-hand rule, for example, is framed in terms of a *section* of conductor: Point the forefinger in the (conventional) direction of the current i and the third finger in the direction of the magnetic induction B; then the thumb will point in the direction of the force F that tends to displace the conductor. Compare this sketch with that in (500., *comment*) to confirm that the respective outcomes are consistent.

500.2 If a parallelogram be drawn, whose sides are parallel and proportional to the strength of the current at any point, and to the magnetic induction at the same point, then the force on unit of length of the conductor is numerically equal to the area of this parallelogram, and is perpendicular to its plane, and acts in the direction in which the motion of turning the handle of a right-handed screw from the direction of the current to the direction of the magnetic induction would cause the screw to move.

500.3 Hence we have a new electromagnetic definition of a line of magnetic induction. It is that line to which the force on the conductor is always perpendicular.

500.4 It may also be defined as a line along which, if an electric current be transmitted, the conductor carrying it will experience no force.

501.] It must be carefully remembered, that the mechanical force which urges a conductor carrying a current across the lines of magnetic force, acts, not on the electric current, but on the conductor which carries it. If the conductor be a rotating disk or a fluid it will move in obedience to this force, and this motion may or may not be accompanied by a change of position of the electric current which it carries. But if the current itself be free to choose any path through a fixed solid conductor or a network of wires, then, when a constant magnetic force is made to act on the system, the path of the current through the conductors is not permanently altered, but after certain transient phenomena, called

500.2 *then the force on unit of length of the conductor is numerically equal to the area of this parallelogram...*: Maxwell is recapitulating a relation he stated earlier; see (490.6–490.8) and comments thereto, where his result is expressed in vector form as $\mathbf{F} = i\,\mathbf{ds} \times \mathfrak{B}$. Since \mathbf{ds} is an element of length of the conductor, the "force on unit length" will be $(i\,\mathbf{ds}/ds) \times \mathfrak{B}$ or $i\mathfrak{B}\sin\theta$; this product is equal to the area of a parallelogram whose respective sides represent the current and the magnetic induction, with θ the angle between them, as Maxwell states.

500.3 *a new electromagnetic definition of a line of magnetic induction*: Maxwell previously defined the line of magnetic induction in Art. 404 as the intersection of two surfaces of no induction.

501. *the mechanical force ... acts, not on the electric current, but on the conductor which carries it*: I questioned this dictum of Maxwell's in the editor's introduction to the present chapter, and again in (500., *comment*).

after certain transient phenomena ... have subsided: Maxwell himself seems to be aware that magnetic force is capable of deflecting the current while the conductor remains fixed; but he thinks such deflection is only momentary and therefore inconsequential. As Hall will later discover, the deflection is *not* momentary; but would not even a momentary deflection challenge Maxwell's insistence that the force on a current-carrying wire exposed to magnetic force acts solely on the conductor, not on the current?

induction currents, have subsided, the distribution of the current will be found to be the same as if no magnetic force were in action.*

501.1 The only force which acts on electric currents is electromotive force, which must be distinguished from the mechanical force which is the subject of this chapter.

*{Mr. Hall has discovered (Phil. Mag. ix. p. 225, x. p. 301, 1880) that a steady magnetic field does slightly alter the distribution of currents in most conductors, so that the statement ... must be regarded as only approximately true.}

501., footnote. The note was supplied by J. J. Thomson, editor for the Third Edition of the *Treatise*. While factually correct, it blithely ignores the main issue, which is the challenge Hall's discovery presents to Maxwell's distinction between the *mechanical force on a conductor* and the *electromotive force on a current* (501.1). In particular, as I noted when introducing the present chapter, the discovery that magnetism acts directly on a current threatens to reduce Maxwell's "mechanical force on the conductor" to the action of a *deflected current* on its own conductor—and so to imply that electric current itself possesses ponderable mass.

Curiously, the telegraph engineer W. H. Preece reported in *Proceedings of the Institution of Electrical Engineers*, vol. 13 (1884) that Hall originally thought that his discovery constituted "an indication that electricity was something separate from matter"—that is, that it actually bolstered Maxwell's opinion. It would be interesting to know what considerations Hall had in mind, if indeed Preece's report is accurate.

Editor's Introduction to
Chapter III

In proceding directly from Chapter I to Chapter III of Part IV we are bypassing an entire chapter devoted to Ampère's investigation of the mutual action of electric currents. In one respect, the very presence of such a chapter in the *Treatise* is anomalous, as Ampère's theory is fully committed to the ideas of action-at-a-distance and active centers of force—ideas very much at variance with Maxwell's own convictions. On the other hand, Maxwell could not compose the *Treatise* as a personal manifesto; it was to serve as the preparatory textbook for the recently-established Natural Sciences Tripos at Cambridge,* and as such it could certainly not neglect Ampère's theory, which held sway as the standard theory of electromagnetic interaction at the time.

It seems clear, however, that Maxwell also entertained a larger purpose. This was to outline the sharp contrast between Ampère's magisterial style of mathematical demonstration and Faraday's more limber and interrogatory approach to nature. The present Chapter III begins with a beautifully appreciative essay that characterizes this contrast. Readers who wish to make the comparison for themselves will enjoy consulting Maxwell's Chapter II of Part IV in a complete edition of the *Treatise*. The chapter is demanding but highly rewarding.

Electromechanical and electromotive force

In the present chapter Maxwell continues attention to the content, no less than the style, of Faraday's experimentation. In Chapter I of Part IV, Maxwell studied Faraday's Rotator (486.), the earliest of Faraday's major electromagnetic discoveries. There Maxwell emphatically distinguished the electromechanical force that moves a conductor from the electromotive force that drives a current (501.1). In the present Chapter III Maxwell will turn to another of Faraday's greatest experimental achievements, the discovery that magnetism is capable of inducing electromotive

*David B. Wilson. "Experimentalists among the mathematicians: Physics in the Cambridge Natural Sciences Tripos, 1851–1900," *Historical Studies in the Physical Sciences* 12:2 (1982).

force. The chapter abounds in fanciful devices, from a railway carriage (532.) whose motion serves no purpose but to generate a feeble electric current capable of deflecting a galvanometer (!) to a gigantic brass arch extending above the earth from pole to equator (533.1)—a structure of truly Babelian proportions. Here is Maxwell's imagination at its most colorful and engaging level.

The "ballistic" galvanometer

Maxwell introduced the galvanometer in Part II Chapter I as a device which, by its angle of deflection, indicates the magnetic action of a current and thereby measures the current itself. The discussion in (240.1, *comment*) applies to a steady current, which is indeed the usual case considered by Maxwell. Occasionally, however, one has to deal not with prolonged currents but with *brief discharges*. A momentary pulse of current does not afford the galvanometer needle enought time to take up a steady position; instead, it throws the needle into oscillation about its rest position—inciting a large angle of swing if the total quantity of electrical discharge is strong, a small angle if weak.

Long before, Faraday had established that the galvanometer's initial maximum deflection is proportional to the quantity of electricity discharged, provided that the electricity fully completes its action before the galvanometer needle has departed very far from its original position.* This requirement is easily met by designing an instrument with high rotational inertia and low restoring torque, thus insuring a long period of oscillation. Perhaps deriving from an image of being "thrown" to a maximum deflection, a galvanometer is today said to be in "ballistic" mode when it is used to detect current pulses that are very brief compared with its period of oscillation.

* *Experimental Researches in Electricity*, Third Series, ¶¶363–367.

CHAPTER III

ON THE INDUCTION OF ELECTRIC CURRENTS.

528.] THE DISCOVERY by Örsted of the magnetic action of an electric current led by a direct process of reasoning to that of magnetization by electric currents, and of the mechanical action between electric currents. It was not, however, till 1831 that Faraday, who had been for some time endeavouring to produce electric currents by magnetic or electric action, discovered the conditions of magneto-electric induction. The method which Faraday employed in his researches consisted in a constant appeal to experiment as a means of testing the truth of his ideas, and a constant cultivation of ideas under the direct influence of experiment. In his published researches we find these ideas expressed in language which is all the better fitted for a nascent science, because it is somewhat alien from the style of physicists who have been accustomed to establish mathematical forms of thought.

528.1 The experimental investigation by which Ampère established the laws of the mechanical action between electric currents is one of the most brilliant achievements in science

528.2 The whole, theory and experiment, seems as if it had leaped, full grown and full armed, from the brain of the 'Newton of electricity.' It is perfect in form, and unassailable in accuracy, and it is summed up in a formula from which all the phenomena may be deduced, and which must always remain the cardinal formula of electro-dynamics.

528.3 The method of Ampère, however, though cast into an inductive form, does not allow us to trace the formation of the ideas which guided it. We can scarcely believe that Ampère really discovered the law of action by means of the experiments which he describes. We are led to suspect, what, indeed, he tells us himself,* that he discovered the law by some process which he has not shewn us, and that when he had afterwards built up a perfect demonstration he removed all traces of the scaffolding by which he had raised it.

*Théorie des phenomènes Électrodynamiques, p. 9.

528. *a constant appeal to experiment as a means of testing the truth of his ideas, and a constant cultivation of ideas under the direct influence of experiment*: Faraday's experimental practices, which David Gooding characterized as "thinking *through* doing as well as thinking *about* doing," provides a central basis for Gooding's landmark study *Experiment and the Making of Meaning* (Kluwer, 1990); see especially his Chapters 5 and 6.

528.4 Faraday, on the other hand, shews us his unsuccessful as well as his successful experiments, and his crude ideas as well as his developed ones, and the reader, however inferior to him in inductive power, feels sympathy even more than admiration, and is tempted to believe that, if he had the opportunity, he too would be a discoverer. Every student should therefore read Ampère's research as a splendid example of scientific style in the statement of a discovery, but he should also study Faraday for the cultivation of a scientific spirit, by means of the action and reaction which will take place between the newly discovered facts as introduced to him by Faraday and the nascent ideas in his own mind.

528.5 It was perhaps for the advantage of science that Faraday, though thoroughly conscious of the fundamental forms of space, time, and force, was not a professed mathematician. He was not tempted to enter into the many interesting researches in pure mathematics which his discoveries would have suggested if they had been exhibited in a mathematical form, and he did not feel called upon either to force his results into a shape acceptable to the mathematical taste of the time, or to express them in a form which mathematicians might attack. He was thus left at leisure to do his proper work, to coordinate his ideas with his facts, and to express them in natural, untechnical language.

528.6 It is mainly with the hope of making these ideas the basis of a mathematical method that I have undertaken this treatise.

529.] We are accustomed to consider the universe as made up of parts, and mathematicians usually begin by considering a single particle, and then conceiving its relation to another particle, and so on. This has generally been supposed the most natural method. To conceive of a particle, however, requires a process of abstraction, since all our perceptions are related to extended bodies, so that the idea of the *all* that is in our consciousness at a given instant is perhaps as primitive an idea as that of any individual thing. Hence there may be a mathematical method in which we proceed from the whole to the parts instead of from the parts to the whole. For example, Euclid, in his first book, conceives a line

528.4 *sympathy*: Here meaning conformity of feelings, of inclinations, or of temperament (Oxford English Dictionary).

529. *Hence there may be a mathematical method in which we proceed from the whole to the parts instead of from the parts to the whole*: Notice Maxwell's use of "hence," and the connection he thereby draws between mathematics and consciousness. For Maxwell, it is *because* consciousness proceeds from whole to part that a mathematical method may do the same. Maxwell will in fact develop such a method in Chapter V below, and we will find that it greatly advances his project of "making [Faraday's] ideas the basis of a mathematical method" (528.6).

as traced out by a point, a surface as swept out by a line, and a solid as generated by a surface. But he also defines a surface as the boundary of a solid, a line as the edge of a surface, and a point as the extremity of a line.

529.1 In like manner we may conceive the potential of a material system as a function found by a certain process of integration with respect to the masses of the bodies in the field, or we may suppose these masses themselves to have no other mathematical meaning than the volume-integrals of $\frac{1}{4\pi}\nabla^2\Psi$, where Ψ is the potential.

529.2 In electrical investigations we may use formulae in which the quantities involved are the distances of certain bodies, and the electrifications or currents in these bodies, or we may use formulae which involve other quantities, each of which is continuous through all space.

529.3 The mathematical process employed in the first method is integration along lines, over surfaces, and throughout finite spaces, those employed in the second method are partial differential equations and integrations throughout all space.

529.4 The method of Faraday seems to be intimately related to the second of these modes of treatment. He never considers bodies as existing with nothing between them but their distance, and acting on one another according to some function of that distance. He conceives all space as a field of force, the lines of force being in general curved, and

Euclid, in his first book, conceives a line as traced out by a point...: Referring to *Elements*, Book I. This is a very loose reading of Definition 3; Euclid is by no means as explicit as Maxwell suggests in this example of proceeding from part to whole.

But he also defines a surface as the boundary of a solid...: in Book XI, Def. 2, an example of proceeding from whole to part.

529.1 *In like manner we may conceive...*: Maxwell continues to develop the distinction introduced in the previous paragraph. If we define (gravitational) potential by integration with respect to individual masses, we proceed in the direction from part to whole. But if we define the masses themselves in terms of the potential, we proceed from whole to part.

$\frac{1}{4\pi}\nabla^2\Psi$: Maxwell discussed Laplace's Operator ∇^2 in Art. 77.

529.2 *we may use ... the distances of certain bodies, and the electrifications or currents in these bodies...*: The quantities mentioned are *discrete*—in fact this seems to have been the hallmark of being a "part" in the foregoing examples.

or we may use ... other quantities, each of which is continuous throughout all space: Continuity throughout all space—the very opposite of discreteness—thus begins to stand forth as the hallmark of being a "whole."

529.3 *the first method ... the second method*: The first method deals with discrete, bounded quantities, the second with continuous ones.

those due to any body extending from it on all sides, their directions being modified by the presence of other bodies. He even speaks* of the lines of force belonging to a body as in some sense part of itself, so that in its action on distant bodies it cannot be said to act where it is not. This, however, is not a dominant idea with Faraday. I think he would rather have said that the field of space is full of lines of force, whose arrangement depends on that of the bodies in the field, and that the mechanical and electrical action on each body is determined by the lines which abut on it.

PHENOMENA OF MAGNETO-ELECTRIC INDUCTION.**

530.]] 1. *Induction by Variation of the Primary Current.*

Let there be two conducting circuits, the Primary and the Secondary circuit, The primary circuit is connected with a voltaic battery by which the primary current may be produced, maintained, stopped, or reversed. The secondary circuit includes a galvanometer to indicate any currents which may be formed in it. This galvanometer is placed at such a distance from all parts of the primary circuit that the primary current has no sensible direct influence on its indications.

*Exp. Res., vol. ii. p. 293; vol. iii. p. 447.

** Read Faraday's *Experimental Researches,* Series i and ii.

529.4 *This, however, is not a dominant idea with Faraday*: Contrary to Maxwell's opinion, a willingness to dissolve bodies into bundles of lines of force becomes an increasingly dominant idea with Faraday. We find it in his *A Speculation touching Electric Conduction and the Nature of Matter* (*Phil. Mag.*, 1844), in the 11th and 28th Series of Experimental Researches, and in his magnificent essay *On the Physical Character of the Lines of Magnetic Force* (*Phil. Mag.*, 1852).

I think he would rather have said...: Maxwell's uncharacteristically presumptuous revision of Faraday's view grants to "bodies" an ontological status at least equal to that of "lines." This seems surprising in light of Maxwell's expressed appreciation for mathematical methods that emphasize the continuous over the discrete. What is there about "bodies" that Maxwell is reluctant to see subverted?

530., footnote ()**: Let us stop to note, and appreciate, Maxwell's unequivocal command, "Read!" When such a word came to St. Augustine, it proved life-changing (*Confessions*, Book VIII). Maxwell is not inviting us to "see" or "consult" Faraday, to confirm a reference or pluck some useful factlet from Faraday's discussion, but to *take it to heart*—as Maxwell himself must have done when he "resolved to read no more mathematics on the subject till I had first read through Faraday's *Experimental Researches in Electricity*" (Maxwell's Preface, page 4 above).

530.1 Let part of the primary circuit consist of a straight wire, and part of the secondary circuit of a straight wire near and parallel to the first, the other parts of the circuits being at a greater distance from each other.

530.2 It is found that at the instant of sending a current through the straight wire of the primary circuit the galvanometer of the secondary circuit indicates a current in the secondary straight wire in the *opposite* direction. This is called the induced current. If the primary current is maintained constant, the induced current soon disappears, and the primary current appears to produce no effect on the secondary circuit. If now the primary current is stopped, a secondary current is observed, which is in the same direction as the primary current. Every variation of the primary current produces electromotive force in the secondary circuit. When the primary current increases, the electromotive force is in the opposite direction to the current. When it diminishes, the electromotive force is in the same direction as the current. When the primary current is constant, there is no electromotive force.

530.3 These effects of induction are increased by bringing the two wires nearer together. They are also increased by forming them into two circular or spiral coils placed close together, and still more by placing an iron rod or a bundle of iron wires inside the coils.

2. *Induction by Motion of the Primary Circuit.*

530.4 We have seen that when the primary current is maintained constant and at rest the secondary current rapidly disappears.

530.5 Now let the primary current be maintained constant, but let the primary straight wire be made to approach the secondary straight wire. During the approach there will be a secondary current in the *opposite* direction to the primary.

530.2 *at the instant of sending a current through the straight wire of the primary circuit the galvanometer ... indicates a current in the secondary straight wire in the opposite direction*: When the switch is closed, current in the primary wire has the direction of the arrow; and a momentary current flows in the secondary wire with the opposite direction, from *b* to *a*. But when the switch is opened and the primary current stopped, a momentary current flows in the secondary wire with the same direction as the primary current, that is, from *a* to *b*.

530.5 *During the approach there will be a secondary current in the opposite direction to the primary*: The primary wire carries current in the direction of the arrow; while it approaches the secondary, current flows in the secondary wire with the opposite direction, from *b* to *a*. In the next paragraph Maxwell will note that the direction of the secondary current will be reversed if the motion of the primary is reversed.

530.6 If the primary circuit be moved away from the secondary, there will be a secondary current in the *same* direction as the primary.

3. *Induction by Motion of the Secondary Circuit.*

530.7 If the secondary circuit be moved, the secondary current is opposite to the primary when the secondary wire is approaching the primary wire, and in the same direction when it is receding from it.

530.8 In all cases the direction of the secondary current is such that the mechanical action between the two conductors is opposite to the direction of motion, being a repulsion when the wires are approaching, and an attraction when they are receding. This very important fact was established by Lenz.*

4. *Induction by the Relative Motion of a Magnet and the Secondary Circuit.*

530.9 If we substitute for the primary circuit a magnetic shell, whose edge coincides with the circuit, whose strength is numerically equal to that of the current in the circuit, and whose austral face corresponds to the positive face of the circuit, then the phenomena produced by the relative motion of this shell and the secondary circuit are the same as those observed in the case of the primary circuit.

531.] The whole of these phenomena may be summed up in one law. When the number of lines of magnetic induction which pass through

*Pogg., Ann. xxxi, p. 483 (1834).

530.8 *This very important fact...*: Note that the "fact" concerns the *direction* (not the magnitude) of the secondary current.

...was established by Lenz: Heinrich Friedrich Emil Lenz (1804–1865) stated what we now call Lenz's Law in the 1834 article cited in Maxwell's footnote. Of course Faraday's experiments—which indeed first established the existence of induced currents—had also revealed their direction in 1831 (*Experimental Researches in Electricity*, First Series); but Lenz elevated such observations to a general law. In (542.1) below, Maxwell will quote Lenz's Law as follows: "If a constant current flows in the primary A, and if, by the motion of A, or of the secondary circuit B, a current is induced in B, the direction of this induced current will be such that, by its electromagnetic action on A, it tends to oppose the relative motion of the circuits."

530.9 *a magnetic shell whose edge coincides with the circuit*: Recall (482.2).

whose strength is numerically equal to that of the current: Recall (482.2, *comment*) and (484.).

531. *The whole of these phenomena*: That is, the results of all the permutations of motions of conductors and changes of currents, and attending not only to the magnitudes and directions of the induced currents, but also the magnitudes and directions of their associated *electromotive forces*. In Art. 534 Maxwell will note that the induced electromotive force is independent of the nature of the conductor.

the secondary circuit in the positive direction is altered, an electromotive force acts round the circuit, which is measured by the rate of decrease of the magnetic induction through the circuit.

532.] For instance, let the rails of a railway be insulated from the earth, but connected at one terminus through a galvanometer, and let the circuit be completed by the wheels and axle of a railway carriage at a distance x from the terminus. Neglecting the height of the axle above the level of the rails, the induction through the secondary circuit is due to the vertical component of the earth's magnetic force, which in northern latitudes is directed downwards. Hence, if b is the gauge of the railway, the horizontal area of the circuit is bx, and the surface-integral of the magnetic induction through it is Zbx, where Z is the vertical component of the magnetic force of the earth. Since Z is downwards, the

an electromotive force ... is measured by the rate of decrease of the magnetic induction: In (490.1) Maxwell considered a circuit carrying current i and enclosing N lines of induction; and we saw that if there is a change ΔN in N, work $W = i\Delta N$ will be done. This work may be performed, as there discussed, by a mechanical force displacing part or all of the circuit; but it may also take the form of an electromotive force arising about the circuit. Now electromotive force has dimensions of *work/charge*, and the dimensions of current are *charge/time* (237.4, *comment*). Divide both sides of $W = i\Delta N$ by *charge*, and we shall have *work/charge* $= \Delta N / time$. Thus the induced electomotive force is certainly proportional to the rate of change of magnetic induction. That it is *negatively* proportional (electromotive force positive when rate of change is negative) is the substance of Lenz's Law of direction.

532. *let the rails of a railway be insulated from the earth, but connected at one terminus through a galvanometer...*: I have tried to represent Maxwell's charming example in the sketch below. Think of the left-hand wheels as being insulated, while the right-hand pair are joined by their axle, thus completing the circuit.

the surface-integral of the magnetic induction ... is Zbx: that is, the product of area bx and the perpendicular component of induction Z. Although Maxwell actually defines Z as the vertical component of the earth's magnetic force, in air it will be numerically equal to the magnetic induction (404.4).

Galvanometer

lower face of the circuit is to be reckoned positive, and the positive direction of the circuit itself is north, east, south, west, that is, in the direction of the sun's apparent diurnal course.

532.1 Now let the carriage be set in motion, then x will vary, and there will be an electromotive force in the circuit whose value is $-Zb\dfrac{dx}{dt}$.

532.2 If x is increasing, that is, if the carriage is moving away from the terminus, this electromotive force is in the negative direction, or north, west, south, east. Hence the direction of this force through the axle is from right to left. If x were diminishing, the absolute direction of the force would be reversed, but since the direction of the motion of the carriage is also reversed, the electromotive force on the axle is still from

532., continued. *the lower face of the circuit is to be reckoned positive*: That is, the positive vertical direction is taken as *downwards*, the direction of Z.

the positive direction of the circuit itself is north, east, south, west: that is, clockwise (looking down upon the horizontal surface). It is the positive direction of rotation in accordance with the right-hand relation between magnetic induction and current (489.7, *comment*).

the sun's apparent diurnal course: The sun's daily motion is from east to south to west in the northern hemisphere; and the shadow cast by the gnomon of a horizontal sundial moves in the same sense, from west to north to east. Such is in fact thought to be the origin of the "clockwise" direction; it is really the "shadow-wise" direction.

532.1 *an electromotive force ... whose value is* $-Zb\,dx/dt$: that is, the rate of decrease of Zbx, the induction through the circuit, in accordance with (531.).

532.2 *If x is increasing... this electromotive force is in the negative direction*: When x increases, Zbx (the number of lines of induction enclosed by the circuit) increases. In accordance with Lenz's Law, the current induced in the circuit will have a direction that opposes this increase. That direction is negative, that is, counterclockwise (looking down upon the horizontal surface)—for by the right-hand grip rule (page 262), the magnetic effect of a counterclockwise current will be opposite to Z and so will oppose the increase in induction through area bx.

Galvanometer

right to left, the observer in the carriage being always supposed to move face forwards. In southern latitudes, where the south end of the needle dips, the electromotive force on a moving body is from left to right.

532.3 Hence we have the following rule for determining the electromotive force on a wire moving through a field of magnetic force. Place, in imagination, your head and feet in the positions occupied by the ends of a compass-needle which point north and south respectively; turn your face in the forward direction of motion, then the electromotive force due to the motion will be from left to right.

533.] As these directional relations are important, let us take another illustration. Suppose a metal girdle laid round the earth at the equator, and a metal wire laid along the meridian of Greenwich from the equator to the north pole.

533.1 Let a great quadrantal arch of metal be constructed [Fig 31], of which one extremity is pivoted on the north pole, while the other is carried round the equator, sliding on the great girdle of the earth, and

Fig. 31.

532.3 *Place ... your head and feet in the positions occupied by the ends of a compass-needle...*: Another body-centered image! At first it may appear to enjoy a close affinity with the railway example, having merely eliminated the carriage while retaining the observer. But notice that the new image also shifts focus away from the *whole circuit* (and the total number of lines of induction it embraces) to a *portion* of it, represented by "left" and "right" in the observer's field of vision.

533. *As these directional relations are important, let us take another illustration*: Perhaps Maxwell is acting pedagogically here, multiplying examples in order to assist the student in mastering these relations. But I think we can detect a larger concern; the illustration Maxwell is about to introduce seems designed to restore that emphasis on the *integral circuit* which was abandoned in (532.3).

We have seen earlier signs that Maxwell is determined to honor the wholeness of electric circuits. The method of magnetic shells (Art. 482) enforces the representation of currents as closed circuits; Maxwell went so far as to treat even the *infinite straight wire* as part of a whole circuit (Art. 485).

533.1 *a great quadrantal arch...*: An arch of 90°, the quadrant of a circle. It is fascinating to see Maxwell progress from an image based on the human body to one that looks down upon the earth and which, therefore, until 12 April 1961 would correspond to *no human experience*. Note that Maxwell's Figure 31 actually shows *two* quadrantal arches, each running from a pole to the equator, along with a semicircular arch running from pole to pole.

following the sun in his daily course. There will then be an electromotive force along the moving quadrant, acting from the pole towards the equator.

533.2 The electromotive force will be the same whether we suppose the earth at rest and the quadrant moved from east to west, or whether we suppose the quadrant at rest and the earth turned from west to east. If we suppose the earth to rotate, the electromotive force will be the same whatever be the form of the part of the circuit fixed in space of which one end touches one of the poles and the other the equator. The current in this part of the circuit is from the pole to the equator.

Fig. 31, repeated

533.1, continued. *following the sun in his daily course*: that is, moving from east to west, presumably with the earth stationary. It is rather awkward that Figure 31 appears to depict the *quadrant* at rest and the *earth* moving from west to east. In the next paragraph Maxwell will point out that both cases eventuate in the same result; nevertheless it may be useful to offer this redrawn figure, which illustrates more directly what Maxwell is describing.

an electromotive force along the moving quadrant, acting from the pole towards the equator: The circuit in this case is formed by *QR*, the meridian wire (533.), axis *RN*, quadrantal arch *NO*, shaft *OP*, and *PQ*, a portion of the metal girdle (533.) supposed to have been laid around the terrestrial equator. Thus the terrestrial lines of induction that are embraced by this circuit will all intersect the spherical triangle *PQR*. Apply either the law of (531.) or the rule of (532.3) to confirm that the electromotive force will indeed act from the pole towards the equator, that is, from *N* to *O*.

533.2 *The electromotive force will be the same whether we suppose the earth at rest...*: Whether *NO* moves westward or *QR* moves towards the east, each will be moving through the earth's lines of induction at the same speed and will therefore experience the same magnitude of electromotive force. Directions *NO* and *QR* are opposite geographically, but they have the same direction with respect to the circuit; the induced electromotive force will have the direction of the unadorned arrows in either case.

Alternatively, whether *NO* moves west or *QR* moves east, the number of lines of induction embraced by circuit *QRNOPQ* (or by triangle *PQR*) will increase at the same rate and will therefore produce the same electromotive force.

533.3 The other part of the circuit, which is fixed with respect to the earth, may also be of any form, and either within or without the earth. In this part the current is from the equator to either pole.

534.] The intensity of the electromotive force of magneto-electric induction is entirely independent of the nature of the substance of the conductor in which it acts, and also of the nature of the conductor which carries the inducing current.

534.1 To shew this, Faraday* made a conductor of two wires of different metals insulated from one another by a silk covering, but twisted together, and soldered together at one end. The other ends of the wires were connected with a galvanometer. In this way the wires were similarly situated with respect to the primary circuit, but if the electromotive force were stronger in the one wire than in the other it would produce a current which would be indicated by the galvanometer. He found, however, that such a combination may be exposed to the most powerful electromotive forces due to induction without the galvanometer being affected. He also found that whether the two branches of the compound conductor consisted of two metals, or of a metal and an electrolyte, the galvanometer was not affected.†

534.2 Hence the electromotive force on any conductor depends only on the form and the motion of that conductor, together with the strength, form, and motion of the electric currents in the field.

535.] Another negative property of electromotive force is that it has of itself no tendency to cause the mechanical motion of any body, but only to cause a current of electricity within it.

<div align="center">* Exp. Res., 195. † Ib., 200.</div>

534. *The intensity of the electromotive force of magneto-electric induction is entirely independent of the nature of the substance of the conductor in which it acts*: It follows from this principle that electromotive force will be developed about any path that encircles a surface through which the magnetic induction changes, *even if that path is not a conductor at all*. This circumstance will become an important element in Maxwell's electromagnetic theory of light.

535. *electromotive force … has of itself no tendency to cause the mechanical motion of any body, but only to cause a current of electricity within it*: Earlier, Maxwell distinguished between mechanical force and electromotive force when he stated that "the mechanical force which urges a conductor carrying a current across the lines of magnetic force, acts, not on the electric current, but on the conductor which carries it" (501.). With the present statement, that distinction becomes absolute. But recall Hall's discovery, which, as we have seen (501., *comment*), militates strongly against Maxwell's distinction.

535.1 If it actually produces a current in the body, there will be mechanical action due to that current, but if we prevent the current from being formed, there will be no mechanical action on the body itself. If the body is electrified, however, the electromotive force will move the body, as we have described in Electrostatics.

* * *

537.1 The theory of the application of the galvanometer to the measurement of transient currents will be considered more at length in Art. 748. At present it is sufficient for our purpose to observe that as long as the galvanometer needle is near its position of equilibrium the deflecting force of the current is proportional to the current itself, and if the whole time of action of the current is small compared with the period of vibration of the needle, the final velocity of the magnet will be proportional to the total quantity of electricity in the current. Hence, if two currents pass in rapid succession, conveying equal quantities of electricity in opposite directions, the needle will be left without any final velocity.

537.2 Thus, to shew that the induction currents in the secondary circuit, due to the closing and the breaking of the primary circuit, are equal in total quantity but opposite in direction, we may arrange the primary circuit in connexion with the battery, so that by touching a key the current may be sent through the primary circuit, or by removing the finger the contact may be broken at pleasure. If the key is pressed down

535.1 *there will be mechanical action due to that current*: Recall that Lenz's Law characterized the current induced in one circuit by another circuit as tending, "by its electromagnetic action," to oppose the relative motion of the circuits (530.8, *comment*). It is this electromagnetic action that Maxwell here characterizes as "mechanical," and acting only on the conductor.

537.1 *the application of the galvanometer to the measurement of transient currents will be considered more at length in Art. 748*: Article 748 is not included in the present selections, but I discussed this application—today called the "ballistic" mode of the galvanometer—in the editor's introduction to the present chapter.

the final velocity of the magnet will be proportional to the total quantity of electricity in the current: When a body oscillates with simple harmonic motion, its initial acceleration, initial velocity, and amplitude of deflection are all proportional to each other. Thus Maxwell's observations concerning the initial velocity of the needle correspond to my remarks, in the editor's introduction, about the angle of maximum deflection.

537.2 *a key*: That is, a momentary switch, as in "telegraph key." According to the Oxford English Dictionary, the earliest instances of this usage were the keys of musical instruments; the term was later extended to any device pressed down by a finger.

for some time, the galvanometer in the secondary circuit indicates, at the time of making contact, a transient current in the opposite direction to the primary current. If contact be maintained, the induction current simply passes and disappears. If we now break contact, another transient current passes in the opposite direction through the secondary circuit, and the galvanometer needle receives an impulse in the opposite direction.

537.3 But if we make contact only for an instant, and then break contact, the two induced currents pass through the galvanometer in such rapid succession that the needle, when acted on by the first current, has not time to move a sensible distance from its position of equilibrium before it is stopped by the second, and, on account of the exact equality between the quantities of these transient currents, the needle is stopped dead.

537.4 If the needle is watched carefully, it appears to be jerked suddenly from one position of rest to another position of rest very near the first.

537.5 In this way we prove that the quantity of electricity in the induction current, when contact is broken, is exactly equal and opposite to that in the induction current when contact is made.

538.] Another application of this method is the following, which is given by Felici in the second series of his *Researches*. It is always possible to find many different positions of the secondary coil B, such that the making or the breaking of contact in the primary coil A produces no induction current in B. The positions of the two coils are in such cases said to be *conjugate* to each other.

538.1 Let B_1 and B_2 be two of these positions. If the coil B be suddenly moved from the position B_1 to the position B_2, the algebraical sum of the transient currents in the coil B is exactly zero, so that the galvanometer needle is left at rest when the motion of B *is* completed.

538.2 This is true in whatever way the coil B is moved from B_1 *to* B_2, and also whether the current in the primary coil A be continued constant, or made to vary during the motion.

538.3 Again, let B' be any other position of B not conjugate to A, so that the making or breaking of contact in A produces an induction current when B is in the position B'.

538.4 Let the contact be made when B is in the conjugate position B_1, there will be no induction current. Move B to B', there will be an

538. *conjugate*: that is, joined together, especially in a pair; for example, the conjugate axes of an ellipse or hyperbola.

induction current due to the motion, but if B is moved rapidly to B', and the primary contact then broken, the induction current due to breaking contact will exactly annul the effect of that due to the motion, so that the galvanometer needle will be left at rest. Hence the current due to the motion from a conjugate position to any other position is equal and opposite to the current due to breaking contact in the latter position.

538.5 Since the effect of making contact is equal and opposite to that of breaking it, it follows that the effect of making contact when the coil B is in any position B' is equal to that of bringing the coil from any conjugate position B_1 to B' while the current is flowing through A.

538.6 If the change of the relative position of the coils is made by moving the primary circuit instead of the secondary, the result is found to be the same.

539.] It follows from these experiments that the total induction current in B during the simultaneous motion of A from A_1 to A_2, and of B from B_1 to B_2, while the current in A changes from γ_1 to γ_2, depends only on the initial state A_1, B_1, γ_1, and the final state A_2, B_2, γ_2, and not at all on the nature of the intermediate states through which the system may pass.

539.1 Hence the value of the total induction current must be of the form

$$F\left(A_2, B_2, \gamma_2\right) - F\left(A_1, B_1, \gamma_1\right),$$

where F is a function of A, B, and γ.

539.2 With respect to the form of this function, we know, by Art. 536, that when there is no motion, and therefore $A_1 = A_2$ and $B_1 = B_2$, the induction current is proportional to the primary current. Hence γ

539. *total induction current*: that is, the *total quantity of electricity* that passes during the time over which the induced current exists. Modern writers would not call this quantity a "current," but rather the *result* of current.

539.1 *the total induction current must be of the form* $F\left(A_2, B_2, \gamma_2\right) - F\left(A_1, B_1, \gamma_1\right)$: A formal expression of what was stated in the previous paragraph—that the total induction current (in coil B) depends only on the initial and final positions of coils A and B, and on the initial and final values of the current in coil A. Note that it shares this property (of depending only on initial and final states) with the line-integral that establishes a *potential* (14.1). Maxwell will make use of this similitude in (539.4) below.

539.2 *we know, by Art. 536, that when there is no motion … the induction current is proportional to the primary current*: that is, to the *change* in primary current, since in the experiment considered the primary current is either made or broken. I have not included Art. 536 in the present selections, but Maxwell there offers a résumé of Felici's experiment establishing the cited result.

enters simply as a factor, the other factor being a function of the form and position of the circuits A and B.

539.3 We also know that the value of this function depends on the relative and not on the absolute positions of A and B, so that it must be capable of being expressed as a function of the distances of the different elements of which the circuits are composed, and of the angles which these elements make with each other.

539.4 Let M be this function, then the total induction current may be written

$$C\big(M_1\gamma_1 - M_2\gamma_2\big),$$

where C is the conductivity of the secondary circuit, and M_1, γ_1 are the original, and M_2, γ_2 the final values of M and γ.

539.5 These experiments, therefore, shew that the total current of induction depends on the change which takes place in a certain quantity, $M\gamma$, and that this change may arise either from variation of the

Hence γ enters simply as a factor, the other factor being a function of the form and position of the circuits: We saw in (539.1) that the total induction current can be expressed as the change in some function $F(A, B, \gamma)$. Now as was just noted, the change in F is proportional to a change in γ when A and B are constant. Thus F must be the product of *at least* two factors, one being γ and another representing the geometrical particulars. As we shall soon see, additional factors are not ruled out.

539.3 *a function of the distances of the different elements ... and of the angles which these elements make with each other*: that is, a function representing the geometrical particulars of the experiment, as noted in the previous comment. We may recall that such a function was defined earlier (492.2); it was designated M, the mutual potential between two closed circuits.

539.4 *Let M be this function, then the total induction current may be written* $C(M_1\gamma_1 - M_2\gamma_2)$: In accordance with the concluding comment to (539.2), Maxwell has formed the product $M\gamma$ and expressed the total induction current in terms of a change in this product. It is tempting to identify the M in this expression with the mutual potential M defined in (492.2); and Maxwell means us to be tempted. Ultimately he will, in fact, identify the two; but for the present he simply lets the uniform notation stand as a tacit sign of his theoretical intention. Notice too that Maxwell has also introduced a third factor, C.

where C is the conductivity of the secondary circuit: Conductivity is the reciprocal of resistance (241.3), expressed per unit cross-sectional area of the conductor, and multiplied by the length of the conductor. It expresses the ability of a given material to conduct electric current. Maxwell gives explicit expression to this quantity because the total induction current must certainly depend on the material of the secondary conductor; recall his discussion of Ohm's Law in Art. 241, above.

primary current γ, or from any motion of the primary or secondary circuit which alters M.

540.] The conception of such a quantity, on the changes of which, and not on its absolute magnitude, the induction current depends, occurred to Faraday at an early stage of his *Researches*.* He observed that the secondary circuit, when at rest in an electromagnetic field which remains of constant intensity, does not shew any electrical effect, whereas, if the same state of the field had been suddenly produced, there would have been a current. Again, if the primary circuit is removed from the field, or the magnetic forces abolished, there is a current of the opposite kind: He therefore recognised in the secondary circuit, when in the electromagnetic field, a 'peculiar electrical condition of matter,' to which he gave the name of the Electrotonic State. He afterwards found that he could dispense with this idea by means of considerations founded on the lines of magnetic force,† but even in his latest *Researches*,‡ he says, 'Again and again the idea of an *electrotonic* state§ has been forced on my mind.'

540.1 The whole history of this idea in the mind of Faraday, as shewn in his published *Researches,* is well worthy of study. By a course of experiments, guided by intense application of thought, but without the aid of mathematical calculations, he was led to recognise the existence of something which we now know to be a mathematical quantity, and which may even be called the fundamental quantity in the theory of electromagnetism. But as he was led up to this conception by a purely experimental path, he ascribed to it a physical existence, and supposed it to be a peculiar condition of matter, though he was ready to abandon this theory as soon as he could explain the phenomena by any more familiar forms of thought.

* *Exp. Res.*, series i. 80. † Ib., series ii. 242. ‡ Ib., 3269.
§ Ib., 60, 1114, 1661, 1729, 1733.

540.1 *he ascribed to it a physical existence ... though was ready to abandon this theory...*: Maxwell is referring to Faraday's acknowledgment that "the reasons which induce me to suppose a particular state in the wire have disappeared" (*Exp. Res.*, Second Series, ¶242). Here Maxwell seems greatly to overestimate Faraday's "readiness" to abandon the electrotonic state as a physical condition in the wire. For in the same passage—a passage cited in Maxwell's own footnote above—Faraday doubts that a wire could be *entirely unaffected* by the nearby presence of a powerful electric current; moreover, as Maxwell noted in (540.), the idea of an electrotonic state held a continuing fascination for Faraday which he was never able to disregard.

I think it highly unlikely that Maxwell is simply overlooking these signs of Faraday's real attachment to the electrotonic state; but he may wish to minimize

540.2 Other investigators were long afterwards led up to the same idea by a purely mathematical path, but, so far as I know, none of them recognised, in the refined mathematical idea of the potential of two circuits, Faraday's bold hypothesis of an electrotonic state. Those, therefore, who have approached this subject in the way pointed out by those eminent investigators who first reduced its laws to a mathematical form, have sometimes found it difficult to appreciate the scientific accuracy of the statements of laws which Faraday, in the first two series of his *Researches,* has given with such wonderful completeness.

540.3 The scientific value of Faraday's conception of an electrotonic state consists in its directing the mind to lay hold of a certain quantity, on the changes of which the actual phenomena depend. Without a much greater degree of development than Faraday gave it, this conception does not easily lend itself to the explanation of the phenomena. We shall return to this subject again in Art. 584.

541.] A method which, in Faraday's hands, was far more powerful is that in which he makes use of those lines of magnetic force which were always in his mind's eye when contemplating his magnets or electric currents, and the delineation of which by means of iron filings he rightly regarded* as a most valuable aid to the experimentalist.

541.1 Faraday looked on these lines as expressing, not only by their direction that of the magnetic force, but by their number and concentration the intensity of that force, and in his later *Researches*† he shews

* *Exp. Res.,* 3234. † Ib., 3122.

Faraday's characterization of it as a condition *in the wire.* I am led to this suggestion inasmuch as Maxwell will himself, in Part IV Chapter VIII, identify Faraday's electrotonic state with the vector-potential **A** (Art. 405)—or, rather, with the loop integral of the same, which measures the *electrokinetic momentum* associated with a current (Art. 590). Maxwell will leave us in no doubt that this momentum is no mere mathematical artifact but a real physical quantity operating in nature. It is, however, not confined to the conductor but is present everywhere in the space surrounding it. Perhaps, then, Maxwell wishes to show that Faraday would not have been so committed to localizing the electrotonic state as to prevent his assent to Maxwell's reinterpretation.

540.2 *none of them recognized, in the refined mathematical idea of the potential of two circuits, Faraday's bold hypothesis of an electrotonic state*: As I mentioned in the preceding comment, Maxwell will link Faraday's electrotonic state to the vector-potential of magnetic induction; he will similarly connect the vector-potential to the quantity M of (539.4) and reveal it as identical to M the mutual potential of two circuits (492.2). Thus the mutual potential and Faraday's electrotonic state are intimately connected, even though several "eminent investigators" failed to notice a relation.

how to conceive of unit lines of force. I have explained in various parts of this treatise the relation between the properties which Faraday recognised in the lines of force and the mathematical conditions of electric and magnetic forces, and how Faraday's notion of unit lines and of the number of lines within certain limits may be made mathematically precise. See Arts. 82, 404, 490.

541.2 In the first series of his *Researches** he shews clearly how the direction of the current in a conducting circuit, part of which is moveable, depends on the mode in which the moving part cuts through the lines of magnetic force.

541.3 In the second series† he shews how the phenomena produced by variation of the strength of a current or a magnet may be explained, by supposing the system of lines of force to expand from or contract towards the wire or magnet as its power rises or falls.

541.4 I am not certain with what degree of clearness he then held the doctrine afterwards so distinctly laid down by him,‡ that the moving conductor, as it cuts the lines of force, sums up the action due to an area or section of the lines of force. This, however, appears no new view of the case after the investigations of the second series§ have been taken into account.

541.5 The conception which Faraday had of the continuity of the lines of force precludes the possibility of their suddenly starting into existence in a place where there were none before. If, therefore, the number of lines which pass through a conducting circuit is made to vary, it can only be by the circuit moving across the lines of force, or else by the lines of force moving across the circuit. In either case a current is generated in the circuit.

541.6 The number of the lines of force which at any instant pass through the circuit is mathematically equivalent to Faraday's earlier conception of the electrotonic state of that circuit, and it is represented by the quantity $M\gamma$.

*Ib., 114. †Ib., 238. ‡Ib., 3082, 3087, 3113.
§*Exp. Res.*, 217, &c.

541.4 *I am not certain with what degree of clearness he then held the view ... that the moving conductor, as it cuts the lines of force, sums up the action...*: Faraday stated this view with unexceptionable clarity in his Twenty-eighth Series. But Maxwell finds its essential elements already in the Second Series, whether Faraday then appreciated their full significance or not.

541.6 *The number of lines of force which at any instant pass through the circuit is mathematically equivalent to ... the electrotonic state...*: Faraday viewed the electrotonic state as a condition of a circuit, any change in which would prompt an

541.7 It is only since the definitions of electromotive force. Arts. 69, 274, and its measurement have been made more precise, that we can enunciate completely the true law of magneto-electric induction in the following terms:—

541.8 The total electromotive force acting round a circuit at any instant is measured by the rate of decrease of the number of lines of magnetic force which pass through it.

541.9 When integrated with respect to the time this statement becomes:—

541.10 The time-integral of the total electromotive force acting round any circuit, together with the number of lines of magnetic force which pass through the circuit, is a constant quantity.

541.11 Instead of speaking of the number of lines of magnetic force, we may speak of the magnetic induction through the circuit, or the surface-integral of magnetic induction extended over any surface bounded by the circuit.

541.12 We shall return again to this method of Faraday.

<p style="text-align:center">* * *</p>

induced current in that circuit. Similarly, as we have seen throughout the earlier portions of this chapter and most recently in (541.5), any change in the number of lines of force embraced by a circuit brings about a proportional induced current in that circuit.

and it is represented by the quantity $M\gamma$: As we saw in (539.5), a change in the product $M\gamma$ is proportional to the total induction current. The number of lines and the product $M\gamma$ being thus both proportional to the total induction current, they are proportional to one another; and the quantity $M\gamma$ may therefore be said to *represent* the number of lines.

541.11 *Instead of speaking of the number of lines of magnetic force, we may speak of the magnetic induction...*: This is true in air or vacuum, where magnetic force \mathfrak{H} and magnetic induction \mathfrak{B} are equal (404.4). In other materials we must bear in mind that $\mathfrak{B} = \mu\mathfrak{H}$ (428.5, *comment*).

541.12 *We shall return again to this method of Faraday*: in Part IV, Chapter VIII, where Maxwell will reconstruct (in thought) Faraday's Moving Wire.

CHAPTER IV.

ON THE INDUCTION OF A CURRENT ON ITSELF.

546.] FARADAY has devoted the ninth series of his *Researches* to the investigation of a class of phenomena exhibited by the current in a wire which forms the coil of an electromagnet.

546.1 Mr. Jenkin has observed that, although it is impossible to produce a sensible shock by the direct action of a voltaic system consisting of only one pair of plates, yet, if the current is made to pass through the coil of an electromagnet, and if contact is then broken between the extremities of two wires held one in each hand, a smart shock will be felt. No such shock is felt on making contact.

546.2 Faraday shewed that this and other phenomena, which he describes, are due to the same inductive action which he had already observed the current to exert on neighbouring conductors. In this case, however, the inductive action is exerted on the same conductor which carries the current, and it is so much the more powerful as the wire itself is nearer to the different elements of the current than any other wire can be.

547.] He observes, however,* that 'the first thought that arises in the mind is that the electricity circulates with something like momentum or inertia in the wire.' Indeed, when we consider one particular wire only,

*Exp. Res., 1077.

546.1 *direct action*: action not mediated by anything else. In this example, think of grasping with the hands the positive and negative terminals of a single voltaic cell; no shock is experienced.

yet, if the current is made to pass through the coil of an electromagnet, and if contact is then broken...: A shock *is* felt under these circumstances. But that seems paradoxical, since neither the electromagnet nor the action of breaking contact would appear capable of augmenting the power of the single cell.

the phenomena are exactly analogous to those of a pipe full of water flowing in a continued stream. If while the stream is flowing we suddenly close the end of the pipe, the momentum of the water produces a sudden pressure, which is much greater than that due to the head of water, and may be sufficient to burst the pipe.

547.1 If the water has the means of escaping through a narrow jet when the principal aperture is closed, it will be projected with a velocity much greater than that due to the head of water, and if it can escape through a valve into a chamber, it will do so, even when the pressure in the chamber is greater than that due to the head of water.

547.2 It is on this principle that the hydraulic ram is constructed, by which a small quantity of water may be raised to a great height by means of a large quantity flowing down from a much lower level.

547. *a sudden pressure, which is much greater than that due to the head of water*: Note the reappearance of a theme noted earlier (546.1), that of an agent whose effect appears to exceed its own capability.

547.2 *the hydraulic ram*: We may apply Maxwell's description in the preceding paragraphs to this diagram of Montgolfier's 1796 hydraulic ram, one of the earliest designs. Suppose that the waste-valve *B* is open and the delivery valve *C* is closed. Then (1) "while the stream is flowing we suddenly close the end of the pipe": This happens when the pressure exerted by the

flowing water on the face of valve *B* causes that valve to close. Next (2) "the momentum of the water produces a sudden pressure, which is much greater that that due to the head of water": The moving column of water in effect collides with the closed valve *B*; the more abruptly it is brought to a halt, the greater the pressure it develops—just as a hammer can develop a force that far exceeds its own weight. Finally (3) "if it can escape through a valve into a chamber, it will do so, even when the pressure in the chamber is greater than that due to the head of water": The pressure in chamber *D* builds up through repeated strokes of the sort described in step 2 and will endeavor to approach the peak pressure developed in each of those strokes. Thus water may be delivered through the riser pipe *E* to a height greater than that of its source.

a small quantity of water may be raised to a great height by means of a large quantity flowing down from a much lower level: Recall that in (35.6) Maxwell expressed the quantity *Energy* in various ways, including the product of mass, gravitation, and height. Under constant gravitation, then, a device that transforms a large mass moving through a small height into a small mass moving through a great height may be understood as transferring a constant amount of energy from the larger mass to the smaller.

548.] These effects of the inertia of the fluid in the tube depend solely on the quantity of fluid running through the tube, on its length, and on its section in different parts of its length. They do not depend on anything outside the tube, nor on the form into which the tube may be bent, provided its length remains the same.

548.1 With a wire conveying a current this is not the case, for if a long wire is doubled on itself the effect is very small, if the two parts are separated from each other it is greater, if it is coiled up into a helix it is still greater, and greatest of all if, when so coiled, a piece of soft iron is placed inside the coil.

548.2 Again, if a second wire is coiled up with the first, but insulated from it, then, if the second wire does not form a closed circuit, the phenomena are as before, but if the second wire forms a closed circuit, an induction current is formed in the second wire, and the effects of self-induction in the first wire are retarded.

549.] These results shew clearly that, if the phenomena are due to momentum, the momentum is certainly not that of the electricity in the wire, because the same wire, conveying the same current, exhibits effects which differ according to its form; and even when its form remains the same, the presence of other bodies, such as a piece of iron or a closed metallic circuit, affects the result.

550.] It is difficult, however, for the mind which has once recognised the analogy between the phenomena of self-induction and those of the motion of material bodies, to abandon altogether the help of this analogy, or to admit that it is entirely superficial and misleading.

548.1 *With a wire conveying a current this is not the case*: Maxwell's "this" refers to the independence displayed by hydraulic phenomena. In contrast, effects such as the shock described in (546.1) are not independent of the shape of the wire.

549. *if the phenomena are due to momentum, the momentum is certainly not that of the electricity in the wire*: since in some crucial respects the electricity fails to exhibit the same behavior as water flowing in a pipe. But why then does Maxwell continue to speak of "momentum" at all? In the face of the observed discrepancies, should he not rather abandon the electric/hydraulic metaphor entirely? That is what Faraday himself did: in the very passage Maxwell quoted in (547.) above, Faraday went on to state that "the idea of momentum cannot apply." In the next paragraph Maxwell will indicate why he departs from Faraday on this occasion.

550. *It is difficult ... to abandon altogether the help of this analogy*: The difficulty is evidently not a manifestation of mere mental habitude, since Maxwell credits the analogy with offering actual "help." What does he count as *help*, and how does this, or any other analogy, provide it?

The fundamental dynamical idea of matter, as capable by its motion of becoming the recipient of momentum, and of energy, is so interwoven with our forms of thought that whenever we catch a glimpse of it in any part of nature, we feel that a path is before us leading, sooner or later, to the complete understanding of the subject.

551.] In the case of the electric current, we find that, when the electromotive force begins to act, it does not at once produce the full current, but that the current rises gradually. What is the electromotive force doing during the time that the opposing resistance is not able to balance it? It is increasing the electric current.

551.1 Now an ordinary force, acting on a body in the direction of its motion, increases its momentum, and communicates to it kinetic energy, or the power of doing work on account of its motion.

551.2 In like manner the unresisted part of the electromotive force has been employed in increasing the electric current. Has the electric current, when thus produced, either momentum or kinetic energy?

550. (continued) *The fundamental dynamical idea of matter ... so interwoven with our forms of thought ... complete understanding of the subject*: I believe Maxwell is saying that when we represent phenomena in terms of *matter in motion* we present them in what is, to us, their most intelligible form. The very length of Maxwell's sentence seems to express a vision in which matter, motion, momentum, energy, and thought stand in a unity that is not to be subdivided.

Perhaps now we can better understand Maxwell's unabashed invocation of momentum in (549): the problem with Faraday's metaphor (547.) was not in the idea of *momentum* but in the assumption that such momentum is *localized in the conductor*.

551. *the electromotive force ... does not at once produce the full current*: No doubt some degree of such sluggishness occurs in every electric circuit; but it is greatly enhanced, and becomes clearly observable, under the same conditions that Maxwell earlier identified as intensifying the electric shock: lengthening the wire, coiling it up, and inserting an iron core (548.1). Thus *the gradual buildup of electric current in a conductor* is to be included, along with the electric shock, in the class of phenomena to which Maxwell is devoting this chapter.

551.1–551.2 *Now an ordinary force...*: In these two paragraphs Maxwell continues to cultivate the analogy Faraday discarded. An ordinary (that is, mechanical) force does not instantly impart a definite velocity to the body to which it is applied but accelerates it gradually, in this respect resembling the electromotive force's production of electric current. A mechanical force increases momentum and kinetic energy in the body to which it is applied; does electromotive force analogously impart either momentum or kinetic energy to electric current?

Notice that since Maxwell is not assuming that either one of these quantities is localized in the conductor, he is *not* reopening the question of *fluid theories of electricity* (36.1–37.5).

551.3 We have already shown that it has something very like momentum, that it resists being suddenly stopped, and that it can exert, for a short time, a great electromotive force.

551.4 But a conducting circuit in which a current has been set up has the power of doing work in virtue of this current, and this power cannot be said to be something very like energy, for it is really and truly energy.

551.5 Thus, if the current be left to itself, it will continue to circulate till it is stopped by the resistance of the circuit. Before it is stopped, however, it will have generated a certain quantity of heat, and the amount of this heat in dynamical measure is equal to the energy originally existing in the current.

551.6 Again, when the current is left to itself, it may be made to do mechanical work by moving magnets, and the inductive effect of these motions will, by Lenz's law, stop the current sooner than the resistance of the circuit alone would have stopped it. In this way part of the energy of the current may be transformed into mechanical work instead of heat.

552.] It appears, therefore, that a system containing an electric current is a seat of energy of some kind; and since we can form no conception of an electric current except as a kinetic phenomenon,* its

* Faraday, *Exp. Res.* 283.

551.3 *something very like momentum*: Maxwell is partially echoing the words of Faraday quoted at (547.) above. He is still using the language of similitude since the "momentum" of an electric current resembles the momentum of a material body only to the extent that electromotive force can be said to resemble ordinary mechanical force.

551.4 [*not*] *something very like energy, for it is really and truly energy*: Where energy is concerned, Maxwell wholeheartedly thrusts aside both simile and metaphor: the electric current's ability to do work is "really and truly"—or, we may say, *literally*—energy.

551.5 *the amount of this heat ... is equal to the energy originally existing in the current*: Since heat is a form of energy, its generation at the expense of the current shows that the current possesses energy. Maxwell discussed the measure of heat in relation to electric current in Art. 242 above.

551.6 *by Lenz's law, stop the current sooner than the resistance of the circuit alone would have stopped it*: See the discussion of Lenz's law in (530.8, *comment*).

part of the energy of the current may be transformed into mechanical work: This mechanical work is in addition to any heat produced, since the current is found to decay more quickly than it would by resistance alone. But just as in the production of heat, the fact that mechanical work is done only at the expense of the current shows that the current possesses energy.

552. *we can form no conception of an electric current except as a kinetic phenomenon*: Maxwell cites Faraday: "By *current*, I mean anything progressive," as though to suggest that if *Faraday* could not conceive current without motion, *none of us* can!

energy must be kinetic energy, that is to say, the energy which a moving body has in virtue of its motion.

552.1 We have already shewn that the electricity in the wire cannot be considered as the moving body in which we are to find this energy, for the energy of a moving body does not depend on anything external to itself, whereas the presence of other bodies near the current alters its energy.

552.2 We are therefore led to enquire whether there may not be some motion going on in the space outside the wire, which is not occupied by the electric current, but in which the electromagnetic effects of the current are manifested.

552.3 I shall not at present enter on the reasons for looking in one place rather than another for such motions, or for regarding these motions as of one kind rather than another.

552.1 *the presence of other bodies near the current alters its energy*: In (551.6), for example, the presence of nearby magnets caused part of the energy of the current to do mechanical work.

552.2 *whether there may not be some motion going on in the space outside the wire*: Maxwell is led to this extraordinary question by inexorable logic. For (i) current manifestly possesses energy; (ii) that energy must be kinetic energy, since motion is conceptually essential to current; but (iii) the seat of that kinetic energy cannot be within the conductor. It must therefore reside *outside* the conductor.

Maxwell's reasoning here transcends its apparently syllogistic character. He is questioning concepts at the deepest level, concepts we barely understand. Nobody knows what current is, really—or, if it indeed involves "motion," what (possibly bizarre) sense of motion that might prove to be. Nor what it would mean to look for motion "external" to a wire in empty space. Formal logic can indeed draw sound conclusions even concerning unknown terms, provided those terms are univocal and consistent. But *current, motion, inside* and *outside*, when applied to our electrical experience, are all too fluid to guarantee either consistency or univocality.

When Maxwell does, in subsequent chapters, indeed establish the existence of "motion" and even "momentum" in the space surrounding a current, we will find those terms being expanded and applied in truly novel ways, ways that cannot be instituted by syllogistic logic alone. Here, I think, Maxwell is using logical analysis not as a tool with which to crank out conclusions but as a vehicle to generate questions. In this he is like Socrates of the dialogues, who famously deploys arguments—in some cases quite flawed ones—to turn the minds of his interlocutors towards certain topics and cultivate insight in chosen directions. In this connection we might recall Maxwell's praise of Faraday's conception of the electrotonic state at (540.3) in the previous chapter. "Without a much greater degree of development than Faraday gave it," he there stated, it does not suffice to explain phenomena. Rather, its scientific value "consists in its directing the mind to lay hold of a certain quantity..." Maxwell too seems intent on "directing the mind" in the present discussion.

552.4 What I propose now to do is to examine the consequences of the assumption that the phenomena of the electric current are those of a moving system, the motion being communicated from one part of the system to another by forces, the nature and laws of which we do not yet even attempt to define, because we can eliminate these forces from the equations of motion by the method given by Lagrange for any connected system.

552.5 In the next five chapters of this treatise I propose to deduce the main structure of the theory of electricity from a dynamical hypothesis of this kind, instead of following the path which has led Weber and

552.4 *a moving system, the motion being communicated from one part of the system to another*: Maxwell is describing a *connected system*, a system of continuity and wholeness, as contrasted with a world of isolated bodies linked by force laws—the supposedly Newtonian world envisaged by action-at-a-distance theorists.

forces, the nature and laws of which we do not yet even attempt to define, because we can eliminate these forces ... by the method given by Lagrange: Why, one might well ask, would one want to "eliminate these forces?" Was it not Newton's great triumph to have established the gravitational force law? Is it not precisely the forces of nature that we wish to know, in order to deduce the motions of nature? "For the whole burden of philosophy seems to consist in this—from the phenomena of motions to investigate the forces of nature, and then from these forces to demonstrate the other phenomena" (Newton, *Principia*, Preface to the First Edition).

The reason we need a method for eliminating unknown quantities is that there must always be some level of analysis at which we simply do not know the inner workings of nature. Ordinary mechanics may still proceed at that level, but only by making specific assumptions—in effect, treating the unknown as though it were known. But such assumptions may prove more burdensome than fruitful at a later stage of development, and yet by then may also have become so ingrained as to require for their correction or replacement nothing short of a revolutionary labor. The "method given by Lagrange," in contrast, never requires us to assume what we do not know. Maxwell will present that method in his next chapter.

552.5 *I propose to deduce ... the theory of electricity from a dynamical hypothesis of this kind*: Although Maxwell calls it a "hypothesis," Lagrange's approach is more like a framework of world-views, flexible enough to accomodate new phenomena without having to fit them to pre-existing categories. Thus, we might say, Lagrangian methods permit nature to speak in her own terms—a ideal pursued by Faraday and much appreciated by Maxwell.

Ordinary mechanics, to be sure, *allows* such an approach—Lagrange indeed developed his methods for the sake of celestial mechanics. But as Maxwell well understands, the phenomena of electricity and magnetism virtually *demand* that we depart from the old framework of isolated bodies and empty space. If Maxwell succeeds in elaborating electromagnetic science in this new way, it will constitute not merely a new science but an veritable exemplar of human knowing.

other investigators to many remarkable discoveries and experiments, and to conceptions, some of which are as beautiful as they are bold. I have chosen this method because I wish to shew that there are other ways of viewing the phenomena, which appear to me more satisfactory, and at the same time are more consistent with the methods followed in the preceding parts of this book than those which proceed on the hypothesis of direct action at a distance.

Editor's Introduction to
Chapter V

With the present chapter we reach what may fairly be regarded as the heart of Maxwell's project in the *Treatise*. Recall that in Chapter III Maxwell had written:

> We are accustomed to consider the universe as made up of parts, and mathematicians usually begin by considering a single particle, and then conceiving its relation to another particle, and so on. This has generally been supposed the most natural method. To conceive of a particle, however, requires a process of abstraction, since all our perceptions are related to extended bodies, so that the idea of the *all* that is in our consciousness at a given instant is perhaps as primitive an idea as that of any individual thing. *Hence there may be a mathematical method in which we proceed from the whole to the parts* instead of from the parts to the whole (Art. 529, emphasis in final sentence added).

Such a method Maxwell found in Lagrange's *Mécanique Analytique,** as expounded by Thomson and Tait in their *Treatise on Natural Philosophy.*† The unique features of Lagrange's method are these: (*i*) it provides a way to represent complex systems solely in terms of the variables which we can observe—it is not necessary to adopt hypotheses concerning the internal and unobservable workings of systems; and (*ii*) it is capable of generalizing beyond conventional mechanical concepts like force, mass, etc.—concepts which, though well-suited to the analysis of machines, are of questionable applicability to electric and magnetic actions in "empty" spaces free of ponderable matter.

This second feature of the Lagrangian method was appreciated and developed more fully by Maxwell than it had been either by Lagrange himself or by Thomson and Tait. I will say more about it later. Maxwell provided a memorable characterization of the first feature in a review published posthumously in 1879; in it he conceived a company of bell-ringers operating, by means of ropes, a set of bells in an inaccessible belfry:

*Joseph-Louis Lagrange, *Mécanique analytique* (Paris: 1788).

†William Thomson and Peter Guthrie Tait, *A Treatise on Natural Philosophy*: Volume 1 (Oxford: 1867); there is no second volume. A second edition in two parts was issued in 1878. A reprint of the final edition (1912) is available as *Principles of Mechanics and Dynamics* (2 vols.; New York: Dover Publications, 1962).

> In an ordinary belfry, each bell has one rope which comes down
> through a hole in the floor to the bellringers' room. But suppose
> that each rope, instead of acting on one bell, contributes to the
> motion of many pieces of machinery, and that the motion of each
> piece is determined not by the motion of one rope alone, but by
> that of several, and suppose, further, that all this machinery is
> silent and utterly unknown to the men at the ropes, who can only
> see as far as the holes in the floor above them....*

In Maxwell's engaging image, the ropes are the variables in terms of
which we attempt to characterize and control the otherwise hidden
system of interconnected bells. A conventional mechanical analysis of
such a carillon might begin by proposing various provisional schemes
for the individual linkages, deriving in each case the consequences for
the system's actual behavior. In contrast, as we shall see, Lagrangian
analysis begins by formulating general relations among the variables,
relations consistent with *any* set of connections, but ones which may be
subsequently refined so far as actual experience permits.

The second distinguishing characteristic of Lagrange's method pro-
vides the basis for what Maxwell terms "dynamical" science. The sciences
having to do with motion had been ranked by Thomson and Tait† (in part
following a suggestion of Ampère) according to the following division:

1. *Kinematics*. This science treats motion in a purely geometrical
manner, expounding the positions and speeds of moving things in rela-
tion to *time*. Ptolemaic astronomy is kinematical; so too is much of
Galileo's work in the Third Day of his *Dialogues Concerning Two New
Sciences*.

2. *Dynamics*. This science treats the action of *forces*; in a mechanical
context it adds to kinematics the concepts of *matter* and *inertia*, as well
as force. As a science of moving systems, then, dynamics goes beyond
kinematics in referring motions not just to formal causes (if we may use
Aristotelian terms) but to efficient causes as well.

Maxwell gave considerable thought to the naming and classification
of the sciences, departing from Thomson and Tait at several points. In
particular he introduced an additional distinction:

3. *"Pure" dynamics*. Here the epithet "pure" may have either of two
senses. In the first sense, dynamics is *pure* when it is not *applied*: Applied
dynamics (or *mechanics*) studies particular mechanisms, while pure
dynamics studies inertial systems in general.

* James Clerk Maxwell, "Thomson and Tait's Natural Philosophy," in *Nature*,
vol. xx. Reprinted in *Scientific Papers*, vol ii, pp. 783–84.

† *Treatise on Natural Philosophy* (1890), vol. i, Preface, p. vi.

In the second sense, dynamics is said to be *pure* when it has been cleansed of the specifically mechanical understanding of such concepts as matter, mass, force, etc. It is this second sense that Maxwell generally employs; and he will show in the present chapter that many mechanical terms are capable of a more general *dynamical*, rather than a conventionally *mechanical*, interpretation.

Additionally, Maxwell took a broader view of the nature of "dynamical" science, whether pure or applied, arguing that its essential concern was *energy* rather than force. As we shall see in the present chapter, concepts like *inertia* in mechanics or *velocity* in kinematics take on metaphorical, "generalized," meanings when they are translated to sciences such as electromagnetism or thermodynamics. But energy, Maxwell believed, is the one concept that retains a literal, univocal meaning in every science, dynamically treated. In an 1864 paper, one of three forerunners to the *Treatise*, he had written:

> [I]n using such words as electric momentum and electric elasticity ..., I wish merely to direct the mind of the reader to mechanical phenomena which will assist him in understanding the electrical ones. All such phrases in the present paper are to be considered as illustrative, not as explanatory.
>
> In speaking of the Energy of the field, however, I wish to be understood literally....*

Two anticipatory expansions

Two steps in Maxwell's development of the Lagrange equations are greatly telescoped; it will be useful, I hope, to offer an expanded account of each of them in advance.

I. In Art. 564 Maxwell undertakes to transform the equation

$$T_p + T_{\dot{q}} - p_1\dot{q}_1 - p_2\dot{q}_2 - \&\text{c.} = 0 \tag{14}$$

to derive

$$\left(\frac{dT_p}{dp_1} - \dot{q}_1\right)\delta p_1 + \left(\frac{dT_p}{dp_2} - \dot{q}_2\right)\delta p_2 + \&\text{c.}$$

$$+ \left(\frac{dT_{\dot{q}}}{d\dot{q}_1} - p_1\right)\delta\dot{q}_1 + \left(\frac{dT_{\dot{q}}}{d\dot{q}_2} - p_2\right)\delta\dot{q}_2 + \&\text{c.} \tag{15}$$

$$+ \left(\frac{dT_p}{dq_1} + \frac{dT_{\dot{q}}}{dq_1}\right)\delta q_1 + \left(\frac{dT_p}{dq_2} + \frac{dT_{\dot{q}}}{dq_2}\right)\delta q_2 + \&\text{c.} = 0.$$

* "A Dynamical Theory of the Electromagnetic Field," *Phil. Trans.* **155,** pp. 459–512; reprinted in *Scientific Papers*, vol. i, pp. 526–571.

Each term of equation (15) is the *variation* of the corresponding term in equation (14), where the variation of a function $f(x)$ is

$$\delta f(x) = \frac{df(x)}{dx}\delta x.$$

To obtain Maxwell's result, take the variation of each term of equation (14) with respect to each of the variables p_1, p_2, q_1, q_2, \dot{q}_1, \dot{q}_2, as follows:

$$\left[\frac{dT_p}{dp_1} + \frac{dT_{\dot{q}}}{dp_1} - \frac{d(p_1\dot{q}_1)}{dp_1} - \frac{d(p_2\dot{q}_2)}{dp_1} - \&c.\right]\delta p_1$$

$$+\left[\frac{dT_p}{dp_2} + \frac{dT_{\dot{q}}}{dp_2} - \frac{d(p_1\dot{q}_1)}{dp_2} - \frac{d(p_2\dot{q}_2)}{dp_2} - \&c.\right]\delta p_2$$

$$+\left[\frac{dT_p}{dq_1} + \frac{dT_{\dot{q}}}{dq_1} - \frac{d(p_1\dot{q}_1)}{dq_1} - \frac{d(p_2\dot{q}_2)}{dq_1} - \&c.\right]\delta q_1$$

$$+\left[\frac{dT_p}{dq_2} + \frac{dT_{\dot{q}}}{dq_2} - \frac{d(p_1\dot{q}_1)}{dq_2} - \frac{d(p_2\dot{q}_2)}{dq_2} - \&c.\right]\delta q_2$$

$$+\left[\frac{dT_p}{d\dot{q}_1} + \frac{dT_{\dot{q}}}{d\dot{q}_1} - \frac{d(p_1\dot{q}_1)}{d\dot{q}_1} - \frac{d(p_2\dot{q}_2)}{d\dot{q}_1} - \&c.\right]\delta\dot{q}_1$$

$$+\left[\frac{dT_p}{d\dot{q}_2} + \frac{dT_{\dot{q}}}{d\dot{q}_2} - \frac{d(p_1\dot{q}_1)}{d\dot{q}_2} - \frac{d(p_2\dot{q}_2)}{d\dot{q}_2} - \&c.\right]\delta\dot{q}_2 = 0.$$

But Maxwell will explain that T_p is not a function of either \dot{q}_1 or \dot{q}_2, nor is $T_{\dot{q}}$ a function of either p_1 or p_2. Thus, as indicated above, the derivatives of T_p with respect to \dot{q}_1 and \dot{q}_2 vanish, as do the derivatives of $T_{\dot{q}}$ with respect to p_1 and p_2. Similarly, the following derivatives also vanish: those of $p_1\dot{q}_1$ with respect to p_2, \dot{q}_2, or any q; and those of $p_2\dot{q}_2$ with respect to p_1, \dot{q}_1, or any q. Finally, derivatives of $p_1\dot{q}_1$ with respect to p_1 and \dot{q}_1 reduce to \dot{q}_1 and p_1, respectively; while derivatives of $p_2\dot{q}_2$ with respect to p_2 and \dot{q}_2 reduce to \dot{q}_2 and p_2, respectively. Collect the resulting terms to obtain equation (15).

II. In paragraph 565.1 Maxwell will formulate one of three expressions for the total kinetic energy of an arbitrary system:

$$T_p = \frac{1}{2}\left(p_1\frac{\partial T_p}{\partial p_1} + p_2\frac{\partial T_p}{\partial p_2} + p_3\frac{\partial T_p}{\partial p_3} + \cdots\right). \tag{i}$$

From this, he asserts, it follows that "T_p is a homogeneous function of the second degree of ... p_1, p_2, &c." In particular, Maxwell ascribes to T_p

the second-degree form

$$2T_p = Q_{11}p_1^2 + Q_{22}p_2^2 + Q_{33}p_3^2 + \cdots$$
$$+ 2Q_{12}p_1p_2 + 2Q_{13}p_1p_3 + 2Q_{23}p_2p_3 + \cdots, \qquad \text{(ii)}$$

in which every term contains a product of two variables, and where the coefficients Q do not contain the variables. To show that the two expressions are equivalent, differentiate equation (ii) with respect to each of the variables respectively to obtain the partial derivatives

$$\frac{\partial T_p}{\partial p_1} = Q_{11}p_1 + Q_{12}p_2 + Q_{13}p_3 + \cdots,$$

$$\frac{\partial T_p}{\partial p_2} = Q_{22}p_2 + Q_{12}p_1 + Q_{23}p_3 + \cdots,$$

$$\frac{\partial T_p}{\partial p_3} = Q_{33}p_3 + Q_{13}p_1 + 2Q_{23}p_2 + \cdots, \text{ etc.}$$

Substituting these derivatives in equation (i), and eliminating T_p between equations (i) and (ii), yields

$$Q_{11}p_1^2 + Q_{22}p_2^2 + Q_{33}p_3^2 + \cdots + 2Q_{12}p_1p_2 + 2Q_{13}p_1p_3 + 2Q_{23}p_2p_3 + \cdots$$
$$= \quad Q_{11}p_1^2 + Q_{12}p_1p_2 + Q_{13}p_1p_3 + \cdots$$
$$+ Q_{22}p_2^2 + Q_{12}p_1p_2 + Q_{23}p_2p_3 + \cdots$$
$$+ Q_{33}p_3^2 + Q_{13}p_1p_3 + Q_{23}p_2p_3 + \cdots.$$

where Q_{12} is the same as Q_{21}, and so on. The equation is an identity; thus we have shown that if T_p has the form (i), it will be a homogeneous second-degree function of the form (ii).

the second-degree form

$$2T_p = Q_{11}\dot{p}_1^2 + Q_{22}\dot{p}_2^2 + Q_{33}\dot{p}_3^2 + \cdots$$
$$+ 2Q_{12}\dot{p}_1\dot{p}_2 + 2Q_{13}\dot{p}_1\dot{p}_3 + 2Q_{23}\dot{p}_2\dot{p}_3 + \cdots, \qquad \text{(ii)}$$

in which every term contains a product of two variables, and where the coefficients Q do not contain the variables. To show that the two expressions are equivalent, differentiate equation (ii) with respect to each of the variables respectively to obtain the partial derivatives

$$\frac{\partial T_p}{\partial p_1} = Q_{11}p_1 + Q_{12}p_2 + Q_{13}p_3 + \cdots,$$

$$\frac{\partial T_p}{\partial p_2} = Q_{21}p_1 + Q_{22}p_2 + Q_{23}p_3 + \cdots,$$

$$\frac{\partial T_p}{\partial p_3} = Q_{31}p_1 + Q_{32}p_2 + Q_{33}p_3 + \cdots, \text{etc.}$$

Substituting these derivatives in equation (i), and eliminating T_p between equations (i) and (ii), yields

$$Q_{11}p_1^2 + Q_{22}p_2^2 + Q_{33}p_3^2 + \cdots + 2Q_{12}p_1p_2 + 2Q_{13}p_2p_3 + 2Q_{32}p_3p_2 + \cdots$$
$$= Q_{11}\dot{p}_1^2 + Q_{12}\dot{p}_1\dot{p}_2 + Q_{13}\dot{p}_1\dot{p}_3 + \cdots$$
$$+ Q_{21}p_2p_1 + Q_{22}\dot{p}_2^2 + Q_{23}\dot{p}_2\dot{p}_3 + \cdots$$
$$+ Q_{31}\dot{p}_3\dot{p}_1 + Q_{32}\dot{p}_3\dot{p}_2 + Q_{33}\dot{p}_3^2 + \cdots.$$

And since Q_{12} is the same as Q_{21}, and so on, this is an identity. Thus we have shown that if T_p has the form (i), it will be a homogeneous second-degree function of the form (ii).

CHAPTER V

ON THE EQUATIONS OF MOTION OF A CONNECTED SYSTEM.

553.] IN the fourth section of the second part of his *Mécanique Analytique,* Lagrange has given a method of reducing the ordinary dynamical equations of the motion of the parts of a connected system to a number equal to that of the degrees of freedom of the system.

553.1 The equations of motion of a connected system have been given in a different form by Hamilton, and have led to a great extension of the higher part of pure dynamics.*

553.2 As we shall find it necessary, in our endeavours to bring electrical phenomena within the province of dynamics, to have our dynamical ideas in a state fit for direct application to physical questions, we shall devote this chapter to an exposition of these dynamical ideas from a physical point of view.

554.] The aim of Lagrange was to bring dynamics under the power of the calculus. He began by expressing the elementary dynamical relations in terms of the corresponding relations of pure algebraical quantities, and from the equations thus obtained he deduced his final equations by a purely algebraical process. Certain quantities (expressing the reactions between the parts of the system called into play by its physical connexions) appear in the equations of motion of the component parts of the system, and Lagrange's investigation, as seen from a

* See Professor Cayley's 'Report on Theoretical Dynamics,' *British Association,* 1857; and Thomson and Tait's *Natural Philosophy.*

553. *degrees of freedom*: Roughly, the number of independent ways in which a system may undergo change. Maxwell will characterize this term more fully in Art. 555 below.

553.1 *pure dynamics*: See the discussion of this term in the editor's introdution to this chapter.

mathematical point of view, is a method of eliminating these quantities from the final equations.

554.1 In following the steps of this elimination the mind is exercised in calculation, and should therefore be kept free from the intrusion of dynamical ideas. Our aim, on the other hand, is to cultivate our dynamical ideas. We therefore avail ourselves of the labours of the mathematicians, and retranslate their results from the language of the calculus into the language of dynamics, so that our words may call up the mental image, not of some algebraical process, but of some property of moving bodies.

554.2 The language of dynamics has been considerably extended by those who have expounded in popular terms the doctrine of the Conservation of Energy, and it will be seen that much of the following statement is suggested by the investigation in Thomson and Tait's *Natural Philosophy,* especially the method of beginning with the theory of impulsive forces.

554.3 I have applied this method so as to avoid the explicit consideration of the motion of any part of the system except the coordinates or variables, on which the motion of the whole depends. It is doubtless important that the student should be able to trace the connexion of the motion of each part of the system with that of the variables, but it is by no means necessary to do this in the process of obtaining the final equations, which are independent of the particular form of these connexions.

The Variables.

555.] The number of degrees of freedom of a system is the number of data which must be given in order completely to determine its position.

554. *eliminating these quantities...*: and thereby dispensing with any attempts to specify the details of the internal connections of the system (552.4, *comment*).

554.3 *I have applied this method so as to avoid the explicit consideration of ... any part of the system except the coordinates or variables*: The variables are thus *the observables*. In Kantian terms they might be said to correspond to the appearances, the phenomena, while the system proper would correspond to things in themselves.

555. *the number of data which must be given in order completely to determine its position*: For example, a body in free space has three degrees of freedom; its position is completely determined by the three Cartesian coordinates x, y, z or, alternatively, by the three spherical coordinates *radius, elevation angle,* and *azimuth*. By contrast, a bead constrained to movement along a fixed curved wire has but one degree of freedom, for its position is fully determined by a single quantity, such as its curvilinear distance from one end of the wire.

Different forms may be given to these data, but their number depends on the nature of the system itself, and cannot be altered.

555.1 To fix our ideas we may conceive the system connected by means of suitable mechanism with a number of moveable pieces, each capable of motion along a straight line, and of no other kind of motion. The imaginary mechanism which connects each of these pieces with the system must be conceived to be free from friction, destitute of inertia, and incapable of being strained by the action of the applied forces. The use of this mechanism is merely to assist the imagination in ascribing position, velocity, and momentum to what appear, in Lagrange's investigation, as pure algebraical quantities.

555.2 Let q denote the position of one of the moveable pieces as defined by its distance from a fixed point in its line of motion. We shall distinguish the values of q corresponding to the different pieces by the suffixes 1, 2, &c. When we are dealing with a set of quantities belonging to one piece only we may omit the suffix.

555.3 When the values of all the variables (q) are given, the position of each of the moveable pieces is known, and, in virtue of the imaginary mechanism, the configuration of the entire system is determined.

The Velocities

556.] During the motion of the system the configuration changes in some definite manner, and since the configuration at each instant is

555.1 *To fix our ideas*: specifically, our ideas of how a whole system can be understood as determined by a limited number of variables.

moveable pieces: Maxwell will later (571.) make clear that these pieces are assumed to be equal in number to the degrees of freedom of the system.

each capable of moving along a straight line, and of no other kind of motion: Think initially of these "moveable pieces" as *indicators*; their readings reflect the positions of the corresponding parts of the system but need not be proportional to them.

to assist the imagination in ascribing position, velocity, and momentum...: Position and velocity are easily manifested by these "indicators"; but since they are massless, their momenta cannot be represented by the ordinary products mv. As we shall see, therefore, Maxwell will take a different approach in ascribing *momentum* to them.

555.3 *the configuration of the entire system is determined*: It is "determined" in the sense that if the indicators are suitably constructed, each configuration of the system corresponds to one and only one set of positions of the indicators. But although the configuration is determined in this sense it is not *known in itself*; the configuration is known only to the extent that the indicators represent it.

fully defined by the values of the variables (q), the velocity of every part of the system, as well as its configuration, will be completely defined if we know the values of the variables (q), together with their velocities dq/dt, or, according to Newton's notation, \dot{q}.

The Forces.

557.] By a proper regulation of the motion of the variables, any motion of the system, consistent with the nature of the connexions, may be produced. In order to produce this motion by moving the variable pieces, forces must be applied to these pieces.

557.1 We shall denote the force which must be applied to any variable q_r by F_r. The system of forces (F) is mechanically equivalent (in virtue of the connexions of the system) to the system of forces, whatever it may be, which really produces the motion.

The Momenta.

558.] When a body moves in such a way that its configuration, with respect to the force which acts on it, remains always the same, (as, for instance, in the case of a force acting on a single particle in the line of its motion,) the moving force is measured by the rate of increase of the momentum. If F is the moving force, and p the momentum,

$$F = \frac{dp}{dt}, \text{ whence } p = \int F dt .$$

558.1 The time-integral of a force is called the Impulse of the force; so that we may assert that the momentum is the impulse of the force which would bring the body from a state of rest into the given state of motion.

556. *the velocity of every part of the system ... will be completely defined*: Just as the positions of the parts of the system are determined by the positions of the indicators, the velocities of the parts are determined by the velocities of the indicators, but they are not known in themselves.

557. *In order to produce this motion by moving the variable pieces, forces must be applied to the pieces*: With this statement of Maxwell's we must no longer regard the "pieces" simply as indicators, for by applying forces to them, the system as a whole is set in motion. The "pieces" are therefore *points of interaction* between the system and ourselves.

557.1 *the system of forces ... which really produces the motion*: that is, the forces that are actually experienced by the respective parts of the system.

558.1 *the momentum is the impulse of the force...*: Conventional mechanics regards impulse $F dt$ and momentum mv as being each a quantity measurable in its own right, and it asserts their equality by means of the "impulse-momentum theorem." But in pure dynamics, that equality is not a theorem but a definition; *momentum*

558.2 In the case of a connected system in motion, the configuration is continually changing at a rate depending on the velocities (\dot{q}), so that we can no longer assume that the momentum is the time-integral of the force which acts on it.

558.3 But the increment δq of any variable cannot be greater than $\dot{q}'\delta t$, where δt is the time during which the increment takes place, and \dot{q}' is the greatest value of the velocity during that time. In the case of a system moving from rest under the action of forces always in the same direction, this is evidently the final velocity.

558.4 If the final velocity and configuration of the system are given, we may conceive the velocity to be communicated to the system in a very small time δt, the original configuration differing from the final configuration by quantities δq_1, δq_2, &c., which are less than $\dot{q}_1 \delta t$, $\dot{q}_2 \delta t$, &c., respectively.

558.5 The smaller we suppose the increment of time δt, the greater must be the impressed forces, but the time-integral, or impulse, of each force will remain finite. The limiting value of the impulse, when the time is diminished and ultimately vanishes, is defined as the *instantaneous* impulse, and the momentum p, corresponding to any variable q, is defined as the impulse corresponding to that variable, when the system is brought instantaneously from a state of rest into the given state of motion.

558.6 This conception, that the momenta are capable of being produced by instantaneous impulses on the system at rest, is introduced

is defined by the impulse of the force that generates it. Notice that this provides a concept of momentum that is free of any reference to, or dependence on, ponderable mass.

558.2 *we can no longer assume that the momentum is the time-integral...*: Since the configuration of the system is changing, the applied forces may have to act at varying angles with respect to the lines of motion of their corresponding "moveable pieces." The mere time-integral of forces would fail to take such changes of angle into account.

558.3 *a system moving from rest under the action of forces always in the same direction*: Sameness of direction is, of course, precisely what we *cannot* in general assume (see the previous note). But we can assume it under suitable limitations, as Maxwell will point out in the next paragraph.

558.4 *a very small time δt*: If the time of action is sufficiently short, we may assume that the configuration of the system—and hence the angle of a force's application—does not change appreciably.

558.5 *when that system is brought instantaneously from ... rest into the given state of motion*: Better, when it is brought into the given state in a vanishingly small time. In (590.5) below, Maxwell will consider the reverse case of *sudden stoppage*.

only as a method of defining the magnitude of the momenta, for the momenta of the system depend only on the instantaneous state of motion of the system, and not on the process by which that state was produced.

558.7 In a connected system the momentum corresponding to any variable is in general a linear function of the velocities of all the variables, instead of being, as in the dynamics of a particle, simply proportional to the velocity.

558.8 The impulses required to change the velocities of the system suddenly from \dot{q}_1, \dot{q}_2 &c. to \dot{q}_1', \dot{q}_2', &c. are evidently equal to $p_1'-p_1$, $p_2'-p_2$, the changes of momentum of the several variables.

Work done by a Small Impulse.

559.] The work done by the force F_1 during the impulse is the space-integral of the force, or

$$W = \int F_1 dq_1,$$
$$= \int F_1 \dot{q}_1 dt.$$

559.1 If \dot{q}_1' is the greatest and \dot{q}_1'' the least value of the velocity \dot{q}_1 during the action of the force, W must be less than

$$\dot{q}_1 \int F dt \quad \text{or} \quad \dot{q}_1'\left(p_1'-p_1\right),$$

and greater than $\quad \dot{q}_1'' \int F dt \quad \text{or} \quad \dot{q}_1''\left(p_1'-p_1\right).$

559.2 If we now suppose the impulse $\int F dt$ to be diminished without limit, the values of \dot{q}_1' and \dot{q}_1'' will approach and ultimately coincide with that of \dot{q}_1, and we may write $p_1'-p_1 = \delta p_1$, so that the work done is ultimately

$$\delta W_1 = \dot{q}_1 \delta p_1,$$

or, *the work done by a very small impulse is ultimately the product of the impulse and the velocity.*

558.7 *a linear function of the velocities...*: Maxwell will formulate the general linear function in (565.5) below.

559.2 *The work done ... is ultimately the product of the impulse and the velocity*: This corresponds to a theorem in conventional mechanics, in which momentum is represented as mv, and the increment of work done in altering the momentum is $mv \cdot dv$ or $v \cdot mdv$—the product of the change in momentum (equal to the impulse) and the velocity. In conventional mechanics we integrate $mv\,dv$ to

Increment of the Kinetic Energy.

560.] When work is done in setting a conservative system in motion, energy is communicated to it, and the system becomes capable of doing an equal amount of work against resistances before it is reduced to rest.

560.1 The energy which a system possesses in virtue of its motion is called its Kinetic Energy, and is communicated to it in the form of the work done by the forces which set it in motion.

560.2 If T be the kinetic energy of the system, and if it becomes $T + \delta T$, on account of the action of an infinitesimal impulse whose components are p_1, p_2, &c., the increment T must be the sum of the quantities of work done by the components of the impulse, or in symbols,

$$\delta T = \dot{q}_1 \delta p_1 + \dot{q}_2 \delta p_2 + \&\text{c.},$$

$$= \sum (\dot{q}\, \delta p). \tag{1}$$

560.3 The instantaneous state of the system is completely defined if the variables and the momenta are given. Hence the kinetic energy, which depends on the instantaneous state of the system, can be expressed in terms of the variables (q), and the momenta (p). This is the mode of expressing T introduced by Hamilton. When T is expressed in this way we shall distinguish it by the suffix p, thus, T_p.

obtain the whole work; it is equal to $\frac{1}{2}mv^2$, which conventional mechanics defines as the kinetic energy. But following the path of "pure" dynamics, Maxwell has expressed momentum without reference to mass; likewise the present formulation of work as the product of impulse and velocity makes no reference to mass. In the next Article Maxwell will go on to express even kinetic energy without making reference to ponderable mass.

560. *conservative system*: defined in (84.11, *comment*).

560.3 *The instantaneous state of the system is completely defined if the variables and the momenta are given*: In bringing a system from rest to its present state (558.4–558.6), it is the impulses that determine the velocities. But the momenta are equal to the impulses since the system was initially at rest (558.1); hence the momenta determine the velocities. And it is the variables *and the velocities* which Maxwell has identified (556.) as the quantities that define the state of a system; therefore the state of the system is defined if the variables and the momenta are given.

Hence the kinetic energy ... can be expressed in terms of the variables (q) and the momenta (p): And since neither q nor p is defined in terms of ponderable mass, kinetic energy too need no longer be understood in terms of ponderable or mechanical mass.

560.4 The complete variation of T_p is

$$\delta T_p = \sum \left(\frac{dT_p}{dp} \delta p \right) + \sum \left(\frac{dT_p}{dq} \delta q \right). \tag{2}$$

560.5 The last term may be written

$$\sum \left(\frac{dT_p}{dq} \dot{q} \delta t \right),$$

which diminishes with t, and ultimately vanishes with it when the impulse becomes instantaneous.

560.6 Hence, equating the coefficients of δp in equations (1) and (2),

we obtain

$$\dot{q} = \frac{dT_p}{dp}, \tag{3}$$

or, *the velocity corresponding to the variable q is the differential coefficient of T_p with respect to the corresponding momentum p.*

560.7 We have arrived at this result by the consideration of impulsive forces. By this method we have avoided the consideration of the change of configuration during the action of the forces. But the instantaneous state of the system is in all respects the same, whether the system was brought from a state of rest to the given state of motion by the transient application of impulsive forces, or whether it arrived at that state in any other manner, however gradual.

560.8 In other words, the variables, and the corresponding velocities and momenta, depend on the actual state of motion of the system at the given instant, and not on its previous history.

560.9 Hence, the equation (3) is equally valid, whether the state of motion of the system is supposed due to impulsive forces, or to forces acting in any manner whatever.

560.10 We may now therefore dismiss the consideration of impulsive forces, together with the limitations imposed on their time of action, and on the changes of configuration during their action.

Hamilton's Equations of Motion.

561.] We have already shewn that

560.4 *complete variation*: When a quantity depends on several variables (as in this case T_p depends on multiple p's and q's), its variation is said to be a *partial* one if it arises from changes in only some of the variables; it is said to be *complete* when it reflects change in all the variables.

560.5 *The last term may be written...*: since $\delta q = \dot{q} \delta t$.

560.6 *the differential coefficient*: Today we say "the derivative."

561. *We have already shewn...*: in the previous equation (3).

$$\frac{dT_p}{dp} = \dot{q}\,. \tag{4}$$

561.1 Let the system move in any arbitrary way, subject to the conditions imposed by its connexions, then the variations of p and q are

$$\delta p = \frac{dp}{dt}\delta t, \quad \delta q = \dot{q}dt\,. \tag{5}$$

561.2 Hence $\qquad \dfrac{dT_p}{dp}\,\delta p \;=\; \dfrac{dp}{dt}\dot{q}\delta t \;=\; \dfrac{dp}{dt}\delta q, \tag{6}$

and the complete variation of T_p is

$$\delta T_p \;=\; \sum\left(\frac{dT_p}{dp}\,\delta p + \frac{dT_p}{dq}\,\delta q\right) \;=\; \sum\left(\left(\frac{dp}{dt} + \frac{dT_p}{dq}\right)\delta q\right). \tag{7}$$

561.3 But the increment of the kinetic energy arises from the work done by the impressed forces, or

$$\delta T_p = \sum(F\delta q)\,. \tag{8}$$

561.4 In these two expressions the variations δq are all independent of each other, so that we are entitled to equate the coefficients of each of them in the two expressions (7) and (8). We thus obtain

$$F_r = \frac{dp_r}{dt} + \frac{dT_p}{dq_r}\,, \tag{9}$$

where the momentum p_r and the force F_r belong to the variable q_r.

561.5 There are as many equations of this form as there are variables. These equations were given by Hamilton. They shew that the force corresponding to any variable is the sum of two parts. The first part is the rate of increase of the momentum of that variable with respect to the time. The second part is the rate of increase of the kinetic energy per unit of increment of the variable, the other variables and all the momenta being constant.

561.4 *the variations δq are all independent of each other*: Since the variables q are equal in number to the degrees of freedom of the system, each corresponds independently to a way in which the system can change; see Arts. 553 and 555 and associated comments.

We thus obtain…: In the Third Edition of the *Treatise*, editor J. J. Thomson inserted a note questioning whether equation (9) legitimately follows from equations (7) and (8). For a defense of Maxwell's reasoning, see Thomas King Simpson, *Maxwell's Mathematical Rhetoric*, chap. 4.

The Kinetic Energy expressed in Terms of the Momenta and Velocities.

562.] Let p_1, p_2, &c. be the momenta, and \dot{q}_1, \dot{q}_2, &c. the velocities at a given instant, and let p_1, p_2, &c., $\dot{\mathrm{q}}_1$, $\dot{\mathrm{q}}_2$, &c. be another system of momenta and velocities, such that

$$\mathrm{p}_1 = np_1, \quad \dot{\mathrm{q}}_1 = n\dot{q}_1, \text{ &c.} \tag{10}$$

562.1 It is manifest that the systems p, $\dot{\mathrm{q}}$ will be consistent with each other if the systems p, \dot{q} are so.

562.2 Now let n vary by δn. The work done by the force F_1 is

$$F_1 \delta \mathrm{q}_1 = \dot{\mathrm{q}}_1 \delta \mathrm{p}_1 = \dot{q}_1 p_1 n \delta n. \tag{11}$$

562.3 Let n increase from 0 to 1, then the system is brought from a state of rest into the state of motion (\dot{q}, p), and the whole work expended in producing this motion is

$$\left(\dot{q}_1 p_1 + \dot{q}_2 p_2 + \text{&c.}\right) \int_0^1 n\, dn. \tag{12}$$

562.4 But
$$\int_0^1 n\, dn = \tfrac{1}{2},$$

and the work spent in producing the motion is equivalent to the kinetic energy. Hence

$$T_{p\dot{q}} = \tfrac{1}{2}\left(p_1 \dot{q}_1 + p_2 \dot{q}_2 + \text{&c.}\right) \tag{13}$$

where $T_{p\dot{q}}$ denotes the kinetic energy expressed in terms of the momenta and velocities. The variables q_1, q_2, &c. do not enter into this expression.

562.5 The kinetic energy is therefore half the sum of the products of the momenta into their corresponding velocities.

562.1 *consistent*: that is, compatible. Compatibility of the momenta p and corresponding velocities \dot{q} cannot be assumed, for momentum and velocity are not independent. In conventional mechanics this is obvious, since momentum is mv and velocity is v. In the new dynamics we no longer have such definite expressions; but we do know (558.7) that "the momentum corresponding to any variable is in general a linear function of all the velocities of all the variables..."; so that momentum and velocity are not independent in the new mechanics either. It is therefore not to be assumed that any system of p's and \dot{q}'s are compatible with each other; but if they *are* compatible, the act of multiplying each member of the system by n cannot destroy their compatibility.

562.3 *Let n increase from 0 to 1...*: Think of n as an index of the growth of the system from its initial state ($n = 0$, necessarily a state of rest) to a final state wherein $\mathrm{p} = 1 \times p$ and $\dot{\mathrm{q}} = 1 \times \dot{q}$. We utilized a similar device previously in (84.12, *comment*).

562.5 *The kinetic energy is therefore half the sum of the products of the momenta into their corresponding velocities*: The formulation $T_{p\dot{q}}$ thus resembles the expression for

562.6 When the kinetic energy is expressed in this way we shall denote it by the symbol $T_{p\dot{q}}$. It is a function of the momenta and velocities only, and does not involve the variables themselves.

563.] There is a third method of expressing the kinetic energy, which is generally, indeed, regarded as the fundamental one. By solving the equations (3) we may express the momenta in terms of the velocities, and then, introducing these values in (13), we shall have an expression for T involving only the velocities and the variables. When T is expressed in this form we shall indicate it by the symbol $T_{\dot{q}}$. This is the form in which the kinetic energy is expressed in the equations of Lagrange.

564.] It is manifest that, since T_p, $T_{\dot{q}}$, and $T_{p\dot{q}}$ are three different expressions for the same thing,

$$T_p + T_{\dot{q}} - 2T_{p\dot{q}} = 0\,,$$

or
$$T_p + T_{\dot{q}} - p_1\dot{q}_1 - p_2\dot{q}_2 - \&\text{c.} = 0\,. \tag{14}$$

564.1 Hence, if all the quantities p, q, and \dot{q} vary,

$$\left(\frac{dT_p}{dp_1} - \dot{q}_1\right)\delta p_1 + \left(\frac{dT_p}{dp_2} - \dot{q}_2\right)\delta p_2 + \&\text{c.}$$

$$+\left(\frac{dT_{\dot{q}}}{d\dot{q}_1} - p_1\right)\delta\dot{q}_1 + \left(\frac{dT_{\dot{q}}}{d\dot{q}_2} - p_2\right)\delta\dot{q}_2 + \&\text{c.} \tag{15}$$

$$+\left(\frac{dT_p}{dq_1} + \frac{dT_{\dot{q}}}{dq_1}\right)\delta q_1 + \left(\frac{dT_p}{dq_2} + \frac{dT_{\dot{q}}}{dq_2}\right)\delta q_2 + \&\text{c.} = 0.$$

kinetic energy that is familiar from conventional mechanics. For momentum is conventionally represented mv, in which case half the sum of the products in question is $\frac{1}{2}mv\cdot v$ or $\frac{1}{2}mv^2$.

563. *a third method of expressing the kinetic energy ... regarded as the fundamental one*: As Maxwell is about to state, this third expression contains *only* the velocities and the variables, and thus constitutes the most completely reduced form. In the next chapter he will select it as the most appropriate form for *electromagnetic* equations.

we shall indicate it by the symbol $T_{\dot{q}}$: Maxwell gives the explicit formulation of $T_{\dot{q}}$ in equation (24) below.

564.1 *Equation* (15): The equation is derived in the editor's introduction to this chapter. Take the variation of each term of equation (14) with respect to each of the variables p_1, p_2, q_1, q_2, \dot{q}_1, \dot{q}_2. But T_p is a function of the momenta p and variables q only (560.3), while $T_{\dot{q}}$ is a function of the velocities \dot{q} and variables q only (563.). Thus the derivatives $dt_{\dot{q}}/dp_1$, $dt_{\dot{q}}/dp_2$, $dT_p/d\dot{q}_1$, $dT_p/d\dot{q}_2$ vanish. Derivatives like $d(p_2\dot{q}_2)/dp_1$ and $d(p_1\dot{q}_1)/dp_2$ likewise vanish; while derivatives like $d(p_1\dot{q}_1)/dp_1$ reduce to \dot{q}_1. The remaining terms yield equation (15).

564.2 The variations δp are not independent of the variations δq and $\delta \dot{q}$, so that we cannot at once assert that the coefficient of each variation in this equation is zero. But we know, from equations (3), that

$$\frac{dT_p}{dp_1} - \dot{q}_1 = 0 \, , \, \&\text{c.,} \tag{16}$$

so that the terms involving the variations δp vanish of themselves.

564.3 The remaining variations $\delta \dot{q}$ and δq are now all independent, so that we find, by equating to zero the coefficients of $\delta \dot{q}_1$, &c.,

$$p_1 = \frac{dT_{\dot{q}}}{d\dot{q}_1}, \quad p_2 = \frac{dT_{\dot{q}}}{d\dot{q}_2}, \, \&\text{c.;} \tag{17}$$

or, *the components of momentum are the differential coefficients of $T_{\dot{q}}$ with respect to the corresponding velocities.*

564.4 Again, by equating to zero the coefficients of δq_1, &c.,

$$\frac{dT_p}{dq_1} + \frac{dT_{\dot{q}}}{dq_1} = 0 \, ; \tag{18}$$

or, *the differential coefficient of the kinetic energy with respect to any variable q_1 is equal in magnitude but opposite in sign when T is expressed as a function of the velocities instead of as a function of the momenta.*

564.5 In virtue of equation (18) we may write the equation of motion (9),

$$F_1 = \frac{dp_1}{dt} - \frac{dT_{\dot{q}}}{dq_1} \, , \tag{19}$$

564.2 *terms involving δp vanish of themselves*: The first line of equation (15) contains all the terms involving δp. When the relations (16) are substituted into those terms the first line disappears.

564.3 *the components of momentum are the differential coefficients of $T_{\dot{q}}$ with respect to the corresponding velocities*: Why does Maxwell state this for $T_{\dot{q}}$ only? Inasmuch as all three formulations of kinetic energy are "expressions for the same thing," as Maxwell wrote in paragraph 564, the same relation must be equally true of T_p and $T_{p\dot{q}}$. However, as we shall see in equation (24) below, $T_{\dot{q}}$ has the advantage of being a function of \dot{q} only, and thus $dT_{\dot{q}}/d\dot{q}$ is fully explicit. In contrast, both $dT_p/d\dot{q}$ and $dT_{p\dot{q}}/d\dot{q}$ implicitly contain additional terms according to the Chain Rule.

Note also that since $p_1 = dT_{\dot{q}}/d\dot{q}_1$, $p_2 = dT_{\dot{q}}/d\dot{q}_2$, etc., all the terms in the *second* line of equation (15) disappear.

564.4 *by equating to zero the coefficients of δq_1, &c.*: Only the third line of equation (15) remains, and its terms sum to zero. But $\delta q_1, \delta q_2$, &c. are not in general equal to zero; therefore the individual multipliers $\dfrac{dT_p}{dq_1} + \dfrac{dT_{\dot{q}}}{dq_1}$, etc., must each be equal to zero, as equation (18) states.

or
$$F_1 = \frac{d}{dt}\frac{dT_{\dot{q}}}{d\dot{q}_1} - \frac{dT_{\dot{q}}}{dq_1}, \qquad (20)$$

which is the form in which the equations of motion were given by Lagrange.

565.] In the preceding investigation we have avoided the consideration of the form of the function which expresses the kinetic energy in terms either of the velocities or of the momenta. The only explicit form which we have assigned to it is

$$T_{p\dot{q}} = \tfrac{1}{2}\left(p_1\dot{q}_1 + p_2\dot{q}_2 + \&c.\right), \qquad (21)$$

in which it is expressed as half the sum of the products of the momenta each into its corresponding velocity.

565.1 We may express the velocities in terms of the differential coefficients of T_p with respect to the momenta, as in equation (3),

$$T_p = \tfrac{1}{2}\left(p_1\frac{dT_p}{dp_1} + p_2\frac{dT_p}{dp_2} + \&c.\right) \qquad (22)$$

565.2 This shews that T_p is a homogeneous function of the second degree of the momenta p_1, p_2, &c.

565.3 We may also express the momenta in terms of $T_{\dot{q}}$, and we find

$$T_{\dot{q}} = \tfrac{1}{2}\left(\dot{q}_1\frac{dT_{\dot{q}}}{d\dot{q}_1} + \dot{q}_2\frac{dT_{\dot{q}}}{d\dot{q}_2} + \&c.\right), \qquad (23)$$

which shews that $T_{\dot{q}}$ is a homogeneous function of the second degree with respect to the velocities \dot{q}_1, \dot{q}_2, &c.

565.4 If we write

$$P_{11} \text{ for } \frac{d^2T_{\dot{q}}}{d\dot{q}_1^2}, \quad P_{12} \text{ for } \frac{d^2T_{\dot{q}}}{d\dot{q}_1\,d\dot{q}_2}, \&c.$$

and
$$Q_{11} \text{ for } \frac{d^2T_p}{dp_1^2}, \quad Q_{12} \text{ for } \frac{d^2T_p}{dp_1\,dp_2}, \&c.;$$

565.2 T_p *is a homogeneous function of the second degree of the momenta p_1, p_2, &c.*: Maxwell will write the specific form of T_p in equation (24) below. A demonstration is provided in the editor's introduction to this chapter.

565.3 *and we find*: by invoking equation (4).

$T_{\dot{q}}$ *is a homogeneous function of the second degree*: Maxwell will write the specific form of $T_{\dot{q}}$ in equation (25) below. It can be derived in the same way as is T_p in the editor's introduction.

565.4 *If we write...*: The demonstration offered in the editor's introduction incorporates these steps implicitly.

then, since $T_{\dot{q}}$ and T_p are functions of the second degree of \dot{q} and p respectively, both the P's and the Q's will be functions of the variables q only, and independent of the velocities and the momenta. We thus obtain the expressions for T,

$$2T_{\dot{q}} = P_{11}\dot{q}_1^2 + 2P_{12}\dot{q}_1\dot{q}_2 + \&c., \tag{24}$$

$$2T_p = Q_{11}p_1^2 + 2Q_{12}p_1p_2 + \&c. \tag{25}$$

565.5 The momenta are expressed in terms of the velocities by the linear equations $p_1 = P_{11}\dot{q}_1 + P_{12}\dot{q}_2 + \&c.,$ (26)
and the velocities are expressed in terms of the momenta by the linear equations $\dot{q}_1 = Q_{11}p_1 + Q_{12}p_2 + \&c.$ (27)

565.6 In treatises on the dynamics of a rigid body, the coefficients corresponding to P_{11}, in which the suffixes are the same, are called Moments of Inertia, and those corresponding to P_{12}, in which the suffixes are different, are called Products of Inertia. We may extend these names to the more general problem which is now before us, in which these quantities are not, as in the case of a rigid body, absolute constants, but are functions of the variables q_1, q_2, &c.

565.7 In like manner we may call the coefficients of the form Q_{11} Moments of Mobility, and those of the form Q_{12} Products of Mobility. It is not often, however, that we shall have occasion to speak of the coefficients of mobility.

* * *

567.] In this outline of the fundamental principles of the dynamics of a connected system, we have kept out of view the mechanism by which the parts of the system are connected. We have not even written down a set of equations to indicate how the motion of any part of the system depends on the variation of the variables. We have confined our attention to the variables, their velocities and momenta, and the forces which act on the pieces representing the variables. Our only assumptions are, that the connexions of the system are such that the time is not explicitly contained in the equations of condition, and that the principle of the conservation of energy is applicable to the system.

567. *time is not explicitly contained in the equations of condition*: This means, roughly, that the structure of the system, whatever it may be, does not change in time. For example, if the "system" is a gasoline engine, we are to assume that the piston rods do not bend.

the principle of the conservation of energy is applicable: Maxwell assumed conservation of energy when, in paragraph 560.2, he treated the kinetic energy of the system as equal to the work done on it—that is, when he declared the system to be a conservative one.

567.1 Such a description of the methods of pure dynamics is not unnecessary, because Lagrange and most of his followers, to whom we are indebted for these methods, have in general confined themselves to a demonstration of them, and, in order to devote their attention to the symbols before them, they have endeavoured to banish all ideas except those of pure quantity, so as not only to dispense with diagrams, but even to get rid of the ideas of velocity, momentum, and energy, after they have been once for all supplanted by symbols in the original equations. In order to be able to refer to the results of this analysis in ordinary dynamical language, we have endeavoured to retranslate the principal equations of the method into language which may be intelligible without the use of symbols.

567.2 As the development of the ideas and methods of pure mathematics has rendered it possible, by forming a mathematical theory of dynamics, to bring to light many truths which could not have been discovered without mathematical training, so, if we are to form dynamical theories of other sciences, we must have our minds imbued with these dynamical truths as well as with mathematical methods.

567.3 In forming the ideas and words relating to any science, which, like electricity, deals with forces and their effects, we must keep constantly in mind the ideas appropriate to the fundamental science of dynamics, so that we may, during the first development of the science, avoid inconsistency with what is already established, and also that when our views become clearer, the language we have adopted may be a help to us and not a hindrance.

567.1 *we have endeavored to retranslate ... into language which may be intelligible without the use of symbols*: Faraday had wished for just such an effort on the part of mathematicians. In a well-known letter to Maxwell dated 13 November, 1857, he wrote: "When a mathematician engaged in investigating physical actions and results has arrived at his own conclusions, may they not be expressed in common language as fully, clearly, and definitely as in mathematical formulae? If so, it would be a great boon...."

567.2 *if we are to form dynamical theories of other sciences*: More than just reconstituting the science of Electricity and Magnetism, the *Treatise* is showing a pathway to new sciences—new areas of knowledge that will no longer be bound by the old archetypes of massive-body mechanics.

567.3 *when our views become clearer, the language we have adopted may be a help to us and not a hindrance*: As we have seen, traditional electricity and magnetism is laden with the baggage of earlier hypotheses, images, and terminology. One of the chief strengths of the dynamical approach is its ability to minimize the accretion of such impedimenta. I will offer additional thoughts on this topic in connection with Chapter XX.

Editor's Introduction to
Chapter VI

In Chapter V Maxwell formulated a set of equations of such purity and far-reaching generality as to apply to *any conceivable dynamical system*. Conventionally, to be sure, such a venture would hardly seem possible—or even desirable. For conventional analysis treats complex systems as assemblages of certain *elements* that we specify; for example, conventional mechanics identifies bodies, together with the forces acting between them, as the elements of physical systems, whether pocket watches or planets in orbit. By contrast, the equations of pure dynamics neither assume the elements of the system nor hypothesize the specific linkages between them; rather, those particulars emerge through a winnowing process in which experimentation reveals nonzero values for some mathematical quantities and null values for others. Maxwell will undertake such a process in the present chapter.

In the preceding chapter Maxwell showed how to express the energy of an abstract dynamical system in three forms. Now he will choose the form he called $T_{\dot{q}}$, set forth in equation (24) of that chapter, as the one best suited to a system consisting of electric currents in conductors:

$$2T_{\dot{q}} = P_{11}\dot{q}_1{}^2 + 2P_{12}\dot{q}_1\dot{q}_2 + \&\text{c.}$$

The quantities \dot{q} represent *velocities*—but in an electrical system the term "velocities" must encompass not only the ordinary velocities of the conductors themselves (supposing them to move), but also the *time rates of change of the electric variables*—that is to say, the *currents* in their respective conductors. In the present chapter, therefore, Maxwell will divide the quantities \dot{q} into two classes: those representing mechanical velocities of the conductors he will call \dot{x}, while those representing electric currents he will style \dot{y}.

In its new form, then, the equation above will contain instances of the following terms:

(*i*) squares of velocities of individual moving conductors ($\dot{x}_1{}^2$),

(*ii*) squares of individual currents ($\dot{y}_1{}^2$),

(*iii*) products of the velocity of one conductor with that of another ($\dot{x}_1\dot{x}_2$),

(*iv*) products of the current in one conductor with that in another ($\dot{y}_1\dot{y}_2$),

(*v*) products of the current in one conductor with the velocity of either its own or another conductor ($\dot{x}_1\dot{y}_1$) or ($\dot{x}_2\dot{y}_1$).

(*i*) Maxwell's suspended coil (*ii*) Maxwell's "electromagnetic top"
(Coil *W*, not shown in Maxwell's own drawing, is mounted within housing *A*)

Each of the terms listed represents a class of possible phenomena; but which of those possible phenomena do electrical systems actually exhibit? That question is answered by *experiment*. In the present chapter Maxwell will describe experiments that aim to measure the coefficients P_{ii} and P_{ij} corresponding to the respective possible terms. If measurement discloses a nonzero value for a coefficient, the corresponding term exists. If the measured value is zero, Maxwell will regard the corresponding term as nonexistent. In this way, the highly abstract and general equation will become more and more specific. Thomas Simpson has called this process the Judgment of the Terms.*

Maxwell's gyroscopic coils

Maxwell will discuss several such experiments in Articles 574–577 below. They involve two devices, illustrated at the top of this page. The "suspended coil" (sketch *i*) is quite simple, while the "electromagnetic top" (sketch *ii*) is complicated both in its construction and in its theory of operation. But I do not think it has been widely noticed that Maxwell's two devices are basically variations of one another, and of the gyroscope. It will be useful, I hope, to point out their common principle here.

The gyroscope, besides being a perenially popular toy, is also a useful laboratory device by which to demonstrate conservation of angular momentum. The gyroscope illustrated on the next page consists of rotor *R* mounted in gimbals—that is, it turns within meridian ring *M*, which is mounted on bearings in the stand *S*, the stand in turn pivoting on a vertical axis at *C*. And first (sketch *iii*) let the instrument be at rest with the rotor axis vertical. Suppose then that a spring motor, or some other

* *Maxwell's Mathematical Rhetoric*, Green Lion Press (2010), pp. 271–289.

(*iii*) Gyroscope, axis vertical (*iv*) Gyroscope, axis horizontal

internal mechanism, sets the rotor R turning in the direction shown;* the rotor will possess angular momentum represented by vector **A**, along the axis of rotation. (The relation between the angular momentum vector and the direction of rotation is given by the right-hand screw rule, which Maxwell discussed in Art. 23.)

But since the total angular momentum of the instrument is constant, any momentum developed by the rotor R must have been lost from the meridian ring M. The ring must therefore turn in an opposite direction, so as to develop a quantity of angular momentum **B** equal and opposite to **A**.

Comparing sketches (*i*) and (*iii*), it will be evident that Maxwell's suspended coil is formally identical to the gyroscope just discussed, with the current in the coil corresponding to rotor R and the coil proper corresponding to meridian ring M. Thus if electric current is associated with mechanical force as a rotating mass is, then a sudden commencement or cessation of current should result in rotation of the coil about its vertical axis.

Next let the rotor axis be horizontal (sketch *iv*), and suppose that all parts of the instrument are at rest except for the rotor, which spins with angular momentum **A**. If we now begin rotating stand S about its vertical axis, we shall be adding angular momentum **B** to the system. The total angular momentum will then be **A+B**; and if the directions are those indicated in the drawing, the rotor axis will incline toward the vertical.

* It is imperative that the source of motion be *internal*—we must not, for example, reach in from the outside with a flick of the finger; otherwise the angular momentum of the system will not be conserved.

(*ii*) Maxwell's "electromagnetic top" (repeated)

(*iv*) Gyroscope, axis horizontal (repeated)

Similarly (sketch *ii*), Maxwell will send a current through coil *W* (mounted within housing *A*) of the electromagnetic top while setting the meridian ring into rotation by means of the pulley and string. If the current exhibits inertial effects like those of a rotating body, Maxwell should detect an alteration in the direction of the coil's axis when the current is started—an inclination like the one seen in the gyroscope (sketch *iv*). A null result, on the other hand, would indicate that in the general equation, terms implying a mechanical interaction between currents and their own conductors are to be omitted.

CHAPTER VI.

DYNAMICAL THEORY OF ELECTROMAGNETISM.

568.] WE have shewn, in Art. 552, that, when an electric current exists in a conducting circuit, it has a capacity for doing a certain amount of mechanical work, and this independently of any external electromotive force maintaining the current. Now capacity for performing work is nothing else than energy, in whatever way it arises, and all energy is the same in kind, how ever it may differ in form. The energy of an electric current is either of that form which consists in the actual motion of matter, or of that which consists in the capacity for being set in motion, arising from forces acting between bodies placed in certain positions relative to each other.

568.1 The first kind of energy, that of motion, is called Kinetic energy, and when once understood it appears so fundamental a fact of nature that we can hardly conceive the possibility of resolving it into anything else. The second kind of energy, that depending on position, is called Potential energy, and is due to the action of what we call forces, that is to say, tendencies towards change of relative position. With respect to

568.1 *forces, that is to say, tendencies towards change of relative position*: Such a characterization of force is bound to startle anyone who has learned to think in terms of Newton's Second Law, according to which the measure of a force is the *change in momentum* it produces (in a given time) rather than the change in position. Nevertheless, there seems no way in which one and the same body can undergo a change in momentum *without* at the same time experiencing a change in position. Our disquietude in the face of Maxwell's description does not arise because it is false, but because it is not convertible: every force may indeed tend towards change of position, but not every change of position requires a force. Force then is a sufficient, but not a necessary, condition for change of position. It remains to be seen what Maxwell has in mind that leads him to offer this unusual formulation.

That he singles out *potential energy* in this connection might suggest that he means the description to apply only to *conservative forces*, since these are the forces associated with a potential; and they are necessarily dependent on position (16.3, 16.6). But Maxwell will bring out the same characterization again in Article 573, in a context that has nothing to do with potential; he really does seem to intend his distinctive formulation to apply to force in general.

these forces, though we may accept their existence as a demonstrated fact, yet we always feel that every explanation of the mechanism by which bodies are set in motion forms a real addition to our knowledge.

569.] The electric current cannot be conceived except as a kinetic phenomenon. Even Faraday, who constantly endeavoured to emancipate his mind from the influence of those suggestions which the words 'electric current' and 'electric fluid' are too apt to carry with them, speaks of the electric current as 'something progressive, and not a mere arrangement.' *

569.1 The effects of the current, such as electrolysis, and the transfer of electrification from one body to another, are all progressive actions which require time for their accomplishment, and are therefore of the nature of motions.

569.2 As to the velocity of the current, we have shewn that we know nothing about it, it may be the tenth of an inch in an hour, or a hundred thousand miles in a second.† So far are we from knowing its absolute value in any case, that we do not even know whether what we call the positive direction is the actual direction of the motion or the reverse.

569.3 But all that we assume here is that the electric current involves motion of some kind. That which is the cause of electric currents has been called Electromotive Force. This name has long been used with great advantage, and has never led to any inconsistency in the language of science. Electromotive force is always to be understood to act on electricity only, not on the bodies in which the electricity resides. It is never to be confounded with ordinary mechanical force, which acts on bodies only, not on the electricity in them. If we ever come to know the formal relation between electricity and ordinary matter, we shall probably also know the relation between electromotive force and ordinary force.

570.] When ordinary force acts on a body, and when the body yields to the force, the work done by the force is measured by the product of the force into the amount by which the body yields. Thus, in the case of

* Exp. Res., 283. † Exp. Res., 1648.

569.2 *As to the velocity of the current, ... we know nothing about it*: In this connection recall (12.5–12.7) and comments.

569.3 *Electromotive force is always to be understood to act on electricity only, not on the bodies in which the electricity resides*: Maxwell here reasserts the doctrine he stated in (501.1), and which I discussed in the editor's introduction to Part IV, Chapter I.

570. *the product of the force into the amount by which the body yields*: For example, the driving weight of a pendulum clock moves incrementally, yet so slowly that

water forced through a pipe, the work done at any section is measured by the fluid pressure at the section multiplied into the quantity of water which crosses the section.

570.1 In the same way the work done by an electromotive force is measured by the product of the electromotive force into the quantity of electricity which crosses a section of the conductor under the action of the electromotive force.

570.2 The work done by an electromotive force is of exactly the same kind as the work done by an ordinary force, and both are measured by the same standards or units.

570.3 Part of the work done by an electromotive force acting on a conducting circuit is spent in overcoming the resistance of the circuit, and this part of the work is thereby converted into heat.

we can consider it as moving with constant but negligible speed. If in the course of a day it descends distance d, then d is "the amount by which the body [the clock mechanism] yields"; thus if the weight is W, the work done is the product Wd. But Maxwell is about to offer a different illustration that takes a more general point of view.

the fluid pressure at the section multiplied into the quantity of water which crosses the section: How is this case relevant, since "fluid pressure" is not a force, and "quantity of water" is not a distance? Bear in mind that pressure is a measure of force *per unit area*; while the quantity of an incompressible fluid must be proportional to its *volume*—that is, to the distance along the pipe multiplied by the pipe's sectional area. Thus Maxwell's product is proportional to

$$\left(\frac{\text{force}}{\text{area}}\right)(\text{distance} \times \text{area}) = \text{force} \times \text{distance},$$

that is, proportional to the corresponding product in the case of a suspended weight. Maxwell has chosen the example of water in a pipe advisedly, since it is analogically related to that of electrical current.

570.1 *In the same way...*: that is, analogically—if we rely on Maxwell's earlier analogy between electric current and the motion of an incompressible fluid (61.). But in fact Maxwell has already obtained the desired result independently of analogy; see the following remark.

...the work done by an electromotive force is ... the product of the electromotive force into the quantity of electricity which crosses a section of the conductor: In (60.10) Maxwell pointed out that "the work spent in augmenting the displacement" is the product of the electromotive force and the amount of the displacement. Although that discussion does not explicitly apply to conduction, Maxwell goes on in (62.7) to draw an intimate connection between electric displacement and conduction current.

570.3 *this part of the work is thereby converted into heat*: See (111.10–111.11) and associated comments.

570.4 Another part of the work is spent in producing the electromagnetic phenomena observed by Ampère, in which conductors are made to move by electromagnetic forces. The rest of the work is spent in increasing the kinetic energy of the current, and the effects of this part of the action are shewn in the phenomena of the induction of currents observed by Faraday.

570.5 We therefore know enough about electric currents to recognise, in a system of material conductors carrying currents, a dynamical system which is the seat of energy, part of which may be kinetic and part potential.

570.6 The nature of the connexions of the parts of this system is unknown to us, but as we have dynamical methods of investigation which do not require a knowledge of the mechanism of the system, we shall apply them to this case.

570.7 We shall first examine the consequences of assuming the most general form for the function which expresses the kinetic energy of the system.

571.] Let the system consist of a number of conducting circuits, the form and position of which are determined by the values of a system of variables x_1 x_2, &c., the number of which is equal to the number of degrees of freedom of the system.

571.1 If the whole kinetic energy of the system were that due to the motion of these conductors, it would be expressed in the form

$$T = \tfrac{1}{2}(x_1 x_1)\dot{x}_1^2 + \&\text{c.} + (x_1 x_2)\dot{x}_1 \dot{x}_2 + \&\text{c.},$$

where the symbols $(x_1 x_1)$, &c. denote the quantities which we have called moments of inertia, and $(x_1 x_2)$, &c. denote the products of inertia.

570.4 *The rest of the work is spent in increasing the kinetic energy of the current*: Increasing it, that is, from *zero* to the current's steady value (551.).

570.6 *we have dynamical methods of investigation which do not require a knowledge of the mechanism of the system*: Maxwell is referring to the methods discussed in the preceding chapter.

571. *the number of which is equal to the number of degrees of freedom of the system*: See Arts. 553, 555 and associated comments.

571.1 *the symbols* $(x_1 x_1)$, &c.: Here $(x_1 x_1)$ does not denote (as it might appear) the product of x_1 by itself. Instead Maxwell is introducing a new notation for *moment of inertia*, which was previously symbolized P_{11} at (565.6) and which is a *function* of x_1. Similarly the symbol $(x_1 x_2)$ represents a *product of inertia*, a function of x_1 and x_2. Bear in mind that "inertia" in these terms is a generalized concept that does not, like mechanical inertia, depend on ponderable mass. Similarly in Arts. 559 and 560 Maxwell developed generalized momentum, work, and kinetic energy without reference to ponderable mass.

571.2 If X' is the impressed force, tending to increase the coordinate x, which is required to produce the actual motion, then, by Lagrange's equation,

$$\frac{d}{dt}\frac{dT}{d\dot{x}} - \frac{dT}{dx} = X'.$$

571.3 When T denotes the energy due to the visible motion only, we shall indicate it by the suffix $_m$, thus, T_m.

571.4 But in a system of conductors carrying electric currents, part of the kinetic energy is due to the existence of these currents. Let the motion of the electricity, and of anything whose motion is governed by that of the electricity, be determined by another set of coordinates y_1, y_2, &c., then T will be a homogeneous function of squares and products of all the velocities of the two sets of coordinates. We may therefore divide T into three portions, in the first of which, T_m, the velocities of the coordinates x only occur, while in the second, T_e, the velocities of the coordinates y only occur, and in the third, T_{me}, each term contains the product of the velocities of two coordinates of which one is an x and the other a y.

571.2 $\dfrac{d}{dt}\dfrac{dT}{d\dot{x}} - \dfrac{dT}{dx} = X'$: Maxwell gave a specific form of this equation in the preceding chapter as eq. (20) in paragraph 564.5, writing it for the variable q_1 and expressing kinetic energy in the form $T_{\dot{q}}$. The present instance is written for a geometrical variable x (571.).

571.3 *the energy due to the visible motion only*: that is, the kinetic energy of the mechanical components of the system—in this case, the conductors.

571.4 *part of the energy is due to the existence of these currents*: These currents represent motions in some sense, though not "visible" motions; and these motions contribute to the total kinetic energy of the system.

Let the motion of the electricity ... be determined by another set of coordinates y_1, y_2, *&c.*: Notice that at this stage, instead of introducing hypotheses about a moving substance that constitutes a current, or where that substance is located, Maxwell introduces unspecified variables y_1, y_2, etc., corresponding to the respective currents.

Then T will be a homogeneous function of squares and products of all the velocities of the two sets of coordinates: We have seen already (565.3) that T can be expressed as a homogeneous second-degree function of the velocities when no distinctions are made among the coordinates; and in (571.1) Maxwell gave an example involving squares and products of the velocities of x-coordinates only.

The same will be true if the coordinates are divided into two classes, x and y; we shall then find squares of like and products of unlike \dot{x} terms, squares of like and products of unlike \dot{y} terms, and products of all combinations of \dot{x} and \dot{y} terms. We may consider these different combinations separately, as Maxwell points out in the remainder of the paragraph.

571.5 We have therefore

$$T = T_m + T_e + T_{me},$$

where
$$T_m = \tfrac{1}{2}(x_1 x_1)\dot{x}_1^2 + \&c. + (x_1 x_2)\dot{x}_1 \dot{x}_2 + \&c.,$$

$$T_e = \tfrac{1}{2}(y_1 y_1)\dot{y}_1^2 + \&c. + (y_1 y_2)\dot{y}_1 \dot{y}_2 + \&c.,$$

$$T_{me} = (x_1 y_1)\dot{x}_1 \dot{y}_1 + \&c.$$

572.] In the general dynamical theory, the coefficients of every term may be functions of all the coordinates, both x and y. In the case of electric currents, however, it is easy to see that the coordinates of the class y do not enter into the coefficients.

572.1 For, if all the electric currents are maintained constant, and the conductors at rest, the whole state of the field will remain constant. But in this case the coordinates y are variable, though the velocities \dot{y} are constant. Hence the coordinates y cannot enter into the expression for T, or into any other expression of what actually takes place.

572.2 Besides this, in virtue of the equation of continuity, if the conductors are of the nature of linear circuits, only one variable is required to express the strength of the current in each conductor. Let the velocities \dot{y}_1, \dot{y}_2, &c. represent the strengths of the currents in the several conductors.

572.3 All this would be true, if, instead of electric currents, we had currents of an incompressible fluid running in flexible tubes. In this case the velocities of these currents would enter into the expression for T, but the coefficients would depend only on the variables x, which determine the form and position of the tubes.

572.4 In the case of the fluid, the motion of the fluid in one tube does not directly affect that of any other tube, or of the fluid in it. Hence, in the value of T, only the squares of the velocities \dot{y}, and not their products, occur, and in T_{me} any velocity \dot{y} is associated only with those velocities of the form \dot{x} which belong to its own tube.

572.2 *equation of continuity*: The equation expressing that electricity cannot be created or annihilated. Maxwell exhibited the fact by means of the experiments of Art. 34 and named the equation in (35.1).

Let the velocities \dot{y}_1, \dot{y}_2, &c. represent the strengths of the currents: Conventionally, these choices might reflect a fluid-flow hypothesis—the "velocities" being interpreted as the rates of locomotion of the fluid. Instead, Maxwell leaves the quantities \dot{y}_1, \dot{y}_2, etc., *uninterpreted* physically. He does not know, and does not need to know, what constitutes the "strength" of a current in physical terms; it is sufficient to have shown, as he has just done, that "only one variable is required to express the strength of the current in each conductor."

572.5 In the case of electrical currents we know that this restriction does not hold, for the currents in different circuits act on each other. Hence we must admit the existence of terms involving products of the form $\dot{y}_1\dot{y}_2$, and this involves the existence of something in motion, whose motion depends on the strength of both electric currents \dot{y}_1 and \dot{y}_2. This moving matter, whatever it is, is not confined to the interior of the conductors carrying the two currents, but probably extends throughout the whole space surrounding them.

573.] Let us next consider the form which Lagrange's equations of motion assume in this case. Let X' be the impressed force corresponding to the coordinate x, one of those which determine the form and position of the conducting circuits. This is a force in the ordinary sense, a tendency towards change of position. It is given by the equation

572.5 *the currents in different circuits act on each other*: This refers to electromagnetic attraction or repulsion between currents. It does *not* refer to current induction, the phenomenon Faraday described in his First Series of *Experimental Researches in Electricity*; see the following remark.

But does not Maxwell's wording here, "the currents in different circuits act on each other," violate his earlier insistence that "the mechanical force which urges a conductor carrying a current across the lines of magnetic force acts, not on the electric current, but on the conductor which carries it" (501.)? It would be easy enough to patch up Maxwell's phraseology to save the earlier doctrine; but is the doctrine really worth saving?

Hence we must admit the existence of terms involving products of the form $\dot{y}_1\dot{y}_2$: As we shall see in (596.), the expression for the electromagnetic force between currents i_1 and i_2 contains a term of the form $i_1 i_2$, which corresponds to $\dot{y}_1\dot{y}_2$ in the present context. Maxwell's reference to the product of currents makes it clear that he does not have current induction in mind, since induction can occur between two circuits even when only one of them carries current.

This moving matter, whatever it is...: As the variable \dot{y} represents a "velocity" only in the generalized sense of a rate of change, so the "matter" whose motion constitutes the current is *generalized* matter, of whose qualities we are ignorant.

...probably extends throughout the whole space: Recall that in Art. 549 Maxwell concluded that an electric current possesses "momentum" in a sense that permits momentum to exist *outside* the conductor—something unimaginable on a conventional fluid-flow hypothesis.

573. *a force in the ordinary sense, a tendency towards change of position*: It is hard to see how Maxwell could have supposed this sense to be "ordinary"; most people probably associate force more with the *effort* involved in shifting a load, than with the shift in position itself. Nevertheless he had already held this opinion for many years when the *Treatise* was composed in 1873. In November, 1857 Faraday wrote to Maxwell: "I perceive that I do not use the word 'force' as you define it, 'the tendency of a body to pass from one place to another.' What I mean by the word is the *source* or *sources* of all possible actions...." Of course Faraday's conception of force, we may remark, is no more "ordinary" than is Maxwell's.

$$X' = \frac{d}{dt}\frac{dT}{d\dot{x}} - \frac{dT}{dx}.$$

573.1 We may consider this force as the sum of three parts, corresponding to the three parts into which we divided the kinetic energy of the system, and we may distinguish them by the same suffixes. Thus

$$X' = X'_m + X'_e + X'_{me}.$$

573.2 The part X'_m is that which depends on ordinary dynamical considerations, and we need not attend to it.

573.3 Since T_e does not contain \dot{x}, the first term of the expression for X'_e is zero, and its value is reduced to

$$X'_e = -\frac{dT_e}{dx}. \tag{1}$$

573.4 This is the expression for the mechanical force which must be applied to a conductor to balance the electromagnetic force, and it

573.2 X'_m ... *depends on ordinary mechanical considerations, and we need not attend to it*: The subscript m identifies quantities belonging exclusively to ordinary mechanics (571.2); nothing about them either bears upon or is altered by Maxwell's *dynamical* approach, so they can safely be left out of discussion.

573.3 *equation* [1]: Maxwell did not number this equation. I have labeled it [1] to facilitate reference.

573.4 *the mechanical force which must be applied to a conductor to balance the electromagnetic force*: From what considerations does Maxwell derive this interpretation? He has already stipulated that the force X'_e is impressed from outside the system (573.). As such, its strength is determined by an outside agent, not by the system itself; it might, for example, be *zero*. Now $X'_e = -dT_e/dx$, according to equation [1]. But dT_e/dx *cannot*, in general, be zero, for that would imply that the energy of the system is indifferent to the positions of its conductors. On the contrary, we find that current-carrying conductors affect one another (572.5). Equation [1] must therefore be not a statement of fact but a *criterion*: some condition must obtain when $X'_e = -dT_e/dx$, which would fail to obtain otherwise. Maxwell identifies that condition as *balance*.

Consider a mechanical analogy. A body of mass m falling through distance x will acquire kinetic energy $T = mgx$. (Since this is a gravitational example, we choose the positive direction *downwards*.) Differentiating T with respect to x, we have $dT/dx = mg$. This value for dT/dx is a property of the gravitational system, not under the control of an outside agent. But it is clear that if an outside agent were to apply a force $X = -dT/dx$, it would balance the gravitational force. By substituting the known value of dT/dx in this case we should find that X must equal $-mg$, that is, it must be an upward force, equal and opposite to the weight of the body.

asserts that it is measured by the rate of *diminution* of the purely electrokinetic energy due to the variation of the coordinate x. The electromagnetic force, X_e, which brings this external mechanical force into play, is equal and opposite to X'_e and is therefore measured by the rate of *increase* of the electrokinetic energy corresponding to an increase of the coordinate x. The value of X_e, since it depends on squares and products of the currents, remains the same if we reverse the directions of all the currents.

573.5 The third part of X' is

$$X'_{me} = \frac{d}{dt}\frac{dT_{me}}{d\dot{x}} - \frac{dT_{me}}{dx}.$$

573.6 The quantity T_{me} contains only products of the form $\dot{x}\dot{y}$, so that $dT_{me}/d\dot{x}$ is a linear function of the strengths of the currents \dot{y}. The first term, therefore, depends on the rate of variation of the strengths of the currents, and indicates a mechanical force on the conductor, which is zero when the currents are constant, and which is positive or negative according as the currents are increasing or decreasing in strength.

573.7 The second term depends, not on the variation of the currents, but on their actual strengths. As it is a linear function with respect to these currents, it changes sign when the currents change sign. Since every term involves a velocity \dot{x}, it is zero when the conductors are at rest. There are also terms arising from the time variations of the coefficients of \dot{y} in $dT_{me}/d\dot{x}$: these remarks apply also to them.

573.8 We may therefore investigate these terms separately. If the conductors are at rest, we have only the first term to deal with. If the currents are constant, we have only the second.

574.] As it is of great importance to determine whether any part of the kinetic energy is of the form T_{me}, consisting of products of ordinary velocities and strengths of electric currents, it is desirable that experiments should be made on this subject with great care.

574. *it is of great importance to determine whether any part of the kinetic energy is of the form T_{me}*: Why is this question "of great importance"? Recall that the *me* subscript identifies energy that arises from interactions not between currents and currents (which action is well known)—nor between conductors and conductors (which, insofar as it exists, is part of ordinary mechanics)—but (571.2) interactions between material conductors and electricity, about whose material or inertial status, Maxwell has reminded us, we know nothing at all (35.). To know whether such interactions exist, then, is important to Maxwell, first, because it is *not* known. I discussed a second, deeper, reason in the editor's introduction to Part IV, Chapter I: that such interactions would challenge the strict separation Maxwell wishes to maintain between the force that acts on a current and the force that acts on a material conductor.

574.1 The determination of the forces acting on bodies in rapid motion is difficult. Let us therefore attend to the first term, which depends on the variation of the strength of the current.

574.2 If any part of the kinetic energy depends on the product of an ordinary velocity and the strength of a current, it will probably be most easily observed when the velocity and the current are in the same or in opposite directions. We therefore take a circular coil of a great many windings, and suspend it by a fine vertical wire, so that its windings are horizontal, and the coil is capable of rotating about a vertical axis, either in the same direction as the current in the coil, or in the opposite direction [Fig. 33].

574.3 We shall suppose the current to be conveyed into the coil by means of the suspending wire, and, after passing round the windings, to complete its circuit by passing downwards through a wire in the same line with the suspending wire and dipping into a cup of mercury.

574.4 Since the action of the horizontal component of terrestrial magnetism would tend to turn this coil round a horizontal axis when the current flows

Fig. 33.

through it, we shall suppose that the horizontal component of terrestrial magnetism is exactly neutralized by means of fixed magnets, or that the experiment is made at the magnetic pole. A vertical mirror is attached to the coil to detect any motion in azimuth.

574.5 Now let a current be made to pass through the coil in the direction N.E.S.W. If electricity were a fluid like water, flowing along the wire, then, at the moment of starting the current, and as long as its velocity is increasing, a force would require to be supplied to produce the angular momentum of the fluid in passing round the coil, and as this must be supplied by the elasticity of the suspending wire, the coil would at first rotate in the opposite direction or W.S.E.N., and this would be detected by means of the mirror. On stopping the current there would be another movement of the mirror, this time in the same direction as that of the current.

574.6 No phenomenon of this kind has yet been observed. Such an action, if it existed, might be easily distinguished from the already known actions of the current by the following peculiarities.

574.2 *most easily observed when the velocity and the current are in the same or in opposite directions*: Since both velocity and current are directional quantities, we would expect their mutual additive or subtractive effect to be a maximum in either of these cases.

574.7 (1) It would occur only when the strength of the current varies, as when contact is made or broken, and not when the current is constant.

574.8 All the known *mechanical* actions of the current depend on the strength of the currents, and not on the rate of variation. The electromotive action in the case of induced currents cannot be confounded with this electromagnetic action.

574.9 (2) The direction of this action would be reversed when that of all the currents in the field is reversed.

574.10 All the known mechanical actions of the current remain the same when all the currents are reversed, since they depend on squares and products of these currents.

574.11 If any action of this kind were discovered, we should be able to regard one of the so-called kinds of electricity, either the positive or the negative kind, as a real substance, and we should be able to describe the electric current as a true motion of this substance in a particular direction. In fact, if electrical motions were in any way comparable with the motions of ordinary matter, terms of the form T_{me} would exist, and their existence would be manifested by the mechanical force X_{me}.

574.12 According to Fechner's hypothesis, that an electric current consists of two equal currents of positive and negative electricity, flowing in opposite directions through the same conductor, the terms of the second class T_{me} would vanish, each term belonging to the positive current being accompanied by an equal term of opposite sign belonging to the negative current, and the phenomena depending on these terms would have no existence.

574.13 It appears to, me, however, that while we derive great advantage from the recognition of the many analogies between the electric current and a current of material fluid, we must carefully avoid making any assumption not warranted by experimental evidence, and that there is,

574.8 *All the known mechanical actions of the current depend on the strength of the current, and not on the rate of variation*: For example, the force exerted between current-carrying conductors (495.1), or the force exerted on a magnet in the vicinity of a current (479.), is constant when each current is constant. But the example of a magnet involves a complication; see the following remark.

574.10 *All the known mechanical actions of the current remain the same when all the currents are reversed*: But the magnetic field about a current (479.)—which is to say, the force exerted on a magnet of unit pole strength—*does* reverse when the current is reversed. In (615.3), however, Maxwell will take note of Ampère's hypothesis that permanent magnets contain molecular electric currents. On that understanding, we will not have met the requirement that *all* currents are reversed unless we also invert the magnet! When we do reverse both the current and the magnet, the force on the magnet indeed remains the same.

as yet, no experimental evidence to shew whether the electric current is really a current of a material substance, or a double current, or whether its velocity is great or small as measured in feet per second.

574.14 A knowledge of these things would amount to at least the beginnings of a complete dynamical theory of electricity, in which we should regard electrical action, not, as in this treatise, as a phenomenon due to an unknown cause, subject only to the general laws of dynamics, but as the result of known motions of known portions of matter, in which not only the total effects and final results, but the whole intermediate mechanism and details of the motion, are taken as the objects of study.

575.] The experimental investigation of the second term of X_{me}, namely dT_{me}/dx, is more difficult, as it involves the observation of the effect of forces on a body in rapid motion.

575.1 The apparatus shewn in Fig. 34, which I had constructed in 1861, is intended to test the existence of a force of this kind. The electromagnet A is capable of rotating about the horizontal axis BB', within a ring which itself revolves about a vertical axis.

575.2 Let A, B, C be the moments of inertia of the electromagnet about the axis of the coil, the horizontal axis BB', and a third axis CC' respectively.

Fig. 34.

575.3 Let θ be the angle which CC' makes with the vertical, φ the azimuth of the axis BB', and ψ a variable on which the motion of electricity in the coil depends.

575.2 *Let A, B, C be the moments of inertia*: If a rigid body spins with angular velocity ω about a specified axis, its angular momentum can be expressed as $L = I\omega$ and its kinetic energy $T = \frac{1}{2}I\omega^2$, where I is called the moment of inertia about the axis in question. Except for cases of obvious symmetry, the moments of inertia for a given body will, in general, be different for different axes. Every body has three "principal" axes, for each of which I is a simple scalar quantity, so that L and ω (which are directed quantities, or vectors) are parallel; thus when a body spins about one of its principal axes, it spins without wobbling. Do not confuse moments of inertia A, B, C with axes AA', BB', CC'.

575.3 *Let θ be the angle which CC' makes with the vertical*: See the sketch on the facing page.

ψ *a variable on which the motion of electricity in the coil depends*: But $\dot{\psi}$ is not the simple measure of current strength, as we will see in (575.6).

575.4 Then the kinetic energy T of the electromagnet may be written

$$2T = A\dot\varphi^2\sin^2\theta + B\dot\theta^2 + C\dot\varphi^2\cos^2\theta + E\big(\dot\varphi\sin\theta + \dot\psi\big)^2,$$

where E is a quantity which may be called the moment of inertia of the electricity in the coil.

575.5 If Θ is the moment of the impressed force tending to increase θ, we have, by the equations of dynamics,

$$\Theta = B\frac{d^2\theta}{dt^2} - \Big\{(A-C)\dot\varphi^2\sin\theta\cos\theta + E\dot\varphi\cos\theta\big(\dot\varphi\sin\theta + \dot\psi\big)\Big\}. \qquad [2]$$

φ *the azimuth of axis BB'*: Since axis BB' is constrained to the horizontal, angle φ is also measured in the horizontal plane—or, equivalently, about the vertical axis. The accompanying drawing shows both angles θ and φ.

575.4 *the kinetic energy T ... may be written*: Each of the terms in Maxwell's expression has the form $I\omega^2$ (575.2, *comment*) for a specified axis. For coil axis AA', $I=A$ and ω is the projection of $\dot\varphi$ onto AA'; as the diagram shows, that projection is $\dot\varphi\sin\theta$. For axis BB', $I=B$ and ω is just $\dot\theta$, since θ is measured about axis BB'. For axis CC', $I=C$ and ω is the projection of $\dot\varphi$ onto CC', or $\dot\varphi\cos\theta$.

The electrical variable ψ also contributes to the kinetic energy associated with the coil axis. Insofar as its time derivative $\dot\psi$ is an angular velocity, the same formalism as before would yield $E\dot\psi^2$. But the coil itself is spinning with angular velocity $\dot\varphi$ about the vertical axis; therefore, even if $\dot\psi$ were equal to zero, the electricity in the coil would still have a component of spin in the direction of AA' in the amount $\dot\varphi\sin\theta$—the projection of $\dot\varphi$ onto AA'. Thus the total *electrical* angular momentum about AA' is $\dot\psi + \dot\varphi\sin\theta$, and the electrokinetic energy is $E(\dot\psi + \dot\varphi\sin\theta)^2$.

575.5 Θ *is the moment of the impressed force tending to increase θ*: I recounted in the editor's introduction how a force tending to increase θ would arise if the current in the coil were to exhibit an inertial reaction correlated with motion of the conductor, as a rotating material body would do. As Maxwell stated in (575.1), it is precisely "a force of this kind" that the present apparatus was designed to detect. The *moment* of that force, Θ, is what modern writers typically call the *torque* about axis BB'.

Equation [2]: Maxwell did not number this equation; I have labeled it [2] to facilitate reference. If a body is moving along a straight path, the derivative of its kinetic energy with respect to distance yields the force that acts on the body; compare (573.4, *comment*). Similarly, if a body is rotating, the derivative of its kinetic energy with respect to the angle of rotation yields the *torque* that acts about the axis of rotation. Thus to find Θ, differentiate the expression for T (575.4) with respect to θ. When differentiating the second term, note that

$$B\dot\theta\frac{d\dot\theta}{d\theta} = B\frac{d\dot\theta}{d\theta}\frac{d\theta}{dt} = B\frac{d\dot\theta}{dt} = B\frac{d^2\theta}{dt^2}.$$

575.6 By making Ψ, the impressed force tending to increase ψ, equal to zero, we obtain

$$\dot{\varphi}\sin\theta + \dot{\psi} = \gamma,$$

a constant, which we may consider as representing the strength of the current in the coil.

575.7 If C is somewhat greater than A, Θ will be zero, and the equilibrium about the axis BB' will be stable when

$$\sin\theta = \frac{E\gamma}{(C-A)\dot{\varphi}}.$$

575.8 This value of θ depends on that of γ, the electric current, and is positive or negative according to the direction of the current.

575.9 The current is passed through the coil by its bearings at B and B', which are connected with the battery by means of springs rubbing on metal rings placed on the vertical axis.

575.10 To determine the value of θ, a disk of paper is placed at C, divided by a diameter parallel to BB' into two parts, one of which is painted red and the other green.

575.6 Ψ, *the impressed force tending to increase* ψ: This is a misprint, as the following remark will explain; read $\dot{\psi}$ in place of ψ.

By making Ψ ... *equal to zero, we obtain*: "Making Ψ ... equal to zero" is not a physical operation but a mathematical investigation into the condition for which $\dot{\psi}$, the time rate of change of ψ, reaches a steady value. If this occurs, we may regard the total electrical angular momentum, $(\dot{\psi} + \dot{\varphi}\sin\theta)$, or γ, as the measure of current strength in the coil.

Thus it is obvious that Ψ cannot be a force that increases ψ itself, for if then Ψ were to become zero and ψ cease to increase, $\dot{\psi}$ would be zero and there would be no further electrical activity in the coil. It is clear from the following equation and subsequent discussion that Maxwell does not contemplate that possibility.

γ, *a constant*: The spinning velocity $\dot{\varphi}$ is set by the apparatus, and we are assuming that $\dot{\psi}$ is constant for the reasons just discussed. Strictly, however, Maxwell cannot yet assert that γ is constant, for he has not yet shown why angle θ may be assumed constant. That argument appears in the next paragraph.

575.7 *If C is somewhat greater than A, Θ will be zero*: A body can execute stable rotation about any of its principal axes, but the stability will be greatest for rotation about the axis having greatest moment of inertia. In Maxwell's apparatus, the coil is free to pivot so that either AA' or CC' lines up—either exactly or approximately—with the vertical axis of rotation. In the first case, angle θ will be nearly $0°$, in the second, nearly $90°$. When Maxwell makes C greater than A (575.12), he causes the coil to favor the first position, $\theta \approx 0°$.

When the rotation is stable, there will be no net torque tending to change θ, and θ will take on a fixed value. Set $\Theta = 0$ and $(\dot{\psi} + \dot{\varphi}\sin\theta) = \gamma$ in equation [2] above to obtain Maxwell's expression for $\sin\theta$.

575.11 When the instrument is in motion a red circle is seen at C when θ is positive, the radius of which indicates roughly the value of θ. When θ is negative, a green circle is seen at C.

575.12 By means of nuts working on screws attached to the electromagnet, the axis CC' is adjusted to be a principal axis having its moment of inertia just exceeding that round the axis A, so as to make the instrument very sensitive to the action of the force if it exists.

575.13 The chief difficulty in the experiments arose from the disturbing action of the earth's magnetic force, which caused the electromagnet to act like a dip-needle. The results obtained

Figure 34, repeated

were on this account very rough, but no evidence of any change in θ could be obtained even when an iron core was inserted in the coil, so as to make it a powerful electromagnet.

575.11 *When the instrument is in motion a red circle is seen at C*: The divided circle at the left in the accompanying sketch corresponds to the bi-colored circle in Figure 34, and point C in the sketch also corresponds to point C in the figure. At the right I have indicated how a clear area appears when the circle rotates about eccentric point C. It should be clear how the radius of the clear area indicates the distance of C from the diameter of the circle and hence provides an estimate of the value of θ in the apparatus.

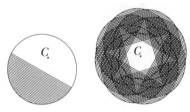

575.12 *so as to make the instrument very sensitive to the action of the force if it exists*: "The force" is the force tending to increase θ (575.5). I noted in (575.7, *comment*) that the coil will favor a position for which $\theta \approx 0°$ if C is greater than A. But if C is only slightly greater than A, that partiality will be less pronounced; thus for a given current γ, angle θ will be large when $(C - A)$ is small—as indeed the expression in (575.7) shows.

575.13 *to act like a dip-needle*: that is, to tend towards alignment with the earth's magnetic field, and so affect the measured value of angle θ. The illustration shows a dip needle design that was widely used in the late 1800's. In 1873 the magnetic "dip" at Kew (about 70 miles south-southwest of Cambridge) was about 67.71 degrees below the horizontal.

575.14 If, therefore, a magnet contains matter in rapid rotation, the angular momentum of this rotation must be very small compared with any quantities which we can measure, and we have as yet no evidence of the existence of the terms T_{me} derived from their mechanical action.

576.] Let us next consider the forces acting on the currents of electricity, that is, the electromotive forces.

576.1 Let Y be the effective electromotive force due to induction, the electromotive force which must act on the circuit from without to balance it is $Y' = -Y$, and, by Lagrange's equation,

$$Y = -Y' = -\frac{d}{dt}\frac{dT}{d\dot{y}} + \frac{dT}{dy}\,.$$

576.2 Since there are no terms in T involving the coordinate y, the second term is zero, and Y is reduced to its first term. Hence, electromotive force cannot exist in a system at rest, and with constant currents.

576.3 Again, if we divide Y into three parts, Y_m, Y_e, and Y_{me}, corresponding to the three parts of T, we find that, since T_m does not contain \dot{y}, $Y_m = 0$.

576.4 We also find
$$Y_e = -\frac{d}{dt}\frac{dT_e}{d\dot{y}}\,.$$

576.5 Here $dT_e/d\dot{y}$ is a linear function of the currents, and this part of the electromotive force is equal to the rate of change of this function. This is the electromotive force of induction discovered by Faraday. We shall consider it more at length afterwards.

577.] From the part of T, depending on velocities multiplied by currents, we find
$$Y_{me} = -\frac{d}{dt}\frac{dT_{me}}{d\dot{y}}\,.$$

577.1 Now $dT_{me}/d\dot{y}$ is a linear function of the velocities of the conductors. If, therefore, any terms of T_{me} have an actual existence, it would be possible to produce an electromotive force independently of all existing currents by simply altering the velocities of the conductors. For instance, in the case of the suspended coil at Art. 574, if, when the coil is at rest, we suddenly set it in rotation about the vertical axis, an

576.2 *electromotive force cannot exist in a system at rest, and with constant currents*: Maxwell is ruling out only electromotive force *due to induction*, as he stated at the beginning of the paragraph. Necessarily, a system with constant current possesses electromotive force associated with that current.

576.5 *This is the electromotive force discovered by Faraday*: Thus it is an observed phenomenon. The purpose of the present experiments is to sort out which possible quantities really exist in nature; and Maxwell has now identified Y_e as one such quantity.

electromotive force would be called into action proportional to the acceleration of this motion. It would vanish when the motion became uniform, and be reversed when the motion was retarded.

577.2 Now few scientific observations can be made with greater precision than that which determines the existence or non-existence of a current by means of a galvanometer. The delicacy of this method far exceeds that of most of the arrangements for measuring the mechanical force acting on a body. If, therefore, any currents could be produced in this way they would be detected, even if they were very feeble. They would be distinguished from ordinary currents of induction by the following characteristics.

577.3 (1) They would depend entirely on the motions of the conductors, and in no degree on the strength of currents or magnetic forces already in the field.

577.4 (2) They would depend not on the absolute velocities of the conductors, but on their accelerations, and on squares and products of velocities, and they would change when the acceleration becomes a retardation, though the absolute velocity is the same.

577.5 Now in all the cases actually observed, the induced currents depend altogether on the strength and the variation of currents in the field, and cannot be excited in a field devoid of magnetic force and of currents. In so far as they depend on the motion of conductors, they depend on the absolute velocity, and not on the change of velocity of these motions.

577.6 We have thus three methods of detecting the existence of the terms of the form T_{me}, none of which have hitherto led to any positive result. I have pointed them out with the greater care because it appears to me important that we should attain the greatest amount of certitude within our reach on a point bearing so strongly on the true theory of electricity.

577.7 Since, however, no evidence has yet been obtained of such terms, I shall now proceed on the assumption that they do not exist, or at least that they produce no sensible effect, an assumption which will considerably simplify our dynamical theory. We shall have occasion, however, in discussing the relation of magnetism to light, to shew that the motion which constitutes light may enter as a factor into terms involving the motion which constitutes magnetism.

577.7 *the motion which constitutes light ... the motion which constitutes magnetism*: Maxwell is referring to the magnetic rotation of light discovered by Faraday; see another reference to this topic in (636.2).

Editor's Introduction to
Chapter VII

In Chapter V Maxwell developed the Lagrangian equations in their most universal form, a form that transcends (but implicitly contains) the particular principles of mechanics, of electricity, and of heat. Indeed going beyond, as they do, the equations of all the special sciences, they might almost be characterized as the equations of Nature herself!

Such universal equations, if supplied with appropriate mechanical concepts such as *force* and *mass*, can be given a form that is equivalent to Newton's laws—and thus constitute the fundamental principles of mechanics. In Chapter VI Maxwell showed how to shape the general equations in a different direction—one in which they became the dynamical principles of electromagnetism.

But while the characteristic concepts of mechanical dynamics are well known, it is not at all clear what should count as the essential concepts of electrodynamics. They might, of course, prove to be identical to the mechanical concepts—and thinkers who regard Newtonian mechanics as the foundation of all the other sciences will proceed upon that very assumption. Maxwell, however, follows a path that is more exploratory and more suited to the Lagrangian spirit. His way is one in which, as we shall see, the fundamental electrodynamical concepts will emerge, almost of themselves, from the general equations.

In this chapter Maxwell takes the first step in fostering that emergence by treating the *electric circuit*. Subsequent chapters will extend his treatment to more general electrodynamical systems.

The electric circuit

Consider the circuit diagrammed here, containing a battery, a switch, and a resistance. With the switch in position 1, electric current develops in the wire; when the switch is in position 2, the battery is effectively removed from the circuit containing resistance R.

We have already viewed electric current from the standpoint of Ohm's Law, $E = CR$ (241.2). But Ohm's Law describes a *steady state* in which electromotive force E, resistance R, and therefore current C, are all constant—what, then, happens when the current starts

and stops? If Ohm's Law applied to those moments, applying electromotive force to the circuit (by placing the switch in position 1) would instantaneously establish a steady current. Similarly, upon moving the switch to position 2 the current would instantly cease. But Maxwell has noted that, on the contrary, "when the electromotive force begins to act, it does not at once produce the full current, but ... the current rises gradually" (551.). Nor does an existing current ever seem to cease instantly but, like a moving body, it "resists suddenly being stopped" (551.2).

The starting and stopping behavior of simple mechanical systems may, therefore, help to reveal what Ohm's Law obscures about electrical circuits—that is their *dynamical* character.

A mechanical analogy

The drawing illustrates a typical 19th-century steam engine. Like other engines of its kind, it boasts an enormous flywheel.

Part of the effort necessary to drive a flywheel is expended in overcoming mechanical resistance in order to maintain a constant speed of rotation—or, we may say,

constant mechanical momentum. If effort is expended in excess of that amount, the momentum increases (the wheel accelerates). On the other hand, if all effort is suddenly withdrawn the wheel does not immediately stop but gradually slows down; nevertheless that decaying rotation continues to represent an overcoming of resistance—now, however, at the expense of the wheel's momentum. Similarly, a circuit containing electrical resistance requires a definite impressed electromotive force (its value given by Ohm's Law) to maintain a constant current, by restoring the electromagnetic momentum that would have been lost had the electromotive force not been applied. If that electromotive force is suddenly withdrawn, the electromagnetic momentum decreases (the current decays)— but gradually, not instantaneously.

In Article 579 of the present chapter, Maxwell discusses a simple electric circuit in terms that suggest a mechanical analogy such as the one just considered. There can be no doubting Maxwell's great affection for analogies generally; and as a pedagogical tool, analogy can assist students who are more familiar with mechanical systems to develop appropriate ways of viewing electrical systems. To that end Maxwell designed a mechanical device that illustrates the behavior of coupled electric circuits; see Art. 584, below. Understand, however, that Maxwell is by no means endeavoring to model electrical science upon mechanical science as a paradigm. Rather, he is engaged in a species of reasoning that is more far-reaching than either.

CHAPTER VII.

THEORY OF ELECTRIC CIRCUITS.

578.] WE may now confine our attention to that part of the kinetic energy of the system which depends on squares and products of the strengths of the electric currents. We may call this the Electrokinetic Energy of the system. The part depending on the motion of the conductors belongs to ordinary dynamics, and we have seen that the part depending on products of velocities and currents does not exist.

578.1 Let A_1, A_2, &c. denote the different conducting circuits. Let their form and relative position be expressed in terms of the variables x_1, x_2, &c. the number of which is equal to the number of degrees of freedom of the mechanical system. We shall call these the Geometrical Variables.

578.2 Let y_1 denote the quantity of electricity which has crossed a given section of the conductor A_1 since the beginning of the time t. The strength of the current will be denoted by \dot{y}_1, the fluxion of this quantity.

578.3 We shall call \dot{y}_1 the actual current, and y_1 the integral current. There is one variable of this kind for each circuit in the system.

578. *Electrokinetic Energy*: As defined here, it denotes energy due to the motion of electricity—distinguished from energy due to mechanical motion of the conductors of electricity. But bear in mind that the "motion" we are attributing to current is motion in a *generalized* sense (572.5, *comment*).

the part depending on products of velocities and currents does not exist: as indicated by the negative results of experiments summarized in (577.6–577.7).

578.1 *degrees of freedom*: Recall the discussion of this term in Arts. 553 and 555.

578.2 *fluxion*: Newton's term, still used widely in Maxwell's day, for what we now call the first derivative of a quantity with respect to time—especially when it is symbolized, as here, by a single dot placed over the quantity. By extension, a double dot denotes the second derivative of the quantity, with respect to time.

578.3 *actual current ... integral current*: Maxwell is freely using the language of flow here, the "actual current" \dot{y} corresponding to the time rate of charge passage, and the "integral current" y denoting (as indeed its symbol requires), the integral of \dot{y} over time—as he states, the quantity of electricity which has crossed a given section of the conductor during a specified period of time. As before, we keep in mind that in the present context, all such flow terminology, like the term "motion" itself, is to be taken in a *generalized*—not to say *metaphorical*—sense.

578.4 Let T denote the electrokinetic energy of the system. It is a homogeneous function of the second degree with respect to the strengths of the currents, and is of the form

$$T = \tfrac{1}{2}L_1\dot{y}_1^2 + \tfrac{1}{2}L_2\dot{y}_2^2 + \&c. + M_{12}\dot{y}_1\dot{y}_2 + \&c., \qquad (1)$$

where the coefficients L, M, &c. are functions of the geometrical variables x_1, x_2, &c. The electrical variables y_1, y_2 do not enter into the expression.

578.5 We may call L_1, L_2, &c. the electric moments of inertia of the circuits A_1, A_2, &c., and M_{12} the electric product of inertia of the two circuits A_1 and A_2. When we wish to avoid the language of the dynamical theory, we shall call L_1 the coefficient of self-induction of the circuit A_1, and M_{12} the coefficient of mutual induction of the circuits A_1 and A_2. M_{12} is also called the potential of the circuit A_1 with respect to A_2. These quantities depend only on the form and relative position of the circuits. We shall find that in the electromagnetic system of measurement they are quantities of the dimension of a line. See Art. 627.

578.6 By differentiating T with respect to \dot{y}_1, we obtain the quantity p_1, which, in the dynamical theory, may be called the momentum corresponding to y_1. In the electric theory we shall call p_1 the electrokinetic momentum of the circuit A_1. Its value is

$$p_1 = L_1\dot{y}_1 + M_{12}\dot{y}_2 + \&c.$$

578.7 The electrokinetic momentum of the circuit A_1 is therefore made up of the product of its own current into its coefficient of self-induction, together with the sum of the products of the currents in the other circuits, each into the coefficient of mutual induction of A_1 and that other circuit.

578.4 *T ... is a homogeneous function of the second degree with respect to the strengths of the currents*: Maxwell had previously expressed T in this form with respect to the generalized velocities \dot{q} in (565.4), eq. (24).

578.5 *When we wish to avoid the language of the dynamical theory*: As he has demonstrated once before already, Maxwell respects his readers' intellectual freedom by declining to enforce a theoretical terminology. In (75.1) he offered an alternative to the language of electrical displacement; here he does the same with respect to the dynamical theory.

self-induction [L] ... *mutual induction* [M]: Engineers and technicians of today use instead the terms self- and mutual *inductance*, the nomenclature favored by Heaviside. But the symbols L and M, which seem to have originated with Maxwell, persist universally with the meanings he gave to them.

578.6 *electrokinetic momentum*: Compare the expression for p_1, electrokinetic momentum, with Maxwell's earlier expression for generalized momentum in

Electromotive Force.

579.] Let E be the impressed electromotive force in the circuit A, arising from some cause, such as a voltaic or thermo-electric battery, which would produce a current independently of magneto-electric induction.

579.1 Let R be the resistance of the circuit, then, by Ohm's law, an electromotive force $R\dot{y}$ is required to overcome the resistance, leaving an electromotive force $E - R\dot{y}$ available for changing the momentum of the circuit. Calling this force Y', we have, by the general equations,

$$Y' = \frac{dp}{dt} - \frac{dT}{dy},$$

but since T does not involve y, the last term disappears.

579.2 Hence, the equation of electromotive force is

$$E - R\dot{y} = Y' = \frac{dp}{dt}, \quad \text{or} \quad E = R\dot{y} + \frac{dp}{dt}. \qquad [2]$$

579.3 The impressed electromotive force E is therefore the sum of two parts. The first, $R\dot{y}$, is required to maintain the current \dot{y} against the resistance R. The second part is required to increase the electro-magnetic momentum p. This is the electromotive force which must be supplied from sources independent of magneto-electric induction. The

(565.5), eq. (26). Here, too, the term "momentum" carries no presumption of ponderable mass.

579. *some cause ... which would produce a current independently of magneto-electric induction*: Not, for example, a typical bicycle lamp dynamo. Maxwell means to enumerate all the relations of self-induction and mutual induction among the circuits to be analyzed; a magnetoelectric generator would introduce additional induction relations into the system.

579.1 *Ohm's Law*: As previously stated (241.2 and *comment*), the electromotive force required to maintain a given current through a resistance is the product of the resistance and the current—in the present case, $R\dot{y}$.

leaving $E - R\dot{y}$ available for changing the momentum of the circuit: If the applied electromotive force E exceeds the amount $R\dot{y}$ that is required to maintain a constant current \dot{y}, the excess $E - R\dot{y}$ will increase the current \dot{y}—or, otherwise stated, increase the electrokinetic momentum of the circuit.

the general equations: Here, Lagrange's equations, specifically equation (19) in (564.5). Maxwell will call another set of equations "general" in Art. 598 of a later chapter; they will not be general in the same sense, however, for Maxwell there crafts them to apply exclusively to cases involving electromotive force.

579.3 *This is the electromotive force which must be supplied*: "This" refers to the sum of both parts of E, as expressed in equation [2]; compare (579.). Note: Maxwell did not number the equations of this chapter beyond (1) in (578.4). I have supplied additional equation numbers in square brackets.

electromotive force arising from magneto-electric induction alone is evidently $-dp/dt$, or, *the rate of decrease of the electrokinetic momentum of the circuit.*

Electromagnetic Force.

580.] Let X be the impressed mechanical force arising from external causes, and tending to increase the variable x. By the general equations

$$X' = \frac{d}{dt}\frac{dT}{d\dot{x}} - \frac{dT}{dx}.$$

580.1 Since the expression for the electrokinetic energy does not contain the velocity (\dot{x}), the first term of the second member disappears, and we find

$$X' = -\frac{dT}{dx}.$$

580.2 Here X' is the external force required to balance the forces arising from electrical causes. It is usual to consider this force as the reaction against the electromagnetic force, which we shall call X, and which is equal and opposite to X'.

580.3 Hence $$X = \frac{dT}{dx},$$ [3]

or, *the electromagnetic force tending to increase any variable is equal to the rate of increase of the electrokinetic energy per unit increase of that variable, the currents being maintained constant.*

580.4 If the currents are maintained constant by a battery during a displacement in which a quantity, W, of work is done by electromotive force, the electrokinetic energy of the system will be at the same time

579.3, continued. *The electromotive force arising from magneto-electric induction alone is evidently* $-dp/dt$: Consider again equation [2] in (579.1). If now the impressed electromotive force is withdrawn so that E becomes zero, the equation will become $R\dot{y} = -dp/dt$, indicating that the electromotive force ($R\dot{y}$) that is overcoming resistance is now being supplied at the expense of p, the electromagnetic momentum of the circuit.

580.1 *the expression for the electrokinetic energy does not contain the velocity* (\dot{x}): reflecting its character as *electro*kinetic energy, which, as its name implies, does not depend on mechanical motion.

580.4 *a displacement in which a quantity, W, of work is done by electromotive force*: Consider, for example, two parallel conductors carrying currents in the same direction, and which therefore attract one another with *electromagnetic* force F proportional to the product of those currents (495.1). If the conductors are allowed to approach one another through distance x, work will be done in the amount $W = \int F dx$. Maxwell characterizes that work as being done by

increased by W. Hence the battery will be drawn upon for a double quantity of energy, or $2W$, in addition to that which is spent in generating heat in the circuit. This was first pointed out by Sir W. Thomson.* Compare this result with the electrostatic property in Art. 93.

Case of Two Circuits.

581.] Let A_1 be called the Primary Circuit, and A_2 the Secondary Circuit. The electrokinetic energy of the system may be written

$$T = \tfrac{1}{2}L\dot{y}_1^2 + M\dot{y}_1\dot{y}_2 + \tfrac{1}{2}N\dot{y}_2^2, \qquad [4]$$

where L and N are the coefficients of self-induction of the primary and secondary circuits respectively, and M is the coefficient of their mutual induction.

* Nichol's *Cyclopaedia of the Physical Sciences*, ed. 1860, article 'Magnetism, Dynamical Relations of.'

electromotive force because the currents, which mutually exert the electromagnetic force, are themselves (like all currents) maintained by electromotive force; thus any work ascribed to the electromagnetic force draws ultimately on the energy represented by electromotive force.

the electrokinetic energy of the system will be ... increased by W: How can this be if, as Maxwell has just stipulated, "the currents are maintained constant"? As we are about to see in the next Article, the electrokinetic energy includes a term $M\dot{y}_1\dot{y}_2$, in which the coefficient M increases when the conductors approach one another (583.3).

the battery will be drawn upon for a double quantity of energy: As Maxwell is about to remind us, we found a similar result for electrostatics in Art. 93.

In the present example, when the two parallel currents are allowed to approach one another, they will do mechanical work of some amount W, which must ultimately be supplied by the battery.

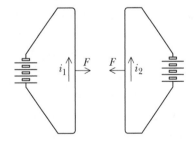

But when currents in the same direction approach one another, each induces in the other a current that diminishes the original (530.5). The batteries, in restoring these currents to their original value, are supplying the additional energy represented in the preceding comment by $M\dot{y}_1\dot{y}_2$; and an argument similar to that of (93., *comment*) shows that the batteries will actually be called upon for a double quantity of energy, $2W$.

581. *The electrokinetic energy of the system may be written*: Maxwell here rewrites equation (1) of (578.4) for the case of two currents.

coefficient of self-induction ... coefficient of mutual induction: These are the more traditional, practical terms; but Maxwell has rooted them in the general, theoretical names for L, M, and N coined in (578.5) above.

581.1 Let us suppose that no electromotive force acts on the secondary circuit except that due to the induction of the primary current. We have then

$$E_2 = R_2 \dot{y}_2 + \frac{d}{dt}\left(M\dot{y}_1 + N\dot{y}_2\right) = 0.$$

581.2 Integrating this equation with respect to t, we have

$$R_2 y_2 + M\dot{y}_1 + N\dot{y}_2 = C, \text{ a constant,}$$

where y_2 is the integral current in the secondary circuit.

581.3 The method of measuring an integral current of short duration will be described in Art. 748, and it is easy in most cases to ensure that the duration of the secondary current shall be very short.

581.4 Let the values of the variable quantities in the equation at the end of the time t be accented, then, if y_2 is the integral current, or the whole quantity of electricity which flows through a section of the secondary circuit during the time t,

$$R_2 y_2 = M\dot{y}_1 + N\dot{y}_2 - \left(M'\dot{y}_1' + N'\dot{y}_2'\right).$$

581.5 If the secondary current arises entirely from induction, its initial value \dot{y}_2 must be zero if the primary current is constant, and the conductors are at rest before the beginning of the time t.

581.1 *We have then*: Maxwell here applies equation [2] to the secondary circuit specifically. The term $R_2\dot{y}_2$, given by Ohm's Law, expresses the electromotive force necessary to maintain current \dot{y}_2 through the resistance of the secondary; $M\dot{y}_1$ and $N\dot{y}_2$ are the contributions to electrokinetic momentum of the secondary circuit made by (1) induction of the primary current and (2) self-induction of the secondary current, respectively.

581.2 y_2 *is the integral current in the secondary circuit*: Integral current was defined in (578.3). In the secondary circuit the integral current will be, as Maxwell will shortly put it, "the whole quantity of electricity which flows through a section of the secondary circuit" during a specified time. As before, the time-integral of \dot{y}_2 (with a dot) is represented as y_2 (without the dot).

581.3 *The method of measuring an integral current of short duration*: The "method" is that of the ballistic galvanometer, discussed in the editor's introduction to Part III, Chapter III; see also (537.1). Maxwell's treatment of it in loving detail appears, unfortunately, in an omitted chapter.

581.4 *Let the values ... be accented*: Here, marked with primes, as M', \dot{y}_2'.

581.5 *its initial value must be zero if the primary current is constant, and the conductors are at rest*: The two circumstances capable of inducing a current in a stationary secondary circuit are *variation of the primary current* (530.) and *motion of the primary circuit* (530.4). If neither of these conditions is the case, there will be no induced current in the secondary.

581.6 If the time t is sufficient to allow the secondary current to die away, $\dot{y}_2{}'$, its final value, is also zero, so that the equation becomes

$$R_2 y_2 = M\dot{y}_1 - M'\dot{y}_1'.$$

581.7 The integral current of the secondary circuit depends in this case on the initial and final values of $M\dot{y}_1$.

Induced Currents.

582.] Let us begin by supposing the primary circuit broken, or $\dot{y}_1 = 0$, and let a current \dot{y}_1 be established in it when contact is made.

582.1 The equation which determines the secondary integral current is

$$R_2 y_2 = -M'\dot{y}_1'.$$

582.2 When the circuits are placed side by side, and in the same direction, M is a positive quantity. Hence, when contact is made in the primary circuit, a negative current is induced in the secondary circuit.

582.3 When the contact is broken in the primary circuit, the primary current ceases, and the induced integral current is y_2, where

$$R_2 y_2 = M\dot{y}_1'.$$

The secondary current is in this case positive.

582.4 If the primary current is maintained constant, and the form or relative position of the circuits altered so that M becomes M', the integral secondary current is y_2, where

$$R_2 y_2 = \left(M - M'\right)\dot{y}_1'.$$

582.5 In the case of two circuits placed side by side and in the same direction M diminishes as the distance between the circuits increases. Hence, the induced current is positive when this distance is increased and negative when it is diminished.

582.2 *when contact is made in the primary circuit, a negative current is induced in the secondary circuit*: Maxwell reported this as an experimental observation in (530.2); now, however, he is in a position to deduce it from the general theory.

582.3 *The secondary current is in this case positive*: that is, when the primary circuit is broken. Maxwell reviewed this effect too in (530.2).

582.4 *the form or relative position of the circuits altered so that M becomes M'*: Recall that M, the coefficient of mutual induction, depends upon the geometrical variables (578.4). It determines the energy attributed jointly to two currents, and therefore expresses the degree of their interaction.

582.5 *M diminishes as the distance between the circuits increases*: This is not deduced from theory but follows from the interpretation of M as a measure of interaction: if the circuits are further apart we presume their interaction is less.

582.6 These are the elementary cases of induced currents described in Art. 530.

Mechanical Action between the Two Circuits.

583.] Let x be any one of the geometrical variables on which the form and relative position of the circuits depend, the electromagnetic force tending to increase x is

$$X = \frac{1}{2}\dot{y}_1^2 \frac{dL}{dx} + \dot{y}_1\dot{y}_2 \frac{dM}{dx} + \frac{1}{2}\dot{y}_2^2 \frac{dN}{dx}.$$ [5]

583.1 If the motion of the system corresponding to the variation of x is such that each circuit moves as a rigid body, L and N will be independent of x, and the equation will be reduced to the form

$$X = \dot{y}_1\dot{y}_2 \frac{dM}{dx}.$$ [6]

583.2 Hence, if the primary and secondary currents are of the same sign, the force X, which acts between the circuits, will tend to move them so as to increase M.

583.3 If the circuits are placed side by side, and the currents flow in the same direction, M will be increased by their being brought nearer together. Hence the force X is in this case an attraction.

583. *the electromagnetic force tending to increase x is*: To obtain equation [5], differentiate equation [4] with respect to x, as was specifed by equation [3].

583.1 *the motion of the system corresponding to the variation of* x: Remember that x is a generalized geometrical variable, not necessarily a Cartesian coordinate. A simple linear increase in x, for example, might correspond physically to non-uniform rectilinear motion, or to motion along a curve, or to a rotation, or to any number of other examples.

if each circuit moves as a rigid body, L and N will be independent of x: L and N each depend on the forms of the circuits, but not on the relations of the circuits to each other.

583.2 *the force X ... will tend to move them so as to increase M*: For if in equation [6] \dot{y}_1 and \dot{y}_2 have the same sign (which normally means they are in the same direction), their product is positive; and force X will have the same sign as dM/dx.

But this inference is not quite as informative as it might seem to be. For the positive direction of X is the direction that increases x (583.). Therefore if X acts in the positive direction dM/dx too will be positive; and M will increase along with x. On the other hand, if X acts in the negative direction dM/dx will be negative; and M will increase as x decreases. Thus, although we can indeed affirm that X acts in a direction that increases M, the equation does not tell us *which* direction that is!

583.3 *M will be increased by their being brought nearer together*: This again is either a perfectly reasonable assumption (582.5, *comment*), or an experimental observation—although, admittedly, that distinction is not a sharp one. In any case it is *not* deduced from equation [6].

584.] The whole of the phenomena of the mutual action of two circuits, whether the induction of currents or the mechanical force between them, depend on the quantity M, which we have called the coefficient of mutual induction. The method of calculating this quantity from the geometrical relations of the circuits is given in Art. 524, but in the investigations of the next chapter we shall not assume a knowledge of the mathematical form of this quantity. We shall consider it as deduced from experiments on induction, as, for instance, by observing the integral current when the secondary circuit is suddenly moved from a given position to an infinite distance, or to any position in which we know that $M = 0$.

NOTE.—{There is a model in the Cavendish Laboratory designed by Maxwell which illustrates very clearly the laws of the induction of currents.

It is represented in Fig. 34a. P and Q are two disks, the rotation of P represents the primary current, that of Q the secondary. These disks are connected together by a differential gearing. The intermediate wheel carries a fly-wheel the moment of inertia of which can be altered by moving weights inwards or outwards. The resistance of the secondary circuit is represented by the friction of a string passing over Q and kept tight by an elastic band. If the disk P is set in rotation (a current started in the primary) the disk Q will turn in the opposite direction (inverse current when the primary in started). When the velocity of rotation of P becomes uniform, Q is at rest (no current in the secondary when the primary current is constant); if the disk P is stopped, Q commences to rotate in the direction in which P was previously moving (direct current in the secondary on breaking the circuit). The effect of an iron core in increasing the induction can be illustrated by increasing the moment of inertia of the fly-wheel.}

Fig. 34a

Hence the force X is ... an attraction: Once we have satisfied ourselves that it is movement in the direction of decreasing x that increases M, we can infer that X acts in that direction, which is the negative direction (583.).

584. *The method of calculating this quantity ... is given in Art. 524*: Article 524 is part of an omitted chapter, but Maxwell previously derived a formula for M as a mathematical quantity in Art. 423. He did not there, however, give to M the interpretation he has so carefully worked out in the present chapter.

584., NOTE. The note and illustration were supplied for the third edition by editor J. J. Thomson. It may be useful to point out to modern readers that the "direct current" he speaks of in the penultimate sentence is not in contradistinction to alternating current; it means a current in the same direction as that which the primary current had before its cessation.

Editor's Introduction to
Chapter VIII

The previous chapter, "Theory of Electric Circuits," did indeed develop a powerful explication of the inductive and electromagnetic relations of multiple circuits. Such a topic is obviously indispensable in a work which, like this one, was to serve as a textbook of electricity and magnetism. The world would no more countenance an electrical treatise that failed to discuss electrical circuits than it would a treatise on mechanics that lacked a chapter on the simple machines.*

Nevertheless, a topic may be indispensable without being fundamental. Electric circuitry, one might suppose, is a specialized technical topic rather than a basic one, a topic belonging to the world of trades and know-how—such as telegraphy and lighting—more than it does to the world of natural philosophy. It may surprise us, then, to read Maxwell's declaration near the outset of the present chapter that he intends to "begin again from a new foundation"—and that the foundation in question is none other than the work of the previous chapter, the theory of electric circuits.

Surprise vanishes, however, once we recognize the significance of Maxwell's title for this chapter, "Exploration of the Field by Means of the Secondary Circuit." The title recalls Faraday's Moving Wire—a specialized contrivance, to be sure, but one that forged for Faraday a fundamentally new view of the field as a real physical entity. Let us briefly review some of Faraday's work with that instrument.

Faraday's Moving Wire

"The moving wire" was Faraday's name for a device, or class of devices, which developed electric currents through *cutting*—so he understood it—lines of magnetic force. One of its earliest forms was the simple wire loop, shown here in a drawing taken from Faraday's Twenty-eighth Series.† Thrust over the end of a bar magnet, it "cuts" the lines of force that emerge from

* Even Newton's *Principia* includes a "chapter" on the simple machines, in the form of Corollary 2 to Law III.

† Faraday, Michael, *Experimental Researches in Electricity*, Volume III, p. 335.

the magnet and in so doing develops, in the form of a current, a quantity of electricity which, Faraday was able to ascertain, is strictly proportional to the number of lines cut. Similarly, by maneuvering it and comparable wire forms in various locations, Faraday was able to map the lines of force that occupied the space surrounding magnets generally. Since it was evident to him that the current so produced represented a determinate amount of power that had been converted from the magnetic lines of force into electrical form, he realized that the Moving Wire *samples*, in effect, a power or activity that characterizes portions of space whether the wire is moving through them or not. In this way he was led to the conception of the magnetic field as a real physical structure, a "system of power" capable of occupying space just as literally as a material body is said to occupy space.

Maxwell's secondary circuit

In the present chapter, Maxwell will consider a system of two circuits. The primary circuit is energized by a battery or other source of electromotive force; it functions like an electromagnet, establishing a magnetic field everywhere in its vicinity. The secondary circuit is moved about like Faraday's exploratory loop, developing an electric current whose strength Maxwell can deduce according to theory in the Case of Two Circuits set out in Article 581 of the preceding chapter.

Maxwell's secondary circuit is equivalent to Faraday's Moving Wire in all but name. Instead of conducting a physical exploration like Faraday's, however, Maxwell will apply the circuit theory of Chapter VII to carry out a mathematical exploration of the field. We might say that Maxwell converts Faraday's physical experiment into a mathematical experiment.

The results of this mathematical investigation quite remarkably corroborate the spirit of Faraday's interpretation of the field, even as they transform it. Faraday's "system of power" becomes, for Maxwell, a store of *electrokinetic momentum*; and this momentum fills space in the same way as Faraday's lines of force filled space. For Faraday, the Moving Wire's action in "cutting" lines of force involved the somewhat mysterious process of conversion of power from magnetic to electrical form.* For Maxwell, on the other hand, there is no conversion or transfiguration of force; when the secondary circuit samples the momentum that occupies space, that momentum is not converted to something else.

* Faraday, of course, would not have called the process "mysterious." For him, conversion of the powers of nature from one form to another was the most fundamental of processes—the sole mode, indeed, in which nature can be said to *act* at all. See Fisher, Howard J., *Faraday's Experimental Researches in Electricity: Guide to a First Reading*, Green Lion Press (2001), pp. 145, 203, 577–578.

Electromotive force arises in the circuit in proportion to the change of momentum it experiences, exemplifying the definitive relation between force and momentum generally, as schematized in the Lagrangian equations of Chapter V. But this relation concerns a *change in quantity*, not a conversion of form.

Refining the railway

Readers will recall—I hope fondly—Maxwell's "railway carriage" vignette of Article 532. Although Maxwell did not provide a visual illustration to accompany it, his descriptive language was itself so realistic and visual, I could not resist supplying an embellished version of this picture, something similar to which, I suppose, every reader must have already constructed in imagination:

In the present chapter, contrastingly, Maxwell does provide a diagram of the Secondary Circuit; it is his Figure 38, shown here in advance. It is clear that the two images, whether we think of them graphically or verbally, are in every essential respect the same. In Maxwell's figure a sliding wire AB corresponds to the rail car's axle; in the figure, loop C closes the circuit just as the galvanometer does in the railway example. Maxwell will even use the word "rails" when he introduces Figure 38 in paragraph 594.1.

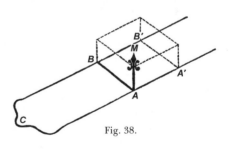

Fig. 38.

In another respect, the two examples differ markedly. The physical apparatus of Figure 38 is utterly simple, having been pared down to the barest essentials; in the earlier example Maxwell called up a complete railway carriage even though nothing but one of its two axles actually performs a functional role. Figure 38 realistically depicts visible elements like the conducting loop and the sliding wire; yet it also includes the mathematical solid formed by $AA'B'B$ and AM, representing, as we shall see, an increment of electrokinetic momentum (594.7)—a quantity that is by no means visible to the optical eye. Maxwell's description of the

railway is, in contrast, realistic in every visual respect—while being at the same time hilariously *un*realistic in conceiving the employment of something as large and lumbering as a railway carriage to produce an electrical effect so delicate as to require a galvanometer for its detection!

Perhaps that is why Maxwell has waited until the present chapter before providing a pictorial illustration. Perhaps he saw no need for a diagram that addresses the optical eye, but a pressing need for one that directs itself to the mind's eye—an eye that is theoretically informed by Maxwell's Chapter V and which is learning to "see" quantities like electrokinetic momentum as physical realities.

CHAPTER VIII.

EXPLORATION OF THE FIELD BY MEANS OF
THE SECONDARY CIRCUIT.

585.] WE have proved in Arts. 582, 583, 584 that the electromagnetic action between the primary and the secondary circuit depends on the quantity denoted by M, which is a function of the form and relative position of the two circuits.

585.1 Although this quantity M is in fact the same as the potential of the two circuits, the mathematical form and properties of which we deduced in Arts. 423, 492, 521, 539 from magnetic and electromagnetic phenomena, we shall here make no reference to these results, but begin again from a new foundation, without any assumptions except those of the dynamical theory as stated in Chapter VII.

585.2 The electrokinetic momentum of the secondary circuit consists of two parts (Art. 578), one, Mi_1, depending on the primary current i_1, while the other, Ni_2, depends on the secondary current i_2. We are now to investigate the first of these parts, which we shall denote by p, where

$$p = Mi_1. \tag{1}$$

585.3 We shall also suppose the primary circuit fixed, and the primary current constant. The quantity p, the electrokinetic momentum of the secondary circuit, will in this case depend only on the form and position of the secondary circuit, so that if any closed curve be taken for the secondary circuit, and if the direction along this curve, which is to be

585.1 *we shall here make no reference to these results, but begin again from a new foundation*: I suggested in the editor's introduction to this chapter that the "new foundation" is, ultimately, the Lagrangian vision outlined in Chapter V. More proximately, it is the Case of Two Circuits of the previous chapter, which incorporates that vision.

585.3 *The quantity p ... will in this case depend on the form and position of the secondary circuit*: The secondary circuit now becomes Faraday's Moving Wire (22.6), that instrument of unparalleled interpretive power. For Faraday the Moving Wire counted lines of force; for Maxwell, at least in Chapter II, it registered lines of magnetic induction (404.7). One consequence of the "new beginning" is that the Moving Wire continues to evolve: here Maxwell will interpret it again—this time as a means of sampling the *electrokinetic momentum* which fills the space surrounding the primary circuit.

reckoned positive, be chosen, the value of p for this closed curve is deter-
minate. If the opposite direction along the curve had been chosen as the
positive direction, the sign of the quantity p would have been reversed.

586.] Since the quantity p depends on the form and position of the
circuit, we may suppose that each portion of the circuit contributes
something to the value of p, and that the part contributed by each por-
tion of the circuit depends on the form and position of that portion
only, and not on the position of other parts of the circuit.

586.1 This assumption is legitimate, because we are not now con-
sidering a *current*, the parts of which may, and indeed do, act on one
another, but a mere *circuit*, that is, a closed curve along which a current
may flow, and this is a purely geometrical figure, the parts of which can-
not be conceived to have any physical action on each other.

586.2 We may therefore assume that the part contributed by the ele-
ment ds of the circuit is $J\,ds$, where J is a quantity depending on the
position and direction of the element ds. Hence, the value of p may be
expressed as a line-integral

$$p = \int J\,ds \qquad (2)$$

where the integration is to be extended once round the circuit.

587.] We have next to determine the form of the quantity J. In the first
place, if ds is reversed in direction, J is reversed in sign. Hence, if two
circuits $ABCE$ and $AECD$ have the arc AEC common,
but reckoned in opposite directions in the two circuits,
the sum of the values of p for the two circuits $ABCE$
and $AECD$ will be equal to the value of p for the circuit
$ABCD$, which is made up of the two circuits.

Fig. 35.

587. *if ds is reversed in direction, J is reversed in sign*: In the drawing, \mathfrak{B} is in the
upward direction; thus in a right-handed system (Art. 23), ab is the direction
associated with increasing \mathfrak{B}. But Maxwell knows experi-
mentally that if \mathfrak{B} should in fact increase, the induced
electromotive force will be in the direction ba (541.8,
541.11). Reversing the coil does not change this.

But the induced electromotive force is also in the direction of decreasing p
(579.2); therefore reversing the coil does not reverse the direction of p. In
equation (2), however, reversing the sign of ds would certainly reverse the sign
of p, unless J also reversed its sign. Maxwell infers that J must, therefore,
depend on ds in such a way as to change its sign whenever ds does.

...the sum of the values of p for the two circuits ... will be equal to the value of p for
the circuit ABCD: Maxwell will reason to this conclusion in the following para-
graphs. It is the same argument I previously discussed in connection with Stokes'
Theorem in the editor's introduction to Art. 24 (section 4 on pages 39–40), and
which Maxwell invoked in Art. 483—as he is about to remind us.

587.1 For the parts of the line-integral depending on the arc AEC are equal but of opposite sign in the two partial circuits, so that they destroy each other when the sum is taken, leaving only those parts of the line-integral which depend on the external boundary of $ABCD$.

587.2 In the same way we may shew that if a surface bounded by a closed curve be divided into any number of parts, and if the boundary of each of these parts be considered as a circuit, the positive direction round every circuit being the same as that round the external closed curve, then the value of p for the closed curve is equal to the sum of the values of p for all the circuits. See Art. 483.

588.] Let us now consider a portion of a surface, the dimensions of which are so small with respect to the principal radii of curvature of the surface that the variation of the direction of the normal within this portion may be neglected. We shall also suppose that if any very small circuit be carried parallel to itself from one part of this portion to another, the value of p for the small circuit is not sensibly altered. This will evidently be the case if the dimensions of the portion of surface are small enough compared with its distance from the primary circuit.

588.1 *If any closed curve be drawn on this portion of the surface, the value of p will be proportional to its area.*

588.2 For the areas of any two circuits may be divided into small elements all of the same dimensions, and having the same value of p. The areas of the two circuits are as the numbers of these elements which they contain, and the values of p for the two circuits are also in the same proportion.

588.3 Hence, the value of p for the circuit which bounds any element dS of a surface is of the form

$$I\,dS,$$

where I is a quantity depending on the position of dS and on the direction of its normal. We have therefore a new expression for p,

$$p = \iint I\,dS. \tag{3}$$

where the double integral is extended over any surface bounded by the circuit.

587.1 *for the parts of the line-integral ... are equal but of opposite sign*: When Maxwell argued this for current in Art. 483 he treated it as an obvious consequence of the nature of current as a directed quantity (483.4). But here Maxwell is reasoning about a line-integral that contains the quantity J, whose physical significance is not known. He therefore had first to establish that the algebraic sign of J depends on the direction of ds, which he did in (587.).

588.3 *I is a quantity depending on the position of dS...*: Maxwell will discover the constitution and significance of I in (591.6) below.

589.] Let $ABCD$ [Fig. 36] be a circuit, of which AC is an elementary portion, so small that it may be considered straight. Let APB and CQB be small equal areas in the same plane, then the value of p will be the same for the small circuits APB and CQB, or

$$p(APB) = p(CQB).$$

Hence

$$p(APBQCD) = p(ABQCD) + p(APB),$$
$$= p(ABQCD) + p(CQB),$$
$$= p(ABCD),$$

Fig. 36.

or the value of p is not altered by the substitution of the crooked line $APQC$ for the straight line AC, provided the area of the circuit is not sensibly altered. This, in fact, is the principle established by Ampère's second experiment (Art. 506), in which a crooked portion of a circuit is shewn to be equivalent to a straight portion provided no part of the crooked portion is at a sensible distance from the straight portion.

589.1 If therefore we substitute for the element ds three small elements, dx, dy, and dz, drawn in succession, so as to form a continuous path from the beginning to the end of the element ds, and if Fdx, Gdy, and Hdz denote the elements of the line-integral corresponding to dx, dy, and dz respectively, then

$$Jds = Fdx + Gdy + Hdz . \qquad (4)$$

589. *the value of p will be the same for the small circuits APB and CQB*: Since circuits having equal area will have the same momentum p, as proved in (588.3).

Ampère's second experiment: It appears in an omitted chapter, but Maxwell's description here is clear enough. In any case he is not relying on experiment but is beginning again "from a new foundation" (585.1).

589.1 *substitute for the element ds three small elements dx, dy, and dz, drawn in succession*: No matter what is the actual shape of path element ds, it can be considered as a simple curve, or even as straight (compare 589.). Its endpoints may be joined by a series of paths dx, dy, dz, as sketched.

Fdx, Gdy, and *Hdz*: Note that Maxwell introduces F, G, and H here as uninterpreted quantities, even though he previously utilized them in Art. 405 to represent the components of \mathfrak{A}, there understood as the magnetic vector-potential. But in keeping with Maxwell's new beginning, the vector \mathfrak{A} itself is about to reveal a new significance—one that enriches, without being dependent on, the former one. Such deliberate cultivation of dialectic among the mathematical quantities is indeed characteristic of Maxwell. In this he honors Faraday, who was surpassingly sensitive to the evolving images under which nature presented herself. (I have written on that aspect of Faraday's work in "The Great Electrical Philosopher," *The College*, July 1979.)

590.] We are now able to determine the mode in which the quantity J depends on the direction of the element ds. For, by (4),

$$J = F\frac{dx}{ds} + G\frac{dy}{ds} + H\frac{dz}{ds} \, . \tag{5}$$

590.1 This is the expression for the resolved part, in the direction of ds, of a vector, the components of which, resolved in the directions of the axes of x, y, and z, are F, G, and H respectively.

590.2 If this vector be denoted by \mathfrak{A}, and the vector from the origin to a point of the circuit by ρ, the element of the circuit will be $d\rho$, and the quaternion expression for Jds will be

$$-S.\mathfrak{A}d\rho \, .$$

590.3 We may now write equation (2) in the form

$$p = \int \left(F\frac{dx}{ds} + G\frac{dy}{ds} + H\frac{dz}{ds} \right) ds \, , \tag{6}$$

or

$$p = -\int S.\mathfrak{A}\, d\rho \, . \tag{7}$$

590.4 The vector \mathfrak{A} and its constituents F, G, H depend on the position of ds in the field, and not on the direction in which it is drawn. They are therefore functions of x, y, z, the coordinates of ds, and not of l, m, n, its direction-cosines.

590.1 *This is the expression for the resolved part, in the direction of ds, of a vector...*: As noted in the editor's introduction to Art. 25, $F\,dx + G\,dy + H\,dz$ is equivalent to the dot product of the two vectors having components F, G, H and dx, dy, dz, respectively; while ds is the magnitude of the latter vector. Their quotient is the resolved part of the first vector in the direction of the second; see page 52, above.

590.2 *the quaternion expression for Jds will be* $-S.\mathfrak{A}d\rho$: Jds was expressed in equation (4) as the sum $F\,dx + G\,dy + H\,dz$. Now Maxwell takes F, G, H as the components of vector \mathfrak{A}; and he is tacitly defining a vector $d\rho$ with magnitude ds and components dx, dy, dz. Then, as in the preceding comment, $F\,dx + G\,dy + H\,dz$ is equivalent to the dot product of \mathfrak{A} and $d\rho$; so that $Jds = \mathfrak{A} \cdot d\rho$. But the dot product of two vectors is the *negative* of the scalar part of the quaternion product, so we have $Jds = -S.\mathfrak{A}d\rho$.

590.3 *Equation (7)*: Equivalent to $p = \int \mathfrak{A} \cdot d\rho$.

590.4 *its constituents depend on the position of ds ... and not on the direction in which it is drawn*: Similarly, as noted in (587., *comment*), reversing the secondary coil does not reverse the direction of p.

l, m, n, its direction-cosines: Recall their character and significance, discussed in (21.2, *comment*).

590.5 The vector \mathfrak{A} represents in direction and magnitude the time-integral of the electromotive intensity which a particle placed at the point (x, y, z) would experience if the primary current were suddenly stopped. We shall therefore call it the Electrokinetic Momentum *at the point* (x, y, z). It is identical with the quantity which we investigated in Art. 405 under the name of the vector-potential of magnetic induction.

590.6 The electrokinetic momentum of any finite line or circuit is the line-integral, extended along the line or circuit, of the resolved part of the electrokinetic momentum at each point of the same.

591.] Let us next determine the value of p for the elementary rectangle $ABCD$, of which the sides are dy and dz, the positive direction being from the direction of the axis of y to that of z.

Fig. 37.

591.1 Let the coordinates of O [Fig. 37], the centre of gravity of the element, be x_0, y_0, z_0, and let G_0, H_0 be the values of G and of H at this point.

591.2 The coordinates of A, the middle point of the first side of the rectangle, are y_0 and $z_0 - \frac{1}{2}dz$. The corresponding value of G is

$$G = G_0 - \frac{1}{2}\frac{dG}{dz}dz + \&c. , \qquad (8)$$

and the part of the value of p which arises from the side of A is approximately

590.5 *if the primary current were suddenly stopped*: The reverse of the process of sudden setting into motion described in (558.1); see also (558.8).

the Electrokinetic Momentum at the point (x, y, z): Why does Maxwell emphasize the phrase *at the point*? Perhaps to drive home that the space surrounding a current is *filled with momentum*, just as it is filled with energy.

it is identical with ... the vector-potential of magnetic induction: The name change is significant, "potential" having mathematical, and "momentum" having physical, significance. In (606.) below, Maxwell will state explicitly that part of the "new foundation" announced in (585.1) will be an effort to denote quantities "by names which indicate their physical relations rather than their mathematical generation."

591. *the positive direction being from the direction of the axis of y to that of z*: Recall the definition of positive rotation (23.). Further on Maxwell will use the phrase "cyclical order" to describe any sequence that preserves the order of the coordinates—either x, y, z or y, z, x or z, x, y.

591.1 *centre of gravity of the element*: Better, *centroid*, since Maxwell is attaching no significance to the weight of the element.

591.2 *The corresponding value of G is*: Representing G as an unknown function of z, Maxwell uses the Taylor expansion, $G(z+h) = G(z) + hG'(z) + \cdots$, with $z = z_0$ and $h = -\frac{1}{2}dz$ and denoting $G(z_0)$ by G_0.

the part of the value of p ... is approximately: The value given by equation (9)

$$G_0 dy - \frac{1}{2}\frac{dG}{dz}dy\,dz \qquad\qquad (9)$$

591.3 Similarly, for B, $H_0 dz + \frac{1}{2}\frac{dH}{dy}dy\,dz$,

for C, $-G_0 dy - \frac{1}{2}\frac{dG}{dz}dy\,dz$,

for D, $-H_0 dz + \frac{1}{2}\frac{dH}{dy}dy\,dz$.

591.4 Adding these four quantities, we find the value of p for the rectangle, viz.

$$p = \left(\frac{dH}{dy} - \frac{dG}{dz}\right)dy\,dz. \qquad\qquad (10)$$

591.5 If we now assume three new quantities, a, b, c, such that

$$\left. \begin{aligned} a &= \frac{dH}{dy} - \frac{dG}{dz}, \\ b &= \frac{dF}{dz} - \frac{dH}{dx}, \\ c &= \frac{dG}{dx} - \frac{dF}{dy}, \end{aligned} \right\} \qquad\qquad \text{(A)}$$

and consider these as the constituents of a new vector \mathfrak{B}, then by Theorem IV, Art. 24, we may express the line-integral of \mathfrak{A} round any circuit in the form of the surface-integral of \mathfrak{B} over a surface bounded by the circuit, thus

is only approximate since Maxwell is ignoring the higher-order terms designated "&c." in equation (8).

591.5 *Equations* (A): This is the first of a series of equations labeled (A) through (L), whose development will be continued in the next chapter. There Maxwell will characterize them as the "General Equations of the Electromagnetic Field."

a new vector \mathfrak{B}: It is not really "new"; Maxwell has used it before to indicate magnetic induction (Art. 400). But as he stated at the outset of this chapter (585.1), Maxwell is not here assuming any of the properties formerly established for \mathfrak{B} and other quantities. He will return to this point again in (592.) below.

Note that since a, b, c are the components of \mathfrak{B} and F, G, H those of \mathfrak{A}, equations (A) have the vector form $\mathfrak{B} = \nabla \times \mathfrak{A}$ or $\mathfrak{B} = \text{curl } \mathfrak{A}$.

Theorem IV, Art. 24: That is, Stokes' Theorem, proved as equation (11) in that Article. Examine the form of that equation and notice that the integrand on the right-hand side is equivalent to $\mathbf{R} \cdot \mathbf{ds}$, while on the left side, ξ, η, ζ are the components of $\nabla \times \mathbf{R}$ or curl \mathbf{R}, and $l\,dS$, $m\,dS$, $n\,dS$ are those of vector \mathbf{dS}. Thus the theorem can be expressed in vector form as $\int \mathbf{R} \cdot \mathbf{ds} = \iint (\nabla \times \mathbf{R}) \cdot \mathbf{dS}$.

$$p = \int \left(F\frac{dx}{ds} + G\frac{dy}{ds} + H\frac{dz}{ds} \right) ds = \iint (la + mb + nc)dS , \qquad (11)$$

or $$\qquad p = \int T.\mathfrak{A}\cos\varepsilon \, ds = \iint T.\mathfrak{B}\cos\eta \, dS , \qquad (12)$$

where ε is the angle between \mathfrak{A} and ds, and ε that between \mathfrak{B} and the normal to dS, whose direction-cosines are l, m, n, and $T.\mathfrak{A}$, $T.\mathfrak{B}$ denote the numerical values of \mathfrak{A} and \mathfrak{B}.

591.6 Comparing this result with equation (3), it is evident that the quantity I in that equation is equal to $\mathfrak{B}\cos\eta$, or the resolved part of \mathfrak{B} normal to dS.

592.] We have already seen (Arts. 490, 541) that, according to Faraday's theory, the phenomena of electromagnetic force and induction in a circuit depend on the variation of the number of lines of magnetic induction which pass through the circuit. Now the number of these lines is expressed mathematically by the surface-integral of the magnetic induction through any surface bounded by the circuit. Hence, we must regard the vector \mathfrak{B} and its components a, b, c as representing what we are already acquainted with as the magnetic induction and its components.

592.1 In the present investigation we propose to deduce the properties of this vector from the dynamical principles stated in the last chapter, with as few appeals to experiment as possible.

592.2 In identifying this vector, which has appeared as the result of a mathematical investigation, with the magnetic induction, the properties of which we learned from experiments on magnets, we do not depart

591.5, continued. *Equation* (11): As in the preceding comments, note that since F, G, H are the components of \mathfrak{A}, while a, b, c are those of \mathfrak{B} or $\nabla\times\mathfrak{A}$, the equation is equivalent to $\int \mathfrak{A}\cdot\mathbf{ds} = \iint \mathfrak{B}\cdot\mathbf{dS} = \iint (\nabla\times\mathfrak{A})\cdot\mathbf{dS}$.

Equation (12): Like the formalisms S. and V. encountered earlier, $T.$ is quaternion notation. We are accustomed to thinking of vectors as possessing both magnitude and direction; in quaternion terminology, the corresponding terms are *tensor* and *versor*. Thus the tensor part of \mathfrak{A}, symbolized $T.\mathfrak{A}$, expresses the unsigned magnitude of \mathfrak{A} or, as Maxwell is about to express it, its "numerical value." (The term "tensor" in modern mathematics has no connection whatever with the tensors of quaternions.) In modern notation, equation (12) states $p = \int A\cos\varepsilon \, ds = \iint B\cos\eta \, dS$, or $p = \int \mathfrak{A}\cdot\mathbf{ds} = \iint \mathfrak{B}\cdot\mathbf{dS}$.

592.1 *as few appeals to experiment as possible*: Recall that this is how Maxwell characterized the "new foundation" for electrodynamic theory (585.1). Although we indeed *test* theories by making experiments, we expect a mature theory to be able to derive expressions for most quantities from a few fundamental ones.

Fig. 24 (repeated) Fig. 25 (repeated)

from this method, for we introduce no new fact into the theory, we only give a name to a mathematical quantity, and the propriety of so doing is to be judged by the agreement of the relations of the mathematical quantity with those of the physical quantity indicated by the name.

592.3 The vector \mathfrak{B}, since it occurs in a surface-integral, belongs evidently to the category of fluxes described in Art. 12, The vector \mathfrak{A}, on the other hand, belongs to the category of forces, since it appears in a line-integral.

593.] We must here recall to mind the conventions about positive and negative quantities and directions, some of which were stated in Art. 23. We adopt the right-handed system of axes, so that if a right-handed screw is placed in the direction of the axis of x, and a nut on this screw is turned in the positive direction of rotation, that is, from the direction of y to that of z, it will move along the screw in the positive direction of x.

593.1 We also consider vitreous electricity and austral magnetism as positive. The positive direction of an electric current, or of a line of electric induction, is the direction in which positive electricity moves or tends to move, and the positive direction of a line of magnetic induction is the direction in which a compass needle points with that end which turns to the north. See Fig. 24, Art. 498, and Fig. 25, Art. 501.

592.2 *we introduce no new fact into the theory*: which relying on experiments at this stage would do.

593. *the right-handed system of axes*: The system was introduced in Art. 23.

593.1 *We also consider vitreous electricity and austral magnetism as positive*: As defined in (27.5), vitreous electricity is that exhibited by glass when rubbed with amber or other resin. Austral magnetism is the magnetism of a north (that is, a north-*seeking*) pole and therefore that of the earth's South pole (393.).

See Fig. 24 ... and Fig. 25: They are reproduced at the top of the page.

593.2 The student is recommended to select whatever method appears to him most effectual in order to fix these conventions securely in his memory, for it is far more difficult to remember a rule which determines in which of two previously indifferent ways a statement is to be made, than a rule which selects one way out of many.

594.] We have next to deduce from dynamical principles the expressions for the electromagnetic force acting on a conductor carrying an electric current through the magnetic field, and for the electromotive force acting on the electricity within a body moving in the magnetic field. The mathematical method which we shall adopt may be compared with the experimental method used by Faraday* in exploring the field by means of a wire, and with what we have already done in Art. 490, by a method founded on experiments. What we have now to do is to determine the effect on the value of p, the electrokinetic momentum of the secondary circuit, due to given alterations of the form of that circuit.

594.1 Let AA', BB' be two parallel straight conductors connected by the conducting arc C, which may be of any form, and by a straight conductor AB, which is capable of sliding parallel to itself along the conducting rails AA' and BB'.

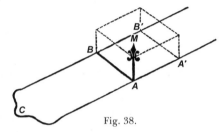

Fig. 38.

594.2 Let the circuit thus formed be considered as the secondary circuit, and let the direction ABC be assumed as the positive direction round it.

594.3 Let the sliding piece move parallel to itself from the position AB to the position $A'B'$. We have to determine the variation of p, the

* Exp. Res., 8082, 3087, 3118.

594. *The mathematical method which we shall adopt may be compared with the experimental method used by Faraday*: What Maxwell is about to describe is nothing less than a mathematical re-creation of Faraday's Moving Wire. We might say (as I suggested in the editor's introduction to this chapter) that he has converted Faraday's physical experiment into a mathematical experiment.

what we have already done… What we have now to do: In Article 490 Maxwell described experiments equivalent to Faraday's Moving Wire and exploratory loop. Now he must derive the same results theoretically, from the principles of electrokinetic momentum. Being so closely modeled on Faraday's procedures, the derivation will bring us, as nearly as a mathematical argument can, into Faraday's own place, experiencing what he experienced. (Thomas K. Simpson explores this theme beautifully in his *Figures of Thought*, pp. 87–89.)

electrokinetic momentum of the circuit, due to this displacement of the sliding piece.

594.4 The secondary circuit is changed from ABC to $A'B'C$, hence, by Art. 587,

$$p\left(A'B'C\right) - p(ABC) = p\left(AA'B'B\right). \tag{13}$$

594.5 We have therefore to determine the value of p for the parallelogram $AA'B'B$. If this parallelogram is so small that we may neglect the variations of the direction and magnitude of the magnetic induction at different points of its plane, the value of p is, by Art. 591, $\mathfrak{B}\cos\varepsilon \cdot AA'B'B$, where \mathfrak{B} is the magnetic induction, and η the angle which it makes with the positive direction of the normal to the parallelogram $AA'B'B$.

594.6 We may represent the result geometrically by the volume of the parallelepiped, whose base is the parallelogram $AA'B'B$, and one of whose edges is the line AM, which represents in direction and magnitude the magnetic induction \mathfrak{B}. If the parallelogram is in the plane of the paper, and if AM is drawn upwards from the paper, or more generally, if the directions of the circuit AB, of the magnetic induction AM, and of the displacement AA', form a right-handed system when taken in this cyclical order, the volume of the parallelepiped is to be taken positively.

594.7 The volume of this parallelepiped represents the increment of the value of p for the secondary circuit due to the displacement of the sliding piece from AB to $A'B'$.

Electromotive Force acting on the Sliding Piece.

595.] The electromotive force produced in the secondary circuit by the motion of the sliding piece is, by Art. 579,

$$E = -\frac{dp}{dt}. \tag{14}$$

594.4 *by Art. 587*: specifically, (587.2).

594.5 *by Art. 591*: specifically, (591.5), where equation (12) states

$$p = \iint T.\mathfrak{B}\cos\eta\,dS,$$

and $T.\mathfrak{B}$ denotes the magnitude of \mathfrak{B} (591.5, *comment*). Then since $AA'B'B$ is "so small" that we can treat \mathfrak{B} and η as constants, integrate to obtain Maxwell's $p = \mathfrak{B}\cos\eta \cdot AA'B'B$ (or, more correctly, $p = T.\mathfrak{B}\cos\eta \cdot AA'B'B$).

594.6 *a right-handed system when taken in cyclical order*: Cyclical order was defined in the editor's introduction to Article 25, page 50.

595. *by Art. 579*: specifically, the last sentence of (579.3).

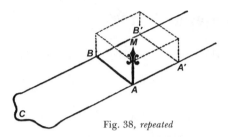

Fig. 38, *repeated*

595.1 If we suppose AA' to be the displacement in unit of time, then AA' will represent the velocity, and the parallelepiped will represent dp, and therefore, by equation (14), the electromotive force in the negative direction BA.

595.2 Hence, the electromotive force acting on the sliding piece AB, in consequence of its motion through the magnetic field, is represented by the volume of the parallelepiped, whose edges represent in direction and magnitude—the velocity, the magnetic induction, and the sliding piece itself, and is positive when these three directions are in right-handed cyclical order.

596.] Let i_2 denote the current in the secondary circuit in the positive direction ABC, then the work done by the electromagnetic force on AB while it slides from the position AB to the position $A'B'$ is $(M'-M)i_1i_2$, where M and M' are the values of M_{12} in the initial and final positions of AB. But $(M'-M)i_1$ is equal to $p'-p$, and this is represented by the volume of the parallelepiped on AB, AM, and AA'. Hence, if we draw a line parallel to AB to represent the quantity $AB \cdot i_2$, the parallelepiped contained by this line, by AM, the magnetic induction, and by AA', the displacement, will represent the work done during this displacement.

595.2 *the parallelepiped, whose edges represent ... the velocity, the magnetic induction, and the sliding piece*: that is, the solid contained by AA', AM', and AB in Figure 38.

596. *the work done ... is* $(M'-M)i_1i_2$: The expression derives from (581.) above, where equation (4) for the energy of a system of two circuits states $T = \frac{1}{2}L\dot{y}_1^2 + M\dot{y}_1\dot{y}_2 + \frac{1}{2}N\dot{y}_2^2$. In the present case i_1 and i_2 correspond to \dot{y}_1 and \dot{y}_2. The first term is constant, since nothing changes in the primary circuit. When the sliding piece moves it alters M, with a resulting energy change equal to $(M'-M)i_1i_2$; this in turn must equal the work done on the sliding piece. Why can Maxwell ignore the contribution of the third term? See (581.6).

if we draw a line parallel to AB to represent the quantity $AB \cdot i_2$, the parallelepiped contained by this line, by AM, the magnetic induction, and by AA': The sketch suggests the parallelepiped in question, with line AD representing the quantity $AB \cdot i_2$.

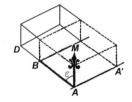

596.1 For a given distance of displacement this will be greatest when the displacement is perpendicular to the parallelogram whose sides are *AB* and *AM*. The electromagnetic force is therefore represented by the area of the parallelogram on *AB* and *AM* multiplied by i_2, and is in the direction of the normal to this parallelogram, drawn so that *AB*, *AM*, and the normal are in right-handed cyclical order.

Four Definitions of a Line of Magnetic Induction.

597.] If the direction *AA'*, in which the motion of the sliding piece takes place, coincides with *AM*, the direction of the magnetic induction, the motion of the sliding piece will not call electromotive force into action, whatever be the direction of *AB*, and if *AB* carries an electric current there will be no tendency to slide along *AA'*.

597.1 Again, if *AB*, the sliding piece, coincides in direction with *AM*, the direction of magnetic induction, there will be no electromotive force called into action by any motion of *AB*, and a current through *AB* will not cause *AB* to be acted on by mechanical force.

597.2 We may therefore define a line of magnetic induction in four different ways. It is a line such that

(1) If a conductor be moved along it parallel to itself it will experience no electromotive force.

(2) If a conductor carrying a current be free to move along a line of magnetic induction it will experience no tendency to do so.

(3) If a linear conductor coincide in direction with a line of magnetic induction, and be moved parallel to itself in any direction, it will experience no electromotive force in the direction of its length.

596.1 *this will be greatest when the displacement is perpendicular to the parallelogram whose sides are AB and AM*: Refer again to the sketch in (596., *comment*). Of all parallelepipeds having parallelogram *AD*, *AM* as base, the greatest is that for which the angle between *AA'* and *AM* is a right angle.

Note that for a similar reason, the parallelepiped contained by *AD*, *AM*, and *AA'* is greatest when all three sides are perpendicular to each other.

the electromagnetic force is therefore represented by the area of the parallelogram on AB and AM multiplied by i_2: equivalent to the parallelogram *AD*, *AM* in the sketch of (596., *comment*). Since the parallelepiped contained by *AD*, *AM*, and *AA'* represents the work done by that force, while *AA'* represents the displacement, parallelogram *AD*, *AM* represents the electromagnetic force itself.

in the direction of the normal to this parallelogram: Experimentally, we already know that the electromagnetic force on a current-carrying conductor in a magnetic field is perpendicular to both the conductor and the magnetic induction (490.6). But if we are to make "as few appeals to experiment as possible" (592.1), we may note that the cross product $AD \times AM$ has magnitude equal to $AB \cdot i_2 AM \sin \varepsilon$, or "the area of the parallelogram on *AB* and *AM* multiplied by i_2," and direction perpendicular to *AB* and *AM*; compare the example in the editor's introduction to Art. 25, page 50 above.

(4) If a linear conductor carrying an electric current coincide in direction with a line of magnetic induction it will not experience any mechanical force.

General Equations of Electromotive Intensity.

598.] We have seen that E, the electromotive force due to induction acting on the secondary circuit, is equal to $-dp/dt$, where

$$p = \int \left(F \frac{dx}{ds} + G \frac{dy}{ds} + H \frac{dz}{ds} \right) ds \; , \tag{1}$$

598.1 To determine the value of E, let us differentiate the quantity under the integral sign with respect to t, remembering that if the secondary circuit is in motion, x, y, and z are functions of the time. We obtain

$$
\begin{aligned}
E = & -\int \left(\frac{dF}{dt}\frac{dx}{ds} + \frac{dG}{dt}\frac{dy}{ds} + \frac{dH}{dt}\frac{dz}{ds} \right) ds \\
& -\int \left(\frac{dF}{dx}\frac{dx}{ds} + \frac{dG}{dx}\frac{dy}{ds} + \frac{dH}{dx}\frac{dz}{ds} \right) \frac{dx}{dt} ds \\
& -\int \left(\frac{dF}{dy}\frac{dx}{ds} + \frac{dG}{dy}\frac{dy}{ds} + \frac{dH}{dy}\frac{dz}{ds} \right) \frac{dy}{dt} ds \\
& -\int \left(\frac{dF}{dz}\frac{dx}{ds} + \frac{dG}{dz}\frac{dy}{ds} + \frac{dH}{dz}\frac{dz}{ds} \right) \frac{dz}{dt} ds \\
& -\int \left(F \frac{d^2x}{ds\,dt} + G \frac{d^2y}{ds\,dt} + H \frac{d^2z}{ds\,dt} \right) ds \; .
\end{aligned}
\tag{2}
$$

598.2 Now consider the second line of the integral, and substitute from equations (A), Art. 591, the values of dG/dx and dH/dx. This line then becomes

$$-\int \left(c\frac{dy}{ds} - b\frac{dz}{ds} + \frac{dF}{dx}\frac{dx}{ds} + \frac{dF}{dy}\frac{dy}{ds} + \frac{dF}{dz}\frac{dz}{ds} \right) \frac{dx}{dt} ds \; ,$$

which we may write

$$-\int \left(c\frac{dy}{ds} - b\frac{dz}{ds} + \frac{dF}{ds} \right) \frac{dx}{dt} ds \; .$$

598. *Equation* (1): It has been Maxwell's practice to begin equation numbers anew in each Article. The present equation (1) is identical to what was labeled equation (6) in Art. 590 above.

598.1 *if the secondary circuit is in motion, x, y, and z are functions of the time*: since x, y, z are the coordinates of an element of the secondary circuit (589.1).

We obtain: Bear in mind that the derivatives in equations (1), (2) and following are partial derivatives, despite Maxwell's lack of a distinctive notation.

598.3 Treating the third and fourth lines in the same way, and collectiug the terms in dx/ds, dy/ds, and dz/ds, remembering that

$$\int\left(\frac{dF}{ds}\frac{dx}{dt}+F\frac{d^2x}{ds\,dt}\right)ds=F\frac{dx}{dt},\tag{3}$$

and therefore that the integral, when taken round the closed curve, vanishes,

$$E=\int\left(c\frac{dy}{dt}-b\frac{dz}{dt}-\frac{dF}{dt}\right)\frac{dx}{ds}\,ds$$

$$+\int\left(a\frac{dz}{dt}-c\frac{dx}{dt}-\frac{dG}{dt}\right)\frac{dy}{ds}\,ds$$

$$+\int\left(b\frac{dx}{dt}-a\frac{dy}{dt}-\frac{dH}{dt}\right)\frac{dz}{ds}\,ds.\tag{4}$$

598.4 We may write this expression in the form

$$E=\int\left(P\frac{dx}{ds}+Q\frac{dy}{ds}+R\frac{dz}{ds}\right)ds,\tag{5}$$

where

$$\left.\begin{array}{l}P=c\dfrac{dy}{dt}-b\dfrac{dz}{dt}-\dfrac{dF}{dt}-\dfrac{d\Psi}{dx},\\[2mm]Q=a\dfrac{dz}{dt}-c\dfrac{dx}{dt}-\dfrac{dG}{dt}-\dfrac{d\Psi}{dy},\\[2mm]R=b\dfrac{dx}{dt}-a\dfrac{dy}{dt}-\dfrac{dH}{dt}-\dfrac{d\Psi}{dz}.\end{array}\right\}\begin{array}{c}\text{Equations of}\\\text{Electromotive}\\\text{Intensity}\end{array}\tag{B}$$

598.3 *Equation* (3): It is an identity. Differentiate both sides with respect to s, as

$$\frac{dF}{ds}\frac{dx}{dt}+F\frac{d^2x}{ds\,dt}=\frac{d}{ds}\left(F\frac{dx}{dt}\right)=\frac{dF}{ds}\frac{dx}{dt}+F\frac{d(dx/dt)}{ds}\text{; the identity is evident.}$$

the integral, when taken round the closed curve, vanishes: "The integral" is the left-hand side of equation (3). It vanishes because dx is the x-component of ds, an element of the circuit (589.1). Since the element does not grow or shrink in time, dx/dt, together with the whole right-hand side of the equation, is zero.

Now in equation (2) above, consider the first and fifth lines. The first term in the first line, plus the first term in the fifth line, are together just the negative of the left-hand side of equation (3); thus they vanish just as it does. We can write identities similar to equation (3) for G and H also, and in this way show that all the terms in the first and fifth lines vanish. We end up with equation (4), the transformed second through fourth lines of equation (2).

598.4 *Equations* (B): Taking (P, Q, R) as the components of electromotive intensity \mathfrak{E}, (a, b, c) those of magnetic induction \mathfrak{B}, (F, G, H) those of electrokinetic momentum \mathfrak{A}, and $(dx/dt, dy/dt, dz/dt)$ as the components of velocity \mathfrak{G}, Maxwell will write these equations as a single quaternion equation in (599.3) below, with vector equivalent $\mathfrak{E}=\mathfrak{G}\times\mathfrak{B}-\dot{\mathfrak{A}}-\nabla\Psi$ (599.3, *comment*).

598.5 The terms involving the new quantity Ψ are introduced for the sake of giving generality to the expressions for P, Q, R. They disappear from the integral when extended round the closed circuit. The quantity Ψ is therefore indeterminate as far as regards the problem now before us, in which the electromotive force round the circuit is to be determined. We shall find, however, that when we know all the circumstances of the problem, we can assign a definite value to Ψ, and that it represents, according to a certain definition, the electric potential at the point (x, y, z).

598.6 The quantity under the integral sign in equation (5) represents the electromotive intensity acting on the element ds of the circuit.

598.7 If we denote by $T.\mathfrak{E}$, the numerical value of the resultant of P, Q, and R, and by ε, the angle between the direction of this resultant and that of the element ds, we may write equation (5),

$$E = \int T.\mathfrak{E}\cos\varepsilon \, ds \, . \tag{6}$$

598.8 The vector \mathfrak{E} is the electromotive intensity at the moving element ds. Its direction and magnitude depend on the position and motion of ds, and on the variation of the magnetic field, but not on the direction of ds. Hence we may now disregard the circumstance that ds forms part of a circuit, and consider it simply as a portion of a moving body, acted on by the electromotive intensity \mathfrak{E}. The electromotive intensity has already been defined in Art. 68. It is also called the resultant electrical intensity, being the force which would be experienced by a unit of positive electricity placed at that point. We have now obtained the most general value of this quantity in the case of a body moving in a magnetic field due to a variable electric system.

598.5 *The terms involving the new quantity Ψ are introduced for the sake of giving generality to the expressions*: By comparison with the equations for L in (16.3, *comment*), it is clear that equation (5) has the form of a difference of potentials, which vanishes over a closed path. If we define P, Q, R simply on the basis of the foregoing argument (which has assumed a closed path) the resulting expressions might prove to vanish over an open path as well. Maxwell adds the terms in Ψ to preserve the possibility of a nonzero open path integral. Notice that the new terms have the same form as the components of a force derived from a potential (16.4, *comment*): $-d\Psi/dx$, $-d\Psi/dy$, $-d\Psi/dz$.

598.7 *If we denote by $T.\mathfrak{E}$, the numerical value of* $[\mathfrak{E}]$: the same notational practice as remarked in (591.5, *comment*). In modern terminology, $T.\mathfrak{E}$ would be called the scalar *magnitude* of \mathfrak{E}. Maxwell has been using R to denote this quantity.

Equation (6): Written $E = \int \mathfrak{E} \cdot \mathbf{ds}$ or $\int R\cos\varepsilon \, ds$ in modern notation.

598.8 *but not on the direction of ds*: Compare the similar observation in (590.4).

598.9 If the body is a conductor, the electromotive force will produce a current; if it is a dielectric, the electromotive force will produce only electric displacement.

598.10 The electromotive intensity, or the force on a particle, must be carefully distinguished from the electromotive force along an arc of a curve, the latter quantity being the line-integral of the former. See Art. 69.

599.] The electromotive intensity, the components of which are defined by equations (B), depends on three circumstances. The first of these is the motion of the particle through the magnetic field. The part of the force depending on this motion is expressed by the first two terms on the right of each equation. It depends on the velocity of the particle transverse to the lines of magnetic induction. If \mathfrak{G} is a vector representing the velocity, and \mathfrak{B} another representing the magnetic induction, then if \mathfrak{E}_1 is the part of the electromotive intensity depending on the motion,

$$\mathfrak{E}_1 = V.\mathfrak{G}\mathfrak{B}, \tag{7}$$

or, the electromotive intensity is the vector part of the product of the magnetic induction multiplied by the velocity, that is to say, the magnitude of the electromotive intensity is represented by the area of the parallelogram, whose sides represent the velocity and the magnetic induction, and its direction is the normal to this parallelogram, drawn so that the velocity, the magnetic induction, and the electromotive intensity are in right-handed cyclical order.

598.9 *If the body is a conductor... if it is a dielectric...*: An electromotive force will arise in both cases—not because the moving element is a body, but because it is moving! Maxwell's argument applies equally well to a mathematically-defined region of space, which is also a dielectric in the sense of supporting electric displacement (60.2). Such a region, moving through a magnetic field, must be the site of an electromotive force.

Modern relativity theory incorporates the preceding observation by saying that a magnetic field, referred to moving axes, exhibits not only a magnetic but an electric component.

598.10 *the force on a particle*: Maxwell is using the word "particle" here as he used "body" in the preceding paragraph: to denote the object of an action, not a quantity of ponderable matter. Electromotive intensity was not defined in relation to matter but to *electrification* specifically (44.3–44.4).

599. *Equation* (7): Its equivalent in modern notation is $\mathfrak{E}_1 = \mathfrak{G} \times \mathfrak{B}$, or $\mathbf{E}_1 = \mathbf{v} \times \mathbf{B}$ using Roman letters. Note: The German \mathfrak{G} actually corresponds to Roman G, but modern writers conventionally employ \mathbf{u} or \mathbf{v} to represent vector velocity in this context.

its direction is normal to the parallelogram: Recall that the cross-product of two vectors is perpendicular to both multiplicands; compare the example in the editor's introduction to Art. 25, page 50 above.

599.1 The third term in each of the equations (B) depends on the time-variation of the magnetic field. This may be due either to the time-variation of the electric current in the primary circuit, or to motion of the primary circuit. Let \mathfrak{E}_2 be the part of the electromotive intensity which depends on these terms. Its components are

$$-\frac{dF}{dt}, \quad -\frac{dG}{dt}, \quad \text{and} \quad -\frac{dH}{dt},$$

and these are the components of the vector, $-\dfrac{d\mathfrak{A}}{dt}$ or $-\dot{\mathfrak{A}}$.

Hence, $\qquad\qquad\qquad\qquad \mathfrak{E}_2 = -\dot{\mathfrak{A}}$. $\qquad\qquad\qquad\qquad$ (8)

599.2 The last term of each equation (B) is due to the variation of the function Ψ in different parts of the field. We may write the third part of the electromotive intensity, which is due to this cause,

$$\mathfrak{E}_3 = -\nabla\Psi . \qquad\qquad (9)$$

599.3 The electromotive intensity, as defined by equations (B) may therefore be written in the quaternion form,

$$\mathfrak{E} = V.\mathfrak{G}\mathfrak{B} - \dot{\mathfrak{A}} - \nabla\Psi . \qquad\qquad (10)$$

$$* \quad * \quad *$$

603.] If the conductor is to be treated not as a line but as a body, we must express the force on the element of length, and the current through the complete section, in terms of symbols denoting the force per unit of volume, and the current per unit of area.

603.1 Let X, Y, Z now represent the components of the force referred to unit of volume, and u, v, w those of the current referred to unit of area. Then, if S represents the section of the conductor, which we shall suppose small, the volume of the element ds will be Sds, and $u = \dfrac{i_2}{S}\dfrac{dx}{ds}$. Hence, equation (7) will become

$$\frac{XSds}{ds} = S(vc - wb), \qquad\qquad (10)$$

or $\qquad\qquad\qquad\qquad X = vc = wb,$

Similarly $\qquad\qquad\qquad\ \ Y = wa - uc,$ \qquad (Equations of Electromagnetic Force) \qquad (C)

and $\qquad\qquad\qquad\qquad\ Z = ub - va.$

599.3 *Equation* (10): Equivalently expressed in vector notation as
$$\mathfrak{E} = \mathfrak{G} \times \mathfrak{B} - \dot{\mathfrak{A}} - \nabla\Psi \quad \text{or} \quad \mathbf{E} = \mathbf{v} \times \mathbf{B} - \dot{\mathbf{A}} - \nabla\Psi.$$

603.2 Here X, Y, Z are the components of the electromagnetic force on an element of a conductor divided by the volume of that element; u, v, w are the components of the electric current through the element referred to unit of area, and a, b, c are the components of the magnetic induction at the element, which are also referred to unit of area.

603.3 If the vector \mathfrak{F} represents in magnitude and direction the force acting on unit of volume of the conductor, and if \mathfrak{C} represents the electric current flowing through it,

$$\mathfrak{F} = V.\mathfrak{C}\mathfrak{B}. \tag{11}$$

603.3 *Equation* (11): Equivalently expressed in vector notation as
$$\mathfrak{F} = \mathfrak{C} \times \mathfrak{B}.$$

Editor's Introduction to
Chapter IX

"Maxwell's Equations"

Readers who have had any exposure to modern treatments of electricity and magnetism will have been introduced to "Maxwell's Equations," so called: *

$$\nabla \cdot \mathbf{D} = 4\pi\rho$$

$$\nabla \times \mathbf{H} = 4\pi\mathbf{J} + \frac{\partial \mathbf{D}}{\partial t}$$

$$\nabla \times \mathbf{E} = -\frac{\partial \mathbf{B}}{\partial t}$$

$$\nabla \cdot \mathbf{B} = 0$$

These four equations are fundamental, powerful, and highly symmetric—especially when expressed, as they are here, in vector form. Nevertheless, echoing the delightful palindrome *Sununu's tonsil is not Sununu's*, we have to acknowledge that *"Maxwell's Equations" are not Maxwell's*. Although versions of them do appear in Maxwell's 1861 essay *On Physical Lines of Force*, it was Oliver Heaviside and, independently, Heinrich Hertz who expressed them as a concise, symmetrical group that would constitute the fundamental axioms of electrodynamics. Those thinkers valued logical economy, just as we do today when we admire it in Newton's three Laws of Motion and in Euclid's Axioms and Postulates. From that point of view, it is inviting to present the four equations as a kind of culmination of electromagnetic science.

Maxwell himself, however, maintains a different stance. In the present chapter he will review twelve, not four, equations (including three that were derived in Chapter VIII), characterizing them in the title as

*The equations are written here in electromagnetic units, with \mathbf{D} representing electric displacement, ρ charge density (charge per unit volume), \mathbf{H} representing magnetic force, \mathbf{J} current density (current per unit area), \mathbf{E} electric field intensity and \mathbf{B} magnetic induction. Maxwell's corresponding symbols are tabulated on page 411 and, in slightly different form, in Art. 618 of the present chapter.

"general" rather than "fundamental." Moreover, in explaining why he has not reduced them to a smaller number (as he knows he can do), he reveals a clear intention that has nothing to do with the logical structure of the science:

> [O]ur object at present is not to obtain compactness in the mathematical formulae, but to express every relation of which we have any knowledge. To eliminate a quantity which expresses a useful idea would be rather a loss than a gain in this stage of our enquiry. (615.)

What, then, distinguishes the General Equations of the Electromagnetic Field featured in this chapter; what qualities set them aside for special consideration? First, they are *expressive*; as Maxwell states, our object is to "express every relation of which we have any knowledge." Second, they are *focused*; they characterize the electromagnetic field as such, thereby distinguishing electromagnetic science from other sciences of energy, such as mechanics and thermodynamics. Third, they are indeed *general*—within the limits of their territory—general enough to express the unity of electromagnetic science. But we should not forget that the most general equations of all are not these, but rather the Lagrange equations discussed in Chapter V.

A garland of quaternions

Having set forth the general equations of the electromagnetic field, Maxwell will proceed to translate the principal ones among them into quaternion form. The section in which he does this, Art. 619, is quite brief; and Maxwell's prefatory remarks concerning his use of quaternions are little short of apologetic (618.). Both these circumstances, I think, obscure his pupose.

In part, to be sure, Maxwell values quaternion expression for its compactness and representational efficiency—the same virtues we laud today in modern vector calculus. But for Maxwell, efficiency has less to do with saving labor and more with empowering thought. Quaternion notation directs our thought to the quantities themselves, not as mediated by an external system of Cartesian axes. Early in his "Preliminary" chapter, Maxwell declared:

> But for many purposes of physical reasoning, as distinguished from calculation, it is desirable to avoid explicitly introducing the Cartesian coordinates, and to fix the mind at once on a point of space instead of its three coordinates, and on the magnitude and direction of a force instead of its three components. This mode of contemplating geometrical and physical quantities is more primitive and more natural than the other, although the ideas connected with it did not receive their full development till Hamilton made

the next great step in dealing with space, by the invention of his Calculus of Quaternions. (10.1)

Perhaps we are in a good position now to practice what Maxwell calls a "more primitive and more natural" way of conceiving physical quantities—to see them as wholes, rather than as ensembles of properties and relations.

Quantities and symbols

This chapter includes Maxwell's compendium of electromagnetic quantities, together with the symbols typically used to represent them in the *Treatise*. His table is by no means a glossary; as I have remarked before (25.6, *comment*), Maxwell is not overly concerned with notational consistency, and he does not everywhere employ the symbology there set out. Still, the table serves as a useful reference. And since I have found it to be more useful while working through Maxwell's derivations than after having completed them, I offer a version of it here, in advance.*

Vector quantities

	Symbol	Constituents
Radius vector of a point	ρ	$x\ y\ z$
Electromagnetic momentum at a point	\mathfrak{A}	$F\ G\ H$
Magnetic induction	\mathfrak{B}	$a\ b\ c$
Total (or "true") electric current density (current per unit area)	\mathfrak{C}	$u\ v\ w$
Electric displacement	\mathfrak{D}	$f\ g\ h$
Electromotive intensity	\mathfrak{E}	$P\ Q\ R$
Mechanical force	\mathfrak{F}	$X\ Y\ Z$
Velocity of a point	\mathfrak{G} or $\dot{\rho}$	$\dot{x}\ \dot{y}\ \dot{z}$
Magnetic force	\mathfrak{H}	$\alpha\ \beta\ \gamma$
Intensity of magnetization	\mathfrak{J}	$A\ B\ C$
Conduction current density (current per unit area)	\mathfrak{K}	$p\ q\ r$

Scalar quantities

Electric potential	Ψ
Magnetic potential	Ω
Charge density (charge per unit volume)	e
Density of magnetic 'matter'	\mathfrak{m}

Physical properties of the medium (at a point)

Conductivity (for electric current)	C
Dielectric inductive capacity	K
Magnetic inductive capacity	μ

*A copy of the table is also provided on the inside rear cover for handy reference.

CHAPTER IX.

GENERAL EQUATIONS OF THE ELECTROMAGNETIC FIELD.

604.] IN our theoretical discussion of electrodynamics we began by assuming that a system of circuits carrying electric currents is a dynamical system, in which the currents may be regarded as velocities, and in which the coordinates corresponding to these velocities do not themselves appear in the equations. It follows from this that the kinetic energy of the system, in so far as it depends on the currents, is a homogeneous quadratic function of the currents, in which the coefficients depend only on the form and relative position of the circuits. Assuming these coefficients to be known, by experiment or otherwise, we deduced, by purely dynamical reasoning, the laws of the induction of currents, and of electromagnetic attraction. In this investigation we introduced the conceptions of the electrokinetic energy of a system of currents, of the electromagnetic momentum of a circuit, and of the mutual potential of two circuits.

604.1 We then proceeded to explore the field by means of various configurations of the secondary circuit, and were thus led to the conception of a vector \mathfrak{A}, having a determinate magnitude and direction at any

604. *In our theoretical discussion of electrodynamics we began...*: Maxwell's "theoretical" discussion began with Chapter VI of Part IV. Earlier stages of the *Treatise*, in contrast, were devoted largely to the acquisition of electrodynamic concepts. In the first few Articles of this chapter Maxwell will review selected stages in the mathematical construction of the theory.

a homogeneous quadratic function of the currents...: See, for example, (571.4).

purely dynamical reasoning: Here, fitting the general equations of Chapter V to the special conditions of electrical circuits and fields—as opposed to deducing consequences from hypotheses. The role of experiment differs dramatically in the two methods; Maxwell's luminous characterization of the way experiment is to function in dynamical reasoning (592.2) will be amplified further in (606.) below.

604.1 *the conception of a vector* \mathfrak{A}: As Maxwell is about to remind us, he first introduced the vector \mathfrak{A} as the "vector potential of magnetic induction" in Art. 405; he rediscovered it through dynamical investigation in (590.5)—but with a new significance, one related to *momentum*.

413

given point of the field. We called this vector the electromagnetic momentum at that point. This quantity may be considered as the time-integral of the electromotive intensity which would be produced at that point by the sudden removal of all the currents from the field. It is identical with the quantity already investigated in Art. 405 as the vector-potential of magnetic induction. Its components parallel to x, y, and z are F, G, and H. The electromagnetic momentum of a circuit is the line-integral of \mathfrak{A} round the circuit.

604.2 We then, by means of Theorem IV, Art. 24, transformed the line-integral of \mathfrak{A} into the surface-integral of another vector, \mathfrak{B}, whose components are a, b, c, and we found that the phenomena of induction due to motion of a conductor, and those of electromagnetic force, can be expressed in terms of \mathfrak{B}. We gave to \mathfrak{B} the name of the magnetic induction, since its properties are identical with those of the lines of magnetic induction as investigated by Faraday.

604.3 We also established three sets of equations: the first set, (A), are those of magnetic induction, expressing it in terms of the electromagnetic momentum. The second set, (B), are those of electromotive intensity, expressing it in terms of the motion of the conductor across the lines of magnetic induction, and of the rate of variation of the electromagnetic momentum. The third set, (C), are the equations of electromagnetic force, expressing it in terms of the current and the magnetic induction.

604.1, continued. *We called this vector the electromagnetic momentum at that point*: Although Maxwell does indeed use the term "electromagnetic momentum," he called it the *electrokinetic* momentum when he renamed vector \mathfrak{A} in (590.5). In general, Maxwell tends to apply the two terms interchangeably—as, for example, at (579.3)

the electromotive intensity which would be produced ... by the sudden removal of all the currents: Maxwell invoked such a thought experiment in (590.5). That it indeed indicates "something like momentum" is what he observed, in the name of Faraday, at (551.3). Compare his discussion of the role of *sudden stoppage* in the operation of the hydraulic ram in Art. 547.

The electromagnetic momentum ... is the line-integral of \mathfrak{A} round the circuit: Maxwell developed the line-integral of \mathfrak{A} in Art. 405; he identified the electromagnetic (or electrokinetic) momentum of a circuit in a different form at (578.6).

604.3 *We also established three sets of equations*: Equations (A) were set forth in (591.5), equations (B) in (598.4), equations (C) in (603.1). They make up the first three of a total of twelve sets of General Equations of the Electromagnetic Field expounded in the present chapter.

604.4 The current in all these cases is to be understood as the actual current, which includes not only the current of conduction, but the current due to variation of the electric displacement.

604.5 The magnetic induction \mathfrak{B} is the quantity which we have already considered in Art. 400. In an unmagnetized body it is identical with the force on a unit magnetic pole, but if the body is magnetized, either permanently or by induction, it is the force which would be exerted on a unit pole, if placed in a narrow crevasse in the body, the walls of which are perpendicular to the direction of magnetization. The components of \mathfrak{B} are a, b, c.

604.6 It follows from the equations (A), by which a, b, c are defined, that

$$\frac{da}{dx} + \frac{db}{dy} + \frac{dc}{dz} = 0.$$

604.7 This was shewn at Art. 403 to be a property of the magnetic induction.

605.] We have defined the magnetic force within a magnet, as distinguished from the magnetic induction, to be the force on a unit pole placed in a narrow crevasse cut parallel to the direction of magnetization. This quantity is denoted by \mathfrak{H}, and its components by α, β, γ. See Art. 398.

604.4 *the actual current, which includes not only ... conduction, but ... displacement*: Maxwell writes here as though he had already introduced the term "actual current" in the sense described. He indeed used that locution in (578.3); but since the discussion there was limited to electrical *circuits* (which are formed of conductors rather than dielectrics), no significant displacement could arise. Later in the present chapter, Arts. 610 and 611, Maxwell will use the term "true current" to denote the sum of conduction and displacement currents; but the present passage marks his first explicit mention of their summation.

604.6 *It follows from equations (A) ... that* $\frac{da}{dx} + \frac{db}{dy} + \frac{dc}{dz} = 0$: Equations (A) were given in (591.5) as $a = \frac{dH}{dy} - \frac{dG}{dz}$, $b = \frac{dF}{dz} - \frac{dH}{dx}$, $c = \frac{dG}{dx} - \frac{dF}{dy}$. Take the derivatives da/dx and db/dy; the terms in H cancel one another; similarly, the terms in F and G cancel pairwise when dc/dz is also taken. We performed a similar transformation in (24.2, *comment*).

This is actually an instance of a general theorem. Since a, b, c are the components of \mathfrak{B}, we recognize $\frac{da}{dx} + \frac{db}{dy} + \frac{dc}{dz}$ as $\nabla \cdot \mathfrak{B}$, the divergence of \mathfrak{B} (25.9, *comment*). Likewise F, G, H being the components of \mathfrak{A} (590.2), equations (A) are equivalent to $\mathfrak{B} = \nabla \times \mathfrak{A}$ or $\mathfrak{B} = \text{curl } \mathfrak{A}$ (591.5, *comment*); thus we have shown that div curl $\mathfrak{A} = 0$. It will be evident that, by the same method, we can prove that the divergence of the curl of *any* vector is identically zero.

605.1 If \Im is the intensity of magnetization, and A, B, C its components, then, by Art. 400,

$$
\left.
\begin{aligned}
a &= \alpha + 4\pi A, \\
b &= \beta + 4\pi B, \\
c &= \gamma + 4\pi C.
\end{aligned}
\right\}
\quad \text{(Equations of Magnetization)} \quad \text{(D)}
$$

605.2 We may call these the equations of magnetization, and they indicate that in the electromagnetic system the magnetic induction \mathfrak{B}, considered as a vector, is the sum, in the Hamiltonian sense, of two vectors, the magnetic force \mathfrak{H}, and the magnetization \Im multiplied by 4π, or

$$\mathfrak{B} = \mathfrak{H} + 4\pi\Im.$$

In certain substances, the magnetization depends on the magnetic force, and this is expressed by the system of equations of induced magnetism given at Arts. 426 and 435.

606.] Up to this point of our investigation we have deduced everything from purely dynamical considerations, without any reference to quantitative experiments in electricity or magnetism. The only use we have made of experimental knowledge is to recognise, in the abstract quantities deduced from the theory, the concrete quantities discovered by experiment, and to denote them by names which indicate their physical relations rather than their mathematical generation.

606.1 In this way we have pointed out the existence of the electromagnetic momentum \mathfrak{A} as a vector whose direction and magnitude vary from one part of space to another, and from this we have deduced, by a mathematical process, the magnetic induction, \mathfrak{B}, as a derived vector. We have not, however, obtained any data for determining either \mathfrak{A} or \mathfrak{B} from the

605.2 *the sum, in the Hamiltonian sense*: that is, the vector sum, the result of joining the tail of one vector to the head of the other, as indicated in the drawing.

606. *to recognize, in the abstract quantities deduced from theory, the concrete quantities discovered by experiment*: with its emphasis on "recognizing," this beautiful characterization makes it clear that experiment, for Maxwell, advances a doubly interpretive activity. By relating abstract mathematical quantities to experience we, first, enrich our experience with meaning and wholeness. Second, we *ground* our mathematical and symbolic knowledge, bestowing upon it a human face that Faraday, above all, would surely have prized: recall Maxwell's remark that "Faraday ... saw lines of force traversing all space where the mathematicians saw centres of force attracting at a distance" (Preface, paragraph 0.21).

denote them by names which indicate their physical relations rather than their mathematical generation: An expression of Maxwell's distaste for what Daniel M. Siegel has called "disembodied mathematics" in his richly rewarding study *Innovation in Maxwell's Electromagnetic Theory*, Cambridge University Press (1991), p. 31.

distribution of currents in the field. For this purpose we must find the mathematical connexion between these quantities and the currents.

606.2 We begin by admitting the existence of permanent magnets, the mutual action of which satisfies the principle of the conservation of energy. We make no assumption with respect to the laws of magnetic force except that which follows from this principle, namely, that the force acting on a magnetic pole must be capable of being derived from a potential.

606.3 We then observe the action between currents and magnets, and we find that a current acts on a magnet in a manner apparently the same as another magnet would act if its strength, form, and position were properly adjusted, and that the magnet acts on the current in the same way as another current. These observations need not be supposed to be accompanied by actual measurements of the forces. They are not therefore to be considered as furnishing numerical data, but are useful only in suggesting questions for our consideration.

606.4 The question these observations suggest is, whether the magnetic field produced by electric currents, as it is similar to that produced by permanent magnets in many respects, resembles it also in being related to a potential?

606.5 The evidence that an electric circuit produces, in the space surrounding it, magnetic effects precisely the same as those produced by a magnetic shell bounded by the circuit, has been stated in Arts. 482–485.

606.6 We know that in the case of the magnetic shell there is a potential, which has a determinate value for all points outside the substance of the shell, but that the values of the potential at two neighbouring points, on opposite sides of the shell, differ by a finite quantity.

606.7 If the magnetic field in the neighbourhood of an electric current resembles that in the neighbourhood of a magnetic shell, the magnetic potential, as found by a line-integration of the magnetic force, will be the same for any two lines of integration, provided one of these lines can be transformed into the other by continuous motion without cutting the electric current.

606.2 *the force acting on a magnetic pole must be capable of being derived from a potential*: Maxwell introduced derivation from a potential generally in (16.4–16.5). For the magnetic case see (395.1) and (398.1, *comment*).

606.3 *They are not ... to be considered as furnishing numerical data, but [as] suggesting questions*: He continues to point out the interpretive role of experiment.

606.6 *the values of the potential ... differ by a finite quantity*: We saw this in (411.).

606.7 *the magnetic potential ... will be the same for any two lines of integration, provided one of these lines can be transformed into the other...*: Maxwell introduced this idea in (481.), but not so explicitly as he states it here. See also (607.2) below.

606.8 If, however, one line of integration cannot be transformed into the other without cutting the current, the line-integral of the magnetic force along the one line will differ from that along the other by a quantity depending on the strength of the current. The magnetic potential due to an electric current is therefore a function having an infinite series of values with a common difference, the particular value depending on the course of the line of integration. Within the substance of the conductor, there is no such thing as a magnetic potential.

607.] Assuming that the magnetic action of a current has a magnetic potential of this kind, we proceed to express this result mathematically.

607.1 In the first place, the line-integral of the magnetic force round any closed curve is zero, provided the closed curve does not surround the electric current.

607.2 In the next place, if the current passes once, and only once, through the closed curve in the positive direction, the line-integral has a determinate value, which may be used as a measure of the strength of the current. For if the closed curve alters its form in any continuous manner without cutting the current, the line-integral will remain the same.

607.3 In electromagnetic measure, the line-integral of the magnetic force round a closed curve is numerically equal to the current through the closed curve multiplied by 4π.

607.4 If we take for the closed curve the rectangle whose sides are dy and dz, the line-integral of the magnetic force round the parallelogram is

$$\left(\frac{d\gamma}{dy} - \frac{d\beta}{dz}\right) dy\, dz,$$

606.8 *Within ... the conductor, there is no such thing as a magnetic potential*: Maxwell voiced an equivalent restriction in (485.1).

607.1 *the line-integral of the magnetic force round any closed curve is zero, provided the closed curve does not surround the electric current*: as stated earlier in (498.3).

607.3 *the line-integral of the magnetic force round a closed curve is numerically equal to the current through the closed curve multiplied by 4π*: To consider a simple case, the magnetic force \mathfrak{H} at distance r from a long straight wire carrying current i is disposed circularly about the wire and has magnitude $T = 2i/r$, as stated at (477.1) and proved in (485.3, *comment*). For a circular path of radius r, the line-integral of \mathfrak{H} will be $2\pi r \cdot 2i/r = 4\pi i$.

607.4 *the line-integral of the magnetic force round the parallelogram*: The drawing shows the parallelogram in question. I discussed a nearly identical case in the editor's introduction to Art. 24, section 2, on page 38 above. Proceeding in the same way, after collecting terms we shall find the line-integral equal to $d\gamma\,dz - d\beta\,dy$. Divide and multiply by $dy\,dz$ to obtain Maxwell's expression.

and if u, v, w are the components of the flow of electricity, the current through the parallelogram is $u\,dy\,dz$.

607.5 Multiplying this by 4π, and equating the result to the line-integral, we obtain the equation

$$4\pi u = \frac{d\gamma}{dy} - \frac{d\beta}{dz},$$

with the similar equations

$$4\pi v = \frac{d\alpha}{dz} - \frac{d\gamma}{dx},$$

$$4\pi w = \frac{d\beta}{dx} - \frac{d\alpha}{dy},$$

(Equations of Electric Currents) (E)

which determine the magnitude and direction of the electric currents when the magnetic force at every point is given.

607.6 When there is no current, these equations are equivalent to the condition that

$$\alpha\,dx + \beta\,dy + \gamma\,dz = -D\Omega,$$

or that the magnetic force is derivable from a magnetic potential in all points of the field where there are no currents.

the current through the parallelogram is $u\,dy\,dz$: Strictly, u, v, and w are the components of the current *density*—current per unit cross section area. Only the component u is perpendicular to the yz-plane; thus the current through area $dy\,dz$ is $u\,dy\,dz$.

607.5 *Equations of Electric Currents* (E): Note that the right-hand sides of the equations together constitute curl \mathfrak{H} or $\nabla\times\mathfrak{H}$; see Art. 25, especially (25.11 and *comment*). For α, β, γ are the components of \mathfrak{H} (605.), while ∇ was defined as

$\left(\mathbf{i}\dfrac{d}{dx} + \mathbf{j}\dfrac{d}{dy} + \mathbf{k}\dfrac{d}{dz}\right)$. Then $\nabla\times\mathfrak{H} = \left(\mathbf{i}\dfrac{d}{dx} + \mathbf{j}\dfrac{d}{dy} + \mathbf{k}\dfrac{d}{dz}\right)\times(\mathbf{i}\alpha + \mathbf{j}\beta + \mathbf{k}\gamma)$. Multi-

plying through, $\nabla\times\mathfrak{H} = \mathbf{i}\left(\dfrac{d\gamma}{dy} - \dfrac{d\beta}{dz}\right) + \mathbf{j}\left(\dfrac{d\alpha}{dz} - \dfrac{d\gamma}{dx}\right) + \mathbf{k}\left(\dfrac{d\beta}{dx} - \dfrac{d\alpha}{dy}\right)$. Since u, v, w

have been identified as the components of current \mathfrak{C} (603.2, 603.3), equations (E) may therefore be expressed in vector form as: $4\pi\mathfrak{C} = \nabla\times\mathfrak{H}$.

607.6 *When there is no current, ...* $\alpha\,dx + \beta\,dy + \gamma\,dz = -D\Omega$: D is Maxwell's symbol for an exact differential, introduced in (16.3). Then since α, β, γ are the components of \mathfrak{H}, the equation is equivalent to $\mathfrak{H} = -\nabla\Omega$; that is, \mathfrak{H} is the negative gradient of Ω, which is therefore a *potential* (17.–17.4 and 17.5, *comment*). Now we just showed that equations (E) may be expressed as $4\pi\mathfrak{C} = \nabla\times\mathfrak{H}$; thus "when there is no current" they reduce to $\nabla\times\mathfrak{H} = 0$ or curl $\mathfrak{H} = 0$. But this will be identically true only if there is some scalar quantity Ω such that $\mathfrak{H} = -\nabla\Omega$; for you can easily prove that the curl of the gradient of any scalar quantity is zero. Thus "the magnetic force is derivable from a potential," Ω, wherever there are no currents.

607.7 By differentiating the equations (E) with respect to x, y, and z respectively, and adding the results, we obtain the equation

$$\frac{du}{dx} + \frac{dv}{dy} + \frac{dw}{dz} = 0,$$

which indicates that the current whose components are u, v, w is subject to the condition of motion of an incompressible fluid, and that it must necessarily flow in closed circuits.

607.8 This equation is true only if we take u, v, and w as the components of that electric flow which is due to the variation of electric displacement as well as to true conduction.

607.9 We have very little experimental evidence relating to the direct electromagnetic action of currents due to the variation of electric displacement in dielectrics, but the extreme difficulty of reconciling the laws of electromagnetism with the existence of electric currents which are not closed is one reason among many why we must admit the existence of transient currents due to the variation of displacement. Their importance will be seen when we come to the electromagnetic theory of light.

608.] We have now determined the relations of the principal quantities concerned in the phenomena discovered by Örsted, Ampère, and Faraday. To connect these with the phenomena described in the former parts of this treatise, some additional relations are necessary.

607.7 *By differentiating the equations (E)...*: Recall that Maxwell performed a similar transformation on equations (A) in (604.6). Here, equations (E) can be expressed as $4\pi\mathfrak{C} = \nabla \times \mathfrak{H}$; taking the divergence ($\nabla\bullet$) of both sides, we immediately deduce that $\nabla\bullet\mathfrak{C} = 0$, since the divergence of $(\nabla\times\mathfrak{H})$ must be zero (604.6, *comment*). And, as before, we recognize $(du/dx + dv/dy + dw/dz)$ as *equivalent* to $\nabla\bullet\mathfrak{C}$.

the current whose components are u, v, w ... must necessarily flow in closed circuits: Otherwise stated, it has the solenoidal property (21.7 and *comment*).

607.8 *true only if we take u, v, w as ... due to the variation of electric displacement as well as to true conduction*: As we have noted already, the conduction current considered in itself is not in every case solenoidal. When charging a capacitor, for example, if we recognize *only* the conduction current, we shall say that electricity accumulates on the plates—a violation of the solenoidal condition (21.7). But when we also take the variation of displacement into account, we understand that electricity does not accumulate anywhere, not even on the plates (61.).

607.9 *the extreme difficulty of reconciling the laws of electromagnetism with ... electric currents that are not closed*: We can find a signal example in equations (E), expressed in vector form as $4\pi\mathfrak{C} = \text{curl } \mathfrak{H}$ (607.7, *comment*). As before, since the divergence of curl \mathfrak{H} is identically zero, the divergence of \mathfrak{C} must also be zero; but div \mathfrak{C} will not be zero at the plates of a charging capacitor, like the one considered in the previous comment—unless we admit the existence of something equivalent to a current in the space between the plates.

the electromagnetic theory of light: to be put forward in Chapter XX.

608.1 When electromotive intensity acts on a material body, it produces in it two electrical effects, called by Faraday Induction and Conduction, the first being most conspicuous in dielectrics, and the second in conductors.

608.2 In this treatise, static electric induction is measured by what we have called the electric displacement, a directed quantity or vector which we have denoted by \mathfrak{D}, and its components by f, g, h.

608.3 In isotropic substances, the displacement is in the same direction as the electromotive intensity which produces it, and is proportional to it, at least for small values of this intensity. This may be expressed by the equation

$$\mathfrak{D} = \frac{1}{4\pi} K\mathfrak{E}, \quad \begin{array}{l}\text{(Equation of Electric} \\ \text{Displacement)}\end{array} \quad \text{(F)}$$

where K is the dielectric capacity of the substance. See Art. 68.

608.4 In substances which are not isotropic, the components f, g, h of the electric displacement \mathfrak{D} are linear functions of the components P, Q, R of the electromotive intensity \mathfrak{E}.

608.5 The form of the equations of electric displacement is similar to that of the equations of conduction as given in Art. 298.

608.6 These relations may be expressed by saying that K is, in isotropic bodies, a scalar quantity, but in other bodies it is a linear and vector function, operating on the vector \mathfrak{E}.

609.] The other effect of electromotive intensity is conduction. The laws of conduction as the result of electromotive intensity were established by Ohm, and are explained in the second part of this treatise, Art. 241. They may be summed up in the equation

$$\mathfrak{K} = C\mathfrak{E}, \quad \text{(Equation of Conductivity)} \quad \text{(G)}$$

where \mathfrak{E} is the electromotive intensity at the point, \mathfrak{K} is the density of the current of conduction, the components of which are $p, q,$ and r, and C is the conductivity of the substance, which in the case of isotropic substances, is a simple scalar quantity, but in other substances becomes a linear and vector function operating on the vector \mathfrak{E}. The form of this function is given in Cartesian coordinates in Art. 298.

610.] One of the chief peculiarities of this treatise is the doctrine which it asserts, that the true electric current \mathfrak{C}, that on which the electromagnetic phenomena depend, is not the same thing as \mathfrak{K}, the current

608.5 *the equations of conduction ... given in Art. 298*: The Article is not included in the present selections, but Maxwell's point is that displacement and conduction can be viewed in comparable terms mathematically.

609. *\mathfrak{K} is the density of the current of conduction*: that is, current per unit cross-section area of the conductor.

of conduction, but that the time-variation of \mathfrak{D}, the electric displacement, must be taken into account in estimating the total movement of electricity, so that we must write,

$$\mathfrak{C} = \mathfrak{K} + \dot{\mathfrak{D}}, \quad \text{(Equation of True Currents)} \qquad \text{(H)}$$

or, in terms of the components,

$$
\left.
\begin{aligned}
u &= p + \frac{df}{dt}, \\
v &= q + \frac{dg}{dt}, \\
w &= r + \frac{dh}{dt}.
\end{aligned}
\right\} \qquad \text{(H*)}
$$

611.] Since both \mathfrak{K} and \mathfrak{D} depend on the electromotive intensity \mathfrak{E}, we may express the true current \mathfrak{C} in terms of the electromotive intensity, thus

$$\mathfrak{C} = \left(C + \frac{1}{4\pi} K \frac{d}{dt} \right) \mathfrak{E}, \qquad \text{(I)}$$

or, in the case in which C and K are constants,

$$
\left.
\begin{aligned}
u &= CP + \frac{1}{4\pi} K \frac{dP}{dt}, \\
v &= CQ + \frac{1}{4\pi} K \frac{dQ}{dt}, \\
w &= CR + \frac{1}{4\pi} K \frac{dR}{dt}.
\end{aligned}
\right\} \qquad \text{(I*)}
$$

612.] The volume-density of the free electricity at any point is found from the components of electric displacement by the equation

$$\rho = \frac{df}{dx} + \frac{dg}{dy} + \frac{dh}{dz}. \qquad \text{(J)}$$

612. *free electricity*: Something of a figure of speech inherited from conventional fluid-flow thinking, the term was used in Art. 36 to denote electricity that is "free" to participate in conduction. It was there distinguished from "combined" or "bound" electricity, which undergoes displacement under the action of electromotive force but returns to its former condition as soon as that force is removed (60.2). But it is misleading to think in terms of *kinds* of electricities, or to think of *any* electricity as being free of force; Maxwell's discussions of conduction and induction have made it quite clear that the essential distinction is rather between forces of tension that easily break down (in a conductor) and those which endure (in a dielectric).

Equation (J): It is a variant of Poisson's equation (77.3). Since f, g, h are the components of displacement \mathfrak{D}, it can also be expressed as $\operatorname{div} \mathfrak{D} = \rho$ or, alternatively, as $\nabla \cdot \mathfrak{D} = \rho$ (77.2, *comment*).

613.] The surface-density of electricity is

$$\sigma = lf + mg + nh + l'f' + m'g' + n'h', \qquad \text{(K)}$$

where l, m, n are the direction-cosines of the normal drawn from the surface into the medium in which f, g, h are the components of the displacement, and l', m', n' are those of the normal drawn from the surface into the medium in which they are f', g', h'.

614.] When the magnetization of the medium is entirely induced by the magnetic force acting on it, we may write the equation of induced magnetization,

$$\mathfrak{B} = \mu\mathfrak{H}, \qquad \text{(L)}$$

where μ is the coefficient of magnetic permeability, which may be considered a scalar quantity, or a linear and vector function operating on \mathfrak{H}, according as the medium is isotropic or not.

615.] These may be regarded as the principal relations among the quantities we have been considering. They may be combined so as to eliminate some of these quantities, but our object at present is not to obtain compactness in the mathematical formulae, but to express every relation of which we have any knowledge. To eliminate a quantity which expresses a useful idea would be rather a loss than a gain in this stage of our enquiry.

613. *the direction-cosines of the normal drawn from the surface into the medium*: Equation (K) gives the charge per unit area that develops on a surface between two different media in the presence of electric displacement. Since the normals are drawn away from the surface into their respective media, they are in opposite directions and their direction-cosines have opposite signs. The products $l'f'$ and lf will similarly have opposite signs. The sketch shows a boundary between two media; the positive x-direction is to the right. Since the net displacement in the direction of the normal, $lf-l'f'$, is directed to the left, the surface-density σ will be negative.

614. $\mathfrak{B} = \mu\mathfrak{H}$: Even though Maxwell has had the materials for writing this relation of magnetic proportionality since Part III Chapter IV, he has not done so until now, where it follows closely upon the electrical proportionality $\mathfrak{D} = \frac{1}{4\pi}K\mathfrak{E}$ (608.3). Is he inviting us to notice a parallelism between these relations? Certainly a formal correspondence is obvious. But physically, magnetism and electricity seem very far from being parallel. The nonexistence of isolated magnetic poles is a major point of difference; the lack of anything like a magnetic "current" is another.

615. *to express every relation of which we have any knowledge*: What impulse drives us to shape truth into speech? While it can have its comical side—"When I think, I must speak!" (Rosalind in *As You Like It*, Act III)—yet there is something quite wonderful about it, worthy even of the heavens: *The heavens declare the glory of God, and the firmament proclaims his handiwork. Day unto day utters speech, and night unto night expresses knowledge* (Psalm 19).

* * *

615.2 If we suppose that no magnets exist in the field except in the form of electric circuits, the distinction which we have hitherto maintained between the magnetic force and the magnetic induction vanishes, because it is only in magnetized matter that these quantities differ from each other.

615.3 According to Ampère's hypothesis, which will be explained in Art. 833, the properties of what we call magnetized matter are due to molecular electric circuits, so that it is only when we regard the substance in large masses that our theory of magnetization is applicable, and if our mathematical methods are supposed capable of taking account of what goes on within the individual molecules, they will discover nothing but electric circuits, and we shall find the magnetic force and the magnetic induction everywhere identical. In order, however, to be able to make use of the electrostatic or of the electromagnetic system of measurement at pleasure we shall retain the coefficient μ, remembering that its value is unity in the electromagnetic system.

* * *

Quaternion Expressions for the Electromagnetic Equations.

618.] In this treatise we have endeavoured to avoid any process demanding from the reader a knowledge of the Calculus of Quaternions.

615.3 *if our mathematical methods are supposed capable of taking account of what goes on within the individual molecules, they will discover nothing but electric circuits*: Is Maxwell acknowledging, however reluctantly, that the "real" physics might lie at the molecular level? That would grant a level of reality to discrete material particles which he has so far conspicuously shunned; continuous space and the aether have been treated as the primary receptacles of energy. Properties such as inductive capacity have been represented as belonging to them *inherently*, not indirectly through being populated by material particles.

What, moreover, would become of the great Lagrangian project of Part IV Chapter V if the molecular view (so we may call it) takes precedence? To be sure, we have characterized the Lagrangian equations as being more schematic than specific—but neither are they mere placeholders. Rather, they express a distinctive view of the cosmos that emphasizes wholeness, connectedness, and priority of energy over matter. Is all that being called into question at this late stage of the *Treatise*?

we shall find the magnetic force and the magnetic induction everywhere identical: Their identity characterizes the *vacuum* (404.4); presumably, vacuum is what we should find were we able to descend into the interior of individual molecules.

the coefficient μ, ... *its value is unity in the electromagnetic system*: Maxwell will set forth the foundations of the electrostatic and electromagnetic systems of

At the same time we have not scrupled to introduce the idea of a vector when it was necessary to do so. When we have had occasion to denote a vector by a symbol, we have used a German letter, the number of different vectors being so great that Hamilton's favourite symbols would have been exhausted at once. Whenever therefore a German letter is used it denotes a Hamiltonian vector, and indicates not only its magnitude but its direction. The constituents of a vector are denoted by Roman or Greek letters.

618.1 The principal vectors which we have to consider are

	Symbol of Vector.	Constituents.
The radius vector of a point	ρ	$x\ y\ z$
The electromagnetic momentum at a point	\mathfrak{A}	$F\ G\ H$
The magnetic induction	\mathfrak{B}	$a\ b\ c$
The (total) electric current	\mathfrak{C}	$u\ v\ w$
The electric displacement	\mathfrak{D}	$f\ g\ h$
The electromotive intensity	\mathfrak{E}	$P\ Q\ R$
The mechanical force	\mathfrak{F}	$X\ Y\ Z$
The velocity of a point	\mathfrak{G} or $\dot{\rho}$	$\dot{x}\ \dot{y}\ \dot{z}$
The magnetic force	\mathfrak{H}	$\alpha\ \beta\ \gamma$
The intensity of magnetization	\mathfrak{J}	$A\ B\ C$
The current of conduction	\mathfrak{K}	$p\ q\ r$

618.2 We have also the following scalar functions:

The electric potential Ψ.

The magnetic potential (where it exists) Ω.

The electric density \mathfrak{e}.

The density of magnetic 'matter' \mathfrak{m}.

618.3 Besides these we have the following quantities, indicating physical properties of the medium at each point:

C, the conductivity for electric currents.

K, the dielectric inductive capacity.

μ, the magnetic inductive capacity.

units in the following chapter. As a consequence of their respective definitions, the magnetic inductive capacity μ of vacuum is unity and dimensionless in the electromagnetic system; in the electrostatic system it has a finite value and the dimensions of *reciprocal velocity* squared. The electrostatic inductive capacity K of vacuum is unity and dimensionless in the electrostatic system; but in the electromagnetic system it has a finite value and the dimensions of *velocity* squared.

618. *Whenever ... a German letter is used it denotes a Hamiltonian vector*: In fact this is Maxwell's practice only for capitals; in (618.2) he will use the lower case German letters \mathfrak{e} and \mathfrak{m} to represent electric density (charge per unit volume) and magnetic density (pole strength per unit volume), both of which are scalar quantities.

618.1 *The (total) electric current*: or, as Maxwell often styles it, the "true" (611.) or "actual" (604.4) current—the sum of the current of conduction \mathfrak{K} and the current due to variation of displacement. But a better term than "current" would be *current density*, as the dimensions of \mathfrak{C} are those of current per unit area.

618.4 These quantities are, in isotropic media, mere scalar functions of ρ, but in general they are linear and vector operators on the vector functions to which they are applied. K and μ are certainly always self-conjugate, and C is probably so also.

619.] The equations (A) of magnetic induction, of which the first is

$$a = \frac{dH}{dy} - \frac{dG}{dz},$$

may now be written

$$\mathfrak{B} = V.\nabla\mathfrak{A},$$

where ∇ is the operator

$$i\frac{d}{dx} + j\frac{d}{dy} + k\frac{d}{dz},$$

and V indicates that the vector part of the result of this operation is to be taken.

619.1 Since \mathfrak{A} is subject to the condition $S.\nabla\mathfrak{A} = 0$, $\nabla\mathfrak{A}$ is a pure vector, and the symbol V is unnecessary.

619.2 The equations (B) of electromotive force, of which the first is

$$P = c\dot{y} - b\dot{z} - \frac{dF}{dt} - \frac{d\Psi}{dx},$$

become

$$\mathfrak{E} = V.\mathfrak{G}\mathfrak{B} - \dot{\mathfrak{A}} - \nabla\Psi.$$

618.4 *K and μ are certainly always self-conjugate*: I have restricted our selections to cases where properties like K and μ are simple scalars. But even when they are vector functions, Maxwell explains, they exhibit a certain form of symmetry; he describes it in the omitted Article 101e: "The most general form of a linear and vector function involves nine independent coefficients. When the coefficients which have the same pair of suffixes are equal, the function is said to be self-conjugate." He gives as an example

$$4\pi f = K_{xx}P + K_{xy}Q + K_{xz}R,$$
$$4\pi g = K_{yx}P + K_{yy}Q + K_{yz}R,$$
$$4\pi h = K_{zx}P + K_{zy}Q + K_{zz}R;$$

where the first suffix of each coefficient K indicates the direction of displacement, and the second that of the electromotive intensity. If $K_{xy} = K_{yx}$, and similarly for other pairs, the function K is self-conjugate.

619. $\mathfrak{B} = V.\nabla\mathfrak{A}$: As we have done previously in (25.4 and 607.5, *comments*), we may write the equation in vector algebra form as $\mathfrak{B} = \nabla\times\mathfrak{A}$ or $\mathfrak{B} = \text{curl}\,\mathfrak{A}$.

619.1 \mathfrak{A} *is subject to the condition* $S.\nabla\mathfrak{A} = 0$: that is, the (quaternion) product $\nabla\mathfrak{A}$ has no scalar part; in modern notation, $-\nabla\bullet\mathfrak{A} = 0$ or $-\text{div}\,\mathfrak{A} = 0$ (the negative signs, though spurious here, remind us that we have translated from the quaternion product). This is the solenoidal condition, which Maxwell obtained for electromagnetic momentum \mathfrak{A} in (406.7).

619.2 *The equations (B) ... become* $\mathfrak{E} = V.\mathfrak{G}\mathfrak{B} - \dot{\mathfrak{A}} - \nabla\Psi$: The magnetic induction \mathfrak{B}, written in component form, is $\mathfrak{B} = a\mathbf{i} + b\mathbf{j} + c\mathbf{k}$; the velocity of the conductor is $\mathfrak{G} = \dot{x}\mathbf{i} + \dot{y}\mathbf{j} + \dot{z}\mathbf{k}$. Take the cross product $\mathfrak{G}\times\mathfrak{B}$; the terms containing \mathbf{i} reduce

619.3 The equations (C) of mechanical force, of which the first is

$$X = cv - bw + eP - m\frac{d\Omega}{dx},^{*}$$

become $\mathfrak{F} = V.\mathfrak{CB} + e\mathfrak{E} - m\nabla\Omega.$

619.4 The equations (D) of magnetization, of which the first is

$$a = \alpha + 4\pi A,$$

become $\mathfrak{B} = \mathfrak{H} + 4\pi\mathfrak{J}.$

619.5 The equations (E) of electric currents, of which the first is

$$4\pi u = \frac{d\gamma}{dy} - \frac{d\beta}{dz},$$

become $4\pi\mathfrak{C} = V.\nabla\mathfrak{H}.$

619.6 The equation of the current of conduction is, by Ohm's Law,

$$\mathfrak{K} = C\mathfrak{E}.$$

619.7 That of electric displacement is

$$D = \frac{1}{4\pi}K\mathfrak{E}.$$

619.8 The equation of the total current, arising from the variation of the electric displacement as well as from conduction, is

$$\mathfrak{C} = \mathfrak{K} + \dot{\mathfrak{D}}.$$

619.9 When the magnetization arises from magnetic induction,

$$\mathfrak{B} = \mu\mathfrak{H}.$$

619.10 We have also, to determine the electric volume-density,

$$\mathfrak{e} = S.\nabla\mathfrak{D}.$$

*{In the 1st and 2nd editions of this work $-d\Psi/dx$ was written for P in this equation. The correction is due to Prof. G. F. Fitzgerald *Trans. R. S. Dublin*, 1883.}

to $c\dot{y} - b\dot{z}$. Again, F is the x-component of \mathfrak{A}, so dF/dt is the x-component of $\dot{\mathfrak{A}}$; similarly, $d\Psi/dx$ is the x-component of $\nabla\Psi$.

Maxwell has given the quaternion form of this equation. Its vector calculus equivalent is $\mathfrak{E} = \mathfrak{G} \times \mathfrak{B} - \dot{\mathfrak{A}} - \nabla\Psi$. The remaining equations are translated in the same way; I will give only their vector equivalents.

619.3 $\mathfrak{F} = V.\mathfrak{CB} + e\mathfrak{E} - m\nabla\Omega$: Equivalently, $\mathfrak{F} = \mathfrak{C} \times \mathfrak{B} + e\mathfrak{E} - m\nabla\Omega$.

619.5 $4\pi\mathfrak{C} = V.\nabla\mathfrak{H}$: Equivalently, $4\pi\mathfrak{C} = \nabla \times \mathfrak{H}$ or $4\pi\mathfrak{C} = \text{curl } \mathfrak{H}$.

619.8 *total current*: Recall that Maxwell also calls it the "true" current (610.) and the "actual" current (604.4).

619.9 *When the magnetization arises from magnetic induction*: that is, when the medium under consideration is not a permanent magnet.

619.10 $\mathfrak{e} = S.\nabla\mathfrak{D}$: Equivalently, $\mathfrak{e} = -\nabla\bullet\mathfrak{D}$ or $\mathfrak{e} = -\text{div } \mathfrak{D}$, representing electric displacement as originating from or terminating in electric charge.

619.11 To determine the magnetic volume-density,

$$\mathfrak{m} = S.\nabla\mathfrak{J}.$$

619.12 When the magnetic force can be derived from a potential,

$$\mathfrak{H} = -\nabla\Omega.$$

619.11 $\mathfrak{m} = S.\nabla\mathfrak{J}$: Equivalently, $\mathfrak{m} = -\nabla\cdot\mathfrak{J}$ or $\mathfrak{m} = -\operatorname{div}\mathfrak{J}$, representing the magnetization of a material as originating from or terminating on magnetic matter. But recall that Maxwell employed the notion of "magnetic matter" only as a temporarily serviceable fiction (386.2). As he pointed out in (604.6), the magnetic induction \mathfrak{B} is everywhere solenoidal ($\operatorname{div}\mathfrak{B} = 0$); and this would not be true if magnetic matter really existed.

Editor's Introduction to
Chapter X

Systems of measurement

In Chapter III of Part IV Maxwell drew an appreciative contrast between two styles of science. Ampère, he noted, gives us finished theory, "perfect in form, and unassailable in accuracy"; yet its very perfection conceals from us the process of its formation. Faraday, to the contrary, offers us not just his finished thoughts but also the origin and course of development of his thinking. Ampère's way serves the professional, the practitioner; Faraday's way invites the seeker.

Systems of measurement may be said to exhibit a similar contrast. Advanced systems, fully developed and comprehensive, build in knowledge and relations that would otherwise have to be kept deliberately before the mind. Their perfection greatly facilitates practical calculations. On the other hand, they can be very hard for the learner to unpack. More elementary systems tend to make their foundations more evident. In preparation for Maxwell's discussion of electrical units in this chapter, then, let us look at some preliminary—perhaps we should say say, incomplete—systems.

"Absolute" measure

In an impassioned and eloquent lecture to the Institution of Civil Engineers in 1883,* William Thomson, Lord Kelvin, extolled the advantages of "absolute" measurement. The term had been Weber's, and it referred to a system of measurement in which all quantities are expressed as combinations of the fundamental dimensions of length, mass, and time—but not restricted to particular units like centimeters, grams, or seconds. Relations that remained unaffected by the particular choice of fundamental units were *absolute* relations. For example, the area of a rectangular surface is the product of its length and width, no matter

*Thomson, Sir William (Lord Kelvin), "Electrical Units of Measurement," *Popular Lectures and Addresses*, Vol. I. London: Macmillan and Co. (1891).

what unit of distance is employed (so long as it is employed consistently); the relation between area and length and width therefore belongs to an absolute system.

Physics generally treats mass, length, and time as the standard fundamental dimensions, but not all sciences require all three of them. Geometry invokes only the dimension of *length*. Kinematics assumes not only length but *time*. Newtonian mechanics demands the addition of *mass* as a third fundamental dimension. What, then, of electrostatics? One of the first quantities defined in that science is *quantity of electricity* (41.1); must *quantity of electricity* constitute a fourth fundamental dimension, or can it be assimilated by a system based on mass, length, and time alone?

As Maxwell describes in Arts. 41 and 42, a suitable definition of quantity of electricity makes such an assimilation possible. On the basis of Coulomb's inverse-square force law we write (41.3)

$$e\,e' = Fr^2,$$

which tells us that the square of the units of electrical quantity must have the units of *force* times the square of the units of *length*. From Newton's second law, $F = ma$, the units of force are those of *mass* times those of *acceleration*; thus we have

[quantity of electricity]$^2 = $ [M L T^{-2}][L^2] or [M L^3 T^{-2}],

from which

[quantity of electricity] $= $ [M$^{1/2}$ L$^{3/2}$ T^{-1}].

(I here use Maxwell's dimensional notation, as employed by him in his Preliminary chapter.)

Electrostatic measure

We see that quantity of electricity can be expressed in terms of the standard dimensions of mass, length, and time. On the other hand, its expression derives ultimately from Coulomb's law of electrostatic attraction. We therefore say that quantity of electricity has been expressed in *electrostatic measure*; and any quantities derived from it will also be said to be in electrostatic measure. Electrostatic measure is thus not an alternative to absolute measure but a subset of it.

We can go further, incorporating more electrical quantities into the scheme of absolute electrostatic measurement. Since the measure of *current* is the quotient of *quantity of electricity* by *time*, its dimensions in electrostatic measure are

[current] $= $ [M$^{1/2}$ L$^{3/2}$ T^{-1}] / [T] or [M$^{1/2}$ L$^{3/2}$ T^{-2}].

The quantity *electromotive intensity* (44.4) is measured by the force per unit of electrical quantity. Its dimensions in electrostatic measure are then

$$[\text{electromotive intensity}] = [\text{M L T}^{-2}] / [\text{M}^{1/2}\,\text{L}^{3/2}\,\text{T}^{-1}]$$
$$= [\text{M}^{1/2}\,\text{L}^{-1/2}\,\text{T}^{-1}].$$

We found (45., *comment*) that *electromotive force* does not have the dimensions of mechanical force but rather of *work*, the product of force by distance, divided by *quantity of electricity*. Thus

$$[\text{electromotive force}] = [\text{M L T}^{-2}][\text{L}] / [\text{M}^{1/2}\,\text{L}^{3/2}\,\text{T}^{-1}]$$
$$= [\text{M}^{1/2}\,\text{L}^{1/2}\,\text{T}^{-1}].$$

From Ohm's Law, the measure of *electrical resistance* is the quotient of *electromotive force* by *current*. So we have the dimensions of resistance:

$$[\text{resistance}] = [\text{M}^{1/2}\,\text{L}^{1/2}\,\text{T}^{-1}] / [\text{M}^{1/2}\,\text{L}^{3/2}\,\text{T}^{-2}]$$
$$= [\text{L}^{-1}\,\text{T}].$$

Note that resistance has the reciprocal dimensions of *velocity*. As with all absolute relations, this one remains true no matter what particular units of length or time we may choose. But it ultimately derives from the law of electrostatic attraction and is, like the other quantities we have discussed so far, part of the electrostatic system.

Electromagnetic measure

So far we have made no mention of magnetism, or of that quantity called *strength of a magnetic pole*.* Can we bring *pole strength* into our scheme, or will its inclusion require a new fundamental dimension? The answer is clear; since pole strength obeys the same inverse-square force law as does quantity of electricity, we can reduce the dimensions of pole strength to mass, length, and time the same way we did for electricity. Moreover, since the magnetic and electric force laws are identical in form, the dimensions of pole strength will be identical to the dimensions already found for quantity of charge, namely,

$$[\text{pole strength}] = [\text{M}^{1/2}\,\text{L}^{3/2}\,\text{T}^{-1}].$$

This expression is no less "absolute" than those of any of the quantities we have been discussing. On the other hand, it cannot be said to belong to the electrostatic system, as it has no connection with the law of electrostatic attraction. We might think of it as the beginning of a new "magnetostatic" system—naming it by analogy with the electrostatic. But as I will note later, there are stronger reasons for calling it the *electromagnetic system*; and such has become its accepted title.

As before, and corresponding to electromotive intensity as the quotient of force by quantity of electricity, we define *magnetic field intensity* as the quotient of *mechanical force* by *pole strength*. It will have the same dimensions, namely

* Also called *quantity of free magnetism* in Chapter II of Part III.

[magnetic field intensity] $= [\text{M L T}^{-2}] / [\text{M}^{1/2} \text{L}^{3/2} \text{T}^{-1}]$

or $= [\text{M}^{1/2} \text{L}^{-1/2} \text{T}^{-1}]$.

These dimensions, too, are no less absolute than any of the former, but they belong only to the electromagnetic system, not the electrostatic. It would seem that the two systems, electrostatic and electromagnetic, are completely separate subdivisions of Absolute measurement, every quantity appearing in one system or the other, but no quantity appearing in both. This neat separation, however, is demolished with Oersted's discovery of the magnetic influence of an electric current.

As Maxwell pointed out (479.), a long straight wire carrying current i is surrounded by a circular magnetic field of intensity T such that

$$Tr = 2i,$$

where the multiplier 2, obtained by a geometrical argument (485.3, *comment*), is a pure number. On the left-hand side appears the *magnetic force, T*, which we have defined exclusively in the electromagnetic system. On the right-hand side is *current, i*, which has been defined exclusively in the electrostatic system. Perhaps that is a sign that the two systems can be unified; let us look at the dimensions of this equation to explore that question.

[magnetic field intensity] [L] = [current]

$[\text{M}^{1/2} \text{L}^{1/2} \text{T}^{-1}]$ = [current].

Thus $[\text{M}^{1/2} \text{L}^{1/2} \text{T}^{-1}]$ are the dimensions of current in the electromagnetic system. But we previously obtained $[\text{M}^{1/2} \text{L}^{3/2} \text{T}^{-2}]$ for the dimensions of current in the electrostatic system; and neither set of dimensions can be reduced to the other. The two systems, electrostatic and electromagnetic, are not merely separate—they are *incompatible* with one another! At the same time it is clear that since there is at least one quantity—current—that appears in both, each system can be extended to include *every* quantity that is defined in the other.

We thus end up with two alternative but incompatible systems, each of which encompasses all of electromagnetic science, both of which take mass, length, and time as the fundamental dimensions, and both of which—inasmuch as they are independent of the particular fundamental units chosen—deserve to be called "absolute."

Numerical factors

Many scientific expressions of quantitative relations contain numerical factors. The significance of those factors becomes clear when the relations are stated in their "absolute" forms. Consider, for example, two formulas for uniform gravitational acceleration at sea level: (1) $s = \frac{1}{2}gt^2$ and (2) $s = 16t^2$.

Equation (1) is written in absolute measure; the factor g is the gravitational acceleration; and provided we apply units of length and time consistently, the equation stands no matter what specific units we elect to use. Knowing this, we understand that the factor $\frac{1}{2}$ is an intrinsic (or, we may say, *absolute*) element *of the relation itself*, not an artefact of the units employed. In equation (2), by contrast, the numerical coefficient 16 is an amalgam of the absolute factor $\frac{1}{2}$ and the relative value 32, the gravitational acceleration at sea level *in feet per second per second*. Equation (2) is not independent of the units employed.

Specific systems

Notwithstanding the benefits of studying relations in their absolute forms, actual measurements necessarily require, and actual calculations necessarily embody, particular choices of units. Thus there have arisen many important *specific* systems—though for the purpose of reading the *Treatise* we will regard them as but variants within the two great classes, electrostatic and electromagnetic.

The choice of *centimeter*, *gram*, and *second* as fundamental units gave the "CGS" systems—electrostatic and electromagnetic, respectively—favored by Kelvin. Weber chose the *millimeter*, *milligram*, and *second* in constructing his system. The British Association adopted *meter*, *gram*, and *second* in its report of 1863; but that system never gained wide adherence. The choice of *meter*, *kilogram*, and *second* yielded the "MKS" system adopted by the 1893 International Electrical Congress in Chicago. Each of these systems has an electrostatic and an electromagnetic version. By enshrining particular units, each invites the development of practical calculation algorithms geared to those units. None of them is "absolute" in the sense of being independent of particular units; but all share the property, so desired by Kelvin, of expressing all physical quantities in terms of mass, length and time alone.

Practical units

The specific systems just outlined share another property, one that is not so desirable from the standpoint of making actual measurements. Their units are in some cases so small, and in others so large, in comparison to the quantities actually encountered in telegraphy, lighting, and other important applications of electricity, that measurements expressed in terms of them would regularly involve either extravagantly large numbers or excessively tiny fractions. A system of "Practical Units" was therefore devised to keep the expressions of actual measurements within a more convenient range. To take one example, the resistance of one nautical mile of the second Atlantic Telegraph Cable was about

4,190,000,000 CGS electromagnetic units.* The practical unit of resistance, the *ohm*, was defined as equal to 10^9 of those units,† so that the same resistance would be far more conveniently expressed as 4.19 ohms. For comparison, this is approximately the resistance of a typical 36-watt automotive headlamp burning at normal brightness.

Symmetries

The logic of Chapter X is compelling but understated. Maxwell begins by calling attention to three pairs of quantities—electrostatic, magnetic, and electrodynamic, respectively (Art. 621); the product of each pair exhibits dimensions of *energy*, thus revealing a unity among the sciences of electric charges, magnets, and currents. This in turn leads to an extensive litany of purely dimensional symmetries (Arts. 622–624), which Maxwell proceeds to explore to an almost obsessional degree of detail and variety.

It may appear that Maxwell has simply become fascinated with the richness of relations that multiple measurement systems present. As we follow his exposition, however, it becomes evident that he is first setting out relations that hold in *any* system of measurement (624.2), next refining these into transitional systems in which either *quantity of electricity* or *magnetic pole strength* appears as an independent unit (623.), and finally erecting, through a further refinement, the electrostatic and electomagnetic systems in a most elegant fashion (Art. 625).

*Everett, John David, *Units and Physical Constants*. London: Macmillan (1886), p.160.

†As Maxwell will note in a table (627.), the dimensions of electrical resistance in electromagnetic measure are those of *velocity*. Thus the CGS electromagnetic unit of resistance was the centimeter per second, and the ohm was to have a value of 10^9 cm/sec. Attempts to produce a physical realization of the ohm having that value proved unexpectedly challenging. I will say more about those challenges when introducing Chapter XIX.

CHAPTER X.

DIMENSIONS OF ELECTRIC UNITS.

620.] EVERY electromagnetic quantity may be defined with reference to the fundamental units of Length, Mass, and Time. If we begin with the definition of the unit of electricity, as given in Art. 65, we may obtain definitions of the units of every other electromagnetic quantity, in virtue of the equations into which they enter along with quantities of electricity. The system of units thus obtained is called the Electrostatic System.

620.1 If, on the other hand, we begin with the definition of the unit magnetic pole, as given in Art. 374, we obtain a different system of units of the same set of quantities. This system of units is not consistent with the former system, and is called the Electromagnetic System.

620.2 We shall begin by stating those relations between the different units which are common to both systems, and we shall then form a table of the dimensions of the units according to each system.

621.] We shall arrange the primary quantities which we have to consider in pairs. In the first three pairs, the product of the two quantities in each pair is a quantity of energy or work. In the second three pairs, the product of each pair is a quantity of energy referred to unit of volume.

<div align="center">

FIRST THREE PAIRS.

Electrostatic Pair.

Symbol
</div>

(1) Quantity of electricity . e
(2) Electromotive force, or electric potential E

<div align="center">

Magnetic Pair.
</div>

(3) Quantity of free magnetism, or strength of a pole . . . m
(4) Magnetic potential . Ω

620. *the definition of the unit of electricity, as given in Art. 65*: Article 65 is omitted in this edition, but an equivalent definition appears in Article 41.

621. *Electromotive force, or electric potential E*: Note that Maxwell also uses the symbols *V* and Ψ to represent electric potentisal.

Electrokinetic Pair.
(5) Electrokinetic momentum of a circuit p
(6) Electric current . C

SECOND THREE PAIRS.
Electrostatic Pair.
(7) Electric displacement (measured by surface-density) $.\mathfrak{D}$
(8) Electromotive intensity . \mathfrak{E}

Magnetic Pair.
(9) Magnetic induction . \mathfrak{B}
(10) Magnetic force . \mathfrak{H}

Electrokinetic Pair.
(11) Intensity of electric current at a point \mathfrak{C}
(12) Vector potential of electric currents \mathfrak{A}

622. The following relations exist between these quantities. In the first place, since the dimensions of energy are $\left[\dfrac{L^2 M}{T^2}\right]$, and those of energy referred to unit of volume $\left[\dfrac{M}{LT^2}\right]$, we have the following equations of dimensions:

$$[eE] = [m\Omega] = [pC] = \left[\frac{L^2 M^2}{T^2}\right], \tag{1}$$

$$[\mathfrak{D}\mathfrak{E}] = [\mathfrak{B}\mathfrak{H}] = [\mathfrak{C}\mathfrak{A}] = \left[\frac{M}{LT^2}\right]. \tag{2}$$

622.1 Secondly, since e, p, and \mathfrak{A} are the time-integrals of C, E, and \mathfrak{E} respectively,

$$\left[\frac{e}{C}\right] = \left[\frac{p}{E}\right] = \left[\frac{\mathfrak{A}}{\mathfrak{E}}\right] = [T]. \tag{3}$$

622.2 Thirdly, since E, Ω, and p are the line-integrals of \mathfrak{E}, \mathfrak{H}, and \mathfrak{A} respectively,

$$\left[\frac{E}{\mathfrak{E}}\right] = \left[\frac{\Omega}{\mathfrak{H}}\right] = \left[\frac{p}{\mathfrak{A}}\right] = [L].^* \tag{4}$$

$^*\{$We have also $\left[\dfrac{\mathfrak{A}}{\mathfrak{B}}\right] = [L].\}$

621, continued. *Electric current* C: Maxwell also uses the symbol i to represent electric current. In particular he will do so in the following chapter.

622.3 Finally, since e, C, and m are the surface-integrals of \mathfrak{D}, \mathfrak{C}, and \mathfrak{B} respectively,

$$\left[\frac{e}{\mathfrak{D}}\right] = \left[\frac{C}{\mathfrak{C}}\right] = \left[\frac{m}{\mathfrak{B}}\right] = \left[L^2\right]. \tag{5}$$

623.] These fifteen equations are not independent, and in order to deduce the dimensions of the twelve units involved, we require one additional equation. If, however, we take either e or m as an independent unit, we can deduce the dimensions of the rest in terms of either of these.

(1)	$[e]$		$= [e]$	$= \left[\dfrac{L^2M}{mT}\right].$
(2)	$[E]$		$= \left[\dfrac{L^2M}{eT^2}\right]$	$= \left[\dfrac{m}{T}\right].$
(3) and (5)	$[p]$	$= [m]$	$= \left[\dfrac{L^2M}{eT}\right]$	$= [m]$
(4) and (6)	$[C]$	$= [\Omega]$	$= \left[\dfrac{e}{T}\right]$	$= \left[\dfrac{L^2M}{mT^2}\right].$
(7)	$[\mathfrak{D}]$		$= \left[\dfrac{e}{L^2}\right]$	$= \left[\dfrac{M}{mT}\right].$
(8)	$[\mathfrak{E}]$		$= \left[\dfrac{LM}{eT^2}\right]$	$= \left[\dfrac{m}{LT}\right].$
(9)	$[\mathfrak{B}]$		$= \left[\dfrac{M}{eT}\right]$	$= \left[\dfrac{m}{L^2}\right].$
(10)	$[\mathfrak{H}]$		$= \left[\dfrac{e}{LT}\right]$	$= \left[\dfrac{LM}{mT^2}\right].$
(11)	$[\mathfrak{C}]$		$= \left[\dfrac{e}{L^2T}\right]$	$= \left[\dfrac{M}{mT^2}\right].$
(12)	$[\mathfrak{A}]$		$= \left[\dfrac{LM}{eT}\right]$	$= \left[\dfrac{m}{L}\right].$

623. *we take either e or m as an independent unit*: In the table that follows, $[e]$ is taken as an independent unit in the next to last column; $[m]$ is taken as an independent unit in the last column.

624.] The relations of the first ten of these quantities may be exhibited by means of the following arrangement:

$$e, \quad \mathfrak{D}, \quad \mathfrak{H}, \quad C \text{ and } \Omega. \quad \Big| \quad E, \quad \mathfrak{E}, \quad \mathfrak{B}, \quad m \text{ and } p.$$
$$m \text{ and } p, \quad \mathfrak{B}, \quad \mathfrak{E}, \quad E. \quad \Big| \quad C \text{ and } \Omega, \quad \mathfrak{H}, \quad \mathfrak{D}, \quad e.$$

624.1 The quantities in the first line are derived from e by the same operations as the corresponding quantities in the second line are derived from m. It will be seen that the order of the quantities in the first line is exactly the reverse of the order in the second line. The first four of each line have the first symbol in the numerator. The second four in each line have it in the denominator.

624.2 All the relations given above are true whatever system of units we adopt.

625.] The only systems of any scientific value are the electrostatic and the electromagnetic systems. The electrostatic system is founded on the definition of the unit of electricity, Arts. 41, 42, and may be deduced from the equation

$$\mathfrak{E} = \frac{e}{L^2},$$

which expresses that the resultant electric intensity \mathfrak{E} at any point, due to the action of a quantity of electricity e at a distance L, is found by dividing e by L^2. Substituting in the equations of dimensions (1) and (8), we find

$$\left[\frac{LM}{eT^2}\right] = \left[\frac{e}{L^2}\right], \qquad \left[\frac{m}{LT}\right] = \left[\frac{M}{mT}\right],$$

whence
$$[e] = \left[L^{\frac{3}{2}}M^{\frac{1}{2}}T^{-1}\right], \qquad [m] = \left[L^{\frac{1}{2}}M^{\frac{1}{2}}\right]$$

in the electrostatic system.

625.1 The electromagnetic system is founded on a precisely similar definition of the unit of strength of a magnetic pole, Art. 374, leading to the equation

$$\mathfrak{H} = \frac{m}{L^2},$$

whence
$$\left[\frac{e}{LT}\right] = \left[\frac{M}{eT}\right], \qquad \left[\frac{LM}{mT^2}\right] = \left[\frac{m}{L^2}\right],$$

and
$$[e] = \left[L^{\frac{1}{2}}M^{\frac{1}{2}}\right], \qquad [m] = \left[L^{\frac{3}{2}}M^{\frac{1}{2}}T^{-1}\right]$$

624.2 *All the relations given above are true whatever system of units we adopt*: They are therefore "absolute" in Kelvin's sense, as discussed in the editor's introduction to this chapter.

in the electromagnetic system. From these results we find the dimensions of the other quantities.

626.] *Table of Dimensions*

	Symbol.	Dimensions in Electrostatic System.	Electromagnetic System.
Quantity of electricity	e	$[L^{3/2}M^{1/2}T^{-1}]$	$[L^{1/2}M^{1/2}]$
Line-integral of electro-motive intensity	E	$[L^{1/2}M^{1/2}T^{-1}]$	$[L^{3/2}M^{1/2}T^{-2}]$
Quantity of magnetism / Electrokinetic momentum of a circuit	$\begin{Bmatrix} m \\ p \end{Bmatrix}$	$[L^{1/2}M^{1/2}]$	$[L^{3/2}M^{1/2}T^{-1}]$
Electric current / Magnetic potential	$\begin{Bmatrix} C \\ \Omega \end{Bmatrix}$	$[L^{3/2}M^{1/2}T^{-2}]$	$[L^{1/2}M^{1/2}T^{-1}]$
Electric displacement / Surface-density	\mathfrak{D}	$[L^{-1/2}M^{1/2}T^{-1}]$	$[L^{-3/2}M^{1/2}]$
Electromotive intensity	\mathfrak{E}	$[L^{-1/2}M^{1/2}T^{-1}]$	$[L^{1/2}M^{1/2}T^{-2}]$
Magnetic induction	\mathfrak{B}	$[L^{-3/2}M^{1/2}]$	$[L^{-1/2}M^{1/2}T^{-1}]$
Magnetic force	\mathfrak{H}	$[L^{1/2}M^{1/2}T^{-2}]$	$[L^{-1/2}M^{1/2}T^{-1}]$
Strength of current at a point	\mathfrak{C}	$[L^{1/2}M^{1/2}T^{-2}]$	$[L^{-3/2}M^{1/2}T^{-1}]$
Vector potential	\mathfrak{A}	$[L^{-1/2}M^{1/2}]$	$[L^{1/2}M^{1/2}T^{-1}]$

627.] We have already considered the products of the pairs of these quantities in the order in which they stand. Their ratios are in certain cases of scientific importance. Thus

	Symbol.	Electrostatic System.	Electromagnetic System.
$\left[\dfrac{e}{E}\right]$ = capacity of an accumulator	q	$[L]$	$\left[\dfrac{T^2}{L}\right]$.
$\left[\dfrac{p}{C}\right]$ = { coefficient of self-induction of a circuit, or electro-magnetic capacity }	L	$\left[\dfrac{T^2}{L}\right]$	$[L]$.
$\left[\dfrac{\mathfrak{D}}{\mathfrak{E}}\right]$ = { specific inductive capacity of dielectric }	K	$[0]$	$\left[\dfrac{T^2}{L^2}\right]$.

$$\left[\frac{\mathfrak{B}}{\mathfrak{H}}\right] = \text{magnetic inductive capacity} \dots \dots \mu \qquad \left[\frac{T^2}{L^2}\right] \qquad [0].$$

$$\left[\frac{E}{C}\right] = \text{resistance of a conductor} \dots \dots R \qquad \left[\frac{T}{L}\right] \qquad \left[\frac{L}{T}\right].$$

$$\left[\frac{\mathfrak{E}}{\mathfrak{C}}\right] = \left\{\begin{array}{c} \text{specific resistance of a} \\ \text{substance} \end{array}\right\} \dots \dots r \qquad [T] \qquad \left[\frac{L^2}{T}\right].$$

628.] If the units of length, mass, and time are the same in the two systems, the number of electrostatic units of electricity contained in one electromagnetic unit is numerically equal to a certain velocity, the absolute value of which does not depend on the magnitude of the fundamental units employed. This velocity is an important physical quantity, which we shall denote by the symbol v.

Number of Electrostatic Units in one Electromagnetic Unit.

628.1 For e, C, Ω, \mathfrak{D}, \mathfrak{H}, \mathfrak{C}, v.

628.2 For m, p, E, \mathfrak{B}, \mathfrak{E}, \mathfrak{A}, $1/v$.

628.3 For electrostatic capacity, dielectric inductive capacity, and conductivity, v^2.

628.4 For electromagnetic capacity, magnetic inductive capacity, and resistance, $1/v^2$.

628.5 Several methods of determining the velocity v will be given in Arts. 768–780.

628.6 In the electrostatic system the specific dielectric inductive capacity of air is assumed equal to unity. This quantity is therefore represented by $1/v^2$ in the electromagnetic system.

628.7 In the electromagnetic system the specific magnetic inductive capacity of air is assumed equal to unity. This quantity is therefore represented by $1/v^2$ in the electrostatic system.

Practical System of Electric Units.

629.] Of the two systems of units, the electromagnetic is of the greater use to those practical electricians who are occupied with electromagnetic telegraphs. If, however, the units of length, time, and mass

628. *This velocity is an important physical quantity, which we shall denote by the symbol v*: Here, Maxwell can call v a "velocity" only in the sense that it has the *dimensions* of velocity—length divided by time. But in Chapter XX v will turn out to be a velocity in the physical sense, specifically the velocity of a traveling electromagnetic wave. When measured, furthermore, its value will prove to be suggestively close to the measured velocity of light.

are those commonly used in other scientific work, such as the metre or the centimetre, the second, and the gramme, the units of resistance and of electromotive force will be so small that to express the quantities occurring in practice enormous numbers must be used, and the units of quantity and capacity will be so large that only exceedingly small fractions of them can ever occur in practice. Practical electricians have therefore adopted a set of electrical units deduced by the electromagnetic system from a large unit of length and a small unit of mass.

629.1 The unit of length used for this purpose is ten million of metres, or approximately the length of a quadrant of a meridian of the earth.

629.2 The unit of time is, as before, one second.

629.3 The unit of mass is 10^{-11} grammes, or one hundred millionth part of a milligramme.

629.4 The electrical units derived from these fundamental units have been named after eminent electrical discoverers. Thus the practical unit of resistance is called the Ohm, and is represented by the resistance-coil issued by the British Association, and described in Art. 340. It is expressed in the electromagnetic system by a velocity of 10,000,000 metres per second.

629.5 The practical unit of electromotive force is called the Volt, and is not very different from that of a Daniell's cell. Mr. Latimer Clark has recently invented a very constant cell, whose electromotive force is almost exactly 1.454 Volts.

629.6 The practical unit of capacity is called the Farad. The quantity of electricity which flows through one Ohm under the electromotive force of one Volt during one second, is equal to the charge produced in a condenser whose capacity is one Farad by an electromotive force of one Volt.

629.7 The use of these names is found to be more convenient in practice than the constant repetition of the words 'electromagnetic units,' with the additional statement of the particular fundamental units on which they are founded.

629.8 When very large quantities are to be measured, a large unit is formed by multiplying the original unit by one million, and placing before its name the prefix *mega*.

629.9 In like manner by prefixing *micro* a small unit is formed, one millionth of the original unit.

629.4 *the resisstance-coil issued by the British Association, and described in Art. 340*: Article 340 is not included in the present selections, but I describe the British Association resistance coil in the introduction to Part IV Chapter XIX below.

629.10 The following table gives the values of these practical units in the different systems which have been at various times adopted.

Fundamental Units	Practical System	B.A. Report, 1863	Thomson.	Weber.
Length, Time, Mass.	Earth's quadrant, Second, 10^{-11} Gramme.	Metre, Second, Gramme.	Centimetre, Second, Gramme.	Millimetre, Second, Milligramme.
Resistance	Ohm	10^7	10^9	10^{10}
Electromotive force	Volt	10^5	10^8	10^{11}
Capacity	Farad	10^{-7}	10^{-9}	10^{-10}
Quantity	Farad (charged to a Volt.)	10^{-2}	10^{-1}	10

629.10 *The following table gives the values of these practical units*: The practical system of units was subsequently expanded to include a unit of *current*, the ampere. The practical unit of *quantity of electricity* is now called the coulomb, defined as the quantity of electricity delivered in one second by a current of one ampere. So defined, it continues to equal the quantity of electricity held by a capacitance of one farad, charged to a potential of one volt.

CHAPTER XI.

ON ENERGY AND STRESS IN THE ELECTROMAGNETIC FIELD.

Electrostatic Energy.

630.] THE energy of the system may be divided into the Potential Energy and the Kinetic Energy.

630.1 The potential energy due to electrification has been already considered in Art. 85. It may be written

$$W = \tfrac{1}{2}\Sigma(e\Psi), \tag{1}$$

where e is the charge of electricity at a place where the electric potential is Ψ, and the summation is to be extended to every place where there is electrification.

630.2 If f, g, h are the components of the electric displacement, the quantity of electricity in the element of volume $dx\,dy\,dz$ is

$$e = \left(\frac{df}{dx} + \frac{dg}{dy} + \frac{dh}{dz}\right) dx\,dy\,dz, \tag{2}$$

and

$$W = \frac{1}{2}\iiint \left(\frac{df}{dx} + \frac{dg}{dy} + \frac{dh}{dz}\right) \Psi\,dx\,dy\,dz, \tag{3}$$

where the integration is to be extended throughout all space.

630.1 *The potential energy ... has been already considered in Art. 85*: Art. 85 is not included in these selections, but Maxwell gave the equivalent of equation (1) in (84.12), there writing the electric potential as V rather than Ψ.

the summation is to be extended to every place where there is electrification: Rhetorically, this implies that places where electricity resides are the *only* places that contribute to the energy of the system.

630.2 *the quantity of electricity in the element of volume dx, dy, dz*: Equation (2) derives from equation (J) in (612.), since the quotient of e by $dx\,dy\,dz$ yields ρ, the volume-density of electricity. Equation (3) follows by substituting for e in equation (1), and converting the sum to an integral.

the integration is to be extended throughout all space: In contrast to equation (1), this implies that *every portion of space* contributes to the total energy of the system. Maxwell will call attention to this contrast—it is a rhetorical contrast, for the equations are mathematically equivalent—in (631.2).

631.] Integrating this expression by parts, and remembering that when the distance, r, from a given point of a finite electrified system becomes infinite, the potential Ψ becomes an infinitely small quantity of the order r^{-1}, and that f, g, h become infinitely small quantities of the order r^{-2}, the expression is reduced to

$$W = -\frac{1}{2} \iiint \left(f\frac{d\Psi}{dx} + g\frac{d\Psi}{dy} + h\frac{d\Psi}{dz} \right) dx\,dy\,dz, \qquad (4)$$

where the integration is to be extended throughout all space.

631.1 If we now write P, Q, R for the components of the electromotive intensity, instead of $-d\Psi/dx$, $-d\Psi/dy$ and $-d\Psi/dz$, we find

631. *Integrating this expression by parts*: To show the pattern, express the right-hand side of equation (3) as the sum of three integrals, the first of which being

$\iint \left[\int \frac{1}{2}\frac{df}{dx}\Psi\,dx \right] dy\,dz$. Let $\frac{1}{2}\Psi$ be v, and $\frac{df}{dx}dx = df$ be du. Integrating "by

parts" we have $\int v\,du = uv - \int u\,dv$, thus $\int \frac{1}{2}\frac{df}{dx}\Psi\,dx = \frac{1}{2}f\Psi - \frac{1}{2}\int f\,d\Psi$. The

integral on the right hand side, rewritten as $-\frac{1}{2}\int f\frac{d\Psi}{dx}\,dx$, corresponds to the

first term of the integrand in equation (4), while the product, $\frac{1}{2}f\Psi$, is to be evaluated at zero distance and infinite distance from a specified point.

when the distance, r ... becomes infinite, the potential Ψ becomes an infinitely small quantity of the order r^{-1}: Since the system is finite, a place infinitely distant from any point in it is also infinitely distant from the totality of its electricity, e; thus the potential at that place will be e/r or er^{-1} (73.), where r increases indefinitely.

and ... f, g, h become infinitely small quantities of the order r^{-2}: Similarly, the electromotive force R at an infinitely distant point will be e/r^2 or er^{-2} (60.7, *comment*), where r increases indefinitely. But f, g, h are the components of displacement, which is in turn proportional to R (68.6) and therefore also to r^{-2}. Thus each component f, g, h vanishes as r increases indefinitely.

the expression is reduced to ...: Returning to the product term $\frac{1}{2}f\Psi$ previously obtained, the preceding remarks explain why it vanishes at infinite distance. We need to show that it also vanishes at zero distance.

To that end, suppose the volume-density of electricity in the region surrounding the specified point is ρ. Then a sphere of radius r about that point will enclose electricity $e' = \frac{4}{3}\pi r^3\rho$. The potential at the surface of the sphere will be the same as if that electricity were concentrated at the sphere's center (60.7), that is, e'/r or $\frac{4}{3}\pi r^2\rho$. Clearly the potential Ψ vanishes as r approaches 0.

Similarly, the displacement will be $\frac{K}{4\pi}e'/r^2$ or $\frac{K}{3}r\rho$. Then each of its components f, g, h vanishes as r shrinks to zero. Proceeding in the same way for g and h, we shall have all three terms on the right-hand side of equation (4).

$$W = \frac{1}{2} \iiint \left(Pf + Qg + Rh \right) dx\, dy\, dz. *\tag{5}$$

631.2 Hence, the electrostatic energy of the whole field will be the same if we suppose that it resides in every part of the field where electrical force and electrical displacement occur, instead of being confined to the places where free electricity is found.

631.3 The energy in unit of volume is half the product of the electromotive force and the electric displacement, multiplied by the cosine of the angle which these vectors include.

631.4 In Quaternion language it is $-\frac{1}{2}S.\mathfrak{E}\mathfrak{D}$.

Magnetic Energy.

632.] We may treat the energy due to magnetization in a way similar to that pursued in the case of electrification, Art. 85. If A, B, C are the components of magnetization and α, β, γ the components of magnetic force, the potential energy of the system of magnets is then, by Art. 389,

* {This expression for the electrostatic energy was deduced in the first volume on the assumption that the electrostatic force could be derived from a potential function. This proof will not hold when part of the electromotive intensity is due to electromagnetic induction. If however we take the view that this part of the energy arises from the polarized state of the dielectric and is per unit volume $\frac{1}{8\pi K}(f^2 + g^2 + h^2)$, the potential energy will then only depend on the polarization of the dielectric no matter how it is produced. Thus the energy will, since

$$\frac{f}{4\pi K} = P, \quad \frac{g}{4\pi K} = Q, \quad \frac{h}{4\pi K} = R,$$

be equal to $\frac{1}{2}(Pf + Qg + Rh)$ per unit volume.}

631.1, footnote. J. J. Thomson's editorial note, useful though it is, tends to distract from Maxwell's main point, which is that equation (5) represents the *same result* as would be obtained by action-at-a-distance reasoning; compare (84.12, *comment*). Maxwell will make the point explicitly in the following paragraph.

631.3 *The energy in unit volume is...*: In equation (5), the energy is one half of $(Pf + Qg + Rh)$ integrated over *volume*; so half of $(Pf + Qg + Rh)$ is the energy per unit volume.

...half the product ... multiplied by the cosine of the angle which those vectors include: since P, Q, R are the components of electromotive intensity \mathfrak{E}, while f, g, h are those of displacement \mathfrak{D}, we recognize $(Pf + Qg + Rh)$ as equal to the dot product $\mathfrak{D}\bullet\mathfrak{E}$, which is also the product of the magnitudes of \mathfrak{D} and \mathfrak{E} times the cosine of the angle between them (page 50, above).

631.4 *In quaternion language*: Recall that the scalar part of the product of quaternions is the negative of the dot product of the corresponding vectors (25.4, *comment*).

$$-\frac{1}{2}\iiint \left(A\alpha + B\beta + C\gamma \right) dx\, dy\, dz, \tag{6}$$

the integration being extended over the space occupied by magnetized matter. This part of the energy, however, will be included in the kinetic energy in the form in which we shall presently obtain it.

* * *

Electrokinetic Energy.

634.] We have already, in Art. 578, expressed the kinetic energy of a system of currents in the form

$$W = \tfrac{1}{2}\Sigma(pi), \tag{12}$$

where p is the electromagnetic momentum of a circuit, and i is the strength of the current flowing round it, and the summation extends to all the circuits.

634.1 But we have proved, in Art. 590, that p may be expressed as a line-integral of the form

$$p = \int \left(F\frac{dx}{ds} + G\frac{dy}{ds} + H\frac{dz}{ds} \right) ds , \tag{13}$$

where F, G, H are the components of the electromagnetic momentum, \mathfrak{A}, at the point (x, y, z), and the integration is to be extended round the closed circuit s. We therefore find

$$T = \tfrac{1}{2}\sum i \int \left(F\frac{dx}{ds} + G\frac{dy}{ds} + H\frac{dz}{ds} \right) ds. \tag{14}$$

634.2 If u, v, w are the components of the density of the current at any point of the conducting circuit, and if S is the transverse section of the circuit, then we may write

$$i\frac{dx}{ds} = uS, \quad i\frac{dy}{ds} = vS, \quad i\frac{dz}{ds} = wS, \tag{15}$$

and we may also write the volume

$$S\, ds = dx\, dy\, dz,$$

634.1 *Equation* (13): See (590.3), equation (6).

634.2 *the transverse section of the circuit*: that is, the cross-section of the conductor whose path makes up the circuit.

Equations (15): Think, for example, of the first member as having been written in the form $i\, dx = uS ds$; since ds is an element of length of the conductor, each side expresses an element of the current in the x-direction.

and we now find

$$T = \tfrac{1}{2} \iiint (Fu + Gv + Hw) \, dx \, dy \, dz, \qquad (16)$$

where the integration is to be extended to every part of space where there are electric currents.

635.] Let us now substitute for u, v, w their values as given by the equations of electric currents (E), Art. 607, in terms of the components α, β, γ of the magnetic force. We then have

$$T = \frac{1}{8\pi} \iiint \left\{ F\left(\frac{d\gamma}{dy} - \frac{d\beta}{dz}\right) + G\left(\frac{d\alpha}{dz} - \frac{d\gamma}{dx}\right) + H\left(\frac{d\beta}{dx} - \frac{d\alpha}{dy}\right) \right\} dx \, dy \, dz, \quad (17)$$

where the integration is extended over a portion of space including all the currents.

635.1 If we integrate this by parts, and remember that, at a great distance r from the system, α, β, and γ are of the order of magnitude r^{-3}, {and that at a surface separating two media, F, G, H, and the tangential magnetic force are continuous,} we find that when the integration is extended throughout all space, the expression is reduced to

$$T = \frac{1}{8\pi} \iiint \left\{ \alpha\left(\frac{dH}{dy} - \frac{dG}{dz}\right) + \beta\left(\frac{dF}{dz} - \frac{dH}{dx}\right) + \gamma\left(\frac{dG}{dx} - \frac{dF}{dy}\right) \right\} dx \, dy \, dz. \quad (18)$$

635.2 By the equations (A), Art. 591, of magnetic induction, we may substitute for the quantities in small brackets the components of magnetic induction a, b, c, so that the kinetic energy may be written

$$T = \frac{1}{8\pi} \iiint (a\alpha + b\beta + c\gamma) \, dx \, dy \, dz, \qquad (19)$$

where the integration is to be extended throughout every part of space in which the magnetic force and magnetic induction have values differing from zero.

635.1 *integrate this by parts*: As in (631., *comment*), examine a single term. In equation (17), $F\frac{d\gamma}{dy} dx \, dy \, dz = [F \, d\gamma] \, dx \, dz$, while in equation (18), $-\gamma\frac{dF}{dy} dx \, dy \, dz = -[\gamma \, dF] \, dx \, dz$. When integrating by parts, $\left[\int F \, d\gamma\right] dx \, dz = \left[F\gamma - \int \gamma \, dF\right] dx \, dz$, that is, a term in (17) becomes a term in (18) plus a term $F\gamma$, which vanishes.

635.2 *the integration is to be extended throughout every part of space in which the magnetic force and magnetic induction have values differing from zero*: Maxwell has transformed equation (17), which represented the *currents* as exclusively contributing to the energy, to an equation which represents *every part of space where magnetic force and magnetic induction exist* as contributing to the energy of the system. He will note this explicitly in (636.).

635.3 The quantity within brackets in this expression is the product of the magnetic induction into the resolved part of the magnetic force in its own direction.

635.4 In the language of quaternions this may be written more simply,

$$-S.\mathfrak{B}\mathfrak{H},$$

where \mathfrak{B} is the magnetic induction, whose components are a, b, c, and \mathfrak{H} is the magnetic force, whose components are α, β, γ.

636.] The electrokinetic energy of the system may therefore be expressed either as an integral to be taken where there are electric currents, or as an integral to be taken over every part of the field in which magnetic force exists. The first integral, however, is the natural expression of the theory which supposes the currents to act upon each other directly at a distance, while the second is appropriate to the theory which endeavours to explain the action between the currents by means of some intermediate action in the space between them. As in this treatise we have adopted the latter method of investigation, we naturally adopt the second expression as giving the most significant form to the kinetic energy.

636.1 According to our hypothesis, we assume the kinetic energy to exist wherever there is magnetic force, that is, in general, in every part of the field. The amount of this energy per unit of volume is $-\frac{1}{8\pi}S.\mathfrak{B}\mathfrak{H}$, and this energy exists in the form of some kind of motion of the matter in every portion of space.

636.2 When we come to consider Faraday's discovery of the effect of magnetism on polarized light, we shall point out reasons for believing that wherever there are lines of magnetic force, there is a rotatory motion of matter round those lines. See Art. 821.

635.3 *the product of the magnetic induction into the resolved part of the magnetic force in its own direction*: Compare (631.3, *comment*); here a, b, c, the components of magnetic induction \mathfrak{B}, and α, β, γ, the components of magnetic force \mathfrak{H}, have the same relation as P, Q, R and f, g, h did in the earlier paragraph. Thus the quantity in parentheses is the product of the magnitudes of \mathfrak{B} and \mathfrak{H}, multiplied by the cosine of the angle between them. But this *is* the product of \mathfrak{B} with the resolved part of \mathfrak{H} (14.1, *comment*). Alternatively, it is the dot product $\mathfrak{B}\cdot\mathfrak{H}$ (page 50).

635.4 *In the language of quaternions*: As noted in (25.4 and 631.4, *comments*), the scalar part of the product of quaternions is the negative of the dot product of the corresponding vectors.

636.1 *the amount of this energy per unit of volume*: Since the total energy is $-\frac{1}{8\pi}$ times the *volume*-integral of $S.\mathfrak{B}\mathfrak{H}$, $-\frac{1}{8\pi}S.\mathfrak{B}\mathfrak{H}$ itself is the energy per unit volume. Compare (631.3, *comment*).

636.2 *Faraday's discovery of the effect of magnetism on polarized light*: Maxwell's chapter on this topic is not included in these selections.

Magnetic and Electrokinetic Energy compared.

637.] We found in Art. 423 that the mutual potential energy of two magnetic shells, of strengths φ and φ', and bounded by the closed curves s and s' respectively, is

$$-\varphi\varphi' \iint \frac{\cos\varepsilon}{r}\, ds\, ds',$$

where ε is the angle between the directions of ds and ds', and r is the distance between them.

637.1 We also found in Art. 521 that the mutual energy of two circuits s and s', in which currents i and i' flow, is

$$-i\,i' \iint \frac{\cos\varepsilon}{r}\, ds\, ds'.$$

637.2 If i, i' are equal to φ, φ', respectively, the mechanical action between the magnetic shells is equal to that between the corresponding electric circuits, and in the same direction. In the case of the magnetic shells the force tends to diminish their mutual potential energy, in the case of the circuits it tends to increase their mutual energy, because this energy is kinetic.

637.3 It is impossible, by any arrangement of magnetized matter, to produce a system corresponding in all respects to an electric circuit, for the potential of the magnetic system is single valued at every point of space, whereas that of the electric system is many-valued.

637.4 But it is always possible, by a proper arrangement of infinitely small electric circuits, to produce a system corresponding in all respects to any magnetic system, provided the line of integration which we follow in calculating the potential is prevented from passing through any of these small circuits. This will be more fully explained in Art. 833.

637.5 The action of magnets at a distance is perfectly identical with that of electric currents. We therefore endeavour to trace both to the same cause, and since we cannot explain electric currents by means of magnets, we must adopt the other alternative, and explain magnets by means of molecular electric currents.

637.2 *the force tends to diminish their mutual potential energy*...: We saw in (16.4) that when a force is derivable from a potential, it acts in the direction of diminishing potential. Maxwell gave a contrasting example of increasing kinetic energy of currents in 580.4.

637.3 *the magnetic system is single-valued ... the electric system is many-valued*: Here, a "magnetic" system is one founded on supposed magnetic matter; Maxwell's discussion of the scalar potential in Art. 383 shows it to be single-valued. An "electric" system is founded on currents; Maxwell showed the potential about a straight wire to be many-valued in Art. 480.

638.] In our investigation of magnetic phenomena, in Part III of this treatise, we made no attempt to account for magnetic action at a distance, but treated this action as a fundamental fact of experience. We therefore assumed that the energy of a magnetic system is potential energy, and that this energy is *diminished* when the parts of the system yield to the magnetic forces which act on them.

638.1 If, however, we regard magnets as deriving their properties from electric currents circulating within their molecules, their energy is kinetic, and the force between them is such that it tends to move them in a direction such that if the strengths of the currents were maintained constant the kinetic energy would *increase*.

638.2 This mode of explaining magnetism requires us also to abandon the method followed in Part III, in which we regarded the magnet as a continuous and homogeneous body, the minutest part of which has magnetic properties of the same kind as the whole.

638.3 We must now regard a magnet as containing a finite, though very great, number of electric circuits, so that it has essentially a molecular, as distinguished from a continuous structure.

638.4 If we suppose our mathematical machinery to be so coarse that our line of integration cannot thread a molecular circuit, and that an immense number of magnetic molecules are contained in our element of volume, we shall still arrive at results similar to those of Part III, but if we suppose our machinery of a finer order, and capable of investigating all that goes on in the interior of the molecules, we must give up the old theory of magnetism, and adopt that of Ampère, which admits of no magnets except those which consist of electric currents.

638.5 We must also regard both magnetic and electromagnetic energy as kinetic energy, and we must attribute to it the proper sign, as given in Art. 635.

638.6 In what follows, though we may occasionally, as in Art. 639, &c., attempt to carry out the old theory of magnetism, we shall find that we obtain a perfectly consistent system only when we abandon that theory and adopt Ampère's theory of molecular currents, as in Art. 644.

638.1 *the strengths of the currents ... maintained constant*: Such was the case in the example discussed in (580.4).

638.3 *We must now regard a magnet as containing ... electric circuits*: Contrast Maxwell's affirmative tone here with that of (615.3), where he merely acknowledged Ampère's molecular theory. His language will rise almost to the level of endorsement in (638.6) below. But is Maxwell truly accepting the new theory, or is he only accepting the necessity of abandoning the old one? I continue to find a fundamental disharmony between the molecular current hypothesis and the Lagrangian vision that seems to lie at the heart of his project (615.3, *comment*).

638.7 The energy of the field therefore consists of two parts only, the electrostatic or potential energy

$$W = \frac{1}{2} \iiint \left(Pf + Qg + Rh \right) dx\, dy\, dz,$$

and the electromagnetic or kinetic energy

$$T = \frac{1}{8\pi} \iiint \left(a\alpha + b\beta + c\gamma \right) dx\, dy\, dz.$$

* * *

Editor's Introduction to
Chapter XIX

The electromagnetic and the electrostatic unit

We are accustomed to regarding the ratios between units of different measurement systems as mere conversion factors. For example, the ratio 2.54 between the inch and the centimeter discloses nothing about nature or about any entity; it simply mediates between two conventions—those of the English and metric systems, respectively. It may seem strange, then, to think of any comparison of units as incorporating actual knowledge of nature. But Maxwell will begin the present chapter with a clear explanation of how the ratio between the electromagnetic and electrostatic units of electricity has that very role. It is, he says, "a natural quantity of definite magnitude, and the measurement of this quantity is one of the most important researches in electricity."

In the course of the chapter, Maxwell discusses six different methods of measuring this ratio; our selections include two of them.

The method of Weber and Kohlrausch

Maxwell will outline this procedure in Article 771. It consists of discharging a Leyden jar through a ballistic galvanometer. The jar's initial charge is calculated in electrostatic units on the basis of its capacitance, which in turn is measured by comparison with an isolated sphere (whose capacitance can be calculated geometrically). The galvanometer indicates that same quantity of electricity, now thrown into the condition of a current and measured in electromagnetic units. The quotient of the first measurement by the second is the ratio of the electromagnetic to the electrostatic unit of electricity.

Unfortunately, efforts to determine the capacity of a Leyden jar are undermined by what Maxwell earlier (Art. 53) called *electric absorption*. As we learned, "absorption" reflects a peculiar behavior observed in solid dielectrics, but not in gases and vacua. If only one could eliminate the glass from the jar—leaving the metal coatings separated by air alone—that error would disappear.

Such an "air capacitor," however, would have had so little capacitance, it could not have held sufficient electrical charge to register on the galvanometers of Weber and Kohlrausch's time. Today we enjoy vastly more sensitive instruments for measuring current, which make the method a practical one, even in conjunction with an air capacitor.*

The method of intermittent current

Maxwell discusses this method in Article 775. Procedurally, it is nearly identical to that of Weber and Kohlrausch, only instead of discharging the capacitor once, we alternately charge and discharge the capacitor continually. Thus the galvanometer, instead of receiving a single discharge, experiences the pulsating equivalent of a steady current. (It is not, therefore, employed in ballistic mode when using this method.)

A second difference lies in the quantity that is directly measured: instead of comparing two measurements of the same quantity of electricity, the method of intermittent current compares two determinations of the same capacitance. The two differences may seem unrelated at first; but we begin to appreciate a relation between them when we realize that, since the capacitor is being discharged many times per second, even a small capacitor can produce measureable currents. We can therefore employ an air capacitor, whose electrostatic measure may be determined by direct geometric calculation.

As we will see in the present chapter, the method depends on the accurate measurement of a *resistance*. Unhappily, the standard resistance assumed by Maxwell and many of his contemporaries would turn out to be a source of unanticipated error and protracted drama—the topic of the next section. Nevertheless, we will find that the method of intermittent current at least mitigates, though it does not escape, the effect of erroneous resistance measurement.

The British Association Unit

Between 1862 and 1884 the British Association for the Advancement of Science (commonly referred to as the British Association, or B. A.), along with other international bodies, put forth a series of reports proposing standard units of measurement, particularly electrical units, for adoption. While units for charge, current, and electromotive force were

* For reasons I will discuss in the editor's introduction to Chapter XX, we no longer regard the ratio between electrostatic and electromagnetic units as a quantity to be measured; it is fixed instead by definition. Nevertheless, the experimental methods reviewed by Maxwell in the present chapter are still performed, not as research experiments but as teaching exercises. As such, they remain both challenging and rewarding.

largely in place by 1864, determination of a unit of electrical resistance proved to be unexpectedly elusive.

As Maxwell will explain in Article 768 below, the quantity called *electrical resistance* has dimensions of length divided by time—identical to the dimensions of *velocity*. The B. A. report of 1869 proposed a unit of resistance to be called the Ohm, which was to be equivalent to 10,000 kilometers per second.*

Expressed as a velocity, however, such a unit has only a theoretical existence.† Actual measurements require a material embodiment of that theoretical quantity, a physical object that can be wired into circuits and used as a standard with which to calibrate practical instruments. Various devices were already being used as physical standards on an *ad hoc* basis. Wheatstone in 1843 had taken for his standard resistance a one-foot length of copper wire weighing 100 grains. Since 1860 Werner Siemens had been employing the equally arbitrary standard of a one-meter column of mercury, and this had gained considerable acceptance among practical electricians and in telegraphy.

In 1865 the British Association produced a coil of platinum-silver alloy wire for use as a physical standard; it was known as the British Association Unit, or alternatively, the B. A. Ohm. Thirteen such coils were distributed. The one photographed here, mounted in its protective case, was purchased by Michael Faraday on behalf of the Royal Institution.

Doubts soon emerged, however, about the actual value of the B. A. Ohm. Joule's measurements yielded a value about 1.3% less than 10,000 kilometers per second; Kohlrausch found it to be a little more. Joule's figures prevailed, and a series of redefinitions of the British Association Unit ensued. In 1875 the Ohm was redefined as equivalent to a mercury column 104.8 cm high; in 1884 the "Legal Ohm" was defined as the resistance of a mercury column 106 cm high; the International Ohm, a mercury column 106.3 cm high, was adopted in 1893. All these were attempts to specify a physical realization of a resistance of 10,000 kilometers per second.

* Sir William Thomson, "Electrical Units of Measurement," in *Popular Lectures and Addresses*, vol. I, p. 91. London (1891).

† As one delegate to the International Conference for the Determination of Electric Units (Paris, October 1882) complained, "What is the ohm? Who can see an ohm? Who can show what an ohm is? Who can measure the resistance of any conductor for us, in this absolute measure...?" *ibid.*, p. 93.

In Chapter XX Maxwell will report the result of his own and others' comparisons of the electric units. All of them relied on the 1865 B.A. Ohm, and were therefore in error to varying degrees. The errors were acknowledged in the second edition of the *Treatise*, and corrected values were supplied in the third.

A bridge between two worlds

As we have seen, the electrostatic system is based on the electric field associated with electricity at rest, the electromagnetic system with the magnetic field associated with a current—that is, with electricity in motion (770.). Thus the two great classes of electrical measurement can be seen as reflecting the cosmos from the viewpoints of rest and motion, respectively. But unless rest and motion can be harmonized, sciences beyond mechanics will be fractured, at least rhetorically, by presenting two competing and equally inclusive representations of the universe; one founded on rest, the other on motion.

Modern thinkers have scant occasion to reflect on the relationship between rest and motion. Since Galileo's mathematization of velocity, we are accustomed to treating that qualitative dichotomy as nothing but quantitative variation; mathematically, rest is merely a degree (the zero degree) of motion. To a more ancient philosophical tradition, however, the problem of defining a reality in which both rest and motion had valid status was one of the most challenging of philosophic problems:

> And it will be in the same case in relation to the other changes, when it passes from being into cessation of being, or from not-being into becoming—then it passes between certain states of motion and rest, and neither is nor is not, nor becomes nor is destroyed. (Plato, *Parmenides*, 157; tr. Jowett; compare *Sophist*, 255a.)

Philosophically, rest and motion have nothing in common.

Comparing electrostatic and electromagnetic units, then, represents an endeavor to join two worlds. Several imagined experiments, as well as some real ones, bridge these worlds. In (770.1) Maxwell will conceive an experiment designed to show whether a charged body, when set in motion, becomes equivalent to a current—a "convection current," so called. This is a fundamental question, touching on the nature of the electric current. It is also a disturbing question, for if the answer is *yes*, it threatens to rehabilitate the old images of current as material in flow.

In ruminating on the design of such an experiment he suggests that one might look for magnetic effects from an electrified disk spinning at high speed. For Maxwell, this idea may have been sheer exuberance; but Henry A. Rowland performed the experiment in 1876, pretty much in

the way Maxwell had envisioned. Maxwell did not live to include Rowland's important confirmatory discovery in the revisions he planned for a second edition of the *Treatise*, but he did pen a celebratory poem:

> The mounted disk of ebonite
> Has whirled before, nor whirled in vain;
> Rowland of Troy,* that doughty knight,
> Convection currents did obtain
> In such a disk, of power to wheedle
> From its loved North the subtle needle.
> 'Twas when Sir Rowland, as a stage
> From Troy to Baltimore,† took rest
> In Berlin,‡ there old Archimage §
> Armed him to follow up this quest;
> Right glad to find himself possessor
> Of the irrepressible Professor....
>
> JAMES CLERK MAXWELL, 1877‖

Aided both by its subject matter, and by its placement within the *Treatise*, the present chapter gains dramatic power as Maxwell welds together the realms of rest and motion through experiments of thought and deed. It seems almost the stuff of Socratic comedy to know that the philosophical problem of rest and motion—for Plato the very exemplar of impasse and obscurity—will issue in the following chapter as progress and Light. Literate and witty as Maxwell was, I cannot think that he crafted this denouement exclusively with a procedural eye.

*Henry Augustus Rowland studied at the Rensselaer Polytechnic Institute in Troy, New York and later returned there as Assistant Professor of Physics.

†He became the first appointee to the Chair of Physics at the Johns Hopkins University in Baltimore, Maryland in 1876.

‡Before taking up his duties at Johns Hopkins, Rowland studied in Berlin under Helmholtz.

§Archimage (archi- + mage, from L. *magus*, magician; hence, "chief magician"), that is, Helmholtz.

‖Campbell, Lewis and William Garnett, *The Life of James Clerk Maxwell*. London: Macmillan (1882), p. 654.

CHAPTER XIX.

COMPARISON OF THE ELECTROSTATIC
WITH THE ELECTROMAGNETIC UNITS.

Determination of the Number of Electrostatic Units of Electricity in one Electromagnetic Unit.

768.] THE absolute magnitudes of the electrical units in both systems depend on the units of length, time, and mass which we adopt, and the mode in which they depend on these units is different in the two systems, so that the ratio of the electrical units will be expressed by a different number, according to the different units of length and time.

768.1 It appears from the table of dimensions, Art. 628, that the number of electrostatic units of electricity in one electromagnetic unit varies inversely as the magnitude of the unit of length, and directly as the magnitude of the unit of time which we adopt.

768.2 If, therefore, we determine a velocity which is represented numerically by this number, then, even if we adopt new units of length and of time, the number representing this velocity will still be the number of electrostatic units of electricity in one electromagnetic unit, according to the new system of measurement.

768.3 This velocity, therefore, which indicates the relation between electrostatic and electromagnetic phenomena, is a natural quantity of definite magnitude, and the measurement of this quantity is one of the most important researches in electricity.

768.1 *the table of dimensions*: The table is actually given in Art. 626. The dimensions there given for quantity of electricity are $[L^{1/2} M^{1/2}]$ in the electromagnetic system and $[L^{3/2} M^{1/2} T^{-1}]$ in the electrostatic system. If we divide the first dimensions by the second we obtain the dimensions of the number of electrostatic units in one electromagnetic unit; those dimensions are $[T/L]$, as Maxwell states here.

768.4 To shew that the quantity we are in search of is really a velocity, we may observe that in the case of two parallel currents the attraction experienced by a length a of one of them is, by Art. 686,

$$F = 2CC'\frac{a}{b},$$

where C, C' are the numerical values of the currents in electromagnetic measure, and b the distance between them. If we make $b = 2a$, then

$$F = CC'.$$

768.5 Now the quantity of electricity transmitted by the current C in the time t is Ct in electromagnetic measure, or nCt in electrostatic measure, if n is the number of electrostatic units in one electromagnetic unit.

768.6 Let two small conductors be charged with the quantities of electricity transmitted by the two currents in the time t, and placed at a distance r from each other. The repulsion between them will be

$$F = \frac{CC'n^2t^2}{r^2}.$$

768.7 Let the distance r be so chosen that this repulsion is equal to the attraction of the currents, then

$$\frac{CC'n^2t^2}{r^2} = CC'.$$

Hence $\qquad\qquad\qquad r = nt;$

or the distance r must increase with the time t at the rate n. Hence n is a velocity, the absolute magnitude of which is the same, whatever units we assume.

768.4 *To shew that the quantity ... is really a velocity*: Two quantities may have the same dimensional analysis, yet be completely different physically. For example, the units of both work and torque comprise the product of length and force; yet a quantity of work is not a torque, nor is any amount of torque equivalent to a quantity of work.

the attraction ... is: Article 686, to which Maxwell refers, is part of an omitted chapter; but Maxwell gave an equivalent expression at (495.1), noting there that parallel currents in the same direction attract one another.

768.6 *The repulsion between them will be*: Maxwell's expression is just that of Art. 40, where the quantities of electricity are equal to nCt and $nC't$ (768.5).

768.7 *Hence n is a velocity*: Recall that Maxwell intended to show that n is "really" a velocity (768.4). Has he done so? With this highly contrived example he has shown that the distance r "must" increase at the rate n, not that it actually does so; in what sense, then, are we to say that n "is" a velocity?

But we misunderstand Maxwell's imaginative contrivances if we take them as aiming to represent settled reality. We have seen such exuberant flourishes

769.] To obtain a physical conception of this velocity, let us imagine a plane surface charged with electricity to the electrostatic surface-density σ, and moving in its own plane with a velocity v. This moving

from Maxwell before, such as his rhetorical transformation of an infinite straight wire into a half-infinite loop (485.3, *comment*), or his mechanical model of induction at the end of Chapter VII (584.). The present example seems to surpass the bounds of literalness to an even greater extent. It is impossible to overlook the element of fantasy, even comedy, that pervades the "apparatus" here described.

Electromagnetic Electrostatic

First, currents C and C' arise in the straight wires, notwithstanding that the currents apparently have no *sources*. Second, although the currents *terminate* in the small spheres (which are joined to the straight conductors by lengths of flexible wire), continuously charging them, some unknown agency maintains the currents at a constant intensity, such that the attractive forces F remain constant. Third, still another unspecified agency registers the attractive forces F between the straight portions and continuously adjusts the repulsive forces between the spheres (by controlling the distance r) so that they too are always equal to F. Obviously this arrangement is not physical, and is not meant to be. No one would try to build such a device. What then is its purpose?

Maxwell, I think, cultivates and even savors thought-devices like this one both to instruct and to free the imagination. By repeatedly violating some of the canons of realism while strictly observing others, we learn to regard our concepts as flexible, as constantly and legitimately responding to competing interpretive themes. Notice that the present device neatly distinguishes between the electromagnetic elements of the situation (on the left), and the electrostatic elements (on the right). No actual circuit could achieve that level of separation; Maxwell has violated the principles of current flow in order to achieve imaginative clarity. At the same time this fable (I don't think that is too strong a word) demands that we *think* electromagnetic and electrostatic action as a unity. We are not to rest content with dividing one number by another; we are to grasp a real relation between the electromagnetic and the electrostatic aspects of the world. The mathematical operation of dividing one unit by another is but a symbol of that deeper demand of reason.

769. *To obtain a physical conception of this velocity*: Maxwell is acknowledging that the previous narrative did *not* represent physical reality.

a plane surface charged with electricity: See the drawing on the following page.

electrified surface will be an electric current-sheet, the strength of the current flowing through unit of breadth of the surface being σv in electrostatic measure, or $\frac{1}{n}\sigma v$ in electromagnetic measure, if n is the number of electrostatic units in one electromagnetic unit. If another plane surface parallel to the first is electrified to the surface-density σ', and moves in the same direction with the velocity v', it will be equivalent to a second current-sheet.

769.1 The electrostatic repulsion between the two electrified surfaces is, by Art. 124, $2\pi\sigma\sigma'$ for every unit of area of the opposed surfaces.

769.2 The electromagnetic attraction between the two current sheets is, by Art. 653, $2\pi u u'$ for every unit of area, u and u' being the surface-densities of the currents in electromagnetic measure.

769.3 But $u = \frac{1}{n}\sigma v$, and $u' = \frac{1}{n}\sigma' v'$, so that the attraction is

$$2\pi\sigma\sigma'\frac{vv'}{n^2}.$$

769., continued. *the strength of the current flowing through unit of breadth of the surface being σv in electrostatic measure*: Let the surface have breadth a and length b. If the surface-density is σ it will bear a total electric charge of σab. If the surface moves with velocity v, then it, along with σab of electricity, moves past the dotted line in time $t = b/v$,
constituting a current of strength $\sigma ab \div b/v$, or σav. Now consider only that much of the surface for which $a = 1$; then "the strength of the current flowing through unit of breadth" is σv.

769.1 *The electrostatic repulsion ... $2\pi\sigma\sigma'$ for every unit of area of the opposed surfaces*: Art. 124, to which Maxwell refers, is part of an omitted chapter, but we can derive Maxwell's expression using the "Gauss box" method (80.4, *comment*). Maxwell does not specify whether the charged surface in the present example is the surface of a conductor (as in 80.4) or a nonconductor. Either will serve, but the nonconducting case is easier to analyze.

Accordingly, consider a charged surface with surface-density σ, and erect a Gauss box through it as shown. As previously in (80.4, *comment*), there will be no induction through the cylindrical walls of the box; but since the surface is not that of a conductor, there will
be induction through *both* circular faces. Then the total induction through all surfaces of the box will be $2RA$, where R is the resultant intensity. But the box encloses charge equal to σA. Then by Gauss's Theorem (76.8, *comment*), $2RA = 4\pi\sigma A$, or $R = 2\pi\sigma$.

If then a quantity q of electricity is placed on one side of the charged surface, it will experience force $F = qR = 2\pi\sigma q$. Let the charge q occupy area A of a second charged surface having surface-density σ'. Then $q = \sigma' A$, and hence $F = 2\pi\sigma\sigma' A$. Then the force per unit area will be $2\pi\sigma\sigma'$.

769.4 The ratio of the attraction to the repulsion is equal to that of vv' to n^2. Hence, since the attraction and the repulsion are quantities of the same kind, n must be a quantity of the same kind as v, that is, a velocity. If we now suppose the velocity of each of the moving planes to be *equal* to n, the attraction will be equal to the repulsion, and there will be no mechanical action between them. Hence we may define the ratio of the electric units to be a velocity, such that two electrified surfaces, moving in the same direction with this velocity, have no mutual action. Since this velocity is about 300000 kilometres per second, it is impossible to make the experiment above described.

770.] If the electric surface-density and the velocity can be made so great that the magnetic force is a measurable quantity, we may at least verify our supposition that a moving electrified body is equivalent to an electric current.

770.1 … The electrified surface may be that of a non-conducting disk revolving in the plane of the magnetic meridian, and the magnet may be placed close to the ascending or descending portion of the disk, and protected from its electrostatic action by a screen of metal. I am not aware that this experiment has been hitherto attempted.*

*{This effect was discovered by Prof. Rowland in 1876. For subsequent experiments on this subject see Rowland and Hutchinson, *Phil. Mag.* 27. 445 (1887); Röntgen, *Wied. Ann.* 40. 93; Himstedt, *Wied. Ann.* 40. 720.}

769.4 *it is impossible to make the experiment above described*: Impossible, for Maxwell, only because he has no means to accelerate the electrified surfaces, along with the necessary measuring instruments, to a speed of 300,000 km/sec —a speed which he knows to be at least of the same order of magnitude as the speed of light. (Since 1905, of course, we are aware of deeper, relativistic, considerations that make the experiment impossible.) But the present narrative was intended to provide "a physical conception of this velocity" (769.). Has it done so? If it has, how did a physically impossible experiment provide a physical conception?

The first edition of the *Treatise* appeared in 1873. Scarcely a quarter century later, in 1896, the young Einstein would imagine an equally impossible "experiment" that, like Maxwell's, asks in effect, "What happens if I carry out an ordinary observation while moving at the speed of light?" Einstein writes:

> After ten years of reflection such a principle resulted from a paradox upon which I had already hit at the age of sixteen: If I pursue a beam of light with a velocity c (the velocity of light in vacuum), I should observe such a beam of light as a spatially oscillatory electromagnetic field at rest.
> (*Autobiographical Notes*, ed. P. A. Schilpp, Open Court, 1979)

770. *our supposition that a moving electrified body is equivalent to an electric current*: Maxwell made this supposition when he treated a moving electrified surface as an electric current sheet (769.).

I. *Comparison of Units of Electricity.*

771.] Since the ratio of the electromagnetic to the electrostatic unit of electricity is represented by a velocity, we shall in future denote it by the symbol v. The first numerical determination of this velocity was made by Weber and Kohlrausch.*

771.1 Their method was founded on the measurement of the same quantity of electricity, first in electrostatic and then in electromagnetic measure.

771.2 The quantity of electricity measured was the charge of a Leyden jar. It was measured in electrostatic measure as the product of the capacity of the jar into the difference of potential of its coatings. The capacity of the jar was determined in comparison with that of a sphere suspended in an open space at a distance from other bodies. The capacity of such a sphere is expressed in electrostatic measure by its radius. Thus the capacity of the jar may be found and expressed as a certain length. See Art. 227.

771.3 The difference of the potentials of the coatings of the jar was measured by connecting the coatings with the electrodes of an electrometer, the constants of which were carefully determined, so that the difference of the potentials, E, became known in electrostatic measure.

771.4 By multiplying this by c, the capacity of the jar, the charge of the jar was expressed in electrostatic measure.

771.5 To determine the value of the charge in electromagnetic measure, the jar was discharged through the coil of a galvanometer. The effect of the transient current on the magnet of the galvanometer communicated to the magnet a certain angular velocity. The magnet then swung round to a certain deviation, at which its velocity was entirely destroyed by the opposing action of the earth's magnetism.

* *Elektrodynamische Maasbestimmungen*; and Pogg., *Ann.*, xcix (Aug. 1856, pp. 10–25).

771.2 *The capacity of such a sphere is expressed ... by its radius*: Article 227, to which Maxwell refers, is part of an omitted chapter; but we can make an equivalent derivation. The electrical capacity C of a conductor was defined as the quotient of its charge e by its potential V, or $C = e/V$ (50.) Let the conductor be a sphere of radius r. We saw that a quantity of electricity e on the surface of the sphere will distribute itself uniformly and act as though it were concentrated at the center (60.7 and *comment*). But the potential at distance r from e is $V = e/r$ (73.). Combining, we have $C = e \div e/r$ or $C = r$.

771.5 *The effect of the transient current*: In this paragraph and the next Maxwell reviews the operation of the ballistic galvanometer in measuring brief ("transient") currents. I discussed that useful instrument in the editor's introduction to Part IV, Chapter III. See also (537.1).

771.6 By observing the extreme deviation of the magnet the quantity of electricity in the discharge may be determined in electromagnetic measure, as in Art. 748....

771.7 The value of v obtained by MM. Weber and Kohlrausch was

$$v = 310740000 \text{ metres per second.}$$

771.8 The property of solid dielectrics, to which the name of Electric Absorption has been given, renders it difficult to estimate correctly the capacity of a Leyden jar. The apparent capacity varies according to the time which elapses between the charging or discharging of the jar and the measurement of the potential, and the longer the time the greater is the value obtained for the capacity of the jar.

771.9 Hence, since the time occupied in obtaining a reading of the electrometer is large in comparison with the time during which the discharge through the galvanometer takes place, it is probable that the estimate of the discharge in electrostatic measure is too high, and the value of v, derived from it, is probably also too high.

* * *

III. *Electrostatic Capacity in Electromagnetic Measure.*

* * *

Intermittent Current.

775.] If the wire of a battery-circuit be broken at any point, and the broken ends connected with the electrodes of a condenser, the current will flow into the condenser with a strength which diminishes as the difference of the potentials of the plates of the condenser increases, so that when the condenser has received the full charge corresponding to the electromotive force acting on the wire the current ceases entirely.

775.1 If the electrodes of the condenser are now disconnected from the ends of the wire, and then again connected with them in the reverse order, the condenser will discharge itself through the wire, and will then become recharged in the opposite way, then a transient current will flow through the wire, the total quantity of which is equal to two charges of the condenser.

771.8 *Electric Absorption*: Discussed in the editor's introduction to this chapter.

775. *the full charge corresponding to the electromotive force acting on the wire*: The capacitor will charge until the difference of potential between its plates is equal to the electromotive force of the battery (49.) Since in a capacitor $e = CV$ (50., 771.2), the "full charge" of a capacitor of capacity C is that value of e for which V equals the electromotive force (the "voltage") of the battery.

775.1 *a transient current will flow ... equal to two charges of the condenser*: The wire first discharges the previous quantity of electricity e, then it delivers a second quantity e to recharge the capacitor. Although the capacitor connections are reversed, both discharge and recharge have the same direction in the wire.

775.2 By means of a piece of mechanism (commonly called a Commutator, or *wippe*) the operation of reversing the connexions of the condenser can be repeated at regular intervals of time, each interval being equal to T. If this interval is sufficiently long to allow of the complete discharge of the condenser, the quantity of electricity transmitted by the wire in each interval will be $2EC$, where E is the electromotive force, and C is the capacity of the condenser.

775.3 If the magnet of a galvanometer included in the circuit is loaded, so as to swing so slowly that a great many discharges of the condenser occur in the time of one free vibration of the magnet, the succession of discharges will act on the magnet like a steady current whose strength is $2EC/T$.

775.4 If the condenser is now removed, and a resistance coil substituted for it, and adjusted till the steady current through the galvanometer produces the same deflexion as the succession of discharges, and if R is the resistance of the whole circuit when this is the case,

$$\frac{E}{R} = \frac{2EC}{T}; \tag{1}$$

$$R = \frac{T}{2C}. \tag{2}$$

775.2 *Commutator*: It is represented schematically in the drawing as a ganged switch. With the switch in the position shown, the upper plate of capacitor C is connected to the positive terminal of the battery; the capacitor will charge. When the switch is moved to its lower position, the lower plate of the capacitor is connected to the positive terminal; the capacitor will discharge its former electrification, and current will continue to pass until the capacitor is again charged to the same degree as before—only now with the lower plate positive.

the quantity of electricity transmitted ... will be $2EC$: Compare (775., *comment*), with V set equal to electromotive force E.

775.3 *swing so slowly that a great many discharges of the condenser occur*: The period of the galvanometer must be large in comparison to the commutation interval T. Although a long period is characteristic of a ballistic galvanometer, the instrument employed here is not acting "ballistically," as it does not swing to a peak value.

775.4 *Equation* (1): This is just Ohm's Law (241.2, *comment*), with current equal to $2EC/T$.

Equation (2): In this case, better written as $C = T/2R$. Measure R in electromagnetic units; then the equation gives C in electromagnetic units.

775.5 We may thus compare the condenser with its commutator in motion to a wire of a certain electrical resistance, and we may make use of the different methods of measuring resistance described in Arts. 345 to 357 in order to determine this resistance.

* * *

775.5 *we may make use of the different methods of measuring resistance described in Arts. 345 to 357*: The Articles referenced, part of an omitted chapter, set out methods for comparing an unknown resistance, such as R in equation (2) above, with a resistance whose value in electromagnetic units is already known. In another omitted chapter Maxwell describes techniques for determining a resistance in electromagnetic units by direct measurement.

Having determined R by one or another of the methods available, we use equation (2) to calculate the electromagnetic value of C; let us call it C_{emu}. This, in turn, we compare with the electrostatic value of C, found by calculation; and let us call it C_{esu}. Since there are v^2 electrostatic units of capacitance in one electromagnetic unit of capacitance (628.1), C_{esu} will be to C_{emu} as v^2 is to 1; thus we find v.

Since C_{emu} is given by equation (2) we write

$$v^2 = \frac{C_{\mathrm{esu}}}{C_{\mathrm{emu}}} = \frac{2RC_{\mathrm{esu}}}{T}$$

$$v = \sqrt{\frac{2RC_{\mathrm{esu}}}{T}}.$$

Notice that the value of v depends on the *square root* of R. Errors in the measurement of R will therefore have less effect on the result than if the value of v had depended on R directly.

Editor's Introduction to
Chapter XX

The central argument

In my Note to Readers I suggested that following a central argument is not the same thing as reaching a single conclusion. But now that we have nearly completed the path through Maxwell's *Treatise* which this book represents, it may be appropriate to revisit that suggestion.

The present chapter—which brings our selections nearly to a close—presents Maxwell's electromagnetic theory of light. Is the electro-magnetic theory of light, then, the culmination of the *Treatise*? Looking back over the course of the argument, I believe we can say that it is—but not, perhaps, because it reveals the nature of light as such; rather because it represents the most concrete form of Maxwell's vision of connectedness and energy flow. For, looking back on Maxwell's narrative, I think we can say that the *Treatise* is, fundamentally, about energy and its communication and transformation throughout the cosmos.

The shape of the inquiry

To be sure, the *Treatise* did not necessarily exhibit a concentration on energy, or exhibit it with prominence, from the beginning. In its first three Parts, and in the first four chapters of Part IV, Maxwell carried out a survey of the fundamental phenomena of electricity and magnetism. His aim there, it seems, was only in part to establish the principal topics of electromagnetic science; it was also to explore the various terms (both mathematical and lexical) that contribute to accurate and adequate descriptions of those phenomena.

During that initial phase of investigation, Maxwell freely employed the terminologies of various hypotheses—often even of competing ones —never favoring one language or symbology to the exclusion of others. At one point, for example, he gave full countenance to the imagery of electricity as a substance separate from the electric field; while in another place he developed *electric displacement*, with its concomitant interpretation of electric charge as being but a structure emergent from, and subordinate to, the field. In discussing magnetization he made free use of the language of magnetic "matter"; yet he also gave a parallel

account in terms of magnetic polarization, making no appeal whatsoever to such "matter." It is as though Maxwell were instructing us by example that, although we have to speak *some* language in order to get started, we should not assume that our starting language is nature's language. We should use any language in frank acknowledgment that it is destined to be refined or replaced.

During the descriptive stage, in short, Maxwell developed the art of employing the specific to represent the general. That is what Euclidean figures do, and for a similar purpose. When Euclid proves that the sum of angles of a triangle is two right angles, the triangle he draws must necessarily be a particular triangle; but when he reasons with its aid he takes no notice of whether the triangle is acute or obtuse, isosceles or scalene. An essential component in the study of Euclid is learning how to make such enlarged use of specific representative objects. So too does Maxwell help us learn to think representatively, without mistaking the representational for the literal. In so doing, one hopes, we may avoid enshrining, at an early stage of thought, language and concepts that at a later stage may prove to have been idiosyncratic.

Maxwell included a cogent expression of this frame of mind at the end of Part IV, Chapter V:

> In forming the ideas and words relating to any science ... we must keep constantly in mind the ideas appropriate to the fundamental science of dynamics, so that we may, during the first development of the science, avoid inconsistency with what is already established, and also that when our views become clearer, the language we have adopted may be a help to us and not a hindrance. (567.3)

As subsequent chapters demonstrated, Maxwell's attitude in Chapter V was not one of merely prudent restraint but one that embodied a positive precept. He states the root of that precept in Article 831, part of an otherwise omitted chapter:

> The problem of determining the mechanism required to establish a given species of connexion between the motions of the parts of a system always admits of an infinite number of solutions. Of these, some may be more clumsy or more complex than others, but *all must satisfy the conditions of mechanism in general*. [emphasis added]

The "conditions of mechanism in general" are precisely the general (Lagrangian) equations of Chapter V; and with that chapter Maxwell's positive purpose became, first, to derive those equations; and next, to refine them to increasing levels of specificity. That is what he has done throughout the remaining chapters of Part IV, from the sixth through the present.

Maxwell's work of refinement itself passes through two phases. The first was completed with his formulation of the General Equations of the Electromagnetic Field (Chapter IX), which were to serve as the foundation of electromagnetic science. The second phase is attained in his Electromagnetic Theory of Light (in the present Chapter), which, as I urged when introducing Chapter XIX, is really about much more than light. Thus we follow Maxwell through four stages of thought in all: (1) acquisitive, descriptive—from the beginning of the *Treatise* through the first four chapters of Part IV; (2) positive, universal—in Chapter V of Part IV; (3) definitive, general—in Chapter IX; and (4) extensive, concrete—in Chapter XX.

I hope that this brief review of the argument of the *Treatise* clarifies in what sense the theory of light is *the* culmination of the work, and in what sense it is but *a* culmination. As I have repeatedly suggested, there remain powerful reasons to call the equations of Chapter V climactic, as being the purest and most universal; and there are equally strong reasons to view the equations of Chapter IX as marking the zenith with respect to a particular science. Finally, the Electromagnetic Theory of Light represents an apex in that it touches, and changes, our world view most intimately.

The nature of light and the speed of light

The theory that light is a wave has much to recommend it, not least that a transverse wave appears to be the *only* constitution of light that is demonstrably consistent with polarization. Nevertheless the wave theory was by no means universally accepted when Maxwell wrote; for every wave theory demands a medium, and the medium posited for light waves (the so-called *luminiferous aether*) was entirely hypothetical. This was considered a serious defect in the undulatory hypothesis.

Maxwell tacitly accepts that light is a wave (781.). But is it an *electromagnetic* wave? The most telling argument is twofold: first, Maxwell will demonstrate in the present chapter that the velocity of an electromagnetic wave must be the quotient of the electromagnetic unit by the electrostatic unit of electricity; second, when this quotient is measured (Chapter XIX), it proves to be tantalizingly close to the measured velocity of light.

Actually, the level of agreement observed between these quantities was not so very great during Maxwell's lifetime; but improved measurements of both over the next forty to fifty years continued to approach one another, and today their equality is no longer regarded as established by measurement but as true *by definition*. Since this point is sometimes misunderstood, let us consider it further.

As Maxwell will remind us in the opening paragraphs of this chapter, the velocity of any wave is a characteristic of its medium. Now in general, if we do not assume two things to be identical, then there is usually no reason to expect any of their characteristic properties to be equal. On the other hand, if such equality is discovered, it may prompt conjecture that the things observed are, in fact, the same. Once we come to regard their sameness as established, the equality becomes a necessity.

For a not entirely fanciful comparison, imagine that a company of astronomers has been for many years engaged in measuring the respective periods of the Morning Star and the Evening Star, and that they have concluded that the periods are equal, at least to within the limits of their instruments. So long as those celestial objects are considered as separate beings, the observed equality is an exciting result. It generates demands for further measurements of still greater precision. At the same time their equality demands an explanation. But the moment our astronomers accept that the "two stars" are really the single planet Venus, they have to regard the "period of Morning Star" and the "period of Evening Star" as *necessarily* equal.

Such has been the case with Maxwell's theory. The Electromagnetic Theory of Light encountered resistance in several quarters, largely because it appeared to demand a questionable medium. Today, the theory (though not the medium) is universally accepted—at least for purposes of classical physics—and we regard the speed of light as equalling the ratio of electric units by definition.

As Maxwell will show in this very chapter, the ratio of electric units can also be expressed as $1/\sqrt{K\mu}$. An indication of how firmly we have come to accept the Electromagnetic Theory is that a standard reference work* lists the determination of $1/\sqrt{K\mu}$ not as a method to measure the ratio of electrical units, but as a method of measuring the speed of light!

*Froome, Keith Davy and Louis Essen. *The Velocity of Light and Radio Waves.* New York: Academic Press (1969)

CHAPTER XX.

ELECTROMAGNETIC THEORY OF LIGHT.

781.] IN several parts of this treatise an attempt has been made to explain electromagnetic phenomena by means of mechanical action transmitted from one body to another by means of a medium occupying the space between them. The undulatory theory of light also assumes the existence of a medium. We have now to shew that the properties of the electromagnetic medium are identical with those of the luminiferous medium.

781.1 To fill all space with a new medium whenever any new phenomenon is to be explained is by no means philosophical, but if the study of two different branches of science has independently suggested the idea of a medium, and if the properties which must be attributed to the medium in order to account for electromagnetic phenomena are of the same kind as those which we attribute to the luminiferous medium in order to account for the phenomena of light, the evidence for the physical existence of the medium will be considerably strengthened.

781.2 But the properties of bodies are capable of quantitative measurement. We therefore obtain the numerical value of some property of the medium, such as the velocity with which a disturbance is propagated through it, which can be calculated from electromagnetic experiments, and also observed directly in the case of light. If it should be found that the velocity of propagation of electromagnetic disturbances is the same as the velocity of light, and this not only in air, but in other transparent media, we shall have strong reasons for believing that light is an electromagnetic phenomenon, and the combination of the optical with the electrical evidence will produce a conviction of the reality of the medium similar to that which we obtain, in the case of other kinds of matter, from the combined evidence of the senses.

782.] When light is emitted, a certain amount of energy is expended by the luminous body, and if the light is absorbed by another body, this body

781.1 *by no means philosophical*: Today we might say "unscientific."

becomes heated, shewing that it has received energy from without. During the interval of time after the light left the first body and before it reached the second, it must have existed as energy in the intervening space.

782.1 According to the theory of emission, the transmission of energy is effected by the actual transference of light-corpuscles from the luminous to the illuminated body, carrying with them their kinetic energy, together with any other kind of energy of which they may be the receptacles.

782.2 According to the theory of undulation, there is a material medium which fills the space between the two bodies, and it is by the action of contiguous parts of this medium that the energy is passed on, from one portion to the next, till it reaches the illuminated body.

782.3 The luminiferous medium is therefore, during the passage of light through it, a receptacle of energy. In the undulatory theory, as developed by Huygens, Fresnel, Young, Green, &c., this energy is supposed to be partly potential and partly kinetic. The potential energy is supposed to be due to the distortion of the elementary portions of the medium. We must therefore regard the medium as elastic. The kinetic energy is supposed to be due to the vibratory motion of the medium. We must therefore regard the medium as having a finite density.

782.4 In the theory of electricity and magnetism adopted in this treatise, two forms of energy are recognised, the electrostatic and the electrokinetic (see Arts. 630 and 636), and these are supposed to have their seat, not merely in the electrified or magnetized bodies, but in every part of the surrounding space, where electric or magnetic force is observed to act. Hence, the theory agrees with the undulatory theory in assuming the existence of a medium which is capable of becoming a receptacle of two forms of energy. *

783.] Let us next determine the conditions of the propagation of an electromagnetic disturbance through a uniform medium, which we shall suppose to be at rest, that is, to have no motion except that which may be involved in electromagnetic disturbances.

> * 'For my own part, considering the relation of a vacuum to the magnetic force and the general character of magnetic phenomena external to the magnet, I am more inclined to the notion that in the transmission of the force there is such an action, external to the magnet, than that the effects are merely attraction and repulsion at a distance. Such an action may be a function of the aether; for it is not at all unlikely that, if there be an aether, it should have other uses than simply the conveyance of radiations.' —Faraday's *Experimental Researches*, 3075.

782. *it must have existed as energy in the intervening space*: by the principle of conservation of energy.

783.1 Let C be the specific conductivity of the medium, K its specific capacity for electrostatic induction, and μ its magnetic 'permeability.'

783.2 To obtain the general equations of electromagnetic disturbance, we shall express the true current \mathfrak{C} in terms of the vector potential \mathfrak{A} and the electric potential Ψ.

783.3 The true current \mathfrak{C} is made up of the conduction current \mathfrak{K} and the variation of the electric displacement \mathfrak{D}, and since both of these depend on the electromotive intensity \mathfrak{E}, we find, as in Art. 611,

$$\mathfrak{C} = \left(C + \frac{1}{4\pi}K\frac{d}{dt}\right)\mathfrak{E}. \tag{1}$$

783.4 But since there is no motion of the medium, we may express the electromotive intensity, as in Art. 599,

$$\mathfrak{E} = -\dot{\mathfrak{A}} - \nabla\Psi. \tag{2}$$

Hence

$$\mathfrak{C} = -\left(C + \frac{1}{4\pi}K\frac{d}{dt}\right)\left(\frac{d\mathfrak{A}}{dt} + \nabla\Psi\right). \tag{3}$$

783.5 But we may determine a relation between \mathfrak{C} and \mathfrak{A} in a different way, as is shewn in Art. 616, the equations (4) of which may be written

$$4\pi\mu\mathfrak{C} = \nabla^2\mathfrak{A} + \nabla J, \tag{4}$$

where

$$J = \frac{dF}{dx} + \frac{dG}{dy} + \frac{dH}{dz}. \tag{5}$$

783.6 Combining equations (3) and (4), we obtain

$$\mu\left(4\pi C + K\frac{d}{dt}\right)\left(\frac{d\mathfrak{A}}{dt} + \nabla\Psi\right) + \nabla^2\mathfrak{A} + \nabla J = 0, \tag{6}$$

which we may express in the form of three equations as follows—

$$\left.\begin{array}{l}
\mu\left(4\pi C + K\dfrac{d}{dt}\right)\left(\dfrac{dF}{dt} + \dfrac{d\Psi}{dx}\right) + \nabla^2 F + \dfrac{dJ}{dx} = 0, \\[12pt]
\mu\left(4\pi C + K\dfrac{d}{dt}\right)\left(\dfrac{dG}{dt} + \dfrac{d\Psi}{dy}\right) + \nabla^2 G + \dfrac{dJ}{dy} = 0, \\[12pt]
\mu\left(4\pi C + K\dfrac{d}{dt}\right)\left(\dfrac{dH}{dt} + \dfrac{d\Psi}{dz}\right) + \nabla^2 H + \dfrac{dJ}{dz} = 0,
\end{array}\right\} \tag{7}$$

783.7 These are the general equations of electromagnetic disturbances.

783.1 *magnetic 'permeability'*: When Maxwell introduced this term in (614.) he did not place it within quotation marks. Why does he use them here?

783.3 *Equation* (1): It appeared earlier as equation (I) in (611.)

783.4 *Equation* (2): Compare (599.3), equation (10), setting velocity equal to zero.

783.8 If we differentiate these equations with respect to x, y, and z, respectively, and add, we obtain

$$\mu\left(4\pi C + K\frac{d}{dt}\right)\left(\frac{dJ}{dt} - \nabla^2\Psi\right) = 0. \tag{8}$$

783.9 If the medium is a non-conductor, $C = 0$, and $\nabla^2\Psi$, which is proportional to the volume-density of free electricity, is independent of t. Hence J must be a linear function of t, or a constant, or zero, and we may therefore leave J and Ψ out of account in considering periodic disturbances.

Propagation of Undulations in a Non-conducting Medium.

784.] In this case, $C = 0$, and the equations become

$$\left.\begin{array}{l} K\mu\dfrac{d^2F}{dt^2} + \nabla^2F = 0, \\[2mm] K\mu\dfrac{d^2G}{dt^2} + \nabla^2G = 0, \\[2mm] K\mu\dfrac{d^2H}{dt^2} + \nabla^2H = 0, \end{array}\right\} \tag{9}$$

784.1 The equations in this form are similar to those of the motion of an incompressible elastic solid, and when the initial conditions are

783.9 *If the medium is a non-conductor,* $C = 0$: since C is the conductivity of the material (609.).

$\nabla^2\Psi$, *which is proportional to the volume-density of free electricity…*: See (77.2) and (77.4).

…is independent of t: If the material is a nonconductor, the distribution of electricity within it does not change in time; so too the volume-density of electricity within it does not change in time.

Hence J must be…: With $C = 0$, equation (8) becomes

$$\mu K\frac{d}{dt}\left(\frac{dJ}{dt} - \nabla^2\Psi\right) = \mu K\frac{d^2J}{dt^2} - \frac{d}{dt}\left(\nabla^2\Psi\right) = 0.$$

But $\nabla^2\Psi$ being (as we just saw) independent of t, its time-derivatives will be zero. It follows that d^2J/dt^2 is also zero; thus dJ/dt is constant. J is therefore either a linear function of t, or constant, or zero.

we may therefore leave J and Ψ out of account in periodic disturbances: A periodic function is *recurrent*; it must have a nonzero first derivative (with respect to time) or it would not vary; and it must have a nonzero second derivative or it could never regain a previous value. Since J and Ψ do not meet these conditions in a nonconductor, they cannot be periodic there.

784. *the equations become…*: In equations (7), set $C = 0$. Then "leave J and Ψ out of account" by ignoring terms that contain them; the result is equations (9).

784.1 *The equations … are similar to those of the motion of an incompressible elastic solid*: that is, they are wave equations, as Maxwell is about to demonstrate.

given, the solution can be expressed in a form given by Poisson,* and applied by Stokes to the Theory of Diffraction.†

784.2 Let us write $\qquad V = 1/\sqrt{K\mu}$. \hfill (10)

784.4 Let O be the point for which we wish to determine the value of F at the time t....

784.5 Then the value of F at the point O, at the time t, is

$$F = \frac{d}{dt}(\bar{F}t) + t\frac{d\bar{F}}{dt}.$$ \hfill (11)

* * *

785.] It appears, therefore, that the condition of things at the point O at any instant depends on the condition of things at a distance Vt and at an interval of time t previously, so that any disturbance is propagated through the medium with the velocity V.

* *Mém. de l'Acad.*, tom. iii. p. 130, et seq.
† *Cambridge Transactions*, vol. ix. pp. 1–62 (1849).

784.2 *Let us write* $V = 1/\sqrt{K\mu}$: Despite Maxwell's suggestive choice of V to symbolize this quantity, he is not assuming that it represents a velocity. Rather it will be his task to *prove* that equations (9) describe a wave and that $1/\sqrt{K\mu}$ is its velocity of propagation.

784.5 *Equation* (11): Maxwell actually writes three equations here, of which I have given the first as an example. Poisson derived it as a solution to the first member of equations (9); Maxwell is citing it without proof. His purpose is to show that equations (9) describe a wave. We will demonstrate the same thing by means of a simpler function that Maxwell himself will consider later.

785. *the condition of things at the point O at any instant depends on the condition of things at a distance Vt and at an interval of time t previously*: This statement is a beautiful definition of wave motion, and in particular of what it means to attribute velocity V to a wave. Maxwell is making this observation about equation (11), but let us consider instead a function $F(z,t)$ having the form $F(z,t) = g(z - Vt)$. I say it conforms to Maxwell's description and therefore represents a wave traveling with velocity V. For suppose $F(z,t)$ has some value, say h, at the point (z_0, t_0); that is, let $F(z_0, t_0) = g(z_0 - Vt_0) = h$. Then "the condition of things" (to use Maxwell's wording) at point z_0 and at the instant t_0 is that $F = h$.

Now consider another point (z_1, t_1), where $z_1 = z_0 - Vt$ and $t_1 = t_0 - t$. Thus (z_1, t_1) is "at a distance Vt and at an interval of time t previously" from (z_0, t_0). Then "the condition of things" at point (z_1, t_1) will be

$$F(z_1, t_1) = g(z_1 - Vt_1) = g[z_0 - Vt - V(t_0 - t)]$$
$$= g(z_0 - Vt - Vt_0 + Vt) = g(z_0 - Vt_0) = h;$$

that is, "the condition of things" at point (z_0, t_0) depends on—indeed, is in this case identical to—"the condition of things at a distance Vt and at an interval of time t previously." F, then, is a *wave function*, and its velocity is V. But we still have to show that F is a solution of equations (9); this we do in (786., *comment*).

<center>* * *</center>

786.] The quantity V, in Art. 784, which expresses the velocity of propagation of electromagnetic disturbances in a non-conducting medium is, by equation (10), equal to $1/\sqrt{K\mu}$.

786.1 If the medium is air, and if we adopt the electrostatic system of measurement, $K = 1$ and $\mu = 1/v^2$, so that $V = v$, or the velocity of propagation is numerically equal to the number of electrostatic units of electricity in one electromagnetic unit. If we adopt the electromagnetic system, $K = 1/v^2$ and $\mu = 1$, so that the equation $V = v$ is still true.

786. *The quantity V, ... which expresses the velocity of propagation of electromagnetic disturbances is ... equal to* $1/\sqrt{K\mu}$: We saw already that $1/\sqrt{K\mu}$ is the velocity of function $F(z,t)$, that is, $g(z-Vt)$; it is the velocity of *electromagnetic waves* if we can show that $F(z,t)$ is a solution of equations (9). The following method is standard:

Substitute $F(z,t)$ into the first of equations (9). Since $F(z,t)$ is a function of z and t only, $\nabla^2 F$ reduces to $-d^2F/dz^2$ (26.) and the equation becomes

$$K\mu \frac{d^2 F}{dt^2} - \frac{d^2 F}{dz^2} = 0 \,. \tag{i}$$

Differentiate $F(z,t)$:

$$\frac{dF}{dz} = \frac{d}{dz} g(z-Vt) = \frac{dg}{du}\frac{du}{dz} = \frac{dg}{du} \,,$$

where $u = (z-Vt)$; and from this,

$$\frac{d^2 F}{dz^2} = \frac{d}{dz}\left(\frac{dg}{du}\right) = \frac{d}{du}\left(\frac{dg}{du}\right)\frac{du}{dz} = \frac{d^2 g}{du^2} \,.$$

Similarly,

$$\frac{dF}{dt} = \frac{d}{dt} g(z-Vt) = \frac{dg}{du}\frac{du}{dt} = -V\frac{dg}{du}$$

and

$$\frac{d^2 F}{dt^2} = \frac{d}{dt}\left(-V\frac{dg}{du}\right) = \frac{d}{du}\left(-V\frac{dg}{du}\right)\frac{du}{dt} = V^2 \frac{d^2 g}{du^2} \,.$$

Substitute these derivatives into equation (i) to obtain

$$K\mu V^2 \frac{d^2 g}{du^2} - \frac{d^2 g}{du^2} = 0 \quad \text{or} \quad K\mu V^2 - 1 = 0 \,,$$

which is an identity if $V = 1/\sqrt{K\mu}$. We may proceed in the same way for the remaining equations. Thus we have shown that a wave function having velocity $V = 1/\sqrt{K\mu}$ solves equations (9) (though it is by no means the most general solution). And equations (9) are equations of electromagnetic force. Therefore they describe electromagnetic waves propagating with velocity $V = 1/\sqrt{K\mu}$.

786.1 *if we adopt the electrostatic system of measurement,* $K=1$ *and* $\mu=1/v^2$: Maxwell reviewed the units of both the electrostatic and the electromagnetic systems in Chapter X above; see especially Article 628.

786.2 On the theory that light is an electromagnetic disturbance, propagated in the same medium through which other electromagnetic actions are transmitted, V must be the velocity of light, a quantity the value of which has been estimated by several methods. On the other hand, v is the number of electrostatic units of electricity in one electromagnetic unit, and the methods of determining this quantity have been described in the last chapter. They are quite independent of the methods of finding the velocity of light. Hence the agreement or disagreement of the values of V and of v furnishes a test of the electromagnetic theory of light.

787.] In the following table, the principal results of direct observation of the velocity of light, either through the air or through the planetary spaces, are compared with the principal results of the comparison of the electric units:—

Velocity of Light (métres per second).		Ratio of Electric Units (métres per second).	
Fizeau	314000000	Weber	310740000
Aberration, etc., and Sun's parallax	308000000	Maxwell	288000000
Foucault	282000000	Thomson	282000000

787.1 It is manifest that the velocity of light and the ratio of the units are quantities of the same order of magnitude. Neither of them can be said to be determined as yet with such a degree of accuracy as to enable us to assert that the one is greater or less than the other. It is to be hoped that, by further experiment, the relation between the magnitudes of the two quantities may be more accurately determined.

786.2 *the methods of determining this quantity have been described in the last chapter*: Maxwell reviewed six methods of measurement in Chapter XIX; our selection included two of them.

787.1 *It is to be hoped that, by further experiment, the relation ... may be more accurately determined*: Maxwell's hope has been realized. Subsequent measurements of the speed of light include:

Foucault (1862)	298,000,000 m/sec
Michelson (1879)	299,910,000
Michelson (1926)	299,796,000
Evanson et al. (1973)	299,792,457.4

Sixteen years after publication of the third edition of the *Treatise*, a new determination of the ratio of electric units yielded:

Rosa, Dorsey (1907) 299,788,000

Today the ratio of electric units and the speed of light are numerically equal by definition. Since the international redefinition of the meter in 1983, the accepted value for both quantities has been 299,792,458 m/sec.

787.2 In the meantime our theory, which asserts that these two quantities are equal, and assigns a physical reason for this equality, is certainly not contradicted by the comparison of these results such as they are.

788.] In other media than air, the velocity V is inversely proportional to the square root of the product of the dielectric and the magnetic inductive capacities. According to the undulatory theory, the velocity of light in different media is inversely proportional to their indices of refraction.

788.1 There are no transparent media for which the magnetic capacity differs from that of air more than by a very small fraction. Hence the principal part of the difference between these media must depend on their dielectric capacity. According to our theory, therefore, the dielectric capacity of a transparent medium should be equal to the square of its index of refraction.

788.2 But the value of the index of refraction is different for light of different kinds, being greater for light of more rapid vibrations. We must therefore select the index of refraction which corresponds to waves of the longest periods, because these are the only waves whose motion can be compared with the slow processes by which we determine the capacity of the dielectric.

789.] The only dielectric of which the capacity has been hitherto determined with sufficient accuracy is paraffin, for which in the solid form MM. Gibson and Barclay found *

$$K = 1.975. \tag{12}$$

789.1 Dr. Gladstone has found the following values of the index of refraction of melted paraffin, sp. g. 0.779, for the lines A, D and H:

* Phil. Trans. 1871, p. 573.

788. *According to the undulatory theory, the velocity of light in different media is inversely proportional to their indices of refraction*: since, as Huygens proposed in his *Treatise on Light*, the refractive index of a transparent medium equals the quotient of the speed of light in vacuum by the speed with which it travels in that medium.

788.2 *the slow processes by which we determine the capacity of the dielectric*: Dielectric capacity is measured under essentially static conditions, corresponding to an oscillation frequency of zero, that is, an infinite wavelength.

789.1 *the lines A, D and H*: that is, spectroscopic lines. Lines A and H correspond to wavelengths of 759.370 and 396.847 nanometers, respectively. "Line D" is now known to consist of two lines, the so-called "sodium doublet," with wavelengths of 589.592 and 588.595 nm.

Temperature	A	D	H
54° C	1.4306	1.4357	1.4499
57° C	1.4294	1.4343	1.4493

from which I find that the index of refraction for waves of infinite length would be about 1.422.

789.2 The square root of K is 1.405.

789.3 The difference between these numbers is greater than can be accounted for by errors of observation, and shews that our theories of the structure of bodies must be much improved before we can deduce their optical from their electrical properties. At the same time, I think that the agreement of the numbers is such that if no greater discrepancy were found between the numbers derived from the optical and the electrical properties of a considerable number of substances, we should be warranted in concluding that the square root of K, though it may not be the complete expression for the index of refraction, is at least the most important term in it.

Plane Waves.

790.] Let us now confine our attention to plane waves, the fronts of which we shall suppose normal to the axis of z. All the quantities, the variation of which constitutes such waves, are functions of z and t only, and are independent of x and y. Hence the equations of magnetic induction, (A), Art. 591, are reduced to

$$a = -\frac{dG}{dz}, \quad b = \frac{dF}{dz}, \quad c = 0, \tag{13}$$

the index of refraction for waves of infinite length would be about 1.422: Modern values do not confirm this result. Using reciprocals of the modern wavelengths just cited, extrapolation to zero (infinite wavelength) yields 1.40773—a value that agrees with the square root of K much more favorably than does Maxwell's.

the square root of K is 1.405: Thus the discrepancy is actually less than 0.2%.

our theories ... must be much improved before we can deduce their optical from their electrical properties: Today the essential identity (for transparent materials) of optical index of refraction and the square root of dielectric capacity K is considered standard physics.

790. *plane waves, the fronts of which* [*are*] *normal to the axis of z*: A wave "front" is a surface at every point of which the values of the oscillating quantity are in phase—simultaneously reaching their peak values, for example. In the special case of a plane wave, the values have to be equal as well as be in phase; otherwise the fronts will not remain planar as the wave propagates.

the quantities, the variation of which constitutes such waves: As we have had occasion to note before (783.9, *comment*), a wave exhibits periodic variation of a quantity; it is that variation, not the quantity itself, that constitutes the wave.

or the magnetic disturbance is in the plane of the wave. This agrees with what we know of that disturbance which constitutes light.

790.1 Putting $\mu\alpha$, $\mu\beta$, and $\mu\gamma$ for a, b and c respectively, the equations of electric currents, Art. 607, become

$$\left.\begin{aligned} 4\pi\mu u &= -\frac{db}{dz} = -\frac{d^2F}{dz^2}, \\ 4\pi\mu v &= \frac{da}{dz} = -\frac{d^2G}{dz^2}, \\ 4\pi\mu w &= 0. \end{aligned}\right\} \qquad (14)$$

Hence the electric disturbance is also in the plane of the wave, and if the magnetic disturbance is confined to one direction, say that of x, the electric disturbance is confined to the perpendicular direction, or that of y.

790.2 But we may calculate the electric disturbance in another way, for if f, g, h are the components of electric displacement in a non-conducting medium,

790., continued. *the magnetic disturbance...*: here, the *variation* of magnetic induction.

...is in the plane of the wave: We see from equations (13) that the magnetic induction has only x- and y-components (a and b); thus its variation will likewise be confined to the xy-plane and is said to be *transverse* with respect to the direction of propagation, z.

This agrees with what we know of ... light: in particular, that light is capable of being polarized, and must therefore be a transverse wave.

790.1 *Putting $\mu\alpha$, $\mu\beta$, $\mu\gamma$ for a, b, c respectively*: Recall that a, b, c are the components of \mathfrak{B} (400.), while α, β, γ are the components of \mathfrak{H} (398.1). Then since $\mathfrak{B} = \mu\mathfrak{H}$ (614.), the components of \mathfrak{B} are, severally, equal to μ times the respective components of \mathfrak{H}.

the equations of electric currents, Art. 607: Maxwell refers to equations (E) in (607.5). Multiplying the first of those equations by μ, we obtain

$$4\pi\mu u = \frac{d(\mu\gamma)}{dy} - \frac{d(\mu\beta)}{dz} \quad \text{or} \quad 4\pi\mu u = \frac{dc}{dy} - \frac{db}{dz} = -\frac{db}{dz} = -\frac{d^2F}{dz^2},$$

which is the first of equations (14); we obtain the others in the same way.

Hence the electric disturbance is also in the plane of the wave: Maxwell here anticipates what he will state explicitly in the next paragraph: that u, v, w—besides being the components of electric current in a conductor (607.4)—indicate the components of *displacement current* in a nonconductor.

if the magnetic disturbance is confined to one direction, ... the electric disturbance is confined to the perpendicular direction: If, for example, the magnetic induction is exclusively in the x-direction, then a is its sole component and $b = 0$. Then in equations (14) db/dz will be zero, and so also $u = 0$. Thus the sole component of electric displacement will be v, which is in the y-direction.

$$u = \frac{df}{dt}, \quad v = \frac{dg}{dt}, \quad w = \frac{dh}{dt}. \tag{15}$$

790.3 If P, Q, R are the components of the electromotive intensity,

$$f = \frac{K}{4\pi} P, \quad g = \frac{K}{4\pi} Q, \quad h = \frac{K}{4\pi} R; \tag{16}$$

and since there is no motion of the medium, equations (B), Art. 598, become

$$P = -\frac{dF}{dt}, \quad Q = -\frac{dG}{dt}, \quad R = -\frac{dH}{dt}. \tag{17}$$

Hence $\quad u = -\dfrac{K}{4\pi}\dfrac{d^2F}{dt^2}, \quad v = -\dfrac{K}{4\pi}\dfrac{d^2G}{dt^2}, \quad w = -\dfrac{K}{4\pi}\dfrac{d^2H}{dt^2}. \tag{18}$

790.4 Comparing these values with those given in equation (14), we find

$$\left.\begin{aligned}
\frac{d^2F}{dz^2} &= K\mu\frac{d^2F}{dt^2}, \\[2mm]
\frac{d^2G}{dz^2} &= K\mu\frac{d^2G}{dt^2}, \\[2mm]
0 &= K\mu\frac{d^2H}{dt^2}.
\end{aligned}\right\} \tag{19}$$

790.5 The first and second of these equations are the equations of propagation of a plane wave, and their solution is of the well-known form

790.3 *equations (B), Art. 598 become*: The first of these Equations of Electromotive Intensity (598.4) reads

$$P = c\frac{dy}{dt} - b\frac{dz}{dt} - \frac{dF}{dt} - \frac{d\Psi}{dx}.$$

But both it and its cohorts were developed for a case in which "the secondary circuit is in motion" (598.1). If, as in the present case, "there is no motion of the medium," then the time-derivatives of x, y, z will be zero. Moreover, as Maxwell argued in (598.5), the quantity Ψ is indeterminate; let it therefore have any convenient value—in this case the value *zero*. The equation then becomes $P = -dF/dt$, the first of equations (17); the remaining equations follow in the same way. Remember that F, G, H are the components of \mathfrak{A}, the electrokinetic momentum.

equations (18): To obtain them, combine equations (15), (16), and (17).

790.5 *equations of propagation of a plane wave*: We obtained the first of equations (19) already in (786., *comment*) and showed that it was a wave equation. It is the equation of a *plane wave* because F and G are functions of z and t only (790.).

the well-known form: See the comment on equations (20) overleaf.

$$F = f_1(z - Vt) + f_2(z + Vt), \Big]$$
$$G = f_3(z - Vt) + f_4(z + Vt). \Big]$$
(20)

790.6 The solution of the third equation is

$$H = A + Bt$$
(21)

where A and B are functions of z. H is therefore either constant or varies directly with the time. In neither case can it take part in the propagation of waves.

791.] It appears from this that the directions, both of the magnetic and the electric disturbances, lie in the plane of the wave. The mathematical form of the disturbance therefore agrees with that of the disturbance which constitutes light, being transverse to the direction of propagation.

791.1 If we suppose $G = 0$, the disturbance will correspond to a plane-polarized ray of light.

791.2 The magnetic force is in this case parallel to the axis of y and

equal to $\dfrac{1}{\mu}\dfrac{dF}{dz}$, and the electromotive intensity is parallel to the axis of x

790.5, continued. *Equations* (20): First consider function F. Notice that each of its terms has the form $g(z - Vt)$ which we considered in (786., *comment*); the velocity of propagation is V in function f_1 and $-V$ in function f_2. Thus each term is itself a solution to the first two equations (19); and by the principle of superposition, their sum is also a solution—a more general solution, moreover, than is either function alone. The same reasoning applies to function G.

(The reader is invited to prove the principle of superposition: If each of two functions f_1 and f_2 is a solution to a homogeneous linear differential equation, their sum or difference will likewise be a solution.)

790.6 *The solution of the third equation*: If $H = A(z) + B(z) \cdot t$, then $dH/dt = B(z)$ and $d^2H/dt^2 = dB(z)/dt = 0$, thus solving the third equation. (Remember that these are partial derivatives, even though Maxwell uses no special notation.)

In neither case can it take part in the propagation: Maxwell utilized a similar argument in (783.9).

791.1 *a plane polarized ray of light*: A plane wave, the transverse vibrations of which are parallel to one another. Here, if $G = 0$, the sole component of electrokinetic momentum will be F, in the x-direction.

791.2 *The magnetic force is in this case ... equal to...*: Equations (13) in (790.) expressed the components a, b, c of the magnetic induction \mathfrak{B} in terms of F, G, H. In the present case where F, the only nonzero component, is a function of z and t only, the equations reduce to $b = dF/dz$. But $\mathfrak{B} = \mu\mathfrak{H}$; then the magnetic force \mathfrak{H} is in the y-direction and is given by $\mathfrak{H} = \dfrac{1}{\mu}b = \dfrac{1}{\mu}\dfrac{dF}{dz}$.

and the electromotive intensity is parallel to the axis of x: Similarly, equations (17) in (790.3) reduce to $P = -dF/dt$, where P is the x-component of electromotive intensity.

and equal to $-dF/dt$. The magnetic force is there-
fore in a plane perpendicular to that which contains
the electric intensity.

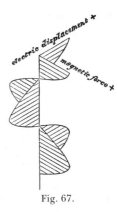

791.3 The values of the magnetic force and of
the electromotive intensity at a given instant at dif-
ferent points of the ray are represented in Fig. 67,
for the case of a simple harmonic disturbance in
one plane. This corresponds to a ray of plane-
polarized light, but, whether the plane of polarization
corresponds to the plane of the magnetic distur-
bance, or to the plane of the electric disturbance,
remains to be seen. See Art. 797.

Fig. 67.

Energy and Stress of Radiation.

792.] The electrostatic energy per unit of volume at any point of the
wave in a non-conducting medium is

$$\tfrac{1}{2} f P = \frac{K}{8\pi} P^2 = \frac{K}{8\pi} \left(\frac{dF}{dt} \right)^2. \tag{22}$$

792.1 The electrokinetic energy at the same point is

$$\frac{1}{8\pi} b \beta = \frac{1}{8\pi\mu} b^2 = \frac{1}{8\pi\mu} \left(\frac{dF}{dz} \right)^2. \tag{23}$$

792.2 In virtue of equation [19] these two expressions are equal for a
single wave, so that at every point of the wave the intrinsic energy of the
medium is half electrostatic and half electrokinetic.

792.3 Let p be the value of either of these quantities, that is, either the
electrostatic or the electrokinetic energy per unit of volume; then, in

791.3 *See Art. 797*: Art. 797 is not included in these selections.

792. *The electrostatic energy per unit volume*: The first member of equations (22)
derives from the integrand of equation (5) of (631.1). Alternatively, we derived
$\tfrac{1}{2}RD$ and $\tfrac{1}{8\pi}KR^2$ in (84.12, *comment*). In the present wave example, displace-
ment D and electromotive intensity R reduce to the components f and P,
respectively. Thus we have the first two members of equations (22). And we
have just seen that $P^2 = (dF/dt)^2$, from which the third member follows.

792.1 *The electrokinetic energy at the same point is*: The first member of equa-
tions (23) derives from the integrand of equation (19) of (635.2). The second
and third members follow from the first since $\beta = \mu b$ (791.2, *comment*).

792.2 *In virtue of equation [19] these two expressions are equal*: The original text
erroneously cited equations (20). From the first of equations (19), it is evident
that the third members of equations (22) and (23) are equal; thus the six mem-
bers of those equations are all equal to one another.

virtue of the electrostatic state of the medium, there is a tension whose magnitude is p, in a direction parallel to x, combined with a pressure, also equal to p, parallel to y and z. See Art. 107.

792.4 In virtue of the electrokinetic state of the medium there is a tension equal to p in a direction parallel to y, combined with a pressure equal to p in directions parallel to x and z. See Art. 643.

792.5 Hence the combined effect of the electrostatic and the electrokinetic stresses is a *pressure* equal to $2p$ in the direction of the propagation of the wave. Now $2p$ also expresses the whole energy in unit of volume.

792.3 *in virtue of the electrostatic state of the medium, there is a tension whose magnitude is p, in a direction parallel to x, combined with a pressure, also equal to p, parallel to y and z*: In (106.12, 106.13) Maxwell showed that electromotive intensity R will produce (in vacuum) longitudinal tension and transverse pressure, each equal to $\frac{1}{8\pi}R^2$. But as noted in (792., *comment*), $p = \frac{1}{8\pi}KR^2$—or $p = \frac{1}{8\pi}R^2$ in vacuum—is also the electrostatic energy per unit volume. Thus the tension and pressure are each equal to the electrostatic energy per unit volume of the medium, p.

In the wave here being considered, the electromotive intensity is in the x-direction (791.2); the longitudinal tension is therefore parallel to x, while the transverse pressure is parallel to y and z. Although Maxwell directs the reader to omitted Art. 107, we have here obtained these results by another route.

792.4 *In virtue of the electrokinetic state of the medium there is a tension equal to p in a direction parallel to y, combined with a pressure equal to p in directions parallel to x and z*: Maxwell derives the pressure and tension due to magnetic action in a lengthy discussion occupying Arts. 639–643. It was not possible to include that protracted treatment in the present selections; but an alternative approach is available.

As Maxwell himself points out (Art. 632), "We may treat the energy due to magnetization in a way similar to that pursued in the case of electrification, Art. 85." That analogy may be extended (though Maxwell himself does not do so) to the calculation of tension and pressure—at least when the medium is a vacuum. Corresponding, then, to the expression $\frac{1}{8\pi}R^2$ (expressed in the *electrostatic* system) for the tension and pressure that develop in a vacuum under electrification (106.12, 106.13), we may anticipate writing $\frac{1}{8\pi}H^2$ (expressed in the *electromagnetic* system) for the longitudinal tension and transverse pressure that develop in a vacuum under the action of magnetic force; and such is indeed the expression Maxwell derives in omitted Art. 643.

In the wave here being considered, the magnetic force H reduces to the component β, so that the tension and pressure will each equal $\frac{1}{8\pi}\beta^2$. But the energy per unit volume, $p = \frac{1}{8\pi}b\beta$ (792.1), likewise reduces to $p = \frac{1}{8\pi}\beta^2$, since $b = \beta$ for vacuum in the electromagnetic system. Thus, at least when the medium is vacuum, the magnetic component of the wave produces a longitudinal tension and a transverse pressure, each equal to the magnetic (electrokinetic) energy per unit volume, p—just as the electrostatic component was found to do.

792.5 *the combined effect*: In the x-direction, the tension arising from electromotive action and the pressure arising from magnetic action mutually cancel. In the y-direction, the tension arising from magnetic action and the pressure

792.6 Hence in a medium in which waves are propagated there is a pressure in the direction normal to the waves, and numerically equal to the energy in unit of volume.

793.] Thus, if in strong sunlight the energy of the light which falls on one square foot is 83.4 foot pounds per second, the mean energy in one cubic foot of sunlight is about 0.0000000882 of a foot pound, and the mean pressure on a square foot is 0.0000000882 of a pound weight. A flat body exposed to sunlight would experience this pressure on its illuminated side only, and would therefore be repelled from the side on which the light falls. It is probable that a much greater energy of radiation might be obtained by means of the concentrated rays of the electric lamp. Such rays falling on a thin metallic disk, delicately suspended in a vacuum, might perhaps produce an observable mechanical effect. When

due to electric action cancel similarly. But in the z-direction the two pressures reinforce one another, totalling $2p$.

Now $2p$ also expresses the whole energy in unit of volume: The energy per unit volume is the sum of p electrostatic and p magnetic (electrokinetic).

793. *the energy … which falls on one square foot is 83.4 foot pound per second*: This figure approximates Pouillet's 1838 determination of 1228 watts/square meter (Dufresne 2008)—equivalent to 84.2 foot-pounds/second/square foot. See also (793.1) below.

the mean energy in a cubic foot of sunlight is…: Suppose light travels at a speed of V feet per second. Then the energy delivered in one second to one square foot of surface is the energy contained in a light column of one square foot cross-section area and V feet in height. Divide that energy by V to obtain the energy in one cubic foot.

Maxwell's figure of 0.0000000882 foot-pounds is therefore the quotient of 83.4 foot-pounds by velocity V; evidently he is assuming for V the value

$$V = 83.4 \div 0.0000000882 \quad \text{or} \quad 9.455 \times 10^8 \text{ feet/second,}$$

which equals 2.882 meters/second. Maxwell, we see, has based his calculation on his own value for the ratio of electric units (787., *table*) rather than any of the optically-measured values for the speed of light.

Such rays falling on a thin metallic disk, delicately suspended in a vacuum: Readers sometimes assume that Maxwell is talking about the device pictured here, Crookes's radiometer, whose pivoting vanes—colored black on one side and white on the other—spin merrily when the bulb is placed in a beam of sunlight. If so, however, the light would exert greater pressure against the reflecting (white) side of the vanes than on the absorbing (black) side; and the vanes would spin with their black sides leading and white sides trailing.

In actuality, the Crookes bulb is only partially evacuated, and the vanes rotate with their white sides leading and black sides trailing. A common explanation of this action—that air adjacent to the black side of each vane becomes hotter, and so exerts greater pressure on the vane, than air adjacent to the white side—was (*continued overleaf*)

a disturbance of any kind consists of terms involving sines or cosines of angles which vary with the time, the maximum energy is double of the mean energy. Hence, if P is the maximum electromotive intensity and β the maximum magnetic force which are called into play during the propagation of light,

$$\frac{K}{8\pi}P^2 = \frac{\mu}{8\pi}\beta^2 = \text{mean energy in unit of volume.} \qquad (22)$$

not favored by Maxwell. In the Appendix to his 1879 essay "On Stresses in Rarefied Gases arising from Inequalities of Temperature" (*Scientific Papers*, Vol. II, pp. 681–712), he noted that "inequalities of temperature at the surface of the solid ... give rise to a force tending to make the gas slide along the surface from colder to hotter places." In Thomson's editorial footnote (†) to (793.1) below, he offers a strictly *electromagnetic* account of the development of radiation pressure.

the maximum energy is double of the mean energy: The total energy of a light wave is double its electrostatic energy and is therefore proportional to $(dF/dt)^2$ (792., 792.2). In the special case of a simple harmonic wave (Fig. 67), F will be sinusoidal. Now the *average of the square* (the "mean square") of a sinusoidal function taken over a whole number of periods is one-half its peak value. (You can prove this by integrating $\sin^2\theta$ from 0 to 2π and dividing by 2π.) Thus the peak energy is double the average energy.

Equations (22): The first term represents the peak electrostatic energy, the second term peak electromagnetic (electrokinetic) energy. Since each of these is one half the peak total energy, each is also equal to the average total energy.

Note: Maxwell's determination of the energy density of a light wave, together with his proof that it exerts pressure on a surface, leads to a consequence that would later prove important in the development of wave mechanics. For consider a portion of light beam having cross-section area A and length l. If the energy per unit volume of the light beam is $2p$ (792.5), then the portion in question contains energy $W = 2pAl$.

Now let the beam illuminate a surface. If the speed of light is c, the portion in question will deliver all its energy to that surface in time l/c. But we also saw that the illuminated surface will experience a pressure equal to $2p$ (792.5); and this is equivalent to a force f distributed over the surface A, so that

$$2p = \frac{f}{A} \quad or \quad f = 2pA.$$

Now the product of a force and the time over which it acts equals the momentum developed by that force. We may therefore write

$$f \cdot \frac{l}{c} = \frac{2pAl}{c} = \frac{W}{c} = momentum.$$

But W is the energy carried by the cylindrical portion of the beam; thus a light

793.1 With Pouillet's data for the energy of sunlight, as quoted by Thomson, *Trans. R. S. E.*, 1854, this gives in electromagnetic measure

$P = 60000000$, or about 600 Daniell's cells per mètre;*

$\beta = 0.193$, or rather more than a tenth of the horizontal magnetic force in Britain.†

* * *

*{I have not been able to verify these numbers; if we assume $v = 3 \times 10^{10}$, the mean energy in one c. c. of sunlight is, according to Pouillet's data, as quoted by Thomson, 3.92×10^{-5} ergs, the corresponding values of P and β as given by (24) are, in C.G.S. units,

$P = 9.42 \times 10^8$ or 9.42 volts per centimetre,

$\beta = .0314$ or rather more than a sixth of the earth's horizontal magnetic force.}

†{We may regard the forces exerted by the incident light on the reflecting surface from a different point of view. Let us suppose that the reflecting surface is metallic, then when the light falls on the surface the variation of the magnetic force induces currents in the metal, and these currents produce opposite inductive effects to the incident light so that the inductive force is screened off from the interior of the metal plate, thus the currents in the plate, and therefore the intensity of the light, rapidly diminish as we recede from the surface of the plate. The currents in the plate are accompanied by magnetic forces at right angles to them, the corresponding mechanical force is at right angles both to the current and the magnetic force, and therefore parallel to the direction of propagation of the light. If the light were passing through a non-absorbent medium this mechanical force would be reversed after half a wave length, and when integrated over a finite time and distance would have no resultant effect. When however the currents rapidly die away as we recede from the surface, the effects due to the currents close to the surface are not counterbalanced by the effects of those at some distance away from it, so that the resultant effect does not vanish....}

wave will possess momentum in the amount

$$momentum = \frac{energy}{c}.$$

In 1924 Louis De Broglie will postulate this same relation for his proposed "matter waves," inasmuch as it had already been shown for electromagnetic waves. Combining it with Planck's hypothesis, that *energy* $= h\nu$ (h being Planck's constant) he will reason that to any material particle endowed with momentum there must be associated a wave such that

$$momentum = \frac{h\nu}{c} = \frac{h}{\lambda},$$

that is, a wave having definite length λ according to the relation above.

793.1 *Thomson's footnote* (*): a modern rule of thumb among solar power designers estimates 100 watts per square foot at the earth's surface. Pouillet's figure of 3.92×10^{-5} erg/cc is equivalent to $3.92 \times 10^{-5} \times 3 \times 10^{10}$ or .117 erg/sec or .117 watt. This is the power falling on surface of 1 sq. cm; then 1 sq. ft. will receive 108.69 watts—nicely in line with the modern rule.

CHAPTER XXIII.

THEORIES OF ACTION AT A DISTANCE.

* * *

865.] There appears to be, in the minds of these eminent men some prejudice, or *à priori* objection, against the hypothesis of a medium in which the phenomena of radiation of light and heat and the electric actions at a distance take place. It is true that at one time those who speculated as to the causes of physical phenomena were in the habit of accounting for each kind of action at a distance by means of a special aethereal fluid, whose function and property it was to produce these actions. They filled all space three and four times over with aethers of different kinds, the properties of which were invented merely to 'save appearances,' so that more rational enquirers were willing rather to accept not only Newton's definite law of attraction at a distance, but even the dogma of Cotes,* that action at a distance is one of the primary properties of matter, and that no explanation can be more intelligible than this fact. Hence the undulatory theory of light has met with much opposition, directed not against its failure to explain the phenomena, but against its assumption of the existence of a medium in which light is propagated.

866.] We have seen that the mathematical expressions for electro-dynamic action led, in the mind of Gauss, to the conviction that a theory of the propagation of electric action in time would be found to be the very keystone of electrodynamics. Now we are unable to conceive of propagation in time, except either as the flight of a material substance through space, or as the propagation of a condition of motion or stress in a medium already existing in space. In the theory of Neumann, the

* Preface to Newton's *Principia*, 2nd edition.

865. *these eminent men*: referring to Gauss, Weber, Neumann, Riemann, and Betti—whose respective theories of action at a distance Maxwell reviewed in Arts. 846–864, omitted here.

mathematical conception called Potential, which we are unable to conceive as a material substance, is supposed to be projected from one particle to another, in a manner which is quite independent of a medium, and which, as Neumann has himself pointed out, is extremely different from that of the propagation of light. In the theories of Riemann and Betti it would appear that the action is supposed to be propagated in a manner somewhat more similar to that of light.

But in all of these theories the question naturally occurs: If something is transmitted from one particle to another at a distance, what is its condition after it has left the one particle and before it has reached the other? If this something is the potential energy of the two particles, as in Neumann's theory, how are we to conceive this energy as existing in a point of space, coinciding neither with the one particle nor with the other? In fact, whenever energy is transmitted from one body to another in time, there must be a medium or substance in which the energy exists after it leaves one body and before it reaches the other, for energy, as Torricelli* remarked, 'is a quintessence of so subtile a nature that it cannot be contained in any vessel except the inmost substance of material things.' Hence all these theories lead to the conception of a medium in which the propagation takes place, and if we admit this medium as an hypothesis, I think it ought to occupy a prominent place in our investigations, and that we ought to endeavour to construct a mental representation of all the details of its action, and this has been my constant aim in this treatise.

* *Lezioni Accademiche* (Firenze, 1715), p. 25.

Bibliography

Bence Jones, Henry. *Life and Letters of Faraday* (1870)

Bork, Alfred M. "Maxwell and the Vector Potential." *Isis* **58**, 2 (Summer 1967) pp. 210–222

Buchwald, Jed Z. *From Maxwell to Microphysics: Aspects of Electromagnetic Theory in the Last Quarter of the Nineteenth Century.* University of Chicago Press (1985)

Crowe, Michael J. *A History of Vector Analysis: The Evolution of the Idea of a Vectorial System.* University of Notre Dame Press (1967)

Darrigol, Olivier. *Electrodynamics from Ampère to Einstein.* Oxford University Press (2000)

Davy, Humphry. *Elements of Chemical Philosophy* (1812)

Densmore, Dana. *Newton's* Principia: *The Central Argument*, 3rd ed. Green Lion Pres (2003)

Dufresne, Jean-Louis. "La détermination de la constante solaire par Claude Matthias Pouillet," *La Météorologie* **60** (February, 2008), pp. 36–43

Einstein, Albert. *Autobiographical Notes*, ed. P. A. Schilpp. Open Court (1979)

Euclid. *Elements*, tr. T. L. Heath. Green Lion Press (2002)

Everett, John David. *Units and Physical Constants.* Macmillan (1886)

Everitt, C. W. F. *James Clerk Maxwell: Physicist and Natural Philosopher. An Expansion of the Article in the Dictionary of Scientific Biography.* Scribner (1976)

Faraday, Michael. *Experimental Researches in Electricity*, 3 vols. (1839, 1844, 1855); facsimile reprint Green Lion Press (2000)

Fisher, Howard J. "The Great Electrical Philosopher," *The College* (St. John's College, Annapolis, Maryland), July 1979

––––––––––. *Faraday's Experimental Researches in Electricity: Guide to a First Reading.* Green Lion Press (2001)

Froome, Keith Davy and Louis Essen. *The Velocity of Light and Radio Waves.* Academic Press (1969)

Gooding, David: *Experiment and the Making of Meaning: Human Agency in Scientific Observation and Experiment*, Kluwer Academic Publishers (1990)

Harman, P. M. *The Natural Philosophy of James Clerk Maxwell*, Cambridge University Press (1998)

––––––––––, ed. *The Scientific Letters and Papers of James Clerk Maxwell*, 3 vols. Cambridge University Press (2002, 2009)

Hertz, Heinrich. *Electric Waves*, trans. D. E. Jones (1893; Dover reprint 1962)

Hunt, Bruce J. *The Maxwellians*. Cornell University Press (1991)

Katz, Victor J. "The History of Stokes' Theorem," *Mathematical Monthly* **52**, 3 (May, 1979), pp. 146–156

Lagrange, Joseph-Louis. *Mécanique analytique* (1788)

Maxwell, James Clerk. *The Electrical Researches of the Honourable Henry Cavendish* Cambridge University Press (1879)

_____. *The Scientific Papers of James Clerk Maxwell*, ed. W. D. Niven, 2 vols. (1890); reprinted Dover Publications, Inc. (1965)

Newton, Isaac. *Sir Isaac Newton's Mathematical Principles of Natural Philosophy* (Philosophiæ Naturalis Principia Mathematica), ed. F. Cajori. University of California Press (1946)

Plato. *Dialogues*, trans. Benjamin Jowett. Macmillan (1892)

Siegel, Daniel M. *Innovation in Maxwell's Electromagnetic Theory: Molecular Vortices, Displacement Current, and Light*, Cambridge University Press (1991)

Simpson, Thomas K. *Figures of Thought*. Green Lion Press (2005)

_____. *Maxwell's Mathematical Rhetoric*. Green Lion Press (2010)

Spiers, I. H. B. and A. G. H., trans. *The Physical Treatises of Pascal*, Coumbia University Press (1937)

Thomson, William (Lord Kelvin). *Popular Lectures and Addresses*, 3 vols. Macmillan (1891)

_____. *Reprint of Papers on Electrostatics and Magnetism* (1872)

_____ and P. G. Tait. *Treatise on Natural Philosophy* (1867, 1878)

Williams, L. P. *The Origins of Field Theory*. Random House (1966)

Index

495

About James Clerk Maxwell

He was born in Edinburgh in 1831 but spent much of his boyhood at Glenlair, the family country estate to which he continued to return throughout his life. His early schooling at the respected Edinburgh Academy began at age ten. There the unsophisticated Maxwell cut an awkward and socially isolated figure, but he did begin two important friendships: with Lewis Campbell, later to be his biographer, and with Peter Guthrie Tait, with whom Maxwell carried on a lifelong correspondence.

At age sixteen Maxwell commenced study at the University of Edinburgh, where his classes included logic and metaphysics, mathematics, and natural philosophy. From there he went to Cambridge, where he attended Peterhouse and then Trinity College, graduating in 1854 with a degree in mathematics and as co-recipient of the prestigious Smith's Prize.

After graduation Maxwell occupied posts at Marischal College in Aberdeen and King's College London. During this period he made advances in color theory and in electricity and magnetism, introducing the concept of displacement current. While in London, Maxwell would frequently visit the Royal Institution; there he met the aging Faraday, for whom Maxwell expressed his surpassing respect in a number of writings, including the *Treatise*.

Maxwell resigned his position at King's College in 1865 and returned to Glenlair, where he continued to produce significant books and research papers. In 1871 he became Cavendish Professor of Physics at Cambridge, supervising the construction and equipping of the Cavendish Laboratory. Eight years later Maxwell died of stomach cancer at age 48.

Maxwell's achievements include having formulated the first comprehensive theory of electricity, magnetism, and light; the theory of electromagnetic waves; and the statistical analysis of molecular behavior in gases. He has been compared to Newton in the importance of his contributions to science; and it expresses a truth of our era that Einstein's momentous 1905 paper on special relativity takes Maxwell's electromagnetic theory, along with Newton's mechanics, as its twin foundations.

About the Editor

Howard J. Fisher has taught for nearly 50 years in the distinctive liberal arts program of St. John's College. The St. John's program, almost uniquely among educational philosophies, sees the liberal arts as encompassing science and mathematics as well as the traditional humanities; and it was in that broadly humanistic community that he first encountered the works of Michael Faraday. Fisher's subsequent decades-long engagement with Faraday's researches and writings led him to the thinking and writing of probably the most profound of Faraday's admirers, James Clerk Maxwell.

With a background in physics and philosophy, Fisher had long before made acquaintance with "the physicists' Maxwell," who was regarded a disciplined and logical theorist, perhaps somewhat hampered by a few quaint 19th-century idiosyncrasies, but masterly in his mathematical skill and physical intuition. Rereading Maxwell's *Treatise* many years later in Faraday's light, and in the context of such thinkers as Plato, Aristotle, Galileo, Newton, Kant—figures never absent from any discussion at St. John's—Fisher discovered a very different Maxwell indeed. Here was a mathematician who continually pressed beyond the formalities of analytic symbolism, notwithstanding Maxwell's own demonstrated mastery of that analytical art. Here too was a physicist who struggled less for technically formalized information and expertise, more for an open and evolving activity of seeking and knowing. "That is the Maxwell," says

Fisher, "I have tried to do justice to in this guide to the *Treatise on Electricity and Magnetism*."

Although St. John's College has been Fisher's permanent academic home, he has enjoyed opportunities to teach and present talks at other institutions in the United States, Canada, and Great Britain, including a memorable year as Visiting Senior Member at Linacre College, Oxford.